Signals and Communication Technology

For further volumes:
http://www.springer.com/series/4748

Shafiullah Khan and
Al-Sakib Khan Pathan (Eds.)

Wireless Networks and Security

Issues, Challenges and Research Trends

Editors
Dr. Shafiullah Khan
Institute of Information Technology
Kohat University of Science and Technology (KUST)
Kohat City
K.P.K Province
Pakistan

Dr. Al-Sakib Khan Pathan
Founding Head, NDC Laboratory, KICT, IIUM
Assistant Professor, Department of Computer Science
Kulliyyah (Faculty) of Information and Communication Technology
International Islamic University Malaysia (IIUM)
Jalan Gombak
Kuala Lumpur
Malaysia

ISSN 1860-4862
ISBN 978-3-642-36168-5 e-ISBN 978-3-642-36169-2
DOI 10.1007/978-3-642-36169-2
Springer Heidelberg New York Dordrecht London

Library of Congress Control Number: 2012956002

Printed on acid-free paper

Springer is part of Springer Science+Business Media (www.springer.com)

Preface

First of all, we would like to express our solemn gratitude to the "Almighty Allah" for giving us time and keeping us fit for work to timely complete this arduous project. Thanks to the authors from different countries who have contributed to the book, also thanks to those authors whose manuscripts we have not been able to include due to the rigorous review-based selection process. The final outcome has come out with total 14 chapters in total ranging various issues on wireless network security.

It is well understood now-a-days that wireless networks have become a part of our daily technical life. Though the impact of wireless networking was more or less assessed since the advent of basic wireless technologies, today's vast and dynamic features of various wireless applications might not have had been accurately envisaged. Today, the types of wireless networks range from cellular network to ad hoc networks, infrastructure-based networks to infrastructure-less networks, short range networks to large range direct communication wireless networks, static wireless networks to mobile networks, and so on. Hence, while initiating this book project, choosing a plain title seemed to be challenging but that also allowed different topics on wireless security to be compiled in a single volume.

This book is mainly targeted for the researchers, post-graduate students in universities, academics, and industry practitioners or professionals. Elementary information about wireless security is not the priority of the book. Hence, some chapters include detailed research works and results on wireless network security. This book provides broad coverage of wireless security issues including cryptographic coprocessors, encryption, authentication, key management, attacks and countermeasures, secure routing, secure medium access control, intrusion detection, epidemics, security performance analysis, security issues in applications. The contributions identify various vulnerabilities in the physical layer, MAC layer, network layer, transport layer, and application layer, and focus on ways of strengthening security mechanisms and services throughout the layers. Instead of simply putting chapters like a regular text book, we mainly have focused on research based outcomes. Hence, while addressing all the relevant issues and works in various layers, we basically lined up the chapters from *easy-to-read* survey type articles to detailed investigation related works.

Though different topics are discussed and addressed in different chapters, the ideas related to security in wireless ad hoc network have taken significant part of the book. Wireless ad hoc network is a combination of computing nodes that can communicate with each other without the presence of a formal central entity (infrastructure-less or semi-infrastructure based) and could be established anytime, anywhere. Each node in an ad hoc network can take the roles of both a host and a router-like device within the network. There might be different forms of ad hoc networks like Mobile Ad hoc Network (MANET), Vehicular Ad hoc Network (VANET), Wireless Mesh Network (WMN), Wireless Sensor Network (WSN), Body Area Network (BAN), Personal Area Network (PAN), etc. Though all of these derive some common features of ad hoc technology, WSN is a network to mention distinctively as this type of network comes with the extra feature that it might have a base station, thus a central entity for processing network packets and all other sensor nodes in the network could be deployed on ad hoc basis. Many of these network structural issues related to security concerns and challenges are presented in some chapters for the general readers.

There are also chapters related to '*not very well known*' topics like: wireless M2M (Machine-to-Machine) systems, Network Coding for Security in Wireless Reconfigurable Networks, Time Synchronization technique to improve the security, Channel Codes for Discrete Variable Quantum Key Distribution (QKD) Applications, Security implementation in EPS/LTE (Evolved Packet System/Long-term Evolution), and so on.

Due to the nature of research works, some of the concepts and future vision may seem to be not fully practical considering the state-of-the-art. Still to capture a snapshot of the current status, past gains, and future possibilities in the fields of wireless network security, the book should be a good and timely collection. This book could also be used as a proper reference material as all the chapters include the citations to the latest research trends and findings.

Because of the unified title of the book, we have opted not to divide the chapters into sections but follow a sequence that the readers may find useful. We hope that this effort of ours would be well appreciated by the readers and practitioners in the relevant fields.

The Editors

Shafiullah Khan
Kohat University of Science and Technology (KUST)
Pakistan
skhan@kust.edu.pk

Al-Sakib Khan Pathan
International Islamic University Malaysia (IIUM)
Malaysia
sakib@iium.edu.my

Contents

Security in Amateur Packet Radio Networks

Miroslav Škorić

University of Novi Sad, Serbia
skoric@uns.ac.rs

Abstract. Computer programs that radio amateurs use in their digital networks give various opportunities for checking user authentication before allowing access to sensitive parts of communication systems. Those systems include not only email servers that handle amateur radio messaging and file exchange, but also include radio-relay networks of digital repeaters that operate in big cities, or in rural and remote locations. This chapter summarizes results of experiments performed in real amateur packet radio networks as well as those provided by simulations with amateur radio software in local area networks. Our intention was to test security in accessing e-mail servers and radio relay systems within the average amateur radio digital infrastructure. This study suggests various methods which aim is to bridge the gap between the improved safety, and eventual discomfort in regular end-user's and system administrator's activities. We focused our work to the following challenges: user authentication in amateur radio email servers; key management, i.e. obtaining, installing, and renewing secret documents ('keys') in between end-users and system administrators; encryption of email content and user passwords; attacks, epidemics, and appropriate countermeasures; and other protective actions that increase the security and satisfaction in average network participants. Described methods will help practitioners, students and teachers in computer science and communication technologies in implementing exciting amateur radio wireless opportunities within educational computer networks, as well as in planning new telecommunication systems.

1 Introduction

Compared to security challenges we face to in our daily Internet-related activities, the amateur radio community does not suffer so much from exposures to known and unknown dangers that may come from their own wireless networks. For such positive situation we can thank to the global and local laws and regulations that require from all amateur radio candidates to pass written and oral, technical and regulatory tests, as well as to pass a basic security background check – before obtaining a license for transmitting radio signals. However, that does mean that the amateur radio digital infrastructure is completely secured and safe for every day's use. Luckily, there are many opportunities in available safety measures, which support the integrity of both user's rights and system administrator's privileges. Personal computers that most of our schools, workplaces and homes are equipped with nowadays are capable to include security features in existing amateur radio programs or to become additionally

S. Khan and A.-S.K. Pathan (Eds.): *Wireless Networks and Security*, SCT, pp. 1–47.
DOI: 10.1007/978-3-642-36169-2_1

enhanced with add-on software. Valuable information about available safety features can be easily obtained in regular amateur radio correspondence with peers, or by using dedicated information channels such as various 'doc' folders in email servers' file repositories. For those who are likely to experiment in an isolated local area network (LAN), consisting of at least two or three machines, there are opportunities to replicate some of the experiments in this chapter.

As described in available sources on the amateur radio simulations [1], a simplest testing scenario might be in a LAN with at least two computers, which is a suitable situation for simulating radio traffic between two different amateur radio facilities, such as digital amateur radio-relay systems - commonly called *digipeaters* (a short of 'digital repeaters'), or *BBS* ('Bulletin Board Systems'; i.e., email servers). By simulating amateur radio traffic, we learn technologies and protocols used in real amateur radio frequency (RF) networks that include thousands radio-relay stations worldwide, as well as radio email servers and various home or work communicating solutions. For those of you who already have experience in dealing with 'ham' (=amateur radio) high frequency (HF), VHF or UHF communications, these home-made simulations will provide useful information on available solutions that are going to improve security of a wireless system that you might be responsible for.

One of the frequently asked questions during author's amateur radio presentations at technical conferences and similar events – is how to ensure the safe access to the end-user email accounts – where the radio waves are the only media for transmitting information. In fact, due to the international regulations, the amateur radio traffic must travel as the open, no ciphered text, which means that all radio amateurs on a frequency are capable to 'read' everything that flows through the channel, by just simple activating his or her antenna, receiver, modem, and appropriate computer software [2]. The same rules also restrict what types of topics and discussions are acceptable in ham radio or not. For example, it is completely common to communicate the following themes: installation of antenna systems, power supply and grounding facilities; building amateur radio receivers and transmitters; programming amateur radio hardware and software; fixing small technical problems with computers and amateur radio stations, etc. That does not mean that general educational topics are not interesting for the local radio amateur community. Discussions about preparing technical conferences, papers and tutorials, or incoming technical expeditions and interesting school projects as well as non-classified details of scientific research or master and doctoral studies are completely suitable for distributing via amateur radio wireless networks. In opposite, it is not acceptable to discuss on things that include political, racial, national, social, sexual, business and similar potentially provocative themes. On the other side, there is a not a strict distinction between more or less priorities in the amateur radio communications. It is obvious that, according to the laws, emergency cases have priority, particularly when it comes to save human's lives or proprieties. But, in any occasion, one can be sure that amateur radio conversations are as 'private' as the talks in, say, public transportation systems, which actually means that there is not much 'privacy' there – if any. As mentioned, every user of a local amateur radio email server should be aware that unknown amateurs could easily read the text of his or her messages – either during an exchange of content with the email server, or during the exchange of content between those 'store & forward' systems, i.e. email servers. In such a relatively open environment, most countries have allowed the amateur radio communications primarily for an exchange of results of radio- and computer-related experiments that do not include commercial discussions, such as advertisements related to selling computers, other home appliances or any other goods. In the other words, the amateur radio laws support the major goal, which is to establish an ordinary '2-way' communication link between two or more wireless enthusiasts who might be the local school's students or teachers, as well as their parents, friends and other relatives. The basic idea is to increase the popularity of engineering and technology in young generations and to motivate them to continue education in technical professions such as electronics, electrical and mechanical engineering, computer science, hardware production, software development, etc. To summarize, when we come to commercial or other topics that are not appropriate for the amateur radio channels, it is the right time for all of us to switch from the amateur radio to commercial email service providers or similar public communicating systems.

1.1 Background

In most cases, the radio amateurs communicate by voice- and computer-related wireless modes, so they are mostly capable to differentiate themselves after a few spoken words or by a few lines of text being sent from their computer systems. In fact, the majority of amateur radio enthusiasts who live in a relatively small geographical area usually know each other, so only the newcomers are unknown to an existing, local 'ham' population. Knowing well the most of local correspondents each other is another good reason why nobody has problems with a fact that all amateur radio communications must travel as clear transmission. In voice communications, the involved correspondents are additionally obliged to repeatedly say their unique identifiers ('*callsigns*' in the radio jargon) every now and then – in order to inform the listeners about their activity on a working frequency. On the other side, the amateur enthusiasts who enjoy working with 'packet radio' – one of the most popular digital communication modes, are also required to identify themselves periodically by sending their callsigns (basic setup with two correspondents, “A” and “B”, is shown in Figure 1).

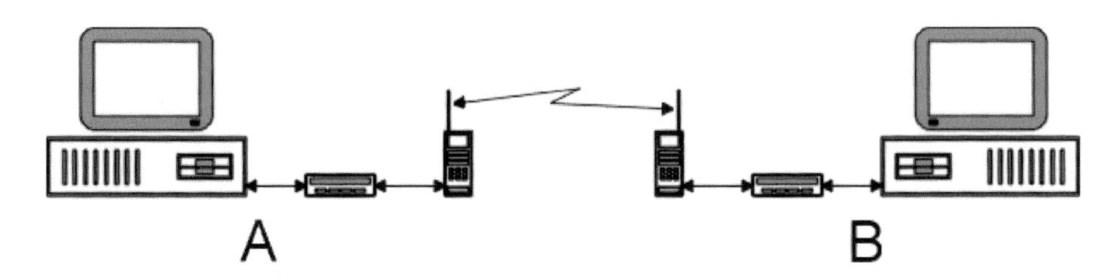

Fig. 1. Amateur radio -related computer communication, called ‘packet-radio’

Communication programs, which are commonly used within the amateur radio community, do that task automatically – provided they are properly configured – so they transmit the callsign every ten to fifteen minutes (depending on the local radio regulations). The same goes for connections to a local 'ham' radio email server. To ensure that feature to be fully functional, the callsign is a parameter of all amateur radio communicating software. It is a requirement for every particular user to ensure that his or her callsign is properly inserted in the program's configuration files. However, there is no logical or any other mechanism in the software and hardware that is capable to control if that callsign is a legal identifier of a particular amateur radio individual or a ham radio club, or not, which means that the software would accept any possible callsign. As a consequence, the callsign put into the configuration file(s) is not only transmitted every dozens of minutes, as a flashlight in the dark, but it is also used as a *username* for accessing the content of an amateur radio mailbox. Furthermore and per default, the email server is going to accept any callsign it receives as a completely valued identifier of an incoming user. Following the connecting procedure, after the first initial contact of an end-user with an email server, his or her callsign becomes the 'primary key' of a new record, added to the user database, which is maintained within the email server software, commonly called WP (‘White Pages’) database, (Jones, 1996). All email server programs are capable to keep the history information of users' activities that, for example, prevents a user to list or read the same messages he or she had read during the previous visits. In addition, the users' database includes not only the callsigns, but also related personal names, their cities' names, postal zip codes, etc. Although these databases are not completely standardized, neighboring email servers often exchange WP information between themselves automatically, with the goal of helping the other users to handle e-mail more easily. As an example, if a Russian radio amateur wanted to post a personal message to an American 'ham', he or she would only need to know the correspondent's callsign, which would go to the "To:" field of a new message header. The WP database would take care to add the recipient local ('home') server's address and location, the shortest path to it, etc.

Having in mind such a friendly environment and legal regulations mentioned earlier, the radio amateurs do not have (significant) problems related to someone’s misusing callsigns, which means the member of the amateur radio community usually perform their email procedures properly. However, that does not mean potential amateur radio 'pirates' do not appear on the horizon from time to time.

1.2 Security Issues and Solutions

If the parameters of an end-user program are not set properly, various problems might occur. In addition, if there is a wrong technical parameter that controls the behavior of the modem or radio station, it might be impossible to establish the communication with other stations at all. When it comes to 'non-technical' parameters, a wrongly inserted user callsign means the mistaken identity, which could lead to malicious misuse the third-party traffic, or in sending non-authored messages to unknown recipients, or including the usage of bad words or some racial and political speech, or deleting unread personal email, etc.

The procedure of obtaining the amateur radio license (the permission to transmit radio signals) is relatively complex. It also keeps motivating people to conform to domestic and international rules and regulations. For example, in some countries it is easier to buy a vehicle than to buy an amateur radio transmitter – regardless the prospective car owner posses a driving license or not. That means a candidate for the ownership of a 'ham' radio station has to prove that he or she is qualified enough to handle the system, which, in turn, means that he or she has to take the amateur radio classes, pass the exam and obtain the license. During the educational part of the courses, one of the most important lessons is to respect the common telecommunication rules and behave properly, because if not – the radio channel might get 'clogged' and unusable for normal communications. The experience confirms that there exist ethically and technically illiterate consumers of transmitting devices who do not hesitate to misuse wireless technologies from time to time. In the amateur radio computer-related communications, malicious users can take actions that in no way contribute to growing and developing our digital systems.

In this chapter, we will discuss about software tools and procedures that extend security level in a) working with *FBB*, which is one of the most popular amateur server programs whose implementation is described in an online user manual [3], as well as in b) working with *4RE*, which is an alternative email server program. We are going to base our examples on the experiments performed in the real wireless networks or in a simulated radio environment. In a MS Windows™ environment, FBB software (called WinFBB) can be configured such as in Figure 2. Here we can see three program windows: the main parameters (on the top), the working frequency monitoring (in the middle), and the situation within all mailbox channels (on the bottom).

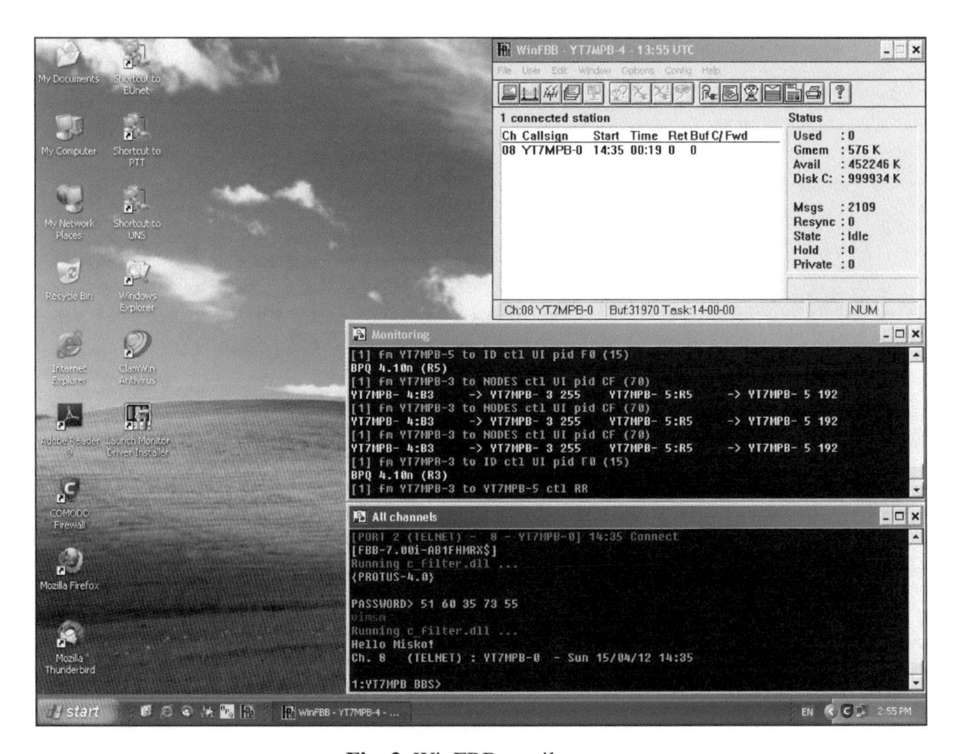

Fig. 2. WinFBB email server

In a Linux environment, FBB software (called here *LinFBB* accordingly) and a node program FPAC can be configured side-by-side, such as in Figure 3. We can see four windows: activating the node (top row, left), the node frequency monitoring (top row, right), activating LinFBB server (bottom row, left), and a local console connection to the server (bottom row, right).

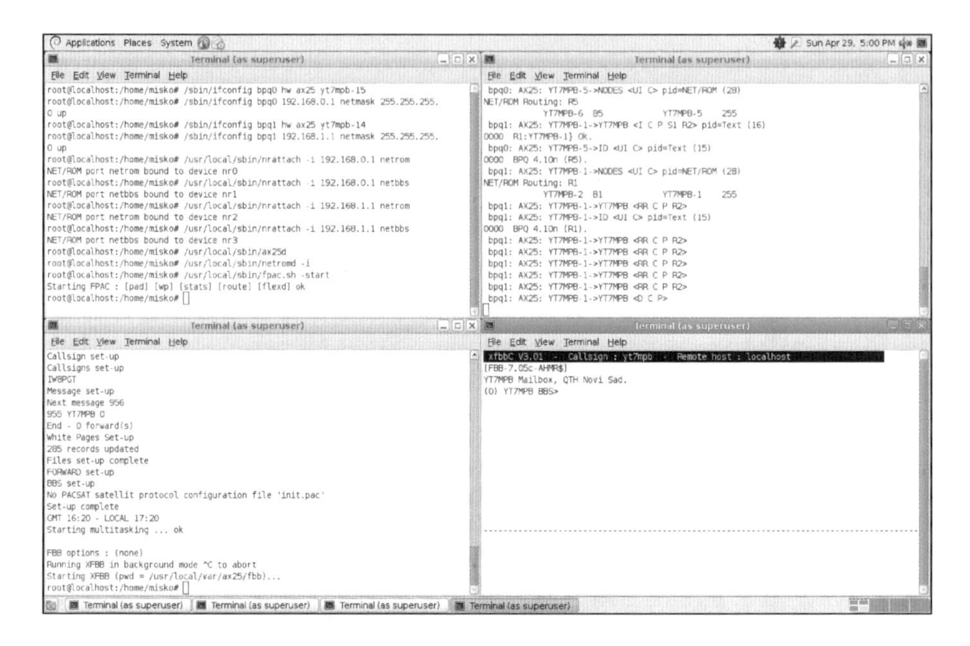

Fig. 3. FPAC node and LinFBB email server

As a 'client' (i.e. an end-user program for accessing the email server) we are going to use WinPack because it supports handling complex passwords, which we will test in advanced steps of this study (E.g. in Experiment 3, etc). A basic configuration of WinPack is shown in Figure 4. For the purpose of this tutorial, we will establish a *telnet* connection between the server- and the client-computer in a wired LAN, what is a perfect simulation of using both FBB and WinPack in real amateur radio wireless networks. (As an alternative, instead of using machines networked in a LAN, it would be also possible to establish a telnet connection by using the Internet, but that is out of scope of this book.)

Regardless which operating system is in use (MS DOS™, MS Windows™, Linux), FBB software allows installing additional tools (commonly called 'servers' or 'PG' programs), which add more functionality and comfort to both system users and administrators. One of those tools is a connection filter (*c_filter*) named *Protus*. By using Protus system administrators are capable to significantly improve the way their users access an email server. That means, in addition to a callsign which plays the role of a *username*, now it is possible to set an optional or mandatory password for each or all users (various combinations are possible). If a password is the secret for all but the

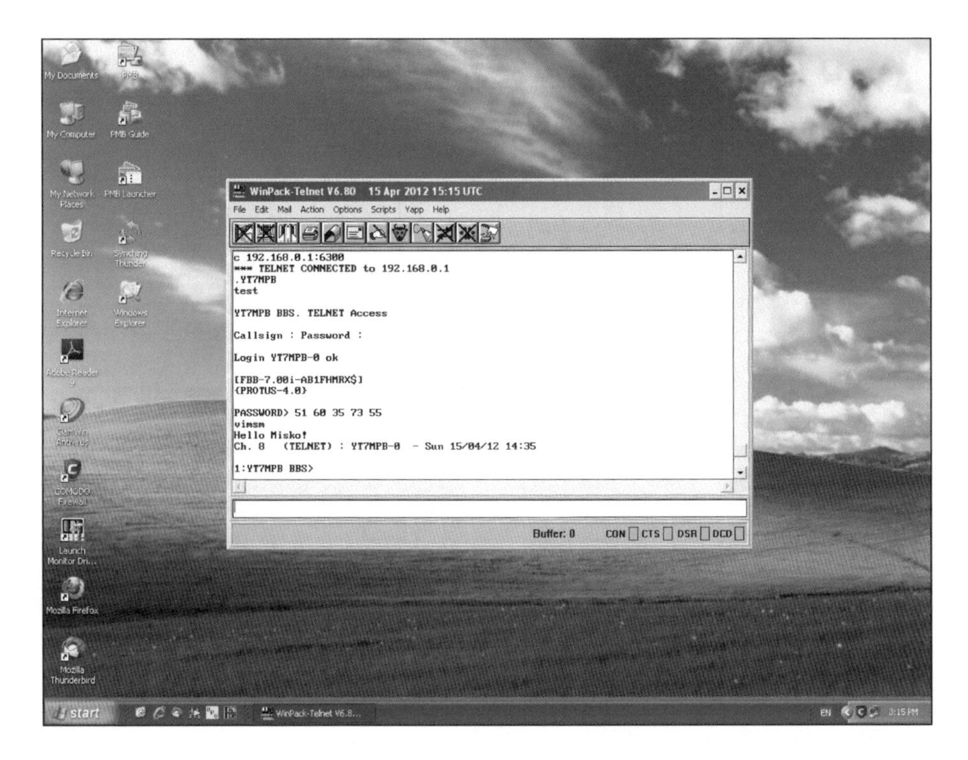

Fig. 4. WinPack email client

particular user and the system operator, the integrity of the user access is significantly better than before. However, the authentication procedure for users and administrators also goes via ‘open’ radio waves such as the content of email messages. Therefore, some additional protective measures must be taken to avoid sensitive passwords traveling in clear text.

2 Securing Access to E-Mail Servers in Windows and DOS

2.1 Experiment 1

The procedure for connecting the email server with Protus password protection is slightly different from the procedure without a password. That means, after an initial connection has been established, the server looks in its password database and checks if there exist a password record for the connecting station. The secret word usually comes in form of a relatively large string (80 characters or more, see Figure 5). That string might be any composition of small and capital letters from the English alphabet and can include numbers 0-9. Be aware that if such password line is not set for an incoming callsign and if the email server is generally set for the 'open access', the incoming user of a callsign will be given full access without further questions.

In opposite, if the callsign is given a password, the server is going to compute a *challenge* (session *A* in Figure 6), which is a random series of five numbers that represent positions of the alphanumeric characters in the secret string. The user's task is to return the proper answer – also called a *response* (session *B* in Figure 6), which consists of five alphanumeric characters, 'translated' from the *challenge*. The easiest way for an end-user to perform this translation is to compare the received *challenge* with the numbered positions of the characters in the string, such as in Figure 5. (Be aware that in this experiment a user calculated, typed and sent the *response* ***vimsm*** manually, such as in Figure 7.)

```
umoransamodkafanavolimpivoirakijuumoransamodkafanavolimpivoirakijuumoransamodkaf

12345678901234567890123456789012345678901234567890123456789012345678901234567890
1        10        20        30        40        50        60        70        80
```

Fig. 5. A sample of a large string

The most important factor in using a password system such as Protus is that the exchange of the large string between any two radio amateurs should be a secret activity for any others. That means the user and administrator have to find other methods in personal communication: E.g. a personal contact with the system administrator, a letter sent via post office, or something else *except* the same type of particular communication mode the password is intended to ensure. Having that in mind, in our case *packet radio* itself would *not* be a proper method for distribution of a password string. However, using passwords as described here is safe for an undefined period – depending on the regularity and frequency in particular users' connections. For those who access their mailboxes, say, ones a day, it should pass a relatively long time to disclose all 80 (or more) characters from the large string. For those who access their mailboxes more frequently, it would be possible that someone performing a thorough surveillance of a radio channel will learn the secret elements of the large string(s) and compromise the passwords much quicker.

Fig. 6. Two phases at the server's c_filter: A) preparing the challenge, B) checking the response

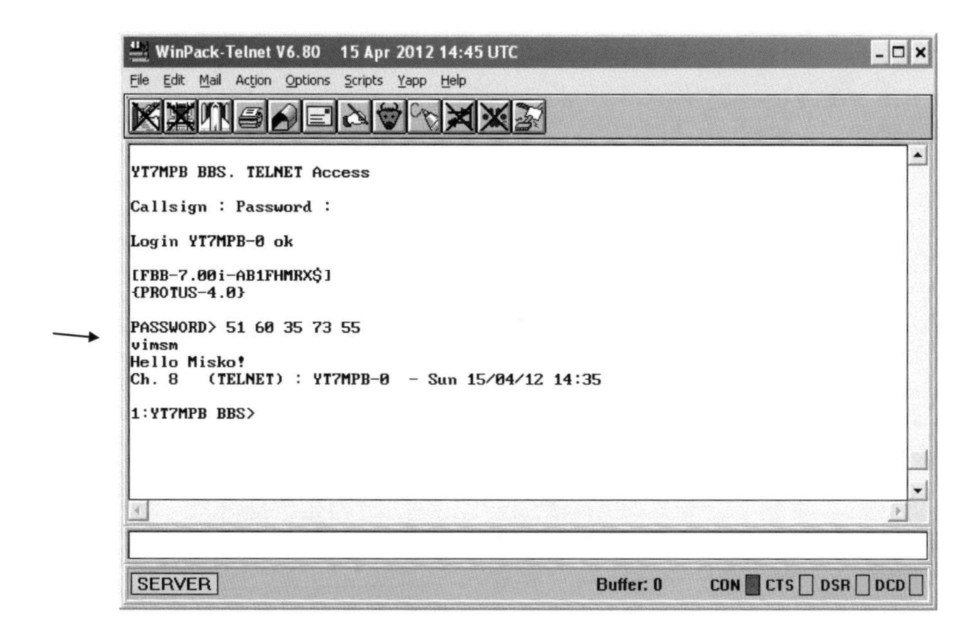

Fig. 7. The end-user's view: Password '**vimsm**' is a response to a challenge '**51 60 35 73 55**'

One of the possible solutions to prevent discovering the elements from the secret string might be to replace the elements in either regular or irregular intervals (E.g. once a week or twice a month). It is important that the users perform such change in a manner that any potential 'pirate' is not aware of a change. The easiest method, which is our proposal for beginners, is to move the leftmost character in the string (i.e. the one at position #1) to the end of the string and the remaining series of 79 characters to move just one place to the left, such as in Figure 8. Using that approach which implementation, in turn, should be negotiated between an administrator and a user, would preserve all original parts of the string but on slightly modified different positions, so a 'pirate' will hardly be capable to recognize any change – even after a prolonged period of surveillance the radio channel. In our example, we performed two simple changes during a period of time, (1) and (2).

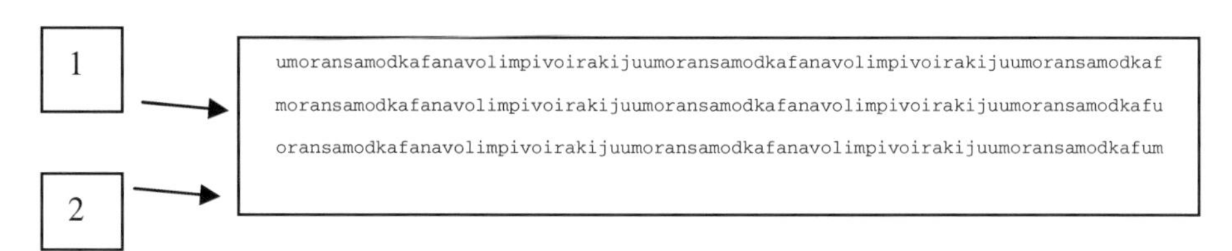

Fig. 8. Two simple consecutive changes within a large string

2.2 Experiment 2

Protus *c_filter* has an interesting option: A specific 'table' having 31 rows of text can replace the secret string. Every row represents a day of the month (assuming possible maximum of 31 days per month) and program is capable to use only the representing row each day. With such a system, the software uses the first row only at the 1st day in a month, the second row only at the 2nd day in a month, etc. This method is going to disclose the secret elements of the rows after a very long time of thoroughly performed radio surveillance. Figure 9 shows a matrix-like table, which basically uses the same characters of the large string, described in the previous example. However, the authenticating procedure is a little bit different with a 'table' password because the *c_filter* does not search through the all characters within the table but only those within the line that corresponds with a particular day in a month. In the following example, we performed an experiment on April 15, and on that day, Protus used only the 15th line within the table, see Figure 10 and Figure 11. The password syntax changed into a format of "**PASSWORD_15**" that meant something similar to "Please, use only the line #15".

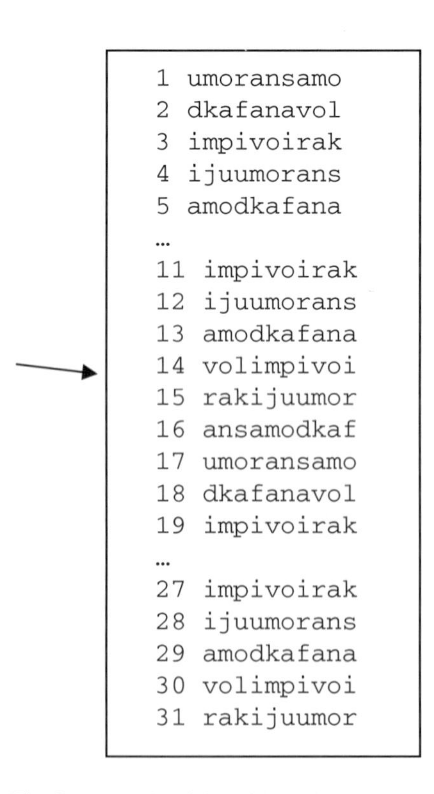

```
1 umoransamo
2 dkafanavol
3 impivoirak
4 ijuumorans
5 amodkafana
...
11 impivoirak
12 ijuumorans
13 amodkafana
14 volimpivoi
15 rakijuumor
16 ansamodkaf
17 umoransamo
18 dkafanavol
19 impivoirak
...
27 impivoirak
28 ijuumorans
29 amodkafana
30 volimpivoi
31 rakijuumor
```

Fig. 9. A sample of the table with many rows

```
All channels
[PORT 2 (TELNET) - 8 - YT7MPB-0] 17:55 Connect
[FBB-7.00i-AB1FHMRX$]
Running c_filter.dll ...
{PROTUS-4.0}

PASSWORD_15> 6 8 5 8 9
umjmo
Running c_filter.dll ...
Hello Misko!
Ch. 8   (TELNET) : YT7MPB-0  - Sun 15/04/12 17:55

1:YT7MPB BBS>
```

Fig. 10. Password '**umjmo**' is a response to a challenge '**6 8 5 8 9**'

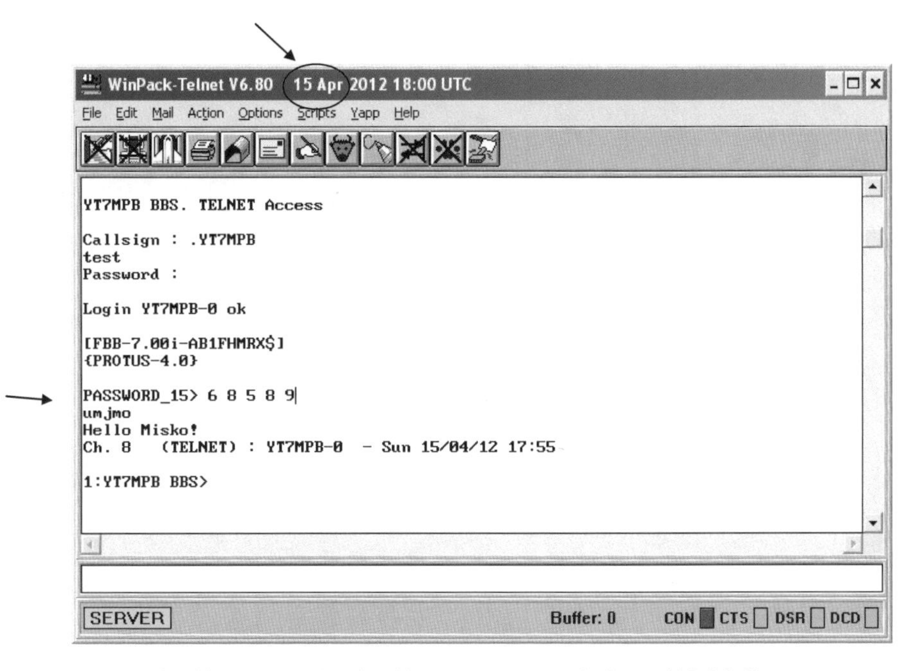

Fig. 11. Password '**umjmo**' is a response to a challenge '**6 8 5 8 9**'

The implementation of a 'secret table' instead of a 'secret string', significantly improves the level of the password safety, because any potential invader would only be capable to register those five *'daily'* characters during a particular day of a month (if we assume that an average legitimate user accesses his or her mailbox once per day). As a result, the users do not need to replace the content of the table so often. In addition, the implementation of the 'secret table' format is not complicated from the point of view of an end-user because if his or her client program eventually does not support an automatic response to a challenge, it would be easy and simple to look at the proper line of a table and calculate the answer quickly.

2.3 Experiment 3

An even better approach the Protus *c_filter* offers is an implementation of the *MD* (*M*essage *D*igest) algorithm. In our test, it looks similar to described Experiment 1 (a single large string), with an addition: The server's *challenge* also has five numbers which represent the positions of alphanumeric characters within the secret string, but now it includes a 10-digit **[square brackets]** sequence, see Figure 12. (In our example, it is **[3654790667]**.) When the client's software (which has to be capable to understand *MD* cipher technology) establishes the link and receives such *challenge*, it uses the sequence in square brackets to compute its *response*, such as in Figure 13. In our case, the *response* is **0aca4c16aff4750e2c4a376fc6d7e0b2**. Then the client's program sends the calculated *response* back to the server, which, in turn, makes a similar computation on its side. Finally, the server compares two results. If the results

```
All channels
[PORT 2 (TELNET) - 8 - YT7MPB-0] 20:00 Connect
[FBB-7.00i-AB1FHMRX$]
Running c_filter.dll ...
{PROTUS-4.0}

PASSWORD> 49 39 80 33 8 [3654790667]
0aca4c16aff4750e2c4a376fc6d7e0b2
Running c_filter.dll ...
Hello Misko!
Ch. 8    (TELNET) : YT7MPB-0   - Sun 15/04/12 20:00

1:YT7MPB BBS>
```

Fig. 12. Password '0aca4c16aff4750e2c4a376fc6d7e0b2' is the response to *the challenge '49 39 80 33 8 [3654790667]'*

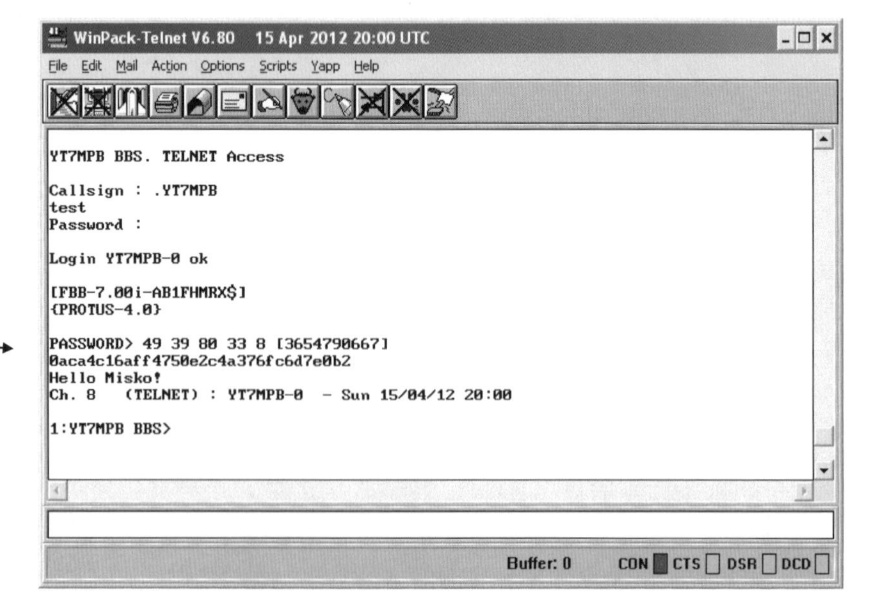

Fig. 13. WinPack enables its user to prepare an MD response automatically

are the same, the server grants free access to the connecting callsign. In opposite, which means if the two computed results are different, an immediate disconnection occurs, following a warning message to the server administrator.

2.4 Experiment 4

In the first three experiments we wanted to protect the access of 'ordinary' users of an email server. However, in most situations such level of security is not needed. In fact, most of the time it would be quite enough to allow all users to access their email accounts without specific password security, but the requirement for additional protection will remain for users with administrative privileges. In that case, we should deactivate Protus c_filter for a particular user because he or she will be granted administrative privileges later; for details see Appendix 4.

Having in mind that FBB itself has capability for sending MD2 password *challenge* (invoked by administrative command *sys*, such as in Figure 14, it is possible to setup WinPack to respond appropriately. In order to simplify this experiment with WinPack, it is possible to create new or adapt existing scripts within the program package. Our first customized script (activated by pressing F2 button on the keyboard) performs the following actions:

1. "Initiate a telnet 'connection' to mailbox YT7MPB" (by using the script line **c 192.168.1.1:6300**)
2. "Wait for the prompt" (**Callsign :**)
3. "Send the callsign to telnet to my local BBS" (**yt7mpb**)
4. "Send the password to telnet to my local BBS" (**test**)

After running the script, we receive the mailbox prompt ***(1) YT7MPB BBS*** > that is the starting point for entering manual commands.

Now we came to a little bit more complex part of a user connection with administrative privileges. Although most of the mailbox commands in FBB are simple one-character key strokes (E.g., **l** for list, **r** for read, etc.), it is not so easy to obtain the status of a system administrator. For security reasons, the administrative privileges always require user authentication. Therefore, if a mailbox visitor wants to activate the administrator's role, he or she has to do the following: a) send **sys** command, b) wait for the mailbox to send an MD2-based password *challenge*, c) compute related MD2-based password *response*, and d) send the response back to the mailbox. That procedure should be done in a timely manner. By using scripts, WinPack is capable to perform the operator's authorization in few seconds.

```
All channels
Running c_filter.dll ...
{PROTUS-4.0}
Hello Misko!
Ch. 8   (TELNET) : YT7MPB-0  - Sun 15/04/12 23:00

1:YT7MPB BBS>
sys
YT7MPB-4>  68 53 12 19 66 [1334527248]
223778f69a7f7e93d1c12ce751f82ffc
Ok

1:YT7MPB BBS>
```

Fig. 14. WinPack enables an authorized user to become an administrator in a safely manner

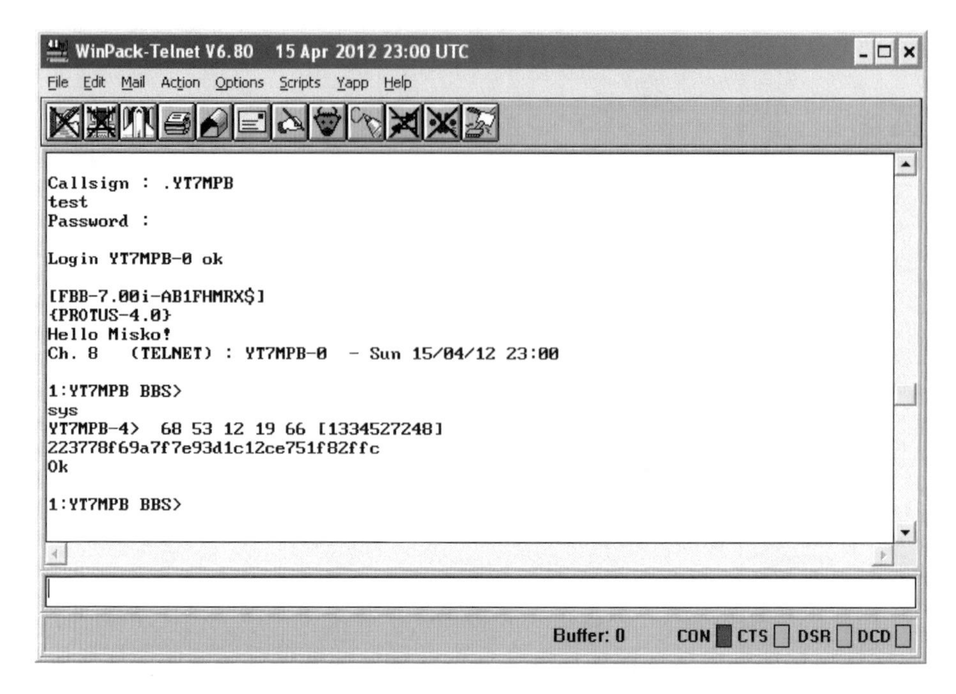

Fig. 15. WinPack enables an authorized user to become an administrator in a safely manner

For that purpose, we customized an existing script called *F3 Send FBB sysop password*, which gave us the result depicted in Figure 15. As shown in the figure, the script (activated by pressing F3 on the keyboard) did the following:

1. It sent the command asking for system administrator's privileges (**sys**).
2. It ran WinPack's subroutine *MD2PASS* (using **68 53 12 19 66 [1334527248]** as the 'key').
3. It sent the calculated response back to the BBS (**223778f69a7f7e93d1c12ce751 f82ffc**).

After successful running the script, we received the confirmation of the system administrator status (**Ok**), followed by the mailbox prompt ***(1) YT7MPB BBS*** > that was the starting point for typing administrator's commands.

As we saw from our four examples, the implementation of Protus connection filter offers system administrators a whole spectrum of various safety mechanisms to ensure the integrity of their users' email activities. In the same time, the level of responsibility in all participants of the amateur radio traffic, including not only the end-users but also the system administrators, is increased.

2.5 Experiment 5

If any two neighboring system administrators use Protus c_filter, it is possible to establish a special automatic BBS-to-BBS protected forwarding session between two servers. (*BBS* stands for the *B*ulletin *B*oard *S*ystem, which is another name for an email server.) That mode does not expect any manual input from the system operators and allows the fast mutual 'recognition' if the two systems implement the same password-authorizing tool. Every next connection results in a completely new *challenge* and *response*, which never contain visible elements of the secret string of alphanumeric characters. In practice, that means such a system is completely satisfying any possible amateur radio safety requirements, because the potential

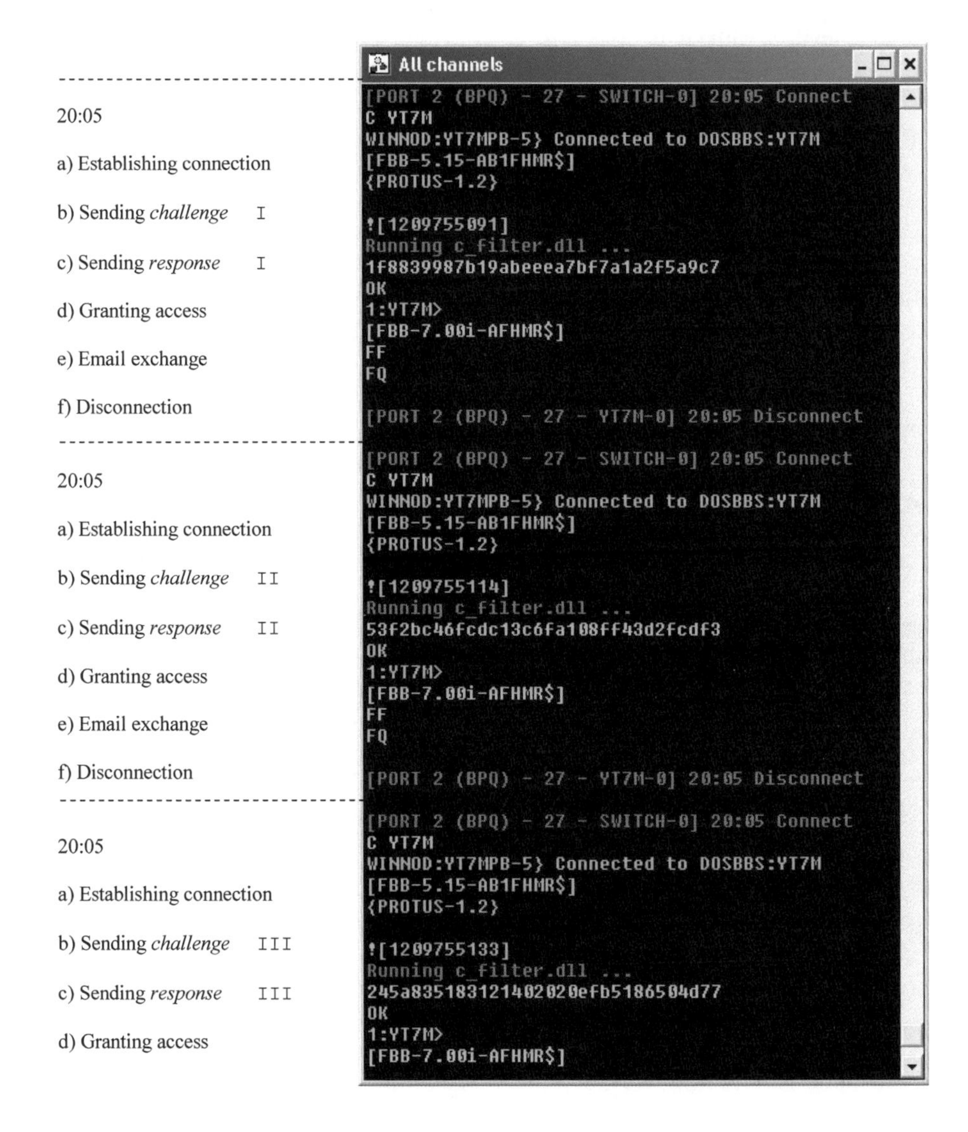

Fig. 16. Three consecutive and successful MD2 authentications between FBB e-mail servers

'pirate' cannot produce the secret key from the recorded transmissions between the two parties. Our next experiment describes three short network communications between two email systems YT7M and YT7MPB. In this example, a 'local' server (YT7M) performs the authentication of an incoming 'remote' server (YT7MPB), by using a shortened version of the *MD2* algorithm. As we can see in Figure 16, the first session's exchange of the *challenge* **![1209755091]** and appropriate *response* **1f8839987b19abeeea7bf7a1a2f5a9c7** is fully automatic and time-efficient, which allows the partnering stations to authorize their credentials for two times more in just a single minute – without any supervisors' interaction!

Be aware that in this experiment we used a version of Protus for DOS operating system. Otherwise, if both correspondents ran Windows or Linux versions of FBB software (and both of them equipped with updated versions of Protus *c_filter*), then an improved *MD5* algorithm would bring even more safety to their email servers. Other differences between versions of Protus for DOS and for Windows/Linux are also known regarding some other features, mainly related to their abilities to understand other parties' *challenges*. For example, Protus for DOS v. 1.2 that we used in this primer seemed not to be capable to compute a *response* to the Protus for Windows' *challenge* – which results in a broken link. Such situations occur when a DosFBB server initiates an outgoing connection request to a WinFBB system. (In the opposite direction, when a WinFBB attempts to establish the link with a DosFBB server, described handshaking goes smoothly, such as in our experiment.)

2.6 Experiment 6

In the next primer, we will test secure connections to alternative e-mail server software, called *AA4RE* (or *4RE* in short), which is capable to run on 'vintage' computers of types PC XT or PC AT, equipped with CPU Intel™ 8086 or 80286 and 640 kilobytes of RAM memory. Despite its maturity and date of initial production – early nineties, the newer versions of AA4RE are "Y2K" compliant, which is a general requirement for reliable e-mail *store & forward* programs. Actually, authors of AA4RE have understood the wishes in many radio amateurs who wanted to keep their old computers in the home amateur laboratories. To be precise, FBB had also solved the "Y2K" issue on time – but only in its newer versions, which were not suitable for installations on older computers mentioned in this chapter. During our tests, we were not capable to confirm eventual compatibilities between systems implementing AA4RE and those running FBB + Protus, what we found as a disadvantage of the former server program. On the other side, AA4RE includes an option that has been in use for accessing amateur radio-relay stations ('digipeaters'). Those stations are often positioned at remote locations (mountaintops etc) that are not always accessible. In the same time, relay stations do not contain computers, so their administrators could re-program the electronic circuits only when on site. That means a frequent replacement of a secret string is almost impossible. To avoid such situations, radio amateurs needed to invent solutions that would ensure safe remote administrative access from radio networks and would not require frequent change of the elements in the secret alphanumeric string. A solution that proved as reliable is similar to the one from Experiment 1, which returns the right answer of five alphanumeric characters, but now the right answer can be inserted into a longer 'word' – so a potential intruder would intercept an unexpectedly longer user's *response*.

Besides that, AA4RE gives an opportunity to accept and analyze not only a single row (a single line) within the answer, but a 'composition' of several lines. Such approach transforms (disguises) a relatively simple phrase to a more complex 'table', described in Figure 17. AA4RE ignores everything excepting the proper part (proper line) within the answer. More precisely, the program *knows* that the real end of such a 'composition' occurs after receiving an empty line, which happens after pressing the *Enter* key twice.

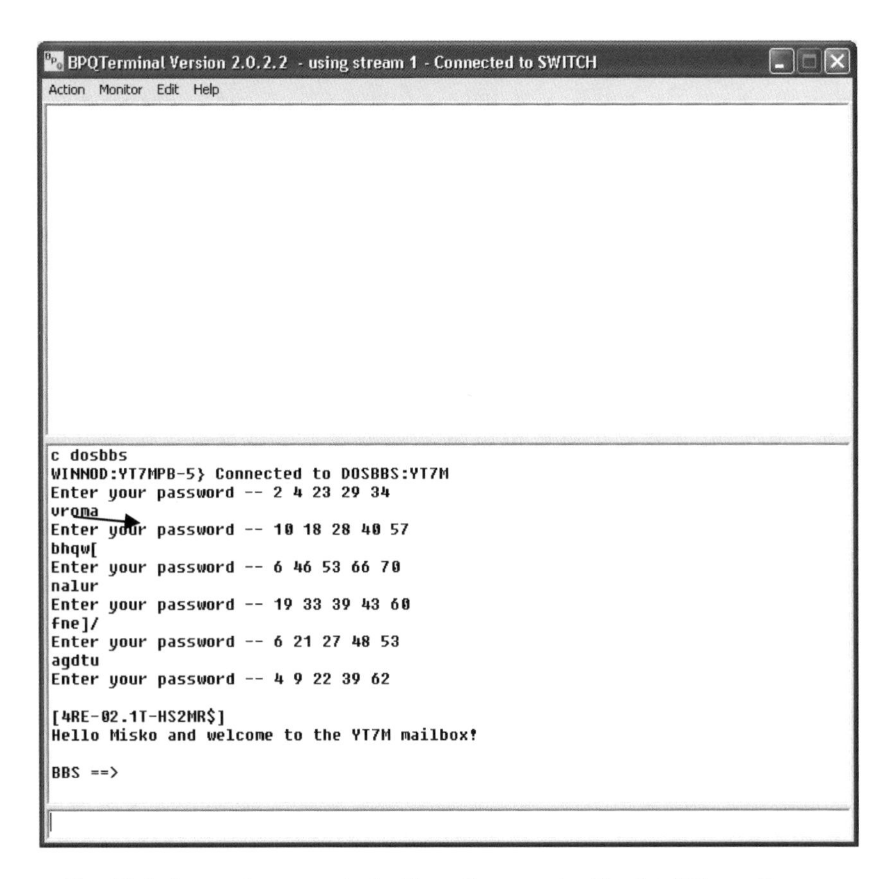

Fig. 17. A disguised response '**nalur**' is easily recognized by AA4RE e-mail server

3 Securing Administrative Access to Relay Systems in Windows and DOS

3.1 Experiment 7

Besides e-mail servers for storing & forwarding messages, the amateur packet-radio networks include their primary end-user access points in form of digital radio-relay systems, commonly called *nodes*. Those nodes can operate within hardware devices named '*t*erminal *n*ode *c*ontrollers' (*TNC*s in short), or within computer equipped with specific node programs. Regardless of type, nodes should operate smoothly and efficiently – almost without any human intervention. However, when it comes to fixing eventual problems in their functioning, such as those after power outages or other failures, it is critical to ensure that only authorized personnel is allowed to check and change node parameters and so on. In addition, it is important to prevent unauthorized 'hams' (radio amateurs) to play with system parameters because such activities could lead to temporary or permanent malfunction of a node. Therefore, all nodes implement a password security. In some cases, handling passwords is made through relatively simple mechanisms, while in other it is made at more complex level.

To simplify experimenting in educational institutions that are not equipped with amateur radio modems and transmitting stations, we will base the next few experiments on software nodes. That means we use appropriate radio-relay programs on computers in a LAN. For personal computers running MS Windows™ operating systems, one of the prevalent node solutions for radio amateurs is BPQ32. Its terminal window enables a user to monitor the working frequency and communicate with a node simultaneously, as described in Figure 18. The upper part of the window shows the radio traffic heard on a designated channel, and the lower part shows only the traffic between the node(s) and the node operator.

In this example, the operator wanted to obtain administrative privileges on a remote node **R1** (while operating at the terminal console of the node **R5**). Therefore, the first step was to initiate a connection between the two nodes, by using the command '*c r1*'. After successful connection, the operator sent the command '*password*' and R1 responded with a challenge '*11 62 36 18 42*' – in the same manner as the email server did in Experiment 1. Then the operator calculated and sent the proper response, in form of '*password daovm*'. Finally, the node confirmed successful authorization by responding with an '*Ok*'.

As you can see in the monitoring part of the window, the whole authentication sequence traveled 'in the air' as clear plain text so the potential intruder was capable to copy both the *challenge* and *response*. Depending on the number of occurrences in similar data exchanges during a period of time, there might be a lower risk or a higher risk of exposing the secret content of the whole 'large string'.

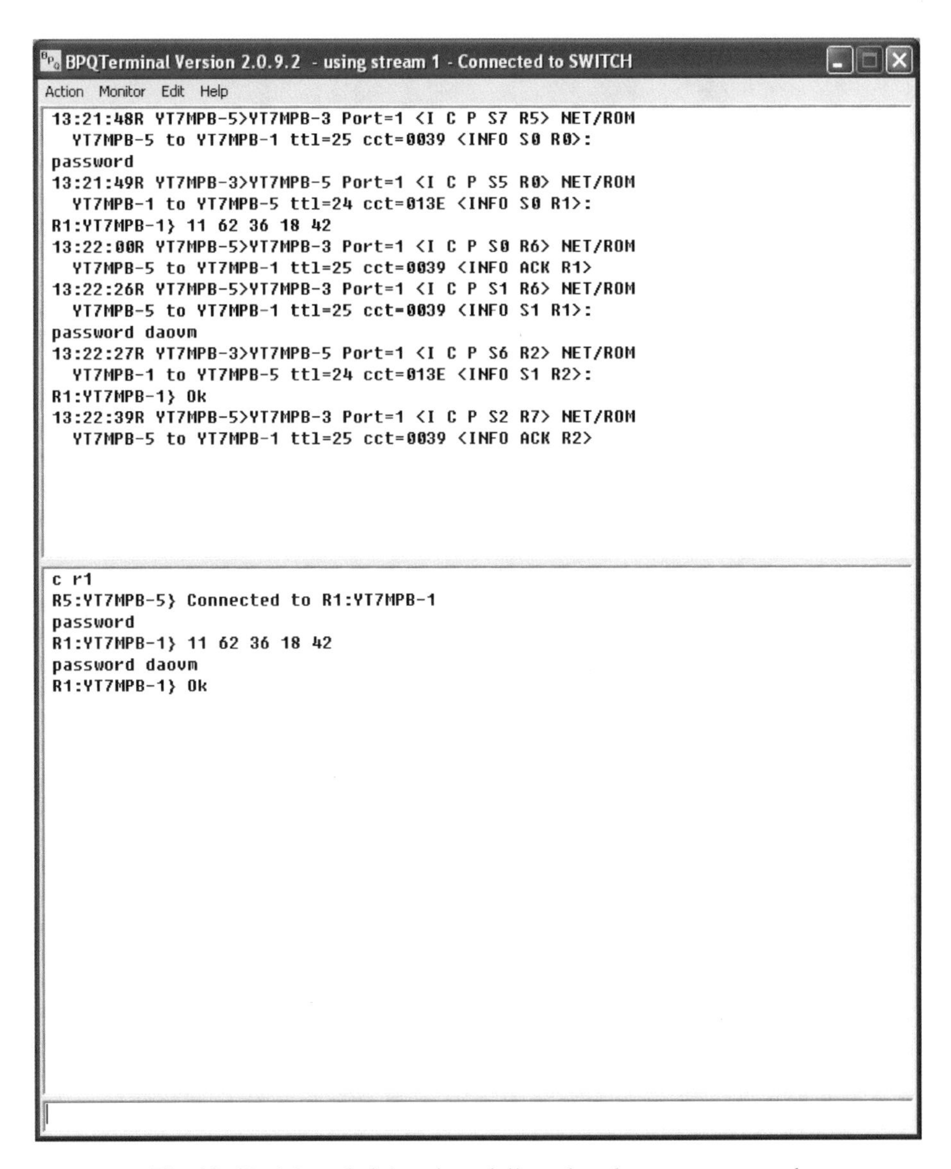

Fig. 18. Obtaining administrative privileges by a less secure approach

3.2 Experiment 8

To decrease the risk of disclosing the string's elements easily, we suggest using the following trick: The operator should send one or more 'rounds' of rather improper (fake) *responses* to the node, by using the same size and format of them, following by the last 'round' (which is actually the proper answer), such as in Figure 19. In that case, a would-be 'pirate' will see more than one answer from the node administrator, but could not be sure which one is the right combination. (Be aware, though, that the

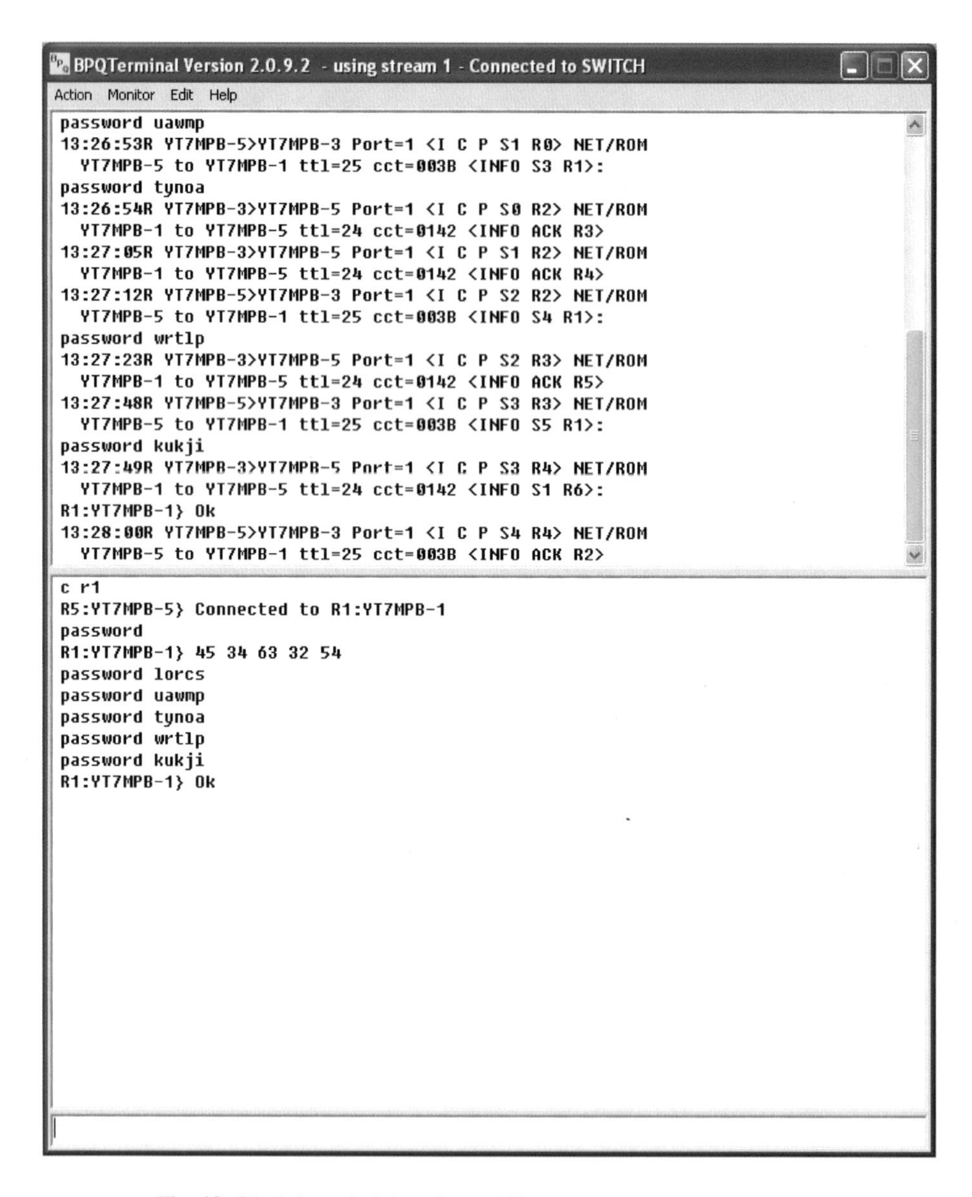

Fig. 19. Obtaining administrative privileges by a more secure approach

node will grant the increased privileges, by sending an '*Ok*', as soon as it gets the right answer. That feature prevents us to insert the right answer somewhere 'in the middle', as we did in Experiment 6.)

3.3 Experiment 9

In the previous experiment we described how to access a BPQ32-based node by using an administrator's console at another BPQ32-equipped computer. It means that both computers run under a proprietary and not so secure MS Windows environment. From the point of view of an open-source inclined educator, a much better and cost-effective solution is to perform the same task by using Linux-based computers. That means, instead of running Windows and BPQ32 program on the second machine – from where we initiate outgoing connection requests, we can easily implement Linux operating system and FPAC node software. In such an occasion, our outgoing terminal console would look similarly to Figure 20.

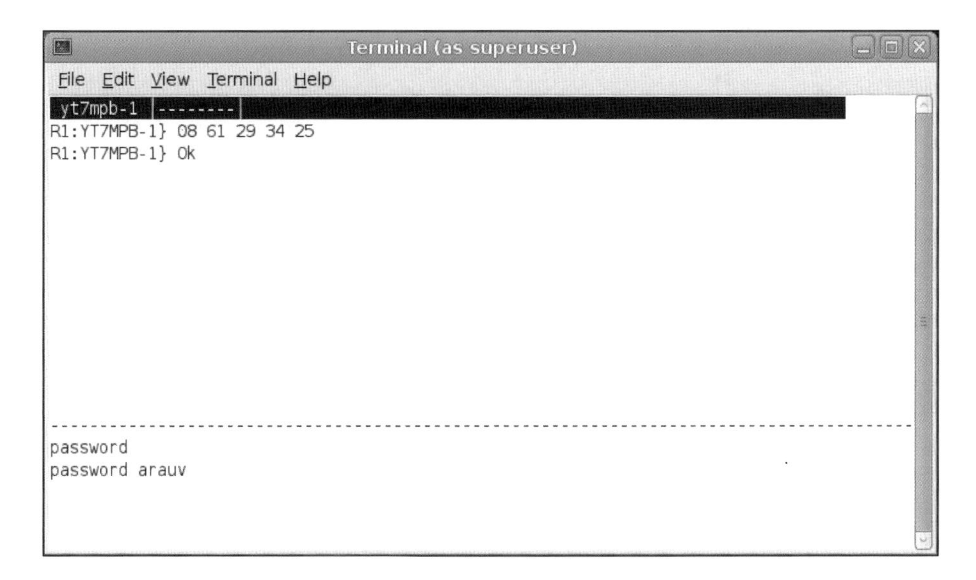

Fig. 20. Obtaining the administrative privileges by less secure approach

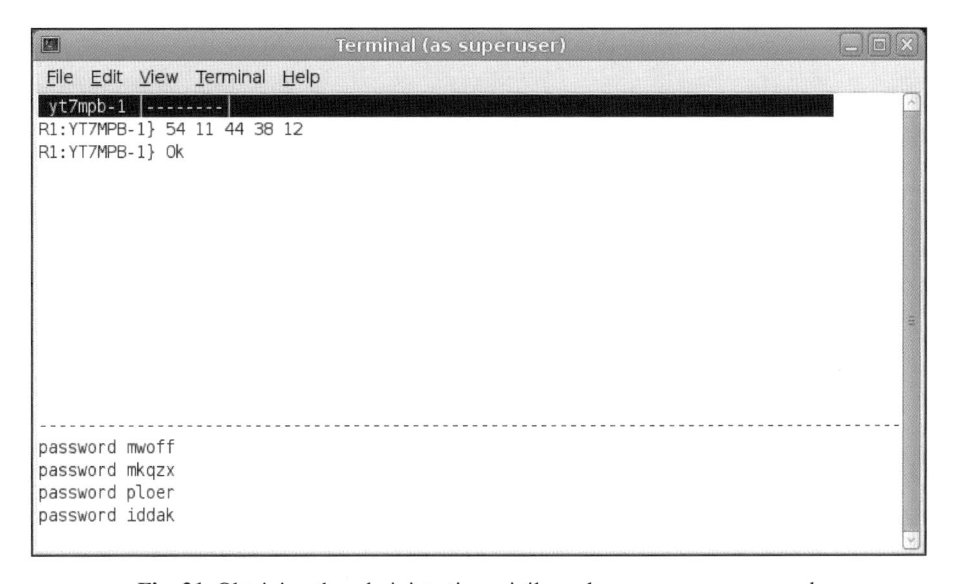

Fig. 21. Obtaining the administrative privileges by more secure approach

In Figure 20 we can see a 2-step procedure in obtaining administrative privileges. This procedure is practically identical to the process described in Experiment 7. In the similar manner, a 'counterpart' to the sequence described in Experiment 8, we can see in Figure 21.

4 Securing User and Administrative Access to E-Mail Servers and Relay Systems in Linux

4.1 Experiment 10

In our next test we are going to replace a remote node and email server environment completely. That means we will use a Linux platform instead of Windows, such as we mentioned in the introductory part of the chapter (consult Figure 3 again).

First of all, we have to activate and configure various parameters in Linux amateur radio and networking subsystem – until we have a running FPAC node, such as in Figure 22. (A detailed description of configuring particular parameters shown in Figure 22 is given in Appendix E.)

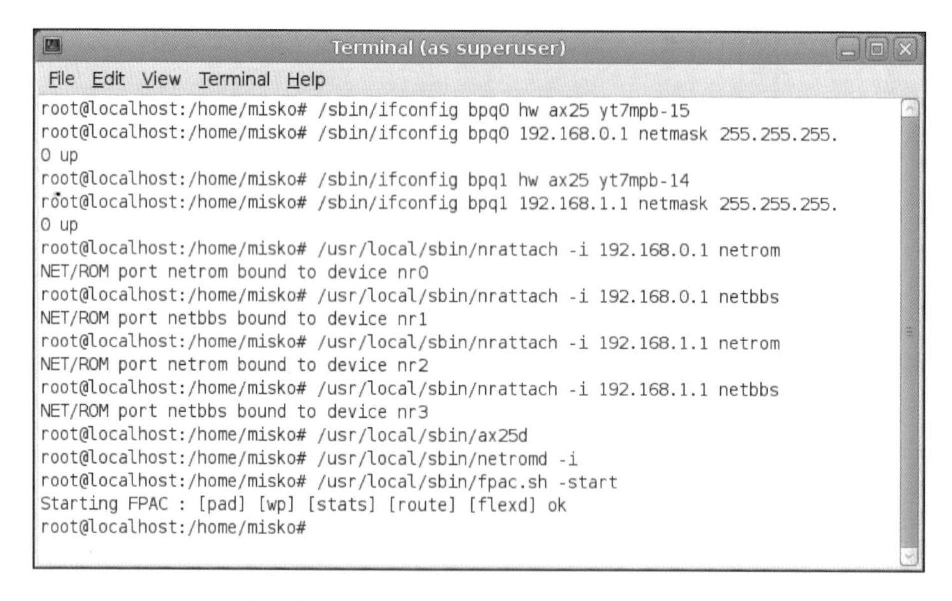

Fig. 22. Steps in activating a FPAC node in Linux

After the first step, we have to activate LinFBB email server, in order to run 'on top' of FPAC node, such as in Figure 23. In the amateur radio jargon, running some software *on top* of another means that the *lower* program (the one that is started first) is mainly intended to provide accessibility to the *higher* program (the one that is executed after the first one). Although in Linux it is quite usual to start several programs one after another in the same terminal window, for the sake of clarity we opened the second terminal for activating the second software, which is XFBB (i.e. LinFBB, a Linux flavor of FBB). In Figure 23 we can only see the last several lines from a relatively long command script that has scrolled down within the second terminal window.

```
Terminal (as superuser)
File Edit View Terminal Help
Callsign set-up
Callsigns set-up
IW8PGT
Message set-up
Next message 956
955 YT7MPB 0
End - 0 forward(s)
White Pages Set-up
285 records updated
Files set-up complete
FORWARD set-up
BBS set-up
No PACSAT satellit protocol configuration file 'init.pac'
Set-up complete
GMT 16:20 - LOCAL 17:20
Starting multitasking ... ok

FBB options : (none)
Running XFBB in background mode ^C to abort
Starting XFBB (pwd = /usr/local/var/ax25/fbb)...
root@localhost:/home/misko#
```

Fig. 23. Final step in activating an XFBB (i.e. LinFBB) server

The next phase of this experiment, after both FPAC node and LinFBB server have been activated, is to start two additional windows – the first one for handling the node-networking, and the other one for accessing email server functions. Those two additional windows are shown in Figure 24.

As it can be seen in Figure 24, although being positioned side-by-side (i.e. top-bottom in this case), those two additional windows operate independently, where the upper one shows a connection to the remote node YT7MPB-1 (performed by an outgoing connection through the 'local' FPAC node) including a request for advanced privileges (system command *password*). On the other side, the lower window shows a local console connection to the LinFBB server, followed by another request for advanced privileges (system command *sys*). As shown in the figure, in both cases the operator had to send a proper answer to the node (***password arauv***) and to the email server (***dnrli***) – before getting confirmations of obtaining administrative rights (***Ok***). Be aware that in this experiment we used only the manual method for sending appropriate responses to the controlled systems, which means that no automatic (scripted) procedures have taken place here. However, even without using more complex procedures, such as MD2 or similar we described earlier, the administrative access to radio nodes and email servers remained secured.

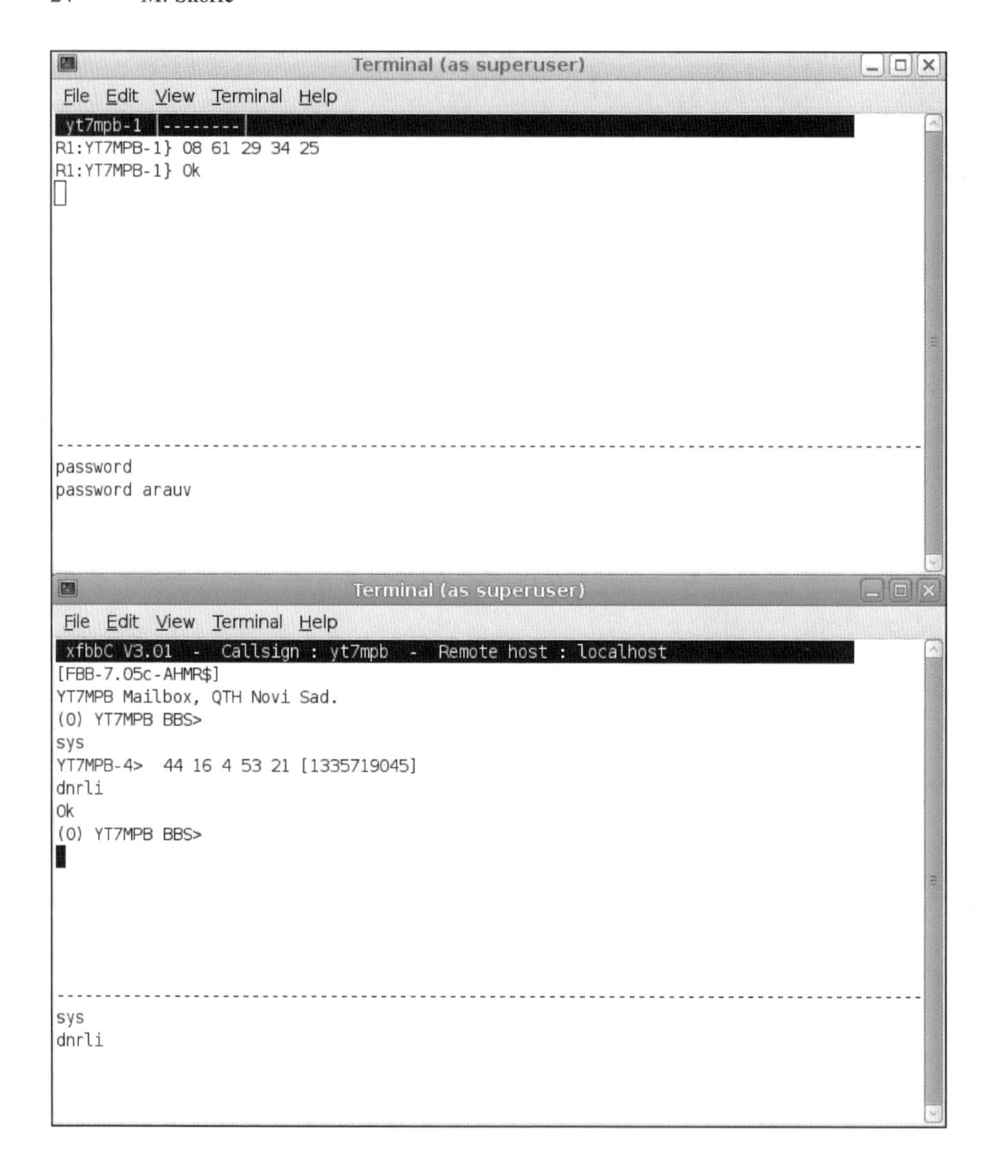

Fig. 24. Final steps in approaching node network and email server

4.2 Experiment 11

In the next test we are going to use WinPack for accessing a remote node and email server that run in Linux environment; see Figure 25 and Figure 26.

In Figure 25 we can see an extended set of commands that are available to a sysop (system administrator), after successfully performed authentication. Among the available commands, there is yet another command named *sys*, which relocates a remote administrator to an internal 'command prompt' of Linux (practically,

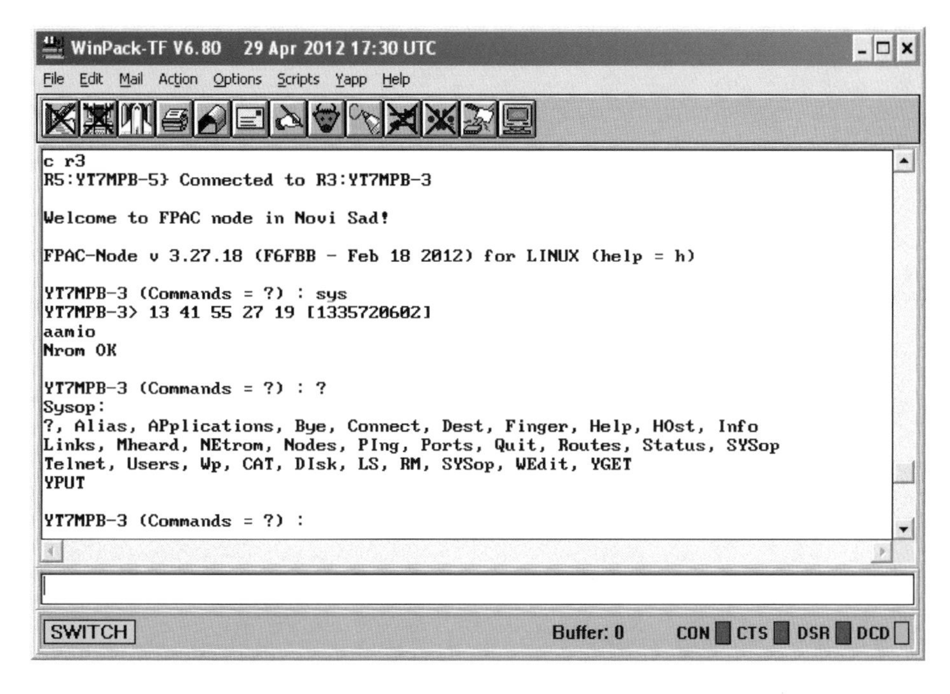

Fig. 25. Performing administrative roles after accessing FPAC radio-node

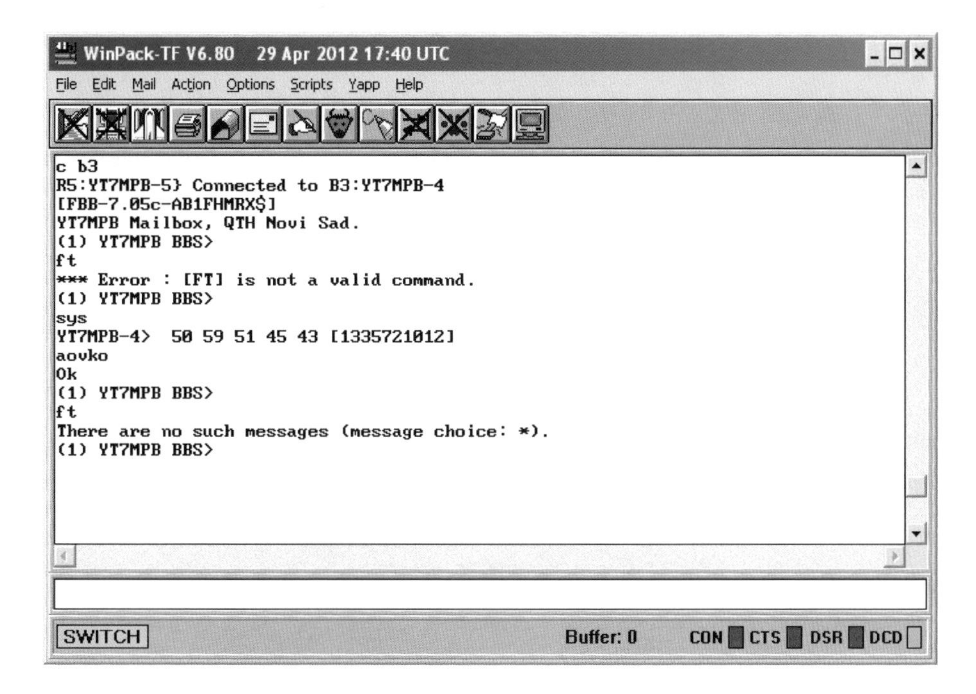

Fig. 26. Performing administrative roles after accessing LinFBB radio-server

a simulated version of another terminal screen, similar to the one in Figure 22). Having in mind that an access to a Linux terminal (command prompt) gives a user almost infinite possibility for using/misusing his or her operating system, it is clear that we have to take care about safe access to the sensitive parts of the computer subsystems that provide remote administration.

When it comes to LinFBB server's administration, described in Figure 26, an example is the command *FT* (means '*f*orwarding *t*otal') that is intended to return a list and volume of messages waiting for exchanging with partnering BBSs. In this example, an ordinary user was informed at first that such a command was "not a valid command" – just because he or she was not approved for administrative rights yet. However, after successful authorization the same command became fully valid but in this particular occasion it responded with "there are no such messages" (i.e. messages for exchanging with neighboring servers). Similarly to WinFBB, LinFBB also allows authorized administrators to read all mails – regardless they are public or private, or to make various restrictions to what ordinary users can do within the server areas, and even to stop the email server functions altogether.

4.3 Experiment 12 (Unfinished)

As a "work in progress" and an idea for further research in this field, we could suggest experimenting with the wireless access to a JNOS amateur radio mailbox, see Figure 27.

There are many similarities and differences between FBB and JNOS but they are mostly out of scope of this book. However, we wanted to point a fact that exciting alternatives are always available for devoted wireless enthusiasts. One of the most interesting things with JNOS is that program users (i.e. prospective system operators of new JNOS mailboxes) can compile very different versions of the program – depending on their particular needs and wishes. In fact, JNOS is also available in the form of a source code, and prospective server administrators are strongly recommended to compile binary executables of that program on their own – instead of installing precompiled versions they can find on the Internet. A reason for this approach is that a 'local compilation' takes care of the characteristics of particular local hardware and software features and is composed accordingly.

For the purpose of satisfying our main goals, which include strengthening in wireless security mechanisms, additional safety features are available when compiling JNOS (see Appendix F). For example, the MD5-based algorithm for telnet user authentication is available, but it is not activated 'per default'. The situation is the same with password authentication for radio users (both those who connect by local *AX.25* radio channels and those who come by *netrom* links). Therefore, the experimenters can activate those options before performing program compilation, which means that the appropriate options should be so-called, 'defined' in the configuration file.

As a result, the MD5-based option for password authentication, which is the sequence **[4fdde1d7]**, is shown in Figure 28.

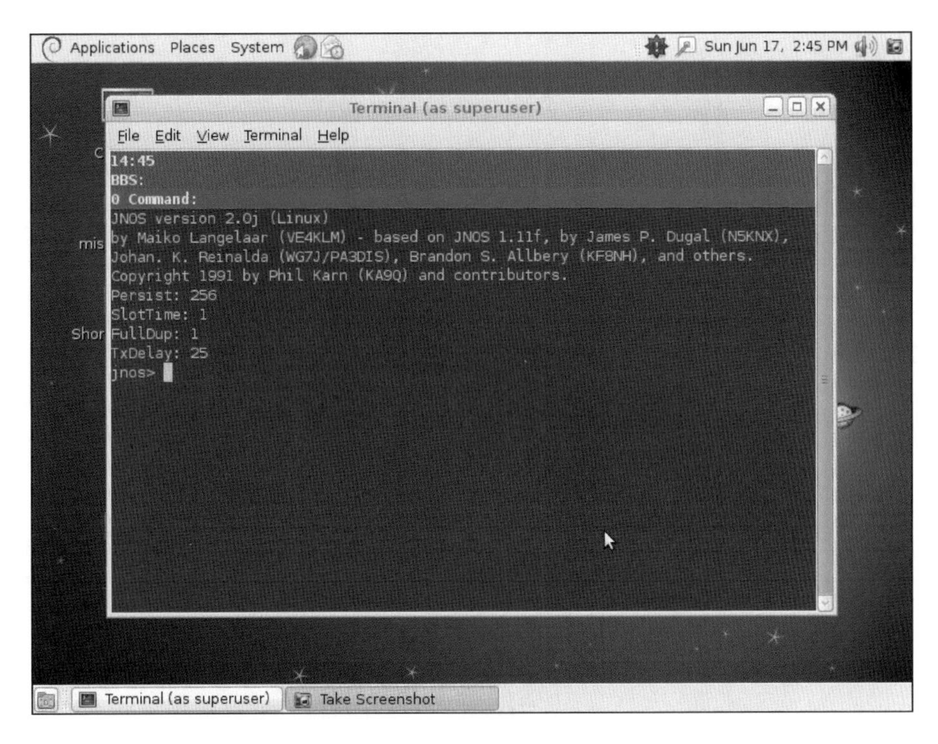

Fig. 27. JNOS amateur radio-server installed in a Linux environment

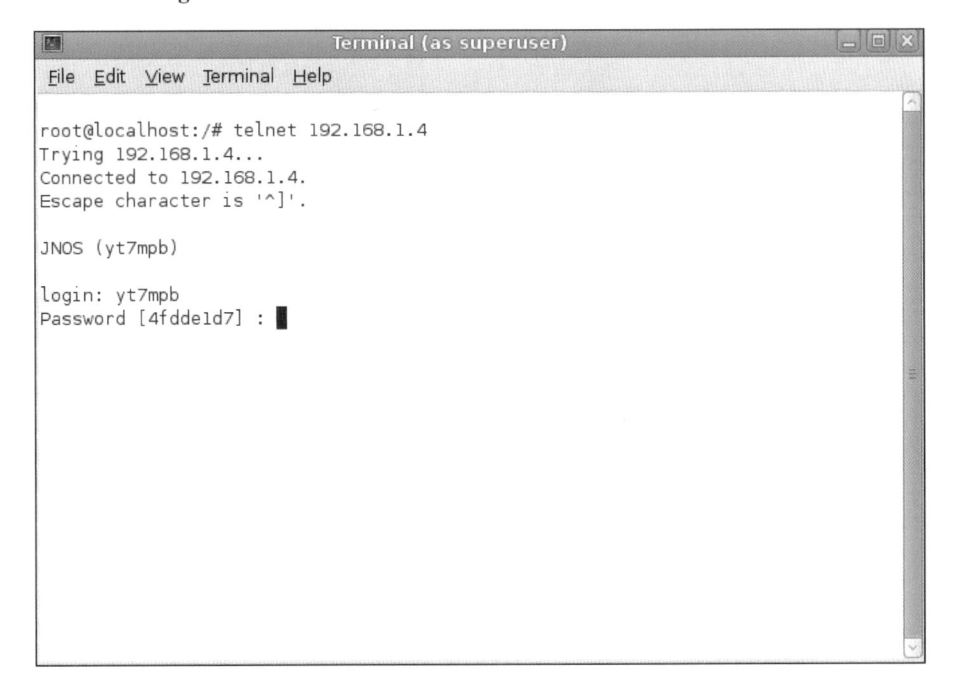

Fig. 28. MD5-based user authentication in a JNOS amateur radio-server

As seen in the figure, by using the telnet command we simulated an incoming user access to a JNOS mailbox. However, this experiment remained unfinished because we have not tested options with AX.25 and netrom radio connections.

5 Discussion

For experiments described in this chapter, we used the following equipment:

Server YT7M
- Computer Intel 80286 CPU clock 12 MHz, 1 MB RAM,
- Operating system MS DOS 5.0,
- Network node software BPQ 4.08a,
- E-mail server software DosFBB 5.15c, AA4RE 2.13t,
- Protus 1.2 and Protus 3.3 for DOS.

Server YT7MPB-1
- Computer AMD Athlon™ CPU clock 1.10 GHz, 512 MB RAM,
- Operating systems MS Windows XP SP3, Linux Debian 6.0.5,
- Network node software BPQ32 4.10n for Windows,
- E-mail server software JNOS 2.0j for Linux.

Server YT7MPB-3
- Computer Intel Celeron™ CPU clock 400 MHz, 224 MB RAM,
- Operating systems MS Windows XP SP3, MS Windows 2000, Linux Debian 6.0.4,
- Network node software BPQ32 4.10d/n for Windows, FPAC 3.27.18 for Linux,
- E-mail server software WinFBB 7.00i for Windows, LinFBB 7.05c for Linux,
- Protus 4.0 for Windows, Protus 4.1b2 for Linux.

Server YT7MPB-5
- Computer Intel Pentium™ Dual-Core CPU clock 3.06 GHz, 3 GB RAM,
- Operating systems MS Windows XP SP3, Linux Ubuntu 10.04.4 LTS,
- Network node software BPQ32 4.10n for Windows, Node 0.3.2-7.1 for Linux,
- E-mail server software WinFBB 7.00i for Windows, LinFBB 7.04j-8.2 for Linux.
- Protus 4.0 for Windows, Protus 4.1b2 for Linux,
- Terminal software WinPack 6.80 for Windows.

With an exception of the relatively new computer that operates YT7MPB-5 subsystem, we did not invest in modern equipment for our tests. Instead, we rather experimented with 'low-level' PC computers that were capable to operate different versions of Windows and Linux operating systems. In addition, we used an outdated 'PC AT' machine for older versions of MS DOS (or PC DOS).

One important remark here: People sometimes consider amateur radio as an expensive 'sport'[1], but that should not be true. In fact, there is no need to waste finances for obtaining top-level computers that are intended for operating amateur radio nodes and email servers. It is always better to invest more in external devices, such as outdoor antennas, in order to expand the coverage of a wireless station.

In our experiments we described various approaches in ensuring safe access to amateur radio wireless infrastructure, focusing to the security in handling user credentials and in obtaining administrative privileges. Most of described amateur radio programs offer full or partial logging of user activities, so the intrusion detection can be possible and system administrators can be informed about unsuccessful connecting attempts, 'mistaken identities' of unknown users, and so on. Besides password protection described in our tests, system administrators are advised to differentiate their users in at least two categories, let us name it 'resticted' (or 'read only') and 'regular' users. What might be a reason for that politics? For example, besides our main goal in Experiment 5, which covered password protection, intended to give additional security to the BBS-to-BBS interactions, the implementation of the automated authentication gives us the opportunity to save our working radio frequencies from overloading and increased traffic. In fact, during an exchange session between two wireless systems ('forwarding') running on the international HF radio waves – mainly intended for the automatic store & forward activities, it is crucial to ensure that on the same channel there are no other interfering stations such as the end-user 'intruders' who might want to access their personal mailboxes manually. There are several reasons for such policy: The HF bands are prone to fading and even a slight fade of signal is enough to cause data loss. Besides that, bad weather conditions, local noise or interference, or even some of them combined, can give a lot of frustration to the system administrators [4]. In that manner, it is possible to set only the passwords for collaborating servers on the otherwise 'closed' systems, so the other connecting stations would be disconnected immediately. In addition to such a rather rigid decision, it might still be possible to reserve some time slots for the 'open' access to the end-users, provided that the automated exchange sessions are finished. However, according to the good practice and common amateur radio rules, there is a variety of available system messages within the program files, including Protus c_filter that inform the users that the safety measures have taken place.

---

[1] Some stories stated that amateur radio in former USSR was called 'Radio sport'. The reasons for such a name they found in a fact that radio amateurs sometimes do use physical efforts in their hobby activities, such as in so called 'fox hunting' – a time-related competition in searching an area for a hidden transmitter. (Nowadays, an official name of such kind of amateur radio activities is ARDF, the Amateur Radio Direction Finding.)

The amateur packet radio is an interesting educational tool for increasing motivation in young generations for studying engineering, computing and related technologies [5]. In regard to that, if the amateur radio infrastructure is linked to a computer network of a school or university, it is important to take care of privileges given to the students. Program FBB is fully capable to differentiate low-risk privileges intended for ordinary '*telnet*'-users within a LAN, from the high-level administrative tasks, Figure 29 and Figure 30. The two slides show that lower privileges give only restricted access to the user's personal email account: ListMine (LM), ReadMine (RM), KillMine (KM) are the most suitable commands for a student (calssign *YU7BDR*, Figure 29), whereas the system administrator (callsign *YT7MPB*, Figure 30) is allowed to do everything else, including removing inappropriate content (if any), accessing radio gateways (if installed), etc. Note that different privileges can be configured not only for the users within a LAN, but also for those who come through the Internet connections, or real wireless channels, and so on.

Furthermore, the content integrity within amateur radio wireless systems can be additionally protected by including various text (i.e. message) filtering mechanisms, commonly named *m_filters*. Such software devices have one function: To screen email server content for 'bad' words or otherwise inappropriate content within public area messages and prevent them from spreading around. It is also possible to make every incoming message invisible for other users, until the server administrator approves that a message is suitable for distribution. That means there are mechanisms whose role is to protect amateur radio wireless infrastructure from proliferating illegal intentions, ideas, and activities. We can only hope that the majority of responsible system operators will follow these instructions and continue building and maintaining wireless systems safe.

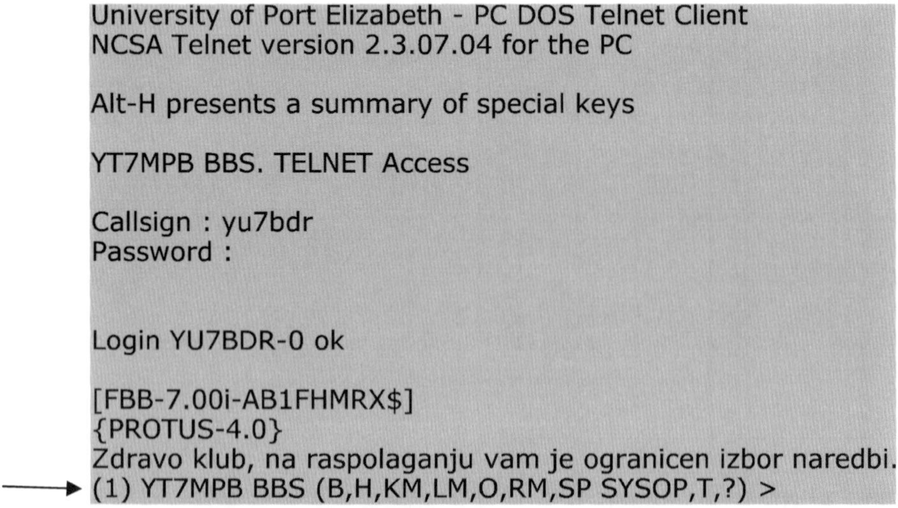

```
University of Port Elizabeth - PC DOS Telnet Client
NCSA Telnet version 2.3.07.04 for the PC

Alt-H presents a summary of special keys

YT7MPB BBS. TELNET Access

Callsign : yu7bdr
Password :


Login YU7BDR-0 ok

[FBB-7.00i-AB1FHMRX$]
{PROTUS-4.0}
Zdravo klub, na raspolaganju vam je ogranicen izbor naredbi.
(1) YT7MPB BBS (B,H,KM,LM,O,RM,SP SYSOP,T,?) >
```

Fig. 29. A low-privileged user has a restricted list of available commands

```
University of Port Elizabeth - PC DOS Telnet Client
NCSA Telnet version 2.3.07.04 for the PC

Alt-H presents a summary of special keys

YT7MPB BBS. TELNET Access

Callsign : yt7mpb
Password :


Login YT7MPB-0 ok

[FBB-7.00i-AB1FHMRX$]
{PROTUS-4.0}
YT7MPB BBS, Novi Sad.
Zdravo Misko! (prethodni QSO je bio: 09-Maj 17:11)
Nove poruke su: 216388 - 220210, aktivnih ima 1913.
Niste procitali nove poruke za vas: 3 kom.
(1) YT7MPB BBS (B,C,D,F,G,H,I,J,K,L,M,N,O,PS,PG,R,S,T,U,V,W,X,Y,Z,?) >
```

Fig. 30. A high-privileged user has a full list of available commands

One of the frequently asked questions is why the amateur radio digital infrastructure does not use encryption of the email content. The answer is simple: According to international and local wireless regulations in most countries, the amateur radio traffic have to be 'open' i.e. an amateur radio station must transmit only the 'clear text'. (The only exception can be the short sequence of user authentication by implementing passwords, such as described in this work.) The reason for that is probably in national security needs, because the governments want to be sure that the amateur radio is not used for illegal, criminal, or terrorist activities, so they can observe it appropriately at all times – whenever needed. In addition, the purpose of giving the wireless radio enthusiasts significant portions of radio spectrum is self-education, amateur research in engineering and technology, improving existing software and hardware, as well as the continual preparation and training for eventual establishing ad-hoc networks in case of emergencies after natural disasters or something like that. Therefore, the most of the radio amateur traffic must remain fully transparent.

Related to that, when it comes to key management for password protection, we remind the readers that the 'secret strings', 'secret tables', etc should never be sent and received by radio waves. Having in mind that the majority of email server users live in a relatively close area, the best way could be that the keys are distributed personally, for example at the amateur radio club gatherings or union conventions, or something similar. For sure, postal mail or Internet email could be used as an alternative too, but you never know if there is a third party or a man-in-the-middle who might be involved in those two types of communication.

Yet another frequently asked question is how the amateur digital wireless services are really, basically safe and if there are significant security issues with existing software applications. As you have realized from the output of the tests provided in this chapter, different approaches have been used by ham radio programmers. Generally, the starting point was that the amateur radio community is just a tiny percentage of the global population and that procedures of obtaining radio licenses usually includes a kind of a security clearance. Having those two facts in mind, as well as the other regulations – partly mentioned in this chapter, the general attitude is that the past experience has taught us that there are not (significantly frequent) incidents related to misusing amateur radio wireless infrastructure. That means, the majority of amateur wireless system administrators are rather inclined to operate 'open' systems, and not enforce 'closed' or 'restricted' features that are under their responsibilities. However, by implementing one or more security measures combined, it is possibly to ensure relatively high levels of safety – regardless of eventual disadvantages in particular applications. For example, if a system administrator thinks that a lack of encrypted password mechanism in his or her email server appliance is a safety issue, then a message 'held' mechanism could be enforced locally, which means that no message would be disseminated without the operator's consent. Similar kinds of synergy can be considered before or after eventual incidents happen.

On the side of end-user programs there is a mixed situation. As we saw, some client programs, such as WinPack, support MD2 procedure, whereas some node/server programs, such as FPAC, support MD5 mode. As a result, using encrypted passwords does not work in all combinations so the workarounds have to be made – by either entering passwords manually, or something else. Therefore, additional efforts in programming new features that would be mutually compatible would be highly appreciated.

6 Conclusion

The examples given in this chapter intended to inform students, teachers and other newcomers to the exciting world of amateur wireless radio that many counter fighting measures can be successfully implemented against potential would-be 'ham pirates'. As you learned here, there exist varieties of methods that help our daily amateur radioactivity to be safer and remain protected. Depending on ideas in the amateur radio programmers' community, there appear various principles for making 'ham' digital communications even more secure. Some approaches might be more suitable with specific server and client software; the others are universal.

Therefore, we want to encourage the readers to explore new horizons by their own, because there is not much literature on this topic available. Final tips and reminders:

System administrators who do not have many users of their email servers (and even less password-protected users), would do the best if implement the simplest security mechanism of a secret 'large string'. Whenever needed – if proven to become compromised or even better in regular intervals, the content of the string can be fully replaced or slightly modified (as described in this chapter).

Those 'sysops' who want to experiment with approaches that are more complex and whose customers use newer communicating programs shall test the feature of monthly 'secret tables'. At first, you can try to use the same 'large string' as a foundation for the daily lines within the secret table. If it proves not to be safe enough (i.e. because of just a few elements in a row), you can try with longer daily lines. Make some more tests and compare the results.

If the password-protected users are equipped with communicating software that integrate automated safety routines (such as *MD* algorithms), use it whenever possible, because the computing process itself would be efficient enough and fully transparent for your end-users.

If you and your collaborating neighboring partner(s) implement newer versions of Protus tools, enable some of available automatic BBS-to-BBS modes for corresponding servers' callsigns. Following that method, you would not need to replace the content of the secret key for very long periods of time.

Keep an eye on daily routines in mailbox activities of your new users, to get familiar with their behavior and practice. In addition to that control, keep them always informed about the existence of the safety facilities you have implemented within your server.

Get familiar with the safety lessons within the amateur radio classes and courses in your area. Advise their lecturers about the safety measures, which are the counter fighting methods against poorly educated beginners in the 'ham' radio community.

Should you, as a scholar or student, consider implementing the amateur radio technology in your educational environment, you do not have to be a computer expert! If you can read operating manuals and follow instructions, you can easily become an active *packeteer*. "Like any other facet of Amateur Radio, you're going to have to learn some new concepts, but that's part of the enjoyment", says Steve, WB8IMY [6]. Other authors noted that too, long ago. As Lucas, Jones and Moore [7] suggested, the amateur wireless radio brought benefits to various social groups and individuals who had links to the educational environment. We focus on stimulating students and teachers for adopting amateur radio in their daily activities with computing and wireless technologies. The best way might be adding radio data communications to our scholarly computer rooms. That means, an amateur radio BBS in your school might provide a useful 'gateway', a corridor to a school's LAN for those parents who already belong to the amateur radio community (or vice versa). By exploring such infrastructure on your own, you would attract other technology enthusiasts from your nearest academic community because 'ham radio' has already found its place at many universities.

Inclusion of described and other amateur radio experiments within an educational environment may help to increase interest in youths and young adults for information and communication technology career. As recent articles reported, in the information economy technical literacy is a prerequisite for many occupations, even beyond technology positions [8]. We can add that serious stepping into the 'ham' radio can be a trigger for advancing existing careers, including widening social circuits among 'radioactive' professionals, students and scholars alike.

Acknowledgements. Teachers cannot be the only ones who are responsible for discovering new talents in worlds of engineering and technology. The process of learning starts within the family so the parents should be able to recognize young potentials at homes. This book chapter is dedicated to the late persons, Mr. Sava Skoric and Mrs. Radmila Skoric, who recognized the exploring nature of their son during his early age. As brave parents, they did not punish him too much for his breaking and destroying various home appliances – while inspecting the secrets of hidden mechanisms *living* inside. Instead, they supported their kid in materializing his imagination and searching for new horizons. Credits also go to all mentioned, known and unknown amateur radio enthusiasts who donated their time and efforts in enriching existing wireless amateur radio technologies with various safety measures.

References

[1] Skoric, M.: Simulation in amateur packet radio networks. In: Al-Bahadili, H. (ed.) Simulation in Computer Network Design and Modeling: Use and Analysis, pp. 216–256. IGI Global, Hershey (2012)
[2] Skoric, M.: The New Amateur Radio University Network – AMUNET. In: Proceedings of the 9th WSEAS International Conference on Computers, pp. 497–717. WSEAS Press, Athens (2005) ISBN: 960-8457-29-7
[3] Skoric, M.: FBB Packet radio BBS mini-HOWTO. The Linux Documentation Project (2000-2010), `http://tldp.org/HOWTO/FBB.html` (retrieved April 15, 2012)
[4] Ford, S.: Your Packet Companion. American Radio Relay League, Newington (1995)
[5] Skoric, M.: The Amateur Radio as a Learning Technology in Developing Countries. In: Proceedings of the 4th IEEE International Conference on Advanced Learning Technologies, pp. 1029–1033. IEEE Computer Society, Los Alamitos (2004)
[6] Ford, S.: Your HF Digital Companion. American Radio Relay League, Newington (1995)
[7] Lucas, L.W., Jones, J.G., Moore, D.L.: Packet Radio: An Educator's Alternative to Costly Telecommunications. Texas Center for Educational Technology, Denton (1992)
[8] Constanza, T.: Youths love tech, but not necessarily tech career. Silicon Republic (2012), `http://www.siliconrepublic.com/careers/item/27746-youths-love-tech-but-not/` (retrieved June 16, 2012)

Additional Reading Section

Blystone, K., Watson, M.: Alternative Information Highways: Networking School BBSs. Texas Center for Educational Technology, Denton (1995)
Corley, A.: Hams in Haiti. IEEE Spectrum (2010), `http://spectrum.ieee.org/telecom/wireless/hams-in-haiti/0` (retrieved October 17, 2010)
Davidoff, M.: The Satellite Experimenter's Handbook. American Radio Relay League, Newington (1994)
Diggens, M.: Enhancing distance education through radio-computer communication. In: Atkinson, R., McBeath, C. (eds.) Open Learning and New Technology: Conference Proceedings, pp. 113–116. Australian Society for Educational Technology, Perth (1990)

Dowie, P.: Presentation: XROUTER Network Infrastructure Software - Paula G8PZT (2002), http://www.g8pzt.pwp.blueyonder.co.uk/fourpak/2002_03.htm (retrieved September 28, 2011)

Edwards, J.: Want to bone up on wireless tech? Try ham radio. Computerworld (2009), http://www.computerworld.com/s/article/9139771/Want_to_bone_up_on_wireless_tech_Try_ham_radio (retrieved October 17, 2011)

Erhardt, W.: Bill's Amateur Radio Page (2010), http://www.k7mt.com/AmateurRadio.htm (retrieved January 16, 2011)

Hill, J.: Amateur Radio—A Powerful Voice in Education. QST 86(12), 52–54 (2002)

Hudspeth, D., Plumlee, R.C.: BBS Uses in Education. Texas Center for Educational Technology, Denton (1994)

Jones, G.: Packet Radio: What? Why? How? Tuckson Amateur Packet Radio, Tuckson (1996)

Kasal, M.: Experimental Satellites Laboratory (2010), http://www.urel.feec.vutbr.cz/esl/ (retrieved January 16, 2011)

Langelaar, M.: JNOS 2.0 - DOS Install. The Easy Way (2009), http://www.langelaar.net/projects/jnos2/documents/install/dos/ (retrieved June 10, 2012)

Lucas, L.: Wide Area Networking Guide for Texas School Districts. Texas Center for Educational Technology, Denton (1997)

Martin, J.: Linux - Jnos Setup and Configuration HOW-TO (2006), http://www.kf8kk.com/packet/jnos-linux/linux-jnos-setup-9.htm (retrieved October 17, 2010)

Martin, J.: JNOS Operators Guide (2006), http://www.nyc-arecs.org/JNOS_OpGuide.pdf (retrieved June 10, 2012)

Martin, J.: Whetting your feet with Jnos (2006), http://legitimate.org/iook/packet/jnos/whetting/whetting.htm (retrieved September 28, 2010)

McCosker, R.: Creating a XRouter Remote Node (2010), http://vk2dot.dyndns.org/XRouter/XRouter.htm (retrieved September 28, 2010)

McDonough, J.: The Michigan Digital Network (2007), http://packet.mi-nts.org/257/MIdigital.pdf (retrieved September 7, 2010)

McLarnon, B.: Packet Radio Technology: An Overview (2008), http://www.friends-partners.org/glosas/Tampere_Conference/Reference_Materials/Packet_Radio_Technology.html (retrieved October 17, 2010)

Moxon, L.: HF Antennas for All Locations. Radio Society of Great Britain, Potters Bar (1993)

Przybylski, J.: Home Web SP1LOP (1997), http://www.sp1lop.ampr.org/ (retrieved January 16, 2011)

Przybylski, J.: All Poland Packet Info Server (2009), http://www.packet.poland.ampr.org/ (retrieved January 16, 2011)

Skoric, M.: The perspectives of the Amateur University Networks – AMUNETs. WSEAS Transactions on Communications 4, 834–845 (2005)

Skoric, M.: The New Amateur Radio University Network – AMUNET (Part 2). In: Proceedings of the 10th WSEAS International Conference on Computers, pp. 45–50. World Scientific and Engineering Academy and Society, Athens (2006)

Skoric, M.: Summer schools on the amateur radio computing. In: Proceedings of the 12th Annual SIGCSE Conference on Innovation and Technology in Computer Science Education, pp. 346–346. Association for Computing Machinery, New York (2007)

Skoric, M.: The New Amateur Radio University Network – AMUNET (Part 3). In: Proceedings of the 12th WSEAS International Conference on Computers: New Aspects of Computers, pp. 432–439. World Scientific and Engineering Academy and Society, Athens (2008)

Skoric, M.: Amateur Radio in Education. In: Song, H., Kidd, T. (eds.) Handbook of Research on Human Performance and Instructional Technology, pp. 223–245. Information Science Reference (IGI-Global), Hershey (2009)

Skoric, M.: The New Amateur Radio University Network – AMUNET (Part 4). In: Recent Advances in Computers: Proceedings of the 13th WSEAS International Conference on Computers, pp. 323–328. WSEAS Press, Athens (2009)

Sumner, D. (ed.): 22nd ARRL and TAPR Digital Communications Conference. American Radio Relay League, Newington (2003)

Sumner, D. (ed.): 26th ARRL and TAPR Digital Communications Conference. American Radio Relay League, Newington (2007)

Wade, I.: NOSintro – TCP/IP over Packet Radio (1992), http://homepage.ntlworld.com/wadei/nosintro/ (retrieved September 28, 2010)

Weiss, R.T.: Ham Radio Operator Heads South To Aid Post-Katrina Communications. Computerworld (2005), http://www.computerworld.com/newsletter/0,4902,104446,00.html?nlid=MW2 (retrieved October 17, 2010)

Weiss, R.T.: Ham radio volunteers help re-establish communications after Katrina. Computerworld (2005), http://www.computerworld.com/securitytopics/security/recovery/story/0,10801,104418,00.html (retrieved October 17, 2010)

Wiseman, J.: BPQAXIP Configuration (2010), http://www.cantab.net/users/john.wiseman/Documents/BPQAXIP%20Configuration.htm (retrieved May 17, 2012)

Wiseman, J.: BPQETHER Ethernet Driver for BPQ32 switch (2010), http://www.cantab.net/users/john.wiseman/Documents/BPQ%20Ethernet.htm (retrieved May 17, 2012)

Worcester Polytechnic Institute Wireless Association (2012), The history of W1YK, http://users.wpi.edu/~wpiwa/about.html (retrieved June 09, 2012)

Key Terms & Definitions

Amateur Radio: Also known as 'ham radio', and similar to terms of 'radio communication' and 'wireless radio'. It is a century-old activity in voluntary experimenting with radio waves, and exploring communications with other parts of the globe on a non-profit basis and by using advantages over commercial services, such as low cost, independence of official infrastructure, etc. Besides that, nowadays the radio amateurs are involved in emergency services world-wide, as a way of establishing ad-hoc communicating systems after natural disasters, such as tornadoes, earthquakes and floods.

AMUNET: This acronym stands for the AMateur radio University computer NETwork, which is a proposed name for the wireless network of an amateur radio bulletin board system at a local university, including one or more amateur radio 'digipeaters', and one or more end-user computers in surrounding schools' computer labs, teachers' offices, and homes of students and their parents. The acronym can replace the terms with similar or corresponding meanings, such as 'packet-radio network', 'ham packet net', etc. AMUNET is associated with amateur radio, packet-radio, and computer communication.

Authentication: A process or a procedure of checking if a user of an e-mail server or other technical infrastructure (such as a radio-relay system, etc., in the context of this chapter), is allowed - or is not allowed to access a remote system and use its resources. Authentication can be implemented for all users (closed systems), or for specific users (open systems). In addition, authentication is performed for checking if incoming "ordinary" users are authorized for advanced and restricted operations within the protected systems, such as system administration.

BBS: An electronic Bulletin Board System, a software that usually operates on a personal computer equipped with one or more telephone lines, amateur radio stations and Internet connections, to provide communication between remote users such as electronic mail, conferences, news, chat, files and databases. Also known as 'message board' or 'mailbox', and is similar to the term of 'e-mail'.

Challenge: The first part in the authenticating process when a remote system asks an incoming user to authorize himself or herself, by sending him or her specific alphanumeric string, which is expected to be responded by sending back a unique and proper "response".

Computer: An electronic device that hosts communicating or other software. Here it is associated with terms of hardware, computer network(s), operating system(s), computer science, computer labs, etc.

Educator: Also known as a teacher, professor, instructor, trainer. That is a person who transfers knowledge to the audience, shares knowledge with peers, experiments with new technologies, etc.

Experiment: A term similar with 'test', 'probe', 'simulation', etc. In the context of this manuscript, it is associated with exploring new approaches to existing or incoming new technologies, etc. Experiments are used to get practical results in school laboratories, for example to test different models in security & privacy in computer networks and wireless systems in order to implement better protective measures, etc. Experiments can be done after school hours, by using inexpensive, older computers.

Gateway: A gateway is a computer that connects two different networks together. The gateway will perform the protocol conversions necessary to go from one network to the other. For example, a gateway could connect a local area network (LAN) of computers in the school to the Internet. In addition, an amateur radio BBS might provide a gateway for the school's LAN to the 'air', or vice versa.

Packet radio: A communication mode between the amateur radio stations where computers control how the radio stations handle the traffic. The computers and attached modems organize information into smaller chunks of it – often referred as 'packets' of data, and route the packets to intended destinations. Also known as ‘AX.25’, although that term references to the ‘Amateur [variant of] X.25’ protocol. It is also similar to the terms such as ‘amateur [digital] radio’, ‘computer [radio] communication’, and ‘AMUNET’.

Password: A secret word, a sentence, or a string of random alphanumerical characters that is usually negotiated between end-users and system operators of email servers or other radio infrastructure. The secret content has to be sent, either as a clear or ciphered text, to the other end of a communication channel, before a user is authored to perform activities within the protected system. Sometimes the password is referred as the term ‘key’, such as in ‘key management’, ‘key exchange’, etc.

Remote Access: The ability to access a technical device (E.g. a computer running as the BBS, etc.) from outside a building in which it is installed. Remote access requires communications hardware, software, and actual physical links. Different users can have different access rights (user permissions) associated with their account on a BBS or a radio-relay system.

Repeater: In radio communications, a repeater is a device that amplifies or regenerates the signal in order to extend the distance of the transmission. Repeaters are available for both analog (voice) and digital (data) signals. 'Digipeater' is the common name for a digital repeater. The term is also known as a ‘radio-relay’ station.

Response: The second part in the authenticating procedure when an incoming user authorizes him or her as a valid user of a remote system, by sending back a specific alphanumeric string, which must uniquely correspond to a “challenge” – a string of characters previously received from the remote station.

Student: Also known as pupil, learner, scholar, etc. That is a person who is associated with schools, universities, learning and studying activities and procedures, and who obtains knowledge and acquires new skills by interacting with teachers, professors, instructors and other educators.

Sysop: It is a short name for Systems Operator. That person operates and maintains an amateur radio BBS or a repeater. Some sources refer to the sysop as "system administrator".

University: A working space intended for learning-studying, teaching, and researching. Similar to the term of ‘academia’, although in this manuscript’s context it represents all kinds of educational environments – where interests of students, teachers, parents and others join together. It is associated with educators, students, experiments, computer labs, etc.

Appendix A

The following content is the functional part of USERS.PRT – the configuration file of the Protus c_filter, such as we used in Experiment 1. Although the same file includes password parameters for various users, for the sake of clarity we left here only the callsign of a user who participated in a particular experiment. Please take attention to the value of parameter in ***bold_italic***.

```
########################################################
#   {PROTUS-4.0}      ###    Fichero de usuarios       #
########################################################
# Cada bloque tiene el siguiente formato:
#   INDICATIVO
#   PUERTO   MODO    PASSWORD (Max. 260 caracteres)
#   --------
#
# El parametro "modo" puede tomar los valores:
#    1 =>  ACCESO LIBRE               (ABIERTO / OPEN)
#    2 =>  MENSAJE Y DESCONEXION      (CERRADO / CLOSED)
#    3 =>  EXCLUIDO                   (EXCLUIDO / EXCLUDED)
#    4 =>  PASSWORD FRASE             (FRASE / FIXED)
#    5 =>  PASSWORD NORMAL            (NORMAL)
#    6 =>  PASSWORD MD2/MD5           (MD / MD2)
#    7 =>  PASSWORD TABLA             (TABLA / TABLE)
#    8 =>  BBS                        (BBS)
#    9 =>  PASSWORD TIPO FLEXNET      (FLEXNET)
#   10 =>  PASSWORD FORWARD BBS MD2   (FORWARD)
#   11 =>  SOLO LECTURA               (SOLOLEER / READONLY)
#   12 =>  MANTENER MENSAJES EN HOLD  (RETENIDOS / HOLD)
#   13 =>  PASSWORD FORWARD BBS MD5   (FORWARD5)
#
# Validos comodines "*" en indicativo y puerto.
########################################################
#
YT7MPB-0
2   5
umoransamodkafanavolimpivoirakijuumoransamodkafanavolimpi
voirakijuumoransamodkaf
--------
#
# Fin del fichero de usuarios.
########################################################
```

The following content is the functional part of BBS.TXT – the script WinPack used for establishing telnet connection to FBB email server, as performed in Experiment 1.

```
; All lines in script files that start with ';' are ignored.
;
; This script file - BBS.TXT - ***MUST*** do nothing else other
; than make a connection to your local BBS and then leave
you at
; the BBS prompt. It is used in auto-connects.
;
; The hot key.
HOTKEY F2
; The title.
TITLE Connect to the local BBS
; Send the command to connect to my local BBS.
SEND c 192.168.0.1:6300
; Wait for the connection to be established. If you get a
very fast
; connection to the BBS, you may find you need to remove
the WAITCON
; otherwise the BBS prompt may be missed.
WAITCON
SEND .$MYCALL
SEND test
```

Appendix B

The following content is the functional part of USERS.PRT – the configuration file of the Protus c_filter, such as we used in Experiment 2. Although the same file includes password parameters for various users, for the sake of clarity we left here only the callsign of a user who participated in a particular experiment. Please take attention to the changed parameter in ***bold_italic***.

```
##########################################################
#    {PROTUS-4.0}       ###     Fichero de usuarios        #
##########################################################
# Cada bloque tiene el siguiente formato:
#   INDICATIVO
#   PUERTO   MODO    PASSWORD (Max. 260 caracteres)
#   --------
#
# El parametro "modo" puede tomar los valores:
#   1 =>  ACCESO LIBRE                 (ABIERTO / OPEN)
#   2 =>  MENSAJE Y DESCONEXION        (CERRADO / CLOSED)
#   3 =>  EXCLUIDO                     (EXCLUIDO / EXCLUDED)
```

```
#    4 =>  PASSWORD FRASE               (FRASE / FIXED)
#    5 =>  PASSWORD NORMAL              (NORMAL)
#    6 =>  PASSWORD MD2/MD5             (MD / MD2)
#    7 =>  PASSWORD TABLA               (TABLA / TABLE)
#    8 =>  BBS                          (BBS)
#    9 =>  PASSWORD TIPO FLEXNET        (FLEXNET)
#   10 =>  PASSWORD FORWARD BBS MD2     (FORWARD)
#   11 =>  SOLO LECTURA                 (SOLOLEER / READONLY)
#   12 =>  MANTENER MENSAJES EN HOLD (RETENIDOS / HOLD)
#   13 =>  PASSWORD FORWARD BBS MD5     (FORWARD5)
#
# Validos comodines "*" en indicativo y puerto.
###########################################################
#
YT7MPB-0
2   7
umoransamodkafanavolimpivoirakijuumoransamodkafanavolimpi
voirakijuumoransamodkaf
--------
#
# Fin del fichero de usuarios.
###########################################################
```

The following content is the functional part of BBS.TXT – the script WinPack used for establishing telnet connection to FBB email server, as performed in Experiment 2.

```
; All lines in script files that start with ';' are ignored.
;
; This script file - BBS.TXT - ***MUST*** do nothing else other
; than make a connection to your local BBS and then leave you at
; the BBS prompt. It is used in auto-connects.
;
; The hot key.
HOTKEY F2
; The title.
TITLE Connect to the local BBS
; Send the command to connect to my local BBS.
SEND c 192.168.0.1:6300
; Wait for the connection to be established. If you get a
very fast
; connection to the BBS, you may find you need to remove
the WAITCON
```

```
; otherwise the BBS prompt may be missed.
WAITCON
SEND .$MYCALL
SEND test
```

Appendix C

The following content is the functional part of USERS.PRT – the configuration file of the Protus c_filter, such as we used in Experiment 3. Although the same file includes password parameters for various users, for the sake of clarity we left here only the callsign of a user who participated in a particular experiment. Please take attention to the changed parameter in ***bold_italic***.

```
########################################################
#    {PROTUS-4.0}       ###      Fichero de usuarios      #
########################################################
# Cada bloque tiene el siguiente formato:
#   INDICATIVO
#   PUERTO   MODO    PASSWORD (Max. 260 caracteres)
#   --------
#
# El parametro "modo" puede tomar los valores:
#   1 =>  ACCESO LIBRE                (ABIERTO / OPEN)
#   2 =>  MENSAJE Y DESCONEXION       (CERRADO / CLOSED)
#   3 =>  EXCLUIDO                    (EXCLUIDO / EXCLUDED)
#   4 =>  PASSWORD FRASE              (FRASE / FIXED)
#   5 =>  PASSWORD NORMAL             (NORMAL)
#   6 =>  PASSWORD MD2/MD5            (MD / MD2)
#   7 =>  PASSWORD TABLA              (TABLA / TABLE)
#   8 =>  BBS                         (BBS)
#   9 =>  PASSWORD TIPO FLEXNET       (FLEXNET)
#  10 =>  PASSWORD FORWARD BBS MD2    (FORWARD)
#  11 =>  SOLO LECTURA                (SOLOLEER / READONLY)
#  12 =>  MANTENER MENSAJES EN HOLD (RETENIDOS / HOLD)
#  13 =>  PASSWORD FORWARD BBS MD5    (FORWARD5)
#
# Validos comodines "*" en indicativo y puerto.
########################################################
#
YT7MPB-0
2  6
umoransamodkafanavolimpivoirakijuumoransamodkafanavolimpi
voirakijuumoransamodkaf
--------
```

```
#
# Fin del fichero de usuarios.
#########################################################
```

The following content is the functional part of BBS.TXT – the script WinPack used for establishing telnet connection to FBB email server, as performed in Experiment 3. Please take attention to the new parameter in ***bold_italic***.

```
; All lines in script files that start with ';' are ignored.
;
; This script file - BBS.TXT - ***MUST*** do nothing else other
; than make a connection to your local BBS and then leave you at
; the BBS prompt. It is used in auto-connects.
;
; The hot key.
HOTKEY F2
; The title.
TITLE Connect to the local BBS
; Send the command to connect to my local BBS.
SEND c 192.168.0.1:6300
; Wait for the connection to be established. If you get a
very fast
; connection to the BBS, you may find you need to remove
the WAITCON
; otherwise the BBS prompt may be missed.
WAITCON
SEND .$MYCALL
SEND test
RUN MD2PASS
```

The following content is the functional part of MD2PASS.CFG – the script that WinPack used for computing a *response* to a FBB's *challenge*, as performed in Experiment 3.

```
;Lines starting with a ';' and blank lines are ignored.
;
;Your MD2 password string.
;RogerBarker79SouthParadeBostonLincsPE217PN
umoransamodkafanavolimpivoirakijuumoransamodkafanavolimpi
voirakijuumoransamodkaf
;Some text from the prompt the BBS sends in the password
line.
```

```
;E.g. My test BBS sent "PASSWORD DE GB7IDE> [1234567890]",
so I used:-
PASSWORD
```

Appendix D

The following content is the functional part of USERS.PRT – the configuration file of the Protus c_filter, such as we used in Experiment 4. Although the same file includes password parameters for various users, for the sake of clarity we left here only the callsign of a user who participated in a particular experiment. Please take attention to the changed (deactivated) parameter in ***bold_italic***.

```
#########################################################
#    {PROTUS-4.0}      ###       Fichero de usuarios    #
#########################################################
# Cada bloque tiene el siguiente formato:
#   INDICATIVO
#   PUERTO   MODO    PASSWORD (Max. 260 caracteres)
#   --------
#
# El parametro "modo" puede tomar los valores:
#   1 =>  ACCESO LIBRE                (ABIERTO / OPEN)
#   2 =>  MENSAJE Y DESCONEXION       (CERRADO / CLOSED)
#   3 =>  EXCLUIDO                    (EXCLUIDO / EXCLUDED)
#   4 =>  PASSWORD FRASE              (FRASE / FIXED)
#   5 =>  PASSWORD NORMAL             (NORMAL)
#   6 =>  PASSWORD MD2/MD5            (MD / MD2)
#   7 =>  PASSWORD TABLA              (TABLA / TABLE)
#   8 =>  BBS                         (BBS)
#   9 =>  PASSWORD TIPO FLEXNET       (FLEXNET)
#  10 =>  PASSWORD FORWARD BBS MD2    (FORWARD)
#  11 =>  SOLO LECTURA                (SOLOLEER / READONLY)
#  12 =>  MANTENER MENSAJES EN HOLD   (RETENIDOS / HOLD)
#  13 =>  PASSWORD FORWARD BBS MD5    (FORWARD5)
#
# Validos comodines "*" en indicativo y puerto.
#########################################################
###############
#
YT7MPB-0
#2   5
umoransamodkafanavolimpivoirakijuumoransamodkafanavolimpi
voirakijuumoransamodkaf
--------
```

```
#
# Fin del fichero de usuarios.
###########################################################
```

The following content is the functional part of BBS.TXT – the script WinPack used for establishing telnet connection to FBB email server, as performed in Experiment 4. Please take attention to the now inactive parameter in ***bold_italic***.

```
; All lines in script files that start with ';' are ignored.
;
; This script file - BBS.TXT - ***MUST*** do nothing else other
; than make a connection to your local BBS and then leave you at
; the BBS prompt. It is used in auto-connects.
;
; The hot key.
HOTKEY F2
; The title.
TITLE Connect to the local BBS
; Send the command to connect to my local BBS.
SEND c 192.168.0.1:6300
; Wait for the connection to be established. If you get a very fast
; connection to the BBS, you may find you need to remove
the WAITCON
; otherwise the BBS prompt may be missed.
WAITCON
SEND .$MYCALL
SEND test
#RUN MD2PASS
```

The following content is the functional part of MD2PASS.CFG – the script that WinPack used for computing a *response* to a FBB's *challenge*, as performed in Experiment 4. Please take attention to the changed parameters in ***bold_italic***.

```
;Lines starting with a ';' and blank lines are ignored.
;
;Your MD2 password string.
;RogerBarker79SouthParadeBostonLincsPE217PN
umoransamodkafanavolimpivoirakijuumoransamodkafanavolimpi
voirakijuumoransamodkaf
;Some text from the prompt the BBS sends in the password line.
;E.g. My test BBS sent "PASSWORD DE GB7IDE> [1234567890]", so
I used:-
;PASSWORD
YT7MPB-4
```

Appendix E

The following content include the steps we used in configuring Linux amateur radio networking with an intention to enable a FPAC node to be visible (and accessible) from within mixed LAN environments that include amateur radio nodes running under MS Windows, such as BPQ32 nodes – examined in our earlier experiments. For the sake of simplicity in this tutorial chapter, we wanted to have a situation in a computer LAN that would simulate a real amateur radio infrastructure, including time delay in between two nodes, etc. To achieve that goal, we decided to implement so-called *bpq* (or precisely *bpqether*) protocol in the LAN, partly because it is a 'native' one for Windows-based computers running BPQ32 nodes. In opposite, FPAC seemed to be dedicated to so-called *rose* protocol that is not fully compatible with BPQ32. However, Linux supports configuring *bpq* ports on Ethernet cards in the LAN, in a way that usual Ethernet interfaces, known as eth0, eth1, etc. can be adopted to become bpq0, bpq1, etc. interfaces accordingly. In our scenario, see the Figure 31, we had two Ethernet cards on our Linux node, so we configured them (by using the first set of commands, *ifconfig*) to exchange amateur radio traffic with two BPQ32-equipped computers, wired by those two networking cards. The second set of commands, *nrattach*, we used to provide automatic visibility of both the node (netrom) and email server (netbbs) in the rest of our Windows-based network. Finally, to make all that working, we started appropriate 'daemons' (background services) in Linux, which are *ax25d*, *netromd*, as well as *fpac.sh* script. Once again, for the sake of simplicity, we entered all those commands one-after-another, although it might be suitable to put all

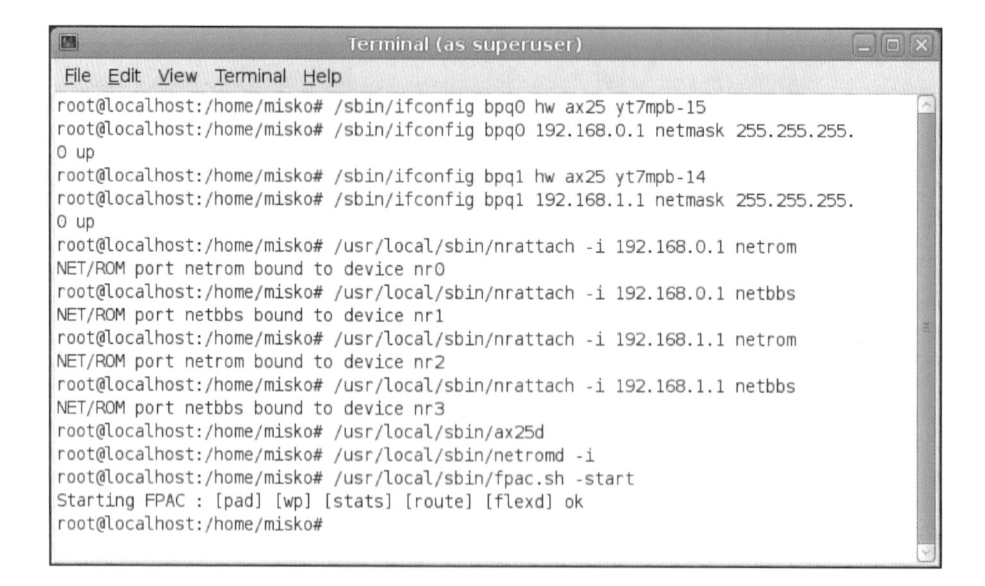

Fig. 31. Configuring amateur wireless networking in Linux environment

of them into a file that activates all of them at the system boot. The only disadvantage, or better to say a 'collateral damage' in our approach we found in eventual disabling the primary functions in our network interfaces eth0, eth1, etc. – so after having newly established amateur radio traffic within the LAN, nothing else remained available (such as the shared Internet access in the same LAN, etc). Therefore, for re-enabling temporarily disabled features mentioned, it would be necessary to shut down Linux and restart it again. (Probably it could be possible to re-enable those functions without system reboot, but investigating that issue is out of scope of this book.)

Appendix F

The following *#define* lines are the password options from *config.h* file that we included in our configuration of JNOS 2.0j amateur radio mailbox. (Previously those three options were not planned to be used, so they had a label *#undef* instead of *#define*.)

```
#define MD5AUTHENTICATE  /* Accept MD5-Authenticated logins */
#define AX25PASSWORD     /* Ask ax.25 users for their passwords */
#define NRPASSWORD       /* Also ask NetRom users for passwords */
```

Security Issues in Mobile Ad Hoc Network

Noman Islam[1] and Zubair Ahmed Shaikh[2]

[1] National University of Computer and Emerging Sciences, Karachi, Pakistan
noman.islam@nu.edu.pk
[2] DHA Suffa University, Karachi
zubair.shaikh@nu.edu.pk

Abstract. Mobile Ad hoc Networks (MANET) are infrastructure-less networks characterized by lack of prior configuration and the hostile environments. Their unique properties make them a natural candidate for situations where unplanned network establishment is required. However this flexibility leads to a number of security challenges. Security in MANET is a very complex job and requires considerations on the issues spanning across all the layers of communication stack. This chapter reviews the security problem in MANET. It provides an updated account of the security solutions for MANET with detailed discussions on secure routing, intrusion detection system and key management problems. The chapter is concluded with a comprehensive security solution for MANET.

Keywords: active attacks, COINS, cross-layer security, intrusion detection, key management, MANET, passive attacks, secure routing, threshold cryptography, trust management.

1 Introduction

Information security has been a significant issue for mankind. During the early days, different kinds of steganographic techniques were used for securing the information. These techniques were based on hiding the data through invisible inks, wax tablets and microdots etc. [1]. With the advent of computer networks, the importance of security grew further. The emergence of wireless networks present new security threats. Different types of security protocols and algorithms were thus developed. The last few years have witnessed the emergence of modern form of computing gadgets. At the same time, advanced forms of connectivity mediums (Blue-tooth, WIMAX etc.) have been proposed. These advancements give rise to the flexibility of network creation on the fly. Such networks called Mobile Ad hoc Network (MANET), are infrastructure-less network comprising of an autonomous collection of mobile routers connected by wireless medium [2]. These networks have dynamic and constantly varying topology. Nodes can join and leave the network any time and the links are created and broken dynamically. The distinctive attributes of MANET has made it useful for a large number of applications. Fig. 1 shows some of the applications of MANET. These applications include various types of commercial (intelligent transportation system, ad hoc gaming, smart agriculture) and non-commercial applications (military application, disaster recovery, wild life monitoring) etc.

S. Khan and A.-S.K. Pathan (Eds.): *Wireless Networks and Security*, SCT, pp. 49–80.
DOI: 10.1007/978-3-642-36169-2_2

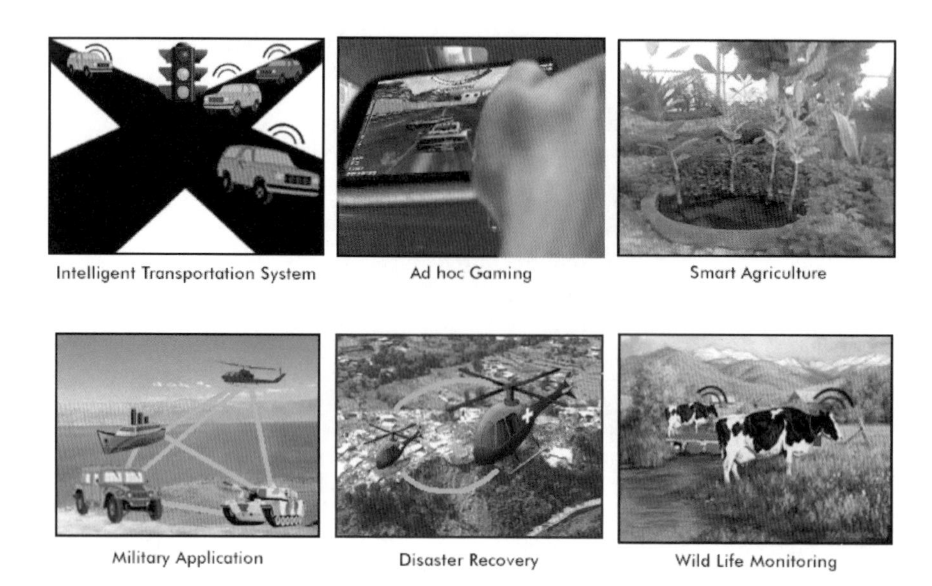

Fig. 1. Applications of MANET

To truly realize the applications of MANET, a large amount of research efforts are underway. These efforts attempt to address the research issues of MANET. For example, at the physical layer there are issues related to MAC protocol design and access control. Due to the wireless nature of the medium, two nodes that are not in each other communication range can simultaneously transmit the message to same source (*hidden node problem*). Similarly, two nodes in each other communication range can unnecessarily delay the transmission, even though their simultaneous transmission doesn't cause any collision (*exposed node problem*). As the nodes and the links in MANET are highly dynamic, novel routing mechanisms are required at network layer that can cope with the frequent topological changes and that utilize techniques to avoid routing loops, false routes etc. Due to the infrastructure-less nature of MANET, addressing protocols are also required that can dynamically assign and manage the address space. At the transport layer, there are protocol design issues, as the TCP congestion avoidance algorithm leads to very poor through put in dynamic networks. There are also issues at application layer like design of novel applications, service discovery and data management problems. A number of cross-layer research issues are also under investigation. Table 1 summarizes various research issues in MANET. Rest of the sections focuses on security problem in MANET. Interested readers can refer to [3] for a detailed discussion on research issues on MANET.

2 A Overview of Security Issues in MANET and Countermeasures

Among the various research issues of MANET, security is considered the most difficult job. It demands attention to a number of new research challenges.

Table 2 summarizes the security challenges in MANET in contrast to traditional systems. To address the security challenges, various solutions have been proposed. Survey of various security solutions for MANET have been provided in [4-7]. However, the need for an updated survey on MANET is desired. This research provides an updated account of currently available security solutions for MANET. The survey specifically ponders over the security solutions for routing, key management and intrusion detection in MANET. In addition, it provides a new classification for secure routing, key management and intrusion detection systems proposed in recent past for MANET.

Table 1. Research Issues in MANET

Physical Layer	Network Layer	Transport Layer	Application Layer
Antenna Design	Routing	Protocol Design	Service Discovery
Access Control	IP Address Assignment		Data Management
Interference	Gateway		
MAC Addressing	Multicasting		
	Clustering		
Power Management			
QoS			
Standards			
Security			

Table 2. Security Challenges in MANET

Challenge	Description
Wireless link	Open and physically accessible to everyone, prone to bit errors/interference
Lack of secure boundaries	Adversaries can easily join and become part of the network
Infrastructure-less	There are no specific infrastructure for addressing, key distribution, certification etc.
Nodes limitation	As the nodes have limited capabilities, their availability can easily be compromised
Link limitation	Cooperation based security algorithms must consider the bandwidth limitation associated with links
Multi-hop routing	As the nodes are dependent on each other for routing, adversaries can generate fabricated routes to create routing loops, false routes etc.

Before proceedings towards the discussion, we start with an overview of known security attacks on MANET. We will discuss active and passive attacks possible at various layers of protocol stack. An *active attack* is a type of attack where an attacker gets access to the medium of communication and modifies or disrupts the transmission. For example, Denial of Service (DoS) attack attempts to overwhelm a target machine (via sending huge number of communication requests), so that the victim can't respond to legitimate requests of other hosts. A *passive attack* only observes the ongoing transmission but doesn't alter or disrupts any activity. A common example is traffic analysis of the snooped data to discover passwords and confidential information of other users. Active attacks usually target integrity and availability of system while passive attack tries to break the confidentiality of the security system. Following section discusses some of the known attacks and the countermeasures for these attacks.

2.1 Physical Layer Attacks

Physical layer is the lowest layer of TCP/IP. It is responsible for encoding the data, converting into signals and transmission at the other end. Being the lowest layer of protocol stack, it is vulnerable to various types of security threats. The easiest attack on this layer is *signal jamming*, where the hacker uses a jamming device to tune to the frequency of the nodes on the network. The jammer then generates a constant and powerful noisy signal that suppresses other messages on the network. To counter this attack, spread spectrum techniques are recommended that changes the frequency of the signal or spread the energy of the signal to a wider spectrum. For example, in Frequency Hopping Spread Spectrum (FHSS), the sender switches the carrier from a set of available carriers based on a pseudo random sequence. The sender modulates the data signal with a sequence of random frequencies. The random frequency changes at regular interval of times. Both the sender and receiver are synchronized such that receiver can reconstruct the original signal. However, an attacker can jam a wide frequency band which makes the FHSS ineffective. In Direct Sequence Spread Spectrum (DSSS), the signal is spread across a frequency band using a spreading code. The complexity associated with DSS however, limits its practicability in networks with low capability devices.

MANET is often deployed in hostile settings like war zone, disaster recovery etc. It is possible that an adversary can get physical access to the device and then *temper* the device. An attacker can also gain physical access to the chip of the node, if the nodes are not locked properly. Using different techniques, the network information maintained in the chip can then be decoded and can be used to set off various security attacks. The physical security of mobile devices can be enforced to some extent by using some security modules like smart cards that requires PIN codes or biometrics for access.

2.2 Link Layer Attacks

IEEE 802.11 is considered as the key enabler for MANET. However, the vulnerabilities associated with this standard can be exploited for different types of security attacks. For example, the malicious node can exploit the exponential back off feature of IEEE 802.11 protocol by sending the data continuously on the medium. This makes the medium busy and other nodes don't get opportunity to send their data. A malicious node can also send RTS/CTS packet with large amount of data for unlimited period of time.

This attack called *indefinite postponement problem* can jeopardize the network, as the nodes on the network are required to wait indefinitely for their turn. Also, if a normal node is sending RTS/CTS packet, malicious node can know that for a particular period of time, another node will be sending data. During that time, an attacker can the inject errors on the medium by means of wireless interference. This problem can be addressed by using error correcting codes. However, their usage leads to more control overhead. Several vulnerabilities of WEP (use of static keys, small key size and small initialization vector) can also be exploited by attackers for compromising the integrity and privacy of the data. Traffic analysis at link layers are also possible that can be prevented by encryption and traffic padding etc.

Various types of DoS attacks are also possible on wireless medium. In the *single adversary attack*, the malicious node attempts to exhaust the battery of the victim node as well as eating up the channel capacity. For this purpose, the attacker sends large volume of data to the victim node which brings down the availability of the victim node. In the *colluding adversary attack*, two or more malicious node sends large volume of data to each other making the transmission channel occupied and unavailable for other nodes on the network. To counter these attacks, a possible approach is to limit the data rate of nodes. A node is allowed to a send a data rate that must not exceed a threshold value. An alternative approach is to use Time Division Multiplexing, where a fixed time slot is allocated for every node to transmit its data. This approach can also solve the indefinite postponement problem in the exponential back off algorithm. Intrusion detection system running at the link layer and monitoring for misbehaviors can also resolve some of the security attacks at the link layer.

2.3 Routing Layer Attacks

One of key distinction of MANET is their multi-hopped routing feature. Nodes are required to route their packets through possibly unknown nodes. Intermediate nodes can add, modify, delete or unnecessarily delay the forwarding of packets. For example, in source routing protocols like DSR, a malicious intermediate node can temper the source route during route discovery/maintenance phase. Hence, several types of attacks are possible during route discovery, maintenance and packet forwarding phase.

For example, a node can launch *routing table overflow attack* by sending huge volume of false routes to overflow the neighbors. The result is that routing tables are occupied with fake routes and new genuine routes are unable to get established. A malicious node can poison the routing cache of neighbors by advertising false routes that can be heard by neighbors to update their routing tables. For example, in AODV, an attacker can advertise a fake route with smallest hop count and with a fresh sequence number. Other nodes considering this as a genuine route can invalidate their old routes. During route maintenance phase, an attacker can send fake route error messages that can cause the initiation of expensive route maintenance operation. Several advanced form of attacks are also possible. These include black hole attack, worm hole and byzantine attacks etc. In *black hole attack*, a node shows its interest in forwarding a packet towards a destination during route discovery phase. The attacker rushes during route discovery to ensure a route is established through it. Later on in the forwarding phase, it drops the packet intended for the destination. A more severe form of attack is when the malicious node tempers the packet as well. The end result is a very low packet delivery ratio. In a *wormhole attack*, an attacker creates a tunnel with another attacking node. All the packets through the first attacker are tunneled to the second attacking node which then sends the packet through normal path ahead. These attacks can compromise the currently on-demand routing protocols. In *byzantine attack*, an individual or a set of colluding nodes works in cooperation to perform attacks like dropping or modifying packets, creating routing loops, poisons the cache etc.

To avoid the network layer attacks, various generalized strategies have been proposed in literature. For example, packet leashes can be added to the routing packet to restrict its transmission beyond some constraints. A temporal leash specifies the time a packet should take to reach to the destination, thus avoiding wormhole attacks. Specialized routing protocols have been proposed to resolve modification (SEAD), replay (SRP) and repudiation attacks (APALLS) etc. in MANET. Various network intrusion detection systems have been proposed that detects malicious actions on the network and isolate the identified intruders on the network.

2.4 Transport Layer Attacks

The transport layer is vulnerable to the same vulnerabilities in MANET as the classical networks. The problem raises severity owing to the limited resources of MANET. Nodes can launch *acknowledgment spoofing attacks* by generating false acknowledgment with large window size. The source will then send data corresponding to the size of window which can cause congestion as well as exhausting the resources of victim. Other forms of attacks can be done by acknowledgment replay, jamming acknowledgments, sequence number alteration, connection request spoofing. To address these problems, there have been several new versions of transport layer protocols proposed in literature. For example, the Transport Layer Security (TLS) / Secure Socket Layer (SSL) are generally recommended for securing transport layer communication. TLS/SSL is based on public key cryptography and it helps in preventing masquerading and replay attacks.

2.5 Application Layer Attacks

Different types of attacks are possible at the application layers as well. There are threats by malicious software (viruses, worms, Trojan) as well as from insider nodes. Use of firewalls can solve some of these issues. However, firewalls are not resistant against repudiation attack, where a malicious node denies an operation it has performed. In addition, insider attacks can't be addressed by the firewalls. Intrusion Detection Systems (IDS) can be used as a second line of defense in such situations. Table 3 summarizes various types of security attacks on MANET at different layers and the countermeasures that can be used to protect against these attacks. Besides the attacks listen in the table, a number of cross-layer security issues also exist for MANET. One such issue is the secure and reliable discovery of service in MANET. Service discovery in MANET can be performed at application layer as well as network layer. It can suffer from various types of attacks (DoS, Byzantine attacks). Different secure service discovery protocols have been proposed in literature [8]. Various types of attacks to the cryptographic mechanisms are also possible [4]. For example, most of the cryptography algorithms are based on random numbers. However, the random numbers generated by computer programs are not truly random. This pseudo randomness can be exploited by attackers to predict the next and future numbers and pose pseudo random attack. The digital signature attack can also be initiated by an attacker. The attacker can use the signature and previously signed message of a victim to fake another message signature.

Table 3. Security attacks on MANET

<table>
<tr><th></th><th colspan="2">Passive Attacks</th><th>Active Attacks</th><th>Solutions</th></tr>
<tr><td>Physical Layer</td><td></td><td rowspan="5">Eavesdropping
Traffic analysis
and monitoring</td><td>Signal Jamming</td><td>Spread Spectrum</td></tr>
<tr><td>Link Layer</td><td></td><td>MAC layer disruption, adversarial attack</td><td>Error Correcting Codes</td></tr>
<tr><td>Network Layer</td><td>Location Disclosure Attack</td><td>Wormhole, Blackhole, Byzantine, Resource Consumption, Routing Table Overflow, Cache Poisoning, Rushing attacks etc.</td><td>Secure Routing Protocols</td></tr>
<tr><td>Transport Layer</td><td></td><td>Session Hijacking</td><td>Securing transport protocol using public key cryptography (TLS, SSL)</td></tr>
<tr><td>Application Layer</td><td></td><td>Repudiation, Viruses</td><td>Firewalls, IDS</td></tr>
</table>

3 Secure Routing in MANET

Routing is one of the major concerns in MANET. The frequent changes in the network make it very difficult to maintain a consistent path from source to destination. Three types of routing protocols for MANET have been proposed in literature. The *proactive protocols* (DSDV, WRP, and OLSR etc.) are based on repeated advertisement of routing information to neighboring nodes. The *reactive protocols* are based on on-demand discovery of routing path (ADOV, DSR, TORA etc.). The *hybrid protocols* (ZRP) are combination of proactive and reactive approaches. Secure routing in MANET is a thorny job and requires dealing with various types of network layer attacks as discussed in previous section. Various secure routing approaches have been proposed in literature. Surveys on secure routing protocols have been provided in [9-12]. However, most of these surveys discuss the routing protocols proposed before 2005. They classify the secure routing protocols as preventive and reactive approaches. An updated classification of currently available security solutions can be done as follows:

- protocols based on exploiting routing header information to identify malicious activities in the network
- protocols based on cryptographic technique to protect routing header
- protocols exploiting redundancy of routing Layers
- protocols based on trust information to identify malicious activities in the network
- protocols that maintain anonymity of routing entities.

3.1 Protocols Based on Routing Header Information

Some secure routing protocols are based on exploiting routing header information to identify malicious activities in the network. In this direction, [13] proposes a solution that identifies routing anomalies using the sequence number field of the routing packet. In routing protocols, a sequence number field defines the last packet received from a node. In case of normal routing operations, subsequent packets must have a higher sequence number. If a packet sequence number is less than previously received packet, misbehavior is suspected. In the proposed approach, two small tables are maintained by every node on the network. These tables help in recognizing the sequence number changes. The first table keeps the sequence number of the last packet sent to other nodes on the network. The second table contains the sequence number of the last received packet by a node. During route discovery, once a RREQ reaches the destination, a RREP is generated. The RREP carries the last packet sequence number received from the source. By comparing the sequence number carried in the reply packet with the one maintained in local tables, any sequence number inconsistencies can be identified.

[14] proposes an anomaly detection approach based on a dynamic training model. The proposal is based on the fact that during a black hole attack, the attacking node changes sequence number of the RREP to a considerably large value. Hence, a black hole attack can be recognized by analyzing the distribution of the sequence number in normal and anomalous state of the network. A feature vector is devised that comprises of number of sent routing requests, number of replies, average difference of sequence number when the request was sent and when it is received. Using a training data set, an attack model is devised. The mean value of the feature vector is calculated using the training data. The Euclidean distance of an input sample from the mean vector is calculated. If the distance is larger than a threshold value, it is classified as a black hole attack. At repeated intervals, the model is updated using previous interval data as a training dataset.

[15] proposes a scheme called DPRAODV to detect black hole attack based on the sequence number of RREP packets. If the sequence number is higher than a threshold, the node is marked as blacklisted. In this case, an ALARM message is generated to notify other nodes. To penalize the black listed node, the routing tables of the node

are neither updated nor are their messages forwarded. To calculate the threshold value, the difference between sequence number of RREP packet and the value in the routing table is first calculated. The average of this difference value is set to the threshold value. The threshold value is updated as soon as a new RREP is received. In this way, the model detects the black hole as well as prevents the attack in some cases.

3.2 Cryptography Based Approaches

The techniques of cryptography have also been used to provide integrity, privacy and non-repudiation of the routing messages. *Secure Ad hoc On-Demand Distance Vector routing (SAODV)* is an asymmetric cryptographic approach that is based on signing the non-mutable fields of AODV routing request headers. Intermediate nodes verify that the fields have not changed before creating a reverse route. After verification, the node broadcasts the request to neighboring nodes. Similar procedure is applied during the RREP message.

Authenticated Routing for Ad hoc Networks (ARAN) is a public key cryptography approach for providing secure routing in MANET. Every node has a certificate issued by a trusted third party. For route discovery, a node generates a request packet called RDP comprising of the IP of the destination, source certificate, a nonce and current time, signed by the source private key. The intermediate nodes verify the signature using the previous node's certificate (that is carried along with the request), sign the received message with their private key and append their own certificated with the message and rebroadcast. The destination generates a reply REP along the reverse route. The REP is signed by a node before it is forwarded to next node. The next node will verify the signature using the certificate of the previous node. *Secure Routing Protocol (SRP)* is a symmetric key approach based on establishing a security association between source and destination using a shared key. The shared key is set up using the other party's public key. All the communications are then encrypted and decrypted using the shared key.

Ariadne provides security over the DSR protocol using TESLA protocol. During route discovery, the source node generates a routing request. The request among other things comprises of a hash chain and a message authentication code (MAC). MAC is computed over the initiator, destination, id of the message and the time interval fields. The hash list is initialized with the MAC. The intermediate node appends its own address to the list of nodes (as in DSR). The node id of intermediate node is then appended to the current hash value and the hash value is recomputed. The MAC code of the whole packet is also recomputed and appended to the MAC field. Using the node list field of DSR, the destination verifies the integrity of the packet by comparing the hash value specified in the packet with its computed value.

endairA[16] is an inspiration of Ariadne protocol that instead of verifying the request, verifies the route reply messages. During route discovery, the request contains the identifier of source, destination and the generated request. Intermediate node appends their identifier to the packet and rebroadcast. When the packet reaches the destination, a route reply is generated and sent back through the reverse route. The reply comprises of the id of source and destination, the accumulated route and a digital signature. Intermediate nodes processing the reply verify the signature and ensure that next and preceding nodes are its neighbors. If the verification is done successfully, the intermediate node also signs the reply packet and forward to the next node. Another inspiration of Ariadne is *APALLS* [17] that is designed for providing non-repudiable secure routing in MANET. The protocol assumes a network with a web server and certifying authority at the backend. During route discovery, RREQ includes digital signature of source, per-hop hash and an optional list of black listed nodes. Intermediate nodes verify the signature of the routing request. The destination node performs various consistency checks and then a reply RREP is generated. In case of any active attack in the network, a non-repudiable proof (signed packet of the attacker) is also generated.

3.3 Protocols Exploiting Redundancy of Routing Layers

These protocols make use of redundancy (multiple routing paths, routing protocols etc.) to ensure the delivery of a routing message through a safe path. [13] propose a solution in which the source node verifies the authenticity of the RREP initiator through network redundancy. It is an extension over AODV protocol. During route discovery, the node waits for more than one RREP through different paths. From the redundant paths, the source extracts common hops and then constructs a safest path to route the message. A slightly different strategy has been used in *SPREAD*[18]. The original routing message is first decomposed into small shares using threshold secret sharing algorithm. Multiple paths towards the source are then determined using an on-demand routing algorithm. The routes are selected keeping into consideration the security levels of the node. The shares of the message are then transmitted towards the destination through these routes. At the destination, different shares of the message are then combined to generate the original message. By using the threshold secret sharing algorithm, it is ensured that if some share gets corrupted by malicious nodes, the whole message can still be reconstructed.

[19] proposes a scheme that employs multiple routing protocols. As different routing protocols are prone to different types of attacks, the idea proposed is to switch the routing protocol upon a particular type of attack detected on the network. The solution detects three types of attacks: black hole, routing table overflow and sleep deprivation torture attack. The black hole attack is detected by watch dog functionality. A node that transmits a message to next node monitors that if it is correctly forwarded by the next node or not. The routing table overflow attack occurs when the nodes along a route sends unnecessarily acknowledgment to the source. By sending a threshold value for the number of acknowledgement packets, this attack can be detected. The sleep deprivation torture occurs when during route discovery a node sends multiple route replies to the source, thus consuming the resources of the source. The approach adopted for the routing overflow attack detection can also be applied for detecting this type of attack. Once an attack is detected, switching of the routing protocol is done. If there is a black hole attack detected on the network, the algorithm tries to switch to On-demand Secure Routing Protocol (OSRP); otherwise Secure Link State Routing Protocol (SLSP) is selected.

3.4 Protocols Based on Trust Models

These approaches are based on maintaining trust information about other nodes on the network. Un-trusted nodes are disregarded during routing operation. [20] proposed an association based routing approach where the most secured route is selected based on the node's trust value. The algorithm extends the DSR protocol. A node maintains the trust value of other nodes based on the packets exchanged and dropped by the nodes. Associations between nodes are thus defined. The association value can be un-known (low trust), known (nodes have exchanged some messages and have moderate trust) and companion (high trust levels as nodes have exchanged lot of message in past). During route discovery, multiple route replies are received from the nodes, as in DSR. The route replies are sorted by trust ratings. The most trusted route is then selected by the source node based on the trust values of the intermediate nodes.

[21] proposed a trust model for secure routing. The trust vector is based on nodes experience, knowledge and recommendation of some other node x in the network. The experience is defined as the ratio of the number of packets forwarded by x to the number of packets transmission x is responsible for. The knowledge parameter is the probability that the data packet will be successfully transmitted between the nodes. The recommendation parameter is based on the recommendation information about x provided by other nodes of the network. Based on these parameters, a trust routing scheme has been proposed. During route discovery, a node sends the trust information about preceding node along with route request. This ensures the spread of trust information across the whole network. Using the available trust information, the proposed approach ensures the selection of a route with the highest trust value.

[22] presents a secure routing scheme using trust levels. The ratio of the 'difference between beacons received and transmitted' to the 'beacons received by the node' is calculated. Based on this ratio, the nodes are sorted in descending order. The first one third of the nodes in the list is classified as ally, the next one third as associate and the last as acquaintance. During routing, a node selects the best neighbor (with the same trust level) and sends it the packet. The neighbor then selects the best node (with the same trust level) and propagates the request ahead. This process continues until the packet is received by the destination.

3.5 Anonymous Routing

Anonymous routing protocols are based on the concept of onion routing. The routing messages are repeatedly encrypted like layers of onions. The intermediate nodes remove a layer of message, see the routing instructions and forward the message to next nodes. In this way, anonymity of the routing entities is preserved. ANODR[23] was the first anonymous routing protocol proposed for MANET. It is assumed that source and destination shares a secret key. During route discovery, the source generates a trapdoor identifier by encrypting a message "you are destination" with the the shared key. The route request contains: sequence number, trapdoor identifier (using secret key), a one-time public key and an onion structure. Onion structure comprises of successive encryption of a message "you are source" using secret keys of intermediate nodes. The intermediate nodes add their one time public key to the message. Every intermediate node stores the public key of previous hop in their table.

This is used during route reply propagation. Destination can recognize if it is the intended destination by successfully able to open the trapdoor. The destination then generates a new link key. The route reply is generated that comprises of a link key encrypted with the public key of previous node and the onion structure encrypted with link key. The node the receives the reply decrypts the link key with its private key and then uses link key to get onion structure. If the node was able to successfully strip away onion ring, the node was in the path of the route. The node then generates reply by encrypting the link key and onion structure. The source will receive the message "you are source" after stripping the onion structure.

[24] proposes ARM, an anonymous routing scheme that is based on a trapdoor generated using one-time pseudonyms. Pseudonyms can be generated using counters or synchronized clocks. During route discovery, the source generates a datagram comprising of the destination ID, a generated secret and a private key, TTL value and pseudonym. A RREQ message is broadcasted on the network. RREQ comprises of pseudonyms, TTL, public key of destination, datagram info and link identifier. Intermediate nodes receiving the RREQ generate a new link identifier and append it to RREQ link identifiers field. Everything is then encrypted with public key of the destination and then re-broadcasted. During route reply RREP, the datagram field of RREQ is decrypted. The node then generates a complete onion encapsulating the intermediate nodes to be traversed. The RREP is then broadcasted on the network. Intermediate nodes perform various types of checks to verify if it is on the anonymous route, strips off one layer of onion and propagates the reply ahead using the link identifiers.

[25] also proposes an efficient anonymous routing protocol called EARP. The protocol is based on onion routing where every roué request and reply message is encrypted with a trust key. A Hello message is sent back to the source during discovery to confirm the node a trusted node to ancestor.

3.6 Other Approaches

We conclude this section with outlining a number of miscellaneous secure routing solutions. [26] provides a security solution for routing and data forwarding operations. It is based on collaborative monitoring performed by nodes. Collaborative monitoring involves nodes overhearing the channel and performs cross checking. For participating in the network operations, every node is granted a token. In case of a malicious node, other nodes must reach a consensus, after which the token of the misbehaving node is revoked. A token renewal algorithm is also proposed such that every node renews its token before expiry. [27] proposes *ODSBR*, an on-demand routing protocol that is based on adaptive probing technique. For a path of length n, if $\log n$ faults have been detected, a misbehaving link is detected. Problematic links are avoided during route discovery. A new metric is proposed that is based on reliability weights assigned to each link. [28] proposes a scheme AODV-WADR for securing AODV protocol against wormhole attacks in emergency scenarios. A wormhole is first identified based on the time interval between route request and its response, hop count value and other parameters. If a black hole is suspected, a cryptographic secret key is exchanged between source and destination. This key can be used for encrypting further communication. Table 4 provides a summary of various secure routing protocols proposed for MANET. Based on this summary, we can identify numerous open issues with currently available protocols. For example, most of the current routing protocol provides security against a specific type of attack. The proposed protocols by and large extend existing routing protocols. Hence, they are useful for a specific type of routing protocol. In addition, the current protocols don't take a global approach towards security. They make various assumptions (availability of a key management system, cooperation among the nodes etc.) that are always not possible in MANET.

Table 4. Summary of various secure routing protocols for MANET

Protocol	Technique	Base Routing Protocol	Attacks addressed	Brief Description
[13]	Sequence Number Inconsistencies, Multiple Routing Paths	AODV	Black Hole attack	a) Identifies anomalies by checking if the sequence number of subsequent sent and received messages are larger than previous values b) constructs safest path based on multiple path information from the received multiple route replies
[14]	Dynamic Learning	AODV	Black Hole attack	An attack model is devised by analyzing the distribution of sequence number difference in normal and anomalous case
DPRAODV	Limit on Sequence Number of RREP	AODV	Black Hole attack	The sequence number of RREP is compared against a threshold to identify black hole attack
ARAN	Sign the request packet	None	Modification, Fabrication and Impersonation	Digitally signs the routing messages using private key that are verified by next node using certificates
SRP	Encryption	ZRP	Modification, Replay and Fabrication attacks	Establish security association using public key and then encrypt the communication using public key
Ariadne	MAC, Hashing	DSR	Worm hole attacks, Modification and Fabrication attacks	Using a hash chain and MAC list, verifies the integrity of the messages using roué request
endairA	MAC and hashing	DSR	Route Modification	Verifies the integrity of route reply messages
APALLS	Digital Signatures	DSR	Non Repudiation	Intermediate nodes do verification using certificates. Non repudiation is achieved by a signed message from the attacker

Table 4. (*continued*)

SEAD	Threshold secret sharing algorithm	None	Eavesdropping, Colluding attack, Modification	Splitting the message in to multiple shares and reconstructs at the destination
[19]	Multiple Routing	SLSP, OSRP	Black hole, Sleep Deprivation and Routing Table Overflow attack.	Decides the routing algorithm based on the type of attack
[20]	Trust Models based on packet drop	DSR	Selective Packet Drop Attacks	An optimal path is chosen based on the degree of association of node with neighbors
[21]	Trust Model based on experience, knowledge recommendation	DSR and AODV	Black Hole	Trust information is carried along with routing request and ensures the selection of the route with highest trust
[22]	Trust model based on beacons	NTP	Black Hole	Trust is maintained based on beacons transmitted/received and trust routing is performed by selecting the best node in the same trust level

4 Key Management in MANET

Cryptography is one of the widely used techniques to provide authentication, authorization and privacy in ad hoc networks. Public Key Infrastructures are normally used to generate keys for employment and execution of various cryptographic operations. Due to the infrastructure-less nature of MANET, there are no prior PKI servers available that can generate public and private keys for the entities of the network. To address this issue, a number of key-management approaches have been proposed in literature. We define *key management* as the process of establishment and maintenance of keying relationship among the entities of the network. A key management solution can employ a centralized certification authority (CA) for key agreement and transport. A distributed CA can also be used where a private key is distributed to a set of nodes on the network, while the public key is known to all the nodes.

Various surveys have been done for key management in MANET. [29, 30] analyses various key management approaches and classify them into approaches based on clustering, identity, routing, offline authority, mobility aware and certificate chaining approaches. [31] classify the schemes as contributory, distributed, certificate based, identity based and symmetric systems. However, existing surveys reviewed the literature mostly published before 2007. This section provides a survey of latest key management approaches for MANET under following classification:

- Approaches based on organizing the nodes as clusters
- Approaches based on identity based cryptography
- Approaches based on certificate chaining
- Approaches exploiting multicasting for key management

4.1 Cluster-Based Approaches

In this approach, the whole network is divided into clusters. This reduces the storage and communication overhead. [32] proposed a partially distributed cluster based approach where every cluster head node maintains a CA information table containing details about the certification authority in the local cluster (and optionally other clusters). Any node in the cluster inquires the cluster head about the whereabouts of the CAs when it wants some certification services. The authors also propose an efficient sequential share update scheme. A set of t servers update their secrets while the remaining nodes update their shares with the help of these t servers.

In [33], the CA private key is distributed to a set of cluster nodes. Whenever any node wants to sign a message, it sends the request to the backbone via its CH. Any CH node that receives the message, signs the message using its own share. When the requesting node has t shares, it can construct the CA signature.

[34] proposes a predictive key management approach for minimizing congestion and traffic overhead. Based on route expiration and nodes velocity, a metric is computed for predicting nodes movement and cluster changes. In addition, a weighted clustering algorithm is proposed based on the weights of the node. The weight is computed based on its speed, distance with neighbors and battery power. Every node calculates its weight and broadcast to other nodes. The node with the minimum weight is elected as the cluster head. Other nodes are then requested to join the cluster head. During cluster maintenance, hello, leave and join request messages are observed by cluster heads to updates the cluster head table accordingly.

[35] proposes a cluster-based key agreement protocol for energy constrained wireless ad hoc networks. A binary tree like cluster of 2-3 nodes is formed. A random node is selected as a leading node that then sends IamAlive message to neighbors. The first two nodes that answer the message will be included in the cluster. This process is repeated until all nodes become part of the cluster. After the cluster formation, the key agreement phase is executed based on elliptic curve cryptography and bilinear maps.

4.2 Identity Based Approaches

The concept of identity based cryptography was first proposed by [36]. In these approaches, the public key of the users are the derived from their identities and thus eliminates the need for public key distribution. A survey of the identity based approaches is provided in [37]. In an identity based system, there is a master public/private key for the whole system. During encryption, the master key of the system, id of the node and the corresponding is provided to get a cipher text. During decryption, the master public key, the private key of the node and the cipher text is provided to get the actual text. [38] proposed a secure and efficient identity based approach. A master key is randomly picked from an algebraic field. Whenever a node k joins the network, it presents its identity id_k to the key generator center. The center applies a hashing function on the *id* to obtain a point on the elliptic curve. The private key for the node is then generated based on this point.

[39] proposed a four-phase secure key management approach. During the initialization phase, each user gets their long term public and private key. A master key is randomly generated by the key generation center. Motivated by the RSA, the public/private key pairs for the users are generated. During registration phase, each user provides its identification to the key generation center to get a certificate. For communication among nodes, a user generates two public keys. The identification of the communicating node and time stamp is used to generate a hash based value. The public keys, time stamp and the generated value are sent to the other node. The generated hash value is verified at the other end during verification phase to ensure the message is sent by the other node. Based on the received public key, session keys are generated for the communication.

[40] proposes a polynomial based approach to key management in hierarchical network based on ID-based threshold cryptography. The identity of a node in the tree is its path from root to that node i.e. $(I_1,I_2, ..)$. A random polynomial is chosen as the master key. If l is the level of the tree then the master key can $F(x_1,y_1,x_2,y_2,\ldots x_L,y_L)$. The secret key of the node at level i is computed as $F(I_1, y_1,x_1, \ldots, I_i, y_i, x_i,\ldots)$.

4.3 Certificate Chaining Based Approaches

In certificate chaining based approaches, each node keeps a repository of certificated nodes in its neighbors. Whenever a node wants to validate the certificate of another node, the nodes combines their certificates and a chain of certificates is searched towards the certifying node. [41] proposes a certificate chain discovery approach based on the existing routing infrastructure. In this approach, the source node sends a message to neighbors that it directly trusts. The directly trusted node appends the certificate of the source to the routing request and forwards it to the node that it directly trusts. This process continues until the message is received by the destination. The receiving node has the whole chain of certificate appended in the message that can be used to recover the public key of the source. The destination then replies the packet through the reverse route. The intermediate nodes append the certificate of the destination node and propagate the message ahead. A similar approach has also been presented in [42]. In this approach, before propagating route requests or reply, the intermediate nodes perform certificate distribution step. A three steps approach is proposed. A direct trust is established by sharing neighboring nodes self certificate. A route request message is only propagated if the node has previous hop certificated. If no certificate is found, certificate exchanges are done among nodes. During stage 2, an indirect trust is also established with the originator and destination. Before forwarding a request, the node asks for the certificate of the owner to the preceding node. During route reply, the certificate of the destination is also piggybacked that can be saved by intermediate nodes as well. The final stage is optimized verification. It is suggested that request messages are forwarded without waiting for the verification process to complete, which is confirmed during the reply route. [43] also presented a certificate chain based scheme to avoid black hole attacks in ODMRP routing protocol. It comprises of route discovery, certification and authentication phase. The route discovery is performed similar to the ODMRP. Once the route is established, the participating node enters into certification phase. Nodes exchange certificates with their neighboring nodes after convinced about their security. The source doesn't transmit the data after the route establishment, but wait for an authenticated message from the destination. Intermediate nodes also append their certificates until it reaches to the source. The source can verify the certificate chain and can thus trust the route for data transmission.

4.4 Multicasting Based Approaches

These key management approaches are based on utilizing multicast structures for key distribution and maintenance in a multicast network. [44] proposes a key management scheme based on threshold cryptography and multicast server groups. The private key of the CA is distributed to a set of shareholders called multicast server groups. The multicast server group comprises of server nodes and forwarding nodes. Together they form a mesh like structure. The server group is formed similar to the ODMRP protocol. The server node initiates a join request packet, which is propagated by the intermediate nodes. The TTL value is signed by nodes thus preventing any malicious activity. Any server group node that receives the request will generate a join reply. Nodes receiving the reply will check if it can be forwarding node and correspondingly update their routing table. The resultant mesh like structure is always connected and handles the certificate generation and distribution in a ticket-based manner. [45] proposed similar approach where a multicast tree is established between members. Two multicast trees are formed called red and black tree, thus ensures fault tolerance. There is a group coordinator (elected based on rotation of members) that computes and distributes intermediate keying material to the other members. All members maintain a logic key tree that are combined to form final group secret. The key tree is initially sorted by the member ids, however further any newly arrived member is added to the right of the logic key tree. Any newly arrived member gets the key based on a blinded key unicasted by the coordinator.

4.5 Other Approaches

We have discussed some of the key management approaches above. There are a number of other approaches to key management. For example, the mobility aware key management schemes instead of relying on the underlying routing structure, proposes to exploit mobility of the nodes for key management [46]. Some approaches were proposed based on pre-distribution of keying information [47]. Different routing based approaches for key management has been also presented [38]. Table 5 summarizes some of the key management approaches proposed in literature. Some key management schemes work in purely distributed fashion while some uses dedicated entities responsible for key management. In addition, some key management approaches are based on integration of network with key management operations. We found that most of the key management approaches don't fully comply with

Table 5. Summary of various key management approaches for MANET

Approach	Basic Technique	Degree of distribution	Network Layer Integration	Multicasting	Description
[32]	Hierarchical clustering	Partial	×	×	Cluster heads are inquired about keys
[35]	Binary Tree	Partial	×	×	Binary trees are maintained. Key management operations are performed using bilinear maps and elliptic curve cryptography
[40]	Polynomial based key management	Full	×	×	Keys are distributed across trees. Keys are generated based on the path of nodes from the root
[39]	Four-phase secure key management approach	Full	×	×	Node provides its ID to key distribution center to get a certificate. A session is then generated for communication among nodes
[41]	Certificate chains formed through routing messages	Full	√	×	Certificates are appended with the routing messages
[42]	Certificate chains established during route discovery	Full	√	×	During route discovery, nodes establishes certificate. RREQ are propagated if the node has previous hop certificate. Similar mechanism during route reply
[43]	Certificate chains via neighbor sharing certificates	Full	√	×	Neighbors exchanges certificates after convinced about security
[44]	Multicast server groups	Partial	√	√	Mesh like structure for certificate generation and distribution

resource limitations and other constraints imposed by MANET. The design of secure and efficient key management system for MANET is still an open area. Future proposal can investigate on cross-layer interaction and multicast communication protocols to design key management system.

5 Intrusion Detection Systems

Intrusion Detection Systems (IDS) are the second line-of-defense once an intruder has entered into the system after breaking the primary security mechanisms. An intrusion is defined as any type of activity considered that attempts to compromise the security objectives. An *IDS* is defined as a system comprising of the mechanisms intended to detect an intrusion, identify the source of intrusion and then isolate this source from the network. Similar to other security issues, designing IDS solution for MANET is a complex job. There are no specialized entities that can take the job of recording network activities and then analyses these activities for intrusion. In addition, the hosts in MANET are constrained by scarcity of resources. The computational, memory and bandwidth demands of an IDS system should be in accordance with the resources of MANET.

The early trend in intrusion detection systems was on designing efficient architectures. The proposal based on this trend can be classified as standalone, cooperative and hierarchical system [8, 48]. In the following section, we briefly discuss the systems based on this trend. Then, recent IDS approaches based on cross-layer design, game theory, theory of evolution etc. will be discussed in detail.

5.1 Standalone, Cooperative and Hierarchical IDS

In standalone systems, a node works without any communication with other nodes and relies on a self-contained approach for detecting malicious activities. [49] proposes a standalone system that is based on monitoring the power consumption of nodes. The power consumption of a node is compared with the power consumption patterns of known attacks. If the power consumption of nodes is similar to the pattern of a known attack, an intrusion is assumed. The system however fails to recognize those attacks that don't effect power consumption. In [50], every node maintains a mis-incident metric that is incremented every time a symptom of an attack is detected. A threshold value is calculated during system initialization. If the mis-incident value goes above the threshold, intrusion is recognized. If during a session, no intrusion is detected, the threshold value is accordingly adjusted.

In a *cooperative system*, every node runs an IDS system. The nodes analyses the behavior locally to identify intrusion. Global Intrusion can be identified by nodes collaborating and sharing information with each other. Various research proposals have been put forwarded in this direction. An example is the system proposed by [51]. The idea is to use local IDS that perform the anomaly detection using nodes profiles. If an intrusion is detected, other nodes are informed. However, in case of suspicion, a node can inquire for global intrusion detection. The requesting node shares its data about suspected intrusion with neighbors. The neighbors also share each other information. A consensus algorithm is thus run. Once, an intrusion is confirmed, intrusion response is initiated. The response includes reroute computation to isolate intruder, re-initialization of communication links etc. [52] adopts similar approach, where malcounts are maintained by nodes for other nodes. If a malcount for a node exceeds a threshold, an alert is initiated. [53] proposed a multi-layer intrusion detection system that operates at all layers of protocol stack. Intrusion detection is performed at every layers of the node. The results from the different layers and neighbors can be combined to identify intrusion. [54] proposed a two tier, two engine based IDS. There are signature-based and anomaly based engines running at each tier. The anomaly engine can be activated if the signature based engine suspects an inconclusive intrusion based on the audit data. [55] proposed agents for intrusion detection based on the performance of the node during intrusion detection phase. Nodes that cooperate satisfactory have the highest reputation. Upon a suspected activity, a set of nodes are selected based on the battery power and nodes reputation. These selected nodes are assigned the task of network monitoring, intrusion decision and intrusion response.

In *hierarchical IDS*, the nodes are structured in a hierarchy and the whole network is divided in to cluster. There is a head for each cluster. An intrusion detected by a node is communicated to its cluster head. A cluster head can launch global response. [56] propose an IDS system, in which cluster heads are selected based on an election algorithm. The nodes run a network detection, local detection, resource management and monitoring module. The first two are meant to perform network level and local level intrusion detection. The resource management module monitors the battery level of cluster head and if it falls below a threshold, a reelection is initiated. The monitoring module ensures that network detection module is active. In [57]., the nodes are structured in the form of a dynamic tree. The upper nodes have the higher authority and can pass directives to below nodes. The lower nodes pass data to upper nodes that are aggregated and useful for detection of the intrusion. The tree is formed in two steps. First a node is randomly selected as a cluster head. Its neighbors become the member of this cluster. Then, a node from the previously cluster member is selected. The selected node is marked as cluster head. All the neighbors of the selected nodes that are not assigned to any other cluster are the member of this new cluster. The process is repeated until a tree is generated. Once a tree is generated, some permutations are performed to enhance the robustness and performance. [58] proposed a system where a cluster head is selected based on voting. Each node votes for the neighbor with the maximum connections and energy. Only cluster heads are responsible for intrusion detection. The cluster head monitors information like: hello, error, request packets, routing table changes and actual packets transmission. Based on the analysis that can be performed for a specific node or cluster level, anomalies are detected.

5.2 Cross Layer Intrusion Detection

A cross-layer IDS system combines information from various layers to perform the identification of intruders. Cross-layer systems are found to be very effective in the detection of intruders. [59] have demonstrated through simulations the need for cross-layer design. Various types of attacks were generated and analyzed with single layer (routing, MAC and physical layer) IDS and a multiple layer IDS. Linear Discriminant Analysis was used for classification purposes. The data collection mechanism can be based on multiple data collection and single data analysis (MCDMA) or multiple data collection and multiple data analysis (MSDMA). In the former, the information from multiple layers are collected and then analyzed at a single layer (MDMA) where as in the later, the data analysis is performed at every layer. Below, we discuss some of the cross-layer techniques.

CRADS [60] is a cross-layer IDS for routing attacks identification. As the cross-layer information from various layers leads to large feature sets, various feature reduction techniques are exploited. Using associations, various features are correlated to give a reduced set of features. Then, feedback-based filtering is used to remove uninformative and redundant information. The resultant features are then trained using Support Vector Machines (SVM). SVM is a non-linear pattern recognition algorithm that outputs the decision boundary between normal and abnormal behavior. The simulation results illustrate the supremacy of the proposed system.

In [61], a novel IDS solution is proposed based on a monitoring process to assess the nodes behavior. The monitoring system keeps into consideration the distance between monitoring and monitored nodes. A cross-layering approach is introduced that utilizes the SNR, ACK and information from other layers at the routing layer. Another novelty of the system is to account for the difference between transmission, interference and sensing region.

[62] proposes a multi-level IDS system for wireless ad hoc sensor networks. In the first stage, an un-supervised algorithm is used to detect anomalous patterns. The detected patterns are then passed to the second stage to identify the type of attack using a supervised pattern classification algorithm. The advantage of this approach is that the un-supervised algorithm is computationally less-expensive. The large number of audit can be handled easily at first stage. Support Vector Machines were used for the detection and identification of intrusions in MANET. The features from network, MAC and physical layers were selected for classification.

[63] presents a cooperative intrusion detection. The cross-layer information across various layers is exploited and data mining has been used to detect misbehaving nodes and various types of network attacks. There is a local data collection module for collecting data packets of physical, MAC and routing layer data streams via the association module. The local detection module detects the attacks based on the association rules. The cooperative detection module is used when it is conclusive that there is an ongoing attack. The majority voting scheme is used in this case.

A cross-layer solution for enhancing the robustness of P2P ad hoc networks has been proposed in [64]. The request messages of the node are analyzed to identify misbehaviors. The packets dropped by a node can be compared with the destination of a packet to identify a malicious node. As P2P systems are based on DHT that provides rules on how the next-hop overlay node is selected. An incorrect request can therefore be identified by comparing the identifier of next node with the proposed next over-lay node. For this purpose, cross-layer information exchange is required.

XIDR[65] is a dynamic, cross-layer intrusion detection and response system. The model comprises of multiple intrusion detection sources. There is an active alert database for various types of intrusion alerts. Each of the alerts has an associated weight calculated based on user's history. The user's history is dependent on the interaction of user with the host's resources. This interaction can be at different layers. Once an intrusion is detected, a response selection engine is used. The responses are selected based on cost-base response selection engine. An operational cost is associated with each response to indicate the severity of the response. The lower-layer's responses are assigned high operational cost as compared to above layers. The selected responses are then deployed using custom scripts running on the server. All the messages are encrypted using cryptography mechanisms to ensure that unauthorized entities can't generate control commands.

5.3 Game Theoretic Approaches

Game theory has also been employed in intrusion detection systems. In a game theory problem, different competing entities interact with each other to achieve their objectives. The game can be: *zero-sum game* where the participant's gain or loss is balanced by gain or loss of other participants of the game or *non-zero-sum game* where aggregate gain or loss of interacting entities can be non-zero. A game can also be cooperative or non-cooperative. In *cooperative game*, the participants form binding commitments. A *pure strategy* is one that player will always follow during the game. A *mixed strategy* is a combination of moves with associated probabilities describing how frequently the move is to be played. A Nash equilibrium is a solution in which incentives of a player is dependent on other player's moves as well. An individual change in strategy of player can't provide any incentive.

This multi-player decision problem can also be mapped to design IDS security solutions for MANET. A number of game theoretic IDS solutions have been proposed in literature. [66] uses game theory to determine if it is essential run an IDS always on nodes of the network. They modeled the interaction between attacker and IDS through a two player non-cooperative, non zero-sum model. The pure strategy for the IDS is to monitor for some percentage of time or not. The pure strategy for the intruder is to attack for some time or not. The game is solved using Nash Equilibrium mixed-strategy pair.

[67] proposes an election algorithm for electing an IDS-leader. The algorithm ensures balanced resource consumption among nodes. The concept of checkers is introduced to ensure that selfish nodes don't provide false information about their resources. A catch and punish approach is provided via a cooperative game theoretic model. To ensure that the elected leader effectively execute the detection service, a zero-sum non-cooperative game theory formulation is done between an intruder and a leader. The game is solved by the Bayesian Nash Equilibrium approach.[68] also

proposes a hybrid model based on game theory for ensuring the security of cluster of nodes in MANET. The nodes play a game using Bayesian game theory. A trust value is calculated by each node. The way intruding nodes are identified and internal intrusion are prevented.

5.4 Evolutionary Approaches

Evolutionary approaches are the light weight solution for intrusion detection. Hence, they are suitable for the resource constrained MANET environments. The idea is to evolve the intrusion detection program automatically using evolutionary techniques, based on the input features. In this direction, [69] proposed an evolutionary solution based on structured GA. A hierarchical structure is maintained that generates multi-shaped detectors in an iterative fashion.

[70] presented genetic programming approach towards IDS in MANET. In genetic programming, a set of candidate solutions are evolved towards the target solution. During each step, the current candidate solutions are cross-over and mutated to generate new solutions. The new solutions are evaluated against a fitness function. Those solutions that passed the fitness criteria are selected as candidate solution and next iteration is iterated. The process is repeated until the termination criterion is satisfied. In [70], the programs are evaluated for each type of attack. The input variables to the genetic programming are features related to mobility and packets (routing control packets statistics). Tournament selection algorithm is used for selecting individuals for cross over. The algorithm is evolved using training data. The fitness function is defined as the difference between detection rate and false positive rate.

5.5 Immune Inspired Approaches

The self-healing property of human immune system has been exploited in some research for detection of intruders. The motivation to use the immune approach is the distributed and autonomous nature of MANET similar to human immune systems. [71] proposed a biologically inspired tactical infrastructure (BTSI). There is a small kernel running on every node. Similar to biology, the notion of damage is introduced. A damage is defined when an application is not getting what is expected from a source. BTSI sends damage notifications to other nodes. A reputation value is thus maintained by every node. Using machine learning techniques, state of the network in future is predicted, based upon reputation and changes in past. [72] presented a framework based on negative selection and danger theory. The first node on the network creates an immune agent. Whenever any node joins the network, the immune agent is replicated on the node. A lymphocytes cell is a substance circulated in human system to detect non-self patterns. The author uses Negative Selection theory to generate detectors similar to the lymphocytes. The negative selection generates random pattern that are compared against self patterns. If it matches the self, then they are not included in the detector patterns. When a non-self pattern is detected, clonal selection process is employed.

5.6 Other Approaches

Besides the approaches outlined above, the researchers have employed different types of innovations for devising an IDS solution. [73] proposes a social network based IDS system. Different SNA algorithms are available for detection of malicious profiles and people on the network .The author suggest use these SNA algorithms for to the detection of malicious activities. Different features are extracted from MAC layer and network layer attributes. Based on these features, socio metrics are built and SNA algorithms are thus executed. [74] presented a solution based on self organizing maps for intrusion detection. Each node shares its security state with neighboring nodes via a map. The global map can thus be created and used to securely route the data. In order to ensure the protection of map during dissemination to other nodes, watermarking technique has been proposed. [7] proposes a fuzzy logic approach for intrusion detection. A node transition probability based routing scheme is proposed.

Table 6. Summary of various IDS Systems for MANET

System	Basic Technique	Punishment Mechanism	Architecture	Cross-layer	Description
CRADS	Unsupervised Learning	None	Stand alone	√	Using SVM on a reduced input feature set, intrusions are detected
[62]	Multi-stage detection	None	Cooperative	√	First anomalous patterns are detected and passed to a pattern classification engine to identify the type of attack
XIDR	Rule based Approach	Cost-base response selection engine	Cooperative	√	Information gathered from various layers are used to classify a request using a set of rules
[67]	Zero-sum non-cooperative game theory	Cooperative punishment model	Hierarchical	×	A game is formulated and the solution to the game is to find Nash Equilibrium
[70]	Genetic Programming	None	Stand alone	×	Set of features are selected and detection algorithm is evolved using mutation and cross over operations
[71]	Danger Theory	None	Cooperative	√	Node maintains reputation based on the network traffic. Machine learning algorithms are used to predict network's future state
[72]	Negative selection and danger theory	clonal selection process	Cooperative	×	Detectors are matured by passing through a negative selection test. These detectors are used to detect any intrusion
[73]	Social Network theory	None	Stand alone	√	Uses SNA analysis (that is used for detection of malicious nodes) for detection of misbehaving nodes

Table 6 provides a summary of some of IDS system for MANET. The research trend suggests the evolution of new techniques (game theory, artificial immune systems etc.) for intrusion detection. Cross-layer design has also been used by some systems. However, our study reveals that most of the IDS have inherent limitations in their application to MANET. Some IDS only consider detection of an intrusion and no details have been provided for intrusion response. There are systems that consider only a specific type of security attack (black hole, wormhole etc.). The performance and accuracy of the intrusion detection also plays a critical in MANET. The design of a generalized and performance efficient IDS solution for MANET still remains an open area of research.

6 COINS: A Correlation Based Intrusion Detection System

Despite the significant research effort, security in MANET still requires extensive research to design new solution that can cope with the challenges of MANET. It is concluded that most of the research proposal focuses on a specific security aspect and a comprehensive security solution is currently lacking for MANET. In this section, we present COINS, a correlation based intrusion detection system for MANET. The proposed system extends the IDS proposed in [75]. COINS is a cross-layer system that identifies the anomalies on the network by monitoring application layer activities. Unlike the other IDS proposed in literature, COINS uses the relationship among the various application layer requests for intrusion detection. Based on the past history,

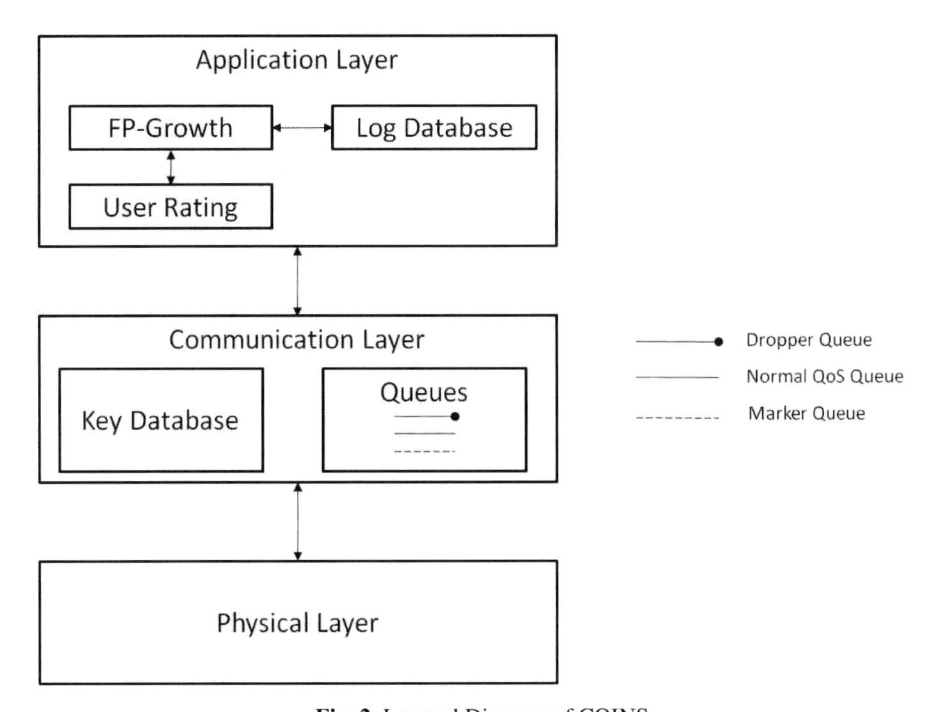

Fig. 2. Layered Diagram of COINS

request patterns are identified and future requests are classified as normal or anomalous based by comparing them with past patterns. During routing, the cross-layer information is utilized to share trust information which is then used for secure delivery of packets across the network. Fig. 2 shows the layered diagram of COINS.

COINS comprises of following layers:

- **Application Layer:** It provides an interface to end user to pose requests for data and resources available on the network. It comprises of a log database that records user activities. Based on these activities, intruding activities are detected on the network. The users are also assigned ratings based on the log database. A user rating can be normal, doubtful or malicious as we will discuss later on.
- **Communication Layer:** It deals with the networking aspects. It provides secure routing and delivery of information across the network. A key database is maintained that stores public keys of other nodes to encrypt routing messages. Several queues are maintained to deliver the routing messages based on the trust ratings.
- **Physical Layer:** The *physical layer* is responsible for the actual transmission of information between two nodes. It encapsulates the link layer and physical layer of OSI protocol stack.

6.1 Key Management

The key management in COINS is inspired from the work presented in [76]. The proposed approach assumes the existence of a secure data point on the network that is responsible for key management. Whenever a set of nodes establish an ad hoc network, a data security point (DSP) is elected from these nodes. The elected node is responsible for configuration, addressing and key management operations on the network.

Any new node that enters in to the network performs the join protocol to become part of the network. For this purpose, it launches a discovery agent on the network. The discovery agent searches for a DSP on the network. After a DSP is found, the node sends a join request to it. DSP uses the identity based cryptography technique to generate an identity and corresponding private key for the new node. The DSP then spawns a topology agent that delivers the topological and key information to the node.

6.2 Intrusion Detection System

There is a monitoring agent running on every node on the network. The monitoring agent logs the user activities in a log database that is maintained by every node on the network. Log database contains the history of the requests of the user in the form of sessions. A new session is started when the time interval between current and past request is greater than a threshold value. Log database keeps the record in the form of a circular list of session records. An FP-Growth algorithm is applied on the log database to discover association rules. These item sets determine what resources are requested together during a normal session. If request of a node during a session doesn't comply with these association rules, a deviation count variable is incremented. Based on the deviation count, a node can be assigned three trust levels i.e. normal, doubtful or misbehaving node. A normal level is assigned when the behavior of a user doesn't deviate at all from the normalcy. While a doubtful level is achieved by a node when its deviation count falls above a threshold. If the deviation count is significantly large, node is assumed to be misbehaving node.

If a secure data point is detected as a misbehaving node, an election algorithm will be re-initiated. After the election of the secure data point, keys are reinitialized for all the nodes on the network.

6.3 Secure Routing

COINS ensures secure transmission of the routing packets by utilizing cross-layer information. To maintain the secrecy of the routing packets, any packet being transmitted is encrypted with the id of the next hop node, so only next hop node can get the content of the packet by decrypting it with its private key. During forwarding a routing packet, a node communicates with its upper layer to identify the trust level of source node. Three types of routing queues are utilized to schedule the delivery of the packets. If a node is a normal behaving node, it is placed into normal QoS delivery queue. The packets of a doubtful node are placed into marker queue where the packet is marked during delivery to alert the next hop node. The packets of a misbehaving node are placed into dropper queue, where it is subsequently dropped.

To ensure the secure delivery of the message, the node also inquires the upper layers about the integrity of the next hope nodes. The priority is given to the next hop nodes with trust level as normal for forwarding of the routing message, while misbehaving nodes are not considered for forwarding of routing packets.

6.4 Election of Secure Data Point

One important question is the how the secure data point is elected from among the set of nodes available on the network. A simple election algorithm is proposed. Every node broadcasts the deviation count information it maintained for every other node on the network based. The receiving nodes accumulate the deviation count values received from all other nodes. The accumulated values are sorted in increasing order and the topmost node is elected as the secure data point.

6.5 Physical Layer Security

Different types of ad hoc networking technologies (blue-tooth, IEEE 802.11 etc.) can be used for the implementation of the physical layer. To ensure the secure communication at physical layer and make it resistant against security attacks various countermeasures can be adopted as discussed earlier in this chapter. For example, to avoid signal jamming spread spectrum techniques are recommended. For countering traffic analysis, traffic padding can be used. For addressing the tempering of the device, biometrics (fingerprints, iris and hand geometry etc.) can be used.

7 Conclusion

This chapter provides an in-depth study of security issues in MANET. A discussion on various types of security threats in MANET has been presented and survey of solutions for secure routing, key management and intrusion detection is provided. The survey of literature on other security issues of MANET has been left as a future exercise. Our study of existing literature highlights that a comprehensive security solution for MANET is still not available. Most of the proposals deliberated on an individual security problem. For example, some solutions focus on providing security for a particular layer of protocol stack or a particular type of security attack. Similarly, there are individual solutions available for secure routing, key management and intrusion detection system for MANET. Hence, a cohesive, multi-fence security solution is thus desired that should address the diverse security issues of MANET. In this direction, this chapter presented COINS, a security solution for routing, key management and intrusion detection. Future work can be done on an extensive evaluation of the performance and accuracy of COINS. Further work is also required to add further components to COINS that can address other security issues of MANET.

References

[1] Cheddad, A., Condell, J., Curran, K., Mc Kevitt, P.: Digital Image Steganography: Survey and Analysis of Current Methods. Signal Processing 90(3), 727–752 (2010)

[2] IETF. IETF MANET Charter, http://datatracker.ietf.org/wg/manet/charter/ (cited May 2012)

[3] Singh, A., Kumar, M., Rishi, R., Madan, D.K.: A Relative Study of MANET and VANET: Its Applications, Broadcasting Approaches and Challenging Issues. Advances in Networks and Communications, 627–632 (2011)

[4] Wu, B., Chen, J., Wu, J., Cardei, M.: A Survey on Attacks and Countermeasures in Mobile Ad Hoc Networks. In: Wireless/Mobile Network Security (2006)

[5] Kärpijoki, V.: Security in Ad Hoc Networks. Helsinki University of Technology (2000)

[6] Cayirci, E., Rong, C.: Security in Wireless Ad hoc and Sensor Networks. Wiley Online Library (2009)

[7] Kumar, S., Narender, M., Ramesh, G.N.: Security Provision For Mobile Ad-Hoc Networks Using Ntp & Fuzzy Logic Techniques. Global Journal of Computer Science and Technology (2010)

[8] Seno, S.A.H., Budiarto, R., Wan, T.-C.: SHSDAP: Secure Hierarchical Service Discovery and Advertisement Protocol in Cluster Based Mobile Ad hoc Network. World Applied Sciences Journal (2009)

[9] Michiardi, P., Molva, R.: Ad hoc Networks Security. Mobile Ad hoc Networking, 275–297 (2004)

[10] Fonseca, E., Festag, A.: A Survey of Existing Approaches for Secure Ad Hoc Routing and Their Applicability to VANETS. NEC Network Laboratories (2006)

[11] Mahapatra, R.P., Katyal, M.: Taxonomy of Routing Security for Ad-Hoc Network. International Journal of Computer Theory and Engineering (2010)

[12] Sreedhar, C., Verma, S.M., Kasiviswanath, N.: A Survey on Security Issues in Wireless Ad hoc Network Routing Protocols. International Journal on Computer Science and Engineering (2010)

[13] Al-Shurman, M., Yoo, S.-M.: Black Hole Attack in Mobile Ad Hoc Networks. In: 42nd Annual Southeast Regional Conference (2004)

[14] Kurosawa, S., Nakayama, H., Kato, N., Jamalipour, A., Nemoto, Y.: Detecting Blackhole Attack on AODV-based Mobile Ad Hoc Networks by Dynamic Learning Method. International Journal of Network Security 5(3) (2007)

[15] Raj, P.N., Swadas, P.B.: DPRAODV: A Dynamic Learning System Against Blackhole Attack in AODV based MANET. IJCSI International Journal of Computer Science Issues 2 (2009)

[16] Buttya, A.G.L., Vajda, I.N.: Provably Secure On-demand Source Routing in Mobile Ad Hoc Networks. IEEE Transactions on Mobile Computing 5(11) (2006)

[17] Kulasekaran, S., Ramkumar, M.: APALLS: A Secure MANET Routing Protocol. Mobile Ad Hoc Networks: Applications (2011)

[18] Lou, W., Liu, W., Zhang, Y., Fang, Y.: SPREAD: Improving Network Security by Multipath Routing in Mobile Ad hoc Networks. Wireless Networks (2009)

[19] Rangara, R.R., Jaipuria, R.S., Yenugwar, G.N., Jawandhiya, P.M.: Intelligent Secure Routing Model For MANET. In: 3rd IEEE International Conference on Computer Science and Information Technology, ICCSIT (2010)

[20] Bhalaji, N.: Reliable Routing against Selective Packet Drop Attack in DSR based MANET. Journal of Software 4(6) (2009)

[21] Gong, W., You, Z., Chen, D., Zhao, X., Gu, M., Lam, K.-Y.: Trust Based Routing for Misbehavior Detection in Ad Hoc Networks. Journal of Networks 5(5) (2010)

[22] Elizabeth., N.E., Radha., S., Priyadarshini., S., Jayasree., S., Swathi, K.N.: SRT-Secure Routing using Trust Levels in MANETs. European Journal of Scientific Research 75(3) (2012)

[23] Hong, J.K.A.X.: ANODR: Anonymous On Demand Routing with Untraceable routes for Mobile Ad-hoc Networks. In: 4th ACM International Symposium on Mobile Ad hoc Networking and Computing, MOBIHOC 2003 (2003)

[24] Seys, S., Preneel, B.: ARM: Anonymous Routing Protocol for Mobile Ad Hoc Networks. International Journal of Wireless and Mobile Computing (2009)

[25] Li, X., Li, H., Ma, J., Zhang, W.: An Efficient Anonymous Routing Protocol for Mobile Ad Hoc Networks. In: Fifth International Conference on Information Assurance and Security (2009)

[26] Yang, H., Shu, J., Meng, X., Lu, S.: SCAN: Self-Organized Network-Layer Security in Mobile Ad Hoc Networks. IEEE Journal on Selected Areas in Communications (2005)

[27] Awerbuch, B., Curtmola, R., Holmer, D., Nita-Rotaru, C., Rubens, H.: ODSBR: An On-Demand Secure Byzantine Resilient Routing Protocol for Wireless Ad Hoc Networks. ACM Transactions on Information and System Security, TISSEC (2008)

[28] Panaousis, E.A., Nazaryan, L., Politis, C.: Securing AODV Against Wormhole Attacks in Emergency MANET Multimedia Communications. In: 5th International ICST Mobile Multimedia Communications Conference (2009)

[29] Van Der Merwe, J., Dawoud, D., McDonald, S.: A survey on peer-to-peer Key Management for Mobile Ad hoc Networks. Computing Surveys 39(1), 1 (2007)

[30] Masdari, M., Jabbehdari, S., Ahmadi, M.R., Hashemi, S.M., Bagherzadeh, J., Khadem-Zadeh, A.: A Survey and Taxonomy of Distributed Certificate Authorities in Mobile Ad hoc Networks. EURASIP Journal on Wireless Communications and Networking (2011)

[31] Hegland, A.M., Winjum, E., Mjølsnes, S.F., Rong, C., Kure, Ø., Spilling, P.: A Survey of Key Managemen in Ad Hoc Networks. IEEE Communications Surveys & Tutorials (2006)

[32] Dong, Y., Sui, A.F., Yiu, S.M., Li, V.O.K., Hui, L.C.K.: Providing Distributed Certificate Authority Service in Cluster-based Mobile Ad hoc Networks. Computer Communications 30(11-12), 2442–2452 (2007)

[33] Chaddoud, G., Martin, K.: Distributed Certificate Authority in Cluster-based Ad hoc Nnetworks. In: Wireless Communications and Networking Conference (2006)

[34] John, S.P., Samuel, P.: A Predictive Clustering Technique for Effective Key Management in Mobile Ad Hoc Networks. Information Security Journal: A Global Perspective 20(4-5), 250–260 (2011)

[35] Konstantinou, E.: Efficient Cluster-based Group Key Agreement Protocols for Wireless Ad hoc Networks. Journal of Network and Computer Applications 34 (2011)

[36] Shamir, A.: Identity-Based Cryptosystems and Signature Schemes. In: Blakely, G.R., Chaum, D. (eds.) CRYPTO 1984. LNCS, vol. 196, pp. 47–53. Springer, Heidelberg (1985)

[37] Silva, E.D., Santos, A.L.D., Albini, A.L.C.P.: Advances in Identity-based Key Management in Mobile Ad hoc Networks: Techniques and Applications. IEEE Wireless Communications (2008)

[38] Balazinska, M., Deshpande, A., Franklin, M.J., Gibbons, P.B., Gray, J., Hansen, M., Liebhold, M., Nath, S., Szalay, A., Tao, V.: Data management in the Worldwide Sensor Web. IEEE Pervasive Computing, 30–40 (2007)

[39] Kapil, A., Rana, S.: Identity-Based Key Management in MANETs using Public Key Cryptography. International Journal of Security (IJS) 3 (2009)

[40] Yu, F.R., Tang, H., Mason, P.C., Wang, F.: A Hierarchical Identity Based Key Management Scheme in Tactical Mobile Ad Hoc Networks. IEEE Transactions On Network And Service Management 7(4) (2010)

[41] Dahshan, H., Irvine, J.: On Demand Self-Organized Public Key Management for Mobile Ad Hoc Networks. In: IEEE 69th Vehicular Technology Conference 2009 (2009)

[42] Gordon, R.L., Dawoud, D.S.: Direct and Indirect Trust Establishment in Ad Hoc Networks by Certificate Distribution and Verification. In: First International Conference on Wireless Communication Society, Vehicular Technology, Information Theory and Aerospace & Electronic Systems Technology (Wireless VITAE 2009) (2009)

[43] Anita, E.A.M., Vasudevan, V.: Black Hole Attack Prevention in Multicast Routing Protocols for Mobile Ad hoc networks using Certificate Chaining. International Journal of Computer Applications (2010)

[44] Wua, B., Wua, J., Fernandeza, E.B., Ilyasa, M., Magliveras, S.: Secure and Efficient Key Management in Mobile Ad hoc Networks. Journal of Network and Computer Applications (2005)

[45] Wu, B., Wu, J., Dong, Y.: An Efficient Group Key management Scheme for Mobile Ad hoc Networks. Int. J. Security and Networks (2008)

[46] Pereira, F., Fraga, J.S., Cust'odio, R.: Self-adaptable and Intrusion Tolerant Certificate Authority for Mobile Ad hoc Networks. In: 22nd International Conference on Advanced Information Networking and Applications (2008)

[47] Eschenauer, L., Gligor: A Key-Management Scheme for Distributed Sensor Networks. In: 9th ACM Conference on Computer and Communication Security (2002)

[48] Xenakis, C., Panos, C., Stavrakakis, I.: A Comparative Evaluation of Intrusion Detection Architectures for Mobile Ad Hoc Networks. Elsevier Computers & Security (2010)

[49] Jacoby, G.A., Davis, N.J.: Mobile Host-Based Intrusion Detection and Attack Identification. IEEE Wireless Communications 14(4) (2007)

[50] Nadkarni, K., Mishra, A.: A Novel Intrusion Detection Approach for Wireless Ad Hoc Networks. In: IEEE Wireless Communications and Networking Conference, WCNC 2004 (2004)

[51] Zhang, Y., Lee, W., Huang, Y.: Intrusion Detection techniques for Mobile Wireless Networks. Wireless Networks Journal (2003)

[52] Bhargava, S., Agrawal, D.: Security Enhancements in AODV protocol for Wireless Ad hoc Networks. In: IEEE Vehicular Technology Conf. (2001)

[53] Bose, S., Bharathimurugan, S., Kannan, A.: Multi-Layer Integrated Anomaly Intrusion Detection System for Mobile Ad Hoc Networks. In: IEEE ICSCN 2007. MIT Campus, Anna University (2007)

[54] Razak, S.A., Furnell, S.M., Clarke, N.L., Brooke, P.J.: Friend-assisted Intrusion Detection and Response mechanisms for Mobile Ad hoc Networks. Ad Hoc Networks (2008)

[55] Ramachandran, C., Misra, M.O.S.: FORK: A Novel two-pronged Sstrategy for an Agentbased Intrusion Detection Scheme in Ad-hoc Networks. Computer Communications 31(16) (2008)

[56] Ma, C., Fang, Z.: A Novel Intrusion Detection Architecture Based on Adaptive Selection Event Triggering for Mobile Ad-hoc Networks. In: IEEE Second International Symposium on Intelligent Information Technology and Security Informatics (2009)

[57] Manousakis, K., Sterne, D., Ivanic, N., Lawler, G., McAuley, A.: A Stochastic Approximation Approach for Improving Intrusion Detection Data Fusion Structures. In: IEEE Military Communications Conference (MILCOM 2008), San Diego, CA (2008)

[58] Deng, H., Xu, R., Li, J., Zhang, F., Levy, R., Lee, W.: Agent-based Cooperative Anomaly Detection for Wireless Ad hoc Networks. In: 12th Conference on Parallel and Distributed Systems (2006)

[59] Joseph, J.F.C., Das, A., Seet, B.-C., Lee, B.-S.: Cross layer Versus Single Layer Approaches for Intrusion Detection in MANETs. In: 15th IEEE International Conference on Networks (ICON). IEEE (2007)

[60] Joseph, J.F.C., Das, A., Seet, B.-C., Lee, B.-S.: CRADS: Integrated Cross layer Approach for Detecting Routing Attacks in MANETs. In: WCNC 2008. IEEE (2008)

[61] Rachedi, A., Benslimane, A.: Toward a Cross-layer Monitoring Process for Mobile Ad hoc Networks. Security and Communication Networks (2008)

[62] Hortos, W.S.: Unsupervised Algorithms for Intrusion Detection and Identification in Wireless Ad hoc Sensor Networks. In: SPIE (2009)

[63] Shrestha, R., Han, K.-H., Choi, D.-Y., Han, S.-J.: A Novel Cross Layer Intrusion Detection System in MANET. In: 4th IEEE International Conference on Advanced Information Networking and Applications (2010)

[64] Gottron, C., König, A.E., Steinmetz, R.: A Cross-Layer Approach for Increasing Robustness of Mobile Peer-to-Peer Networks. In: Proceedings of the Security in NGNs and the Future Internet Workshop (2010)

[65] Svecs, I., Sarkar, T., Basu, S., Wong, J.S.: XIDR: A Dynamic Framework Utilizing Cross-Layer Intrusion Detection for Effective Response Deployment. In: 34th Annual IEEE Computer Software and Applications Conference Workshops (2010)

[66] Marchang, N., Tripathi, R.: A Game theoretical approach for Efficient Deployment of Intrusion Detection System in Mobile Ad Hoc Networks. In: 15th Int. Conf. on Advanced Computing and Communications, Guwahati, Assam (2007)

[67] Otrok, H., Mohammed, N., Wang, L., Debbabi, M., Bhattacharya, P.: A Game-Theoretical Intrusion Detection Model for Mobile Ad Hoc Networks. Computer Communications 31 (2008)
[68] Rafsanjani, M.K., Aliahmadipour, L., Javidi, M.M.: An Optimal Method for Detecting Internal and External Intrusion in MANET. In: Kim, T.-h., Vasilakos, T., Sakurai, K., Xiao, Y., Zhao, G., Ślęzak, D. (eds.) FGCN 2010. CCIS, vol. 120, pp. 71–82. Springer, Heidelberg (2010)
[69] Balachandran, S., Dasgupta, D., Wang, L.: A Hybrid Approach for Misbehavior Detection in Wireless Ad-Hoc Networks. Computer and Information Science (2006)
[70] Sen, S., Clark, J.A.: Evolutionary computation techniques for intrusion detection in mobile ad hoc networks. Computer Networks (2011)
[71] Ford, R., Carvalho, M., Allen, W.H., Ham, F.: Adaptive Security for MANETs via Biology. In: 2nd Cyberspace Research Workshop (2009)
[72] Mohamed, Y.A., Abdullah, A.B.: Immune Inspired Framework for Ad Hoc Network Security. In: 7th IEEE International Conference on Control & Automation (ICCA 2009), Christchurch, New Zealand (2009)
[73] Wang, W., Man, H., Liu, Y.: A framework for intrusion detection systems by social network analysis methods in ad hoc networks. Security and Communication Networks (2009)
[74] Mitrokotsa, A., Komninos, N., Douligeris, C.: Protection of an Intrusion Detection Engine with Watermarking in Ad Hoc Networks. International Journal of Network Security (2010)
[75] Islam, N., Shaikh, Z.A., Aqeel-ur-Rehman: COINS: Towards a Correlation Based Intrusion Detection System for Mobile Ad hoc Network. Sindh University Research Journal (2011)
[76] Shaikh, R.A., Shaikh, Z.A.: A Security Architecture for Multihop Mobile Ad hoc Networks with Mobile Agents. In: 9th International Multitopic Conference, IEEE INMIC 2005 (2005)

Secure AODV Routing Protocol Based on Trust Mechanism

Harris Simaremare[1], Abdelhafid Abouaissa[1], Riri Fitri Sari[2], and Pascal Lorenz[1]

[1] Universire de Haute Alsace, France
{harris.simaremare,abdelhafid.abouaissa,pascal.lorenz}@uha.fr,
lorenz@ieee.org
[2] Universitas Indonesia, Indonesia
riri@ui.ac.id

Abstract. A mobile ad hoc network (MANET) is a wireless network with high of mobility, no fixed infrastructure and no central administration. These characteristics make MANET more vulnerable to attack. In ad hoc network, active attack i.e. DOS, and blackhole attack can easily occur. These attacks could decrease the performance of the routing protocol. In this chapter, we proposed new trust mechanism that has the ability to detect and prevent the potentials attacks into a wireless ad hoc network especially for Denial of Service (DOS) and blackhole attacks. We have proposed some modifications of AODV routing protocol with implemented a trust level calculation. Our proposed mechanism will detect the attack by calculate local and global trust parameters. When a node is suspected as an attacker, the security mechanism will isolate it from the network before communication established. To perform the trust calculation, each node should get all the activity information from his neighbor. In order to ensure the nodes can hear all the activities of his neighbors, each node will run in promiscuous mode. Simulation has been conducted using NS-2 to evaluate our proposed protocol under dos and blackhole attack. We compare the performance of our proposed protocol with existing secure routing protocol such as TCLS [14], LLSP [16], and RSRP [17]. The simulation result shows that our proposed protocols outperform other secure protocols under DOS and blackhole attack in term of packet delivery ratio, end to end delay and routing overhead. We demonstrate that the proposed protocol improves significantly the performance of secure routing protocol.

Keywords: AODV, Blackhole attack, DOS attack, MANET, Promiscuous, Security, Trust mechanism.

1 Introduction

A mobile ad hoc network (MANET) is a wireless network with autonomic, provisional, no fixed infrastructure and no central administration. It is widely used in military information system of battle field, civil emergency search, rescue operations and other occasions. Nodes in the network usually have limited resources such as processor, bandwidth, memory, and energy. Because nodes are mobile and the

S. Khan and A.-S.K. Pathan (Eds.): *Wireless Networks and Security*, SCT, pp. 81–105.
DOI: 10.1007/978-3-642-36169-2_3

topology of the network varies. In traditional wireless networks, a base station or access point facilitate communications between nodes within or outside the network [1]. In contrast, MANET is an infrastructure-less network where every node acts as a router for establishing the connection between sources to destinations. Since there is no administrative node to control the network, every node participating in the network is responsible for the reliable operation of the whole network. The dynamic nodes mobility makes the network topology changes rapidly and unpredictable. When the network topology is changing, the connections need to re-establish. In addition, the features of ad hoc networks are similar to normal wireless network. All the natural behavior in wireless ad hoc network makes security problem become more complex.

In general, routing protocols for the wireless ad hoc networks can be classified into three types, based on the routing information update mechanism. The three types are reactive protocol (on demand), proactive protocol (table driven) and hybrid protocol [2].

Proactive protocol or table driven protocol is a type of protocol where the nodes will keep network topology information, and changes routing information periodically. This mechanism can make flood the network with active request information. A disadvantage from this method is increased overhead inside the network, when finding the path communication to destination, routing information will be generally flooded in the whole network.

Reactive protocol is on demands protocols that discover the route once needed. Reactive protocol does not maintain the network topology information and do not exchange routing information periodically.

The last type is hybrid protocol. This protocol combines the feature of proactive and reactive protocol. In this type, a node communicates with its neighbors using a proactive routing protocol, and uses a reactive protocol to communicate with nodes farther away. In other words, the nodes will choose the best way when establish the communication in the network.

The Internet Engineering Task Force (IETF) working group on [3] produced four experimental Requests for Comments (RFCs) for four routing protocols, i.e. Ad Hoc on Demand Distance Vector (AODV), Optimized Link State Routing (OLSR), Topology Dissemination Based on Reverse-Path Forwarding (TBRPF), and Dynamic Source Routing (DSR).

In this chapter, we will discuss more about reactive protocol. As we explain before, the reactive protocols are on demand protocols that discover the route once needed. Examples of reactive protocol are AODV, DSR [4], SAODV [5], and SAR [6]. The reactive protocols display considerable bandwidth and overhead advantages over proactive protocols. AODV routing protocol offers quick adaptation to dynamic link conditions, low processing, low memory overheads, and low network utilization.

In the standard AODV routing protocol assume that there are no malicious nodes participating in routing operations. This assumption cannot be applied in real MANET because the nature of the MANET such as high of mobility, no central coordination mechanism, open network and it communicates by collaboration between nodes, makes MANET more vulnerable from attack. Some research about security issues of AODV protocol, which is one of the most popular routing protocols in ad hoc networks, is done. In this chapter, we will address the security aspect to enhance AODV routing protocol.

Attacks on MANET can be classified as the active attacks and passive attacks. Passive attacks do not disrupt the operation of a routing protocol or influence the functionality of connection, but only attempts to discover valuable information by listening to the routing traffic. This type of attack is difficult to detect. Active attacks attempts to improperly modify data, destroy data, gain authentication, or procure authorization by inserting false packets into the data stream or modifying packets transition through the network.

In order to prevent network from passive or active attack, secure protocol must follow the general security objectives like:

a) Confidentiality: Only the intended receivers should be able to interpret the transmitted data. For example using digital signature mechanism.
b) Integrity: Data should not change during the transmission process. Sent packet must be the same with the received packet.
c) Availability: Network services should be available all the time when it needed.
d) Authentication: Received data must be authenticated and must initially send by the legitimate node.
e) Non-repudiation is having a proof that the announced author really wrote the message, and such the proof can be verified even without the consent of the said author: the author must not be able to *repudiate* his message.

2 Security Issues in MANET

High of mobility, no central coordination mechanism, and open network makes MANET more vulnerable to attack. Thus the secure solution has to be also dynamic. Some of security problem issues are; first, with many of different nodes inside the network, it is also hard to detect the exact malicious node. Second, some of the nodes may have limited memory space, computational power and battery powered. Thus time and power consuming applications become hindrance for the whole ad-hoc network. Last, active/passive link attacks like spoofing, denial of service, masquerading, eavesdropping, impersonation attacks are possible. The security issues in each layer have been identifies [7] that shown in Table 1.

Table 1. Security issue related each layer in protocol stack [7]

Layer	Security Issues
Application	Prevention, detection of viruses, worms, malicious codes, application abuses
Transport	Authentication and end to end data security through encryption techniques
Network	Security of ad hoc routing protocols and associated parameters.
Physical	Preventing signal jamming, denial of service attacks and other active attacks

Next we will identify attack that can perform in MANET. The latter type of attack is including black hole, wormhole attack, denial of service (DOS),Sleep deprivation torture, rushing attack, Byzantine attack, malign attack, partition attack, detour attack, routing table poisoning, packet replication, session hijacking and impersonation attack have been addressed in literature[1].

Attack definition:

a) Blackhole attack: Malicious node sends a forged RREP packet to a source node that initiates the route discovery in order to pretend to be a destination node itself or a node immediate neighbor of the destination. Source node will forward all of its data packets to the malicious node; which were intended for the destination.
b) Wormhole attack: A malicious node uses a path outside the network to route messages to another compromised node at some other location in the network. For illustration, using a pair of attacker nodes A and B linked via a private network connection. A will forwarded all the packet through wormhole to B.
c) Denial of Service Attack: An adversary tries to disturb the communication in a network, for example by flooding the network with a huge amount of packages. Services offered by the network are not working as usual, slow down or even stop. Ad hoc wireless networks are more affected than wired networks, because there are more possibilities to perform such an attack. Depending on the layer an adversary starts an attack. It could disturb transmissions on physical layer, manipulate the routing process on network layer or bring down important service on application level.
d) Sleep deprivation torture : A malicious node request the services a certain node offers, over and over again, so it cannot go into an idle or power preserving state, thus depriving it of its sleep (hence the name). This can be very devastating to networks with nodes that have limited resources, for example battery power.
e) Rushing attack: A malicious node will attempt to tamper with route request packets, modifying the node list, and hurrying this packet to the next node.
f) Byzantine attacks: In which two or more routers collude to drop, fabricate, modify, or misroute packets in an attempt to disrupt the routing services.
g) Malign attack: Watchdog and path-rater are used in ad hoc routing protocols to keep track of perceived malicious nodes in a blacklist. An attacker may blackmail a good node, causing other good nodes to add that node to their blacklists, thus avoiding that node in routes.
h) Partition attack: An attacker may try to partition the network by injecting forged routing packets to prevent one set of nodes from reaching another.
i) Detour attack: An attacker may attempt to cause a node to use detours through suboptimal routes. Also compromised nodes may try to work together to create a routing loop.
j) Routing table poisoning: The publication and advertisement of fictitious routes.
k) Packet replication: The replication of stale packets. The aim is to consume additional resources such as bandwidth.
l) Session hijacking: One weak point is that most authentications processes are only carried out once when a session starts. An adversary could try to appear as an authentic node and hijack the session. (Transport Layer Attack)

m) Impersonation attack: Also called spoofing attacks. The attacker assumes the identity of another node in the network, thus receiving messages directed to the node it faked. Usually this would be one of the first steps to intrude a network with the aim of carrying out further attacks to disrupt operation.

Types of attacks against ad hoc networks will continue to develop rapidly in new forms and methods. In this chapter, we focus to active attack i.e. blackhole attack and DOS attack. And we propose a mechanism to prevent both of them.

3 AODV Routing Rotocol

AODV (Ad hoc On Demand Distance Vector) [8] has been considered as one of the most popular and promising on-demand routing protocol, which has been standardized by IETF and attracts many concerns because of its lower network overhead and algorithm complexity. The routing procedure of AODV is described as follows.

3.1 Route Creating

Route creating procedure is a mechanism of AODV routing protocol to find and create a new path communication between sources to destination. AODV mechanism performs as follows.

1. The source node broadcasts Route Request (RREQ) message.
2. Once the intermediate node receives the RREQ message, a reverse route towards the upstream node that sends the RREQ message is built. If the node has a fresh route to the destination, it will send Route Reply (RREP) message along the reverse route to the source node, else the RREQ message will be forwarded one by one.
3. The destination node sends RREP message to the source node through reverse route after it receives RREQ message.
4. All nodes on the reverse route update their routing tables, in which a route to the destination node will be built.
5. Once RREP reaches the source node, the route searching process is terminated. A new route is built in its routing table by which the transmission can be done.

3.2 Route Maintaining

Route maintaining is a procedure in AODV routing protocol to cover broken link problem during communication. If a node detects a link break for the next hop of an active route in its routing table, it deletes all route entries including this broken link, and broadcasts Route Errors (RERR) message to notify those upstream nodes to delete the corresponding entries in their routing tables. A node may detect connectivity with others by broadcasting local Hello messages periodically and broadcast RERR message when a link break is detected.

4 Variant of Secure AODV Routing Protocol

In this section we will explain about variant secure protocol based on AODV. Many researchers have proposed new variant and new mechanism to make AODV routing protocol more secure. There are two mainstreams to enhance the security aspect in AODV routing protocol i.e. cryptographic mechanism and trust based mechanism.

4.1 Secure AODV Routing Protocol Using Cryptographic Mechanism

This method use cryptographic mechanism to protect exchanging packet data, route creating, and route maintenance during the communication. It will guarantee the confidentiality and integrity aspect. Many types of cryptography algorithms had applied to secure the packet.

Zapata proposed Secured AODV (SAODV) [5] to protect the route discovery mechanism of AODV using a signature. SAODV added secure based on public key cryptography. Before the data packet sent, firstly performed the encryption then the recipient will be able to read back the data with the specified public key. The Secure-AODV scheme is based on the assumption that each node has certified public keys of all network nodes. Node that generates a routing message signs it with its private key and the nodes that receive this message verify the signature using the sender's public key. The next challenge for researchers is the problem of public key distribution. How we can distribute the public keys to the whole network.

In AODV when node is destination, intermediate nodes must reply with Route reply. SAODV includes a kind of delegation feature that allows intermediate nodes to reply the RREQ messages. This is called the double signature [3]: when a node generates a RREQ message, in addition to the regular signature, it can include a second signature. Intermediate nodes can store this second signature in their routing table, along with other routing information related to node source. The intermediate node generates the RREP message, includes the signature of node that it previously cached, and signs the message with its own private key. RREP packet size will be increase because it must be accompanied by a security key.

Messages that exchange in SAODV become significantly bigger because of added digital signature in each messages and additional signature for reply packet. New problem comes, SAODV requires heavyweight asymmetric cryptographic operations: every time a node generates a routing message, it must generate a signature, and every time it receives a routing message (also as an intermediate node), it must verify a signature. This gets worse when the double signature mechanism is used, because this may require the generation or verification of two signatures for a single message. To mitigate this problem, Cerri and Ghioni proposed A-SAODV [3]. This protocol based on AODV-UU. AODV-UU is a modified protocol based on AODV that add gateway module, and can implement in real world scenario. In SAODV, the intermediate nodes send route reply with signature, so intermediate node generates a cryptographic signature. Nodes may spend much time in computing these signatures and become overloaded. Moreover, if intermediate nodes have a long queue of routing messages that must be cryptographically processed, the resulting delay may be longer for the packet to reaches the destination node. If the double signature mechanism is removed, protocol becomes un-collaborative, in which only the destination node is allowed to reply to a RREQ message. To solve that problem, A-SAODV has adaptive mechanism. The mechanism is makes intermediate nodes reply the request only if they are not overloaded. This option will choose based on value queue length. If queue length lower than threshold, the nodes generate a RREP with signature, if not, RREQ will forward without replying request to source node. This algorithm can optimize performance of SAODV.

Eichler and Roman proposed AODV-SEC protocol [9] which is the development of the AODV routing protocol. AODV-SEC is an improved version of the SAODV. The approaches used in this protocol are the certificate and public key infrastructure. To secure data package, the mechanism added a new certificate called mCert. Two security functions are added to the library and certificates types. For cryptographic library using libcrypto because it has a faster performance. For the certificates AODV-SEC use mCert. These certificates have a new design with a smaller size and simpler in order to reduce overhead. The AODV-SEC protocol tries to secure all possible aspects of the route discovery process. This includes the authentication of the two end nodes as well as the intermediate nodes. Further, it excludes not trusted nodes from the discovered routes. The length of the discovered route is protected in a way that intermediate nodes cannot advertise a potentially shorter route than actually exists. The major difference compared to SAODV is the inclusion of a last-hop authentication mechanism and the defined certificate usage. No certificates have to be distributed before operation for the AODV-SEC protocol. Only the CA certificate needs to be known to the nodes. To improve the performance and capabilities of the protocol, AODV-SEC defined the data container. This additional data field can be used to run a key agreement protocol in parallel to the route setup process. This feature reduces the connection setup time.

4.2 Secure AODV Routing Protocol Using Trust Based Mechanism

In general, this mechanism builds a trust relationship between nodes before perform communication. Trust parameter nodes are represented by opinion, which is an item derived from some definition from normal behavior communication. The opinions are dynamic and updated frequently as the protocol specification: If one node performs normal communications, the opinion from other nodes points of view can be increased. Otherwise, if one node performs some malicious behaviors, it will be ultimately denied by the whole network. A trust recommendation mechanism is also designed to exchange trust information among nodes. With trust relation between nodes, next no need for a node to request and verify certificates all the time. In this protocol, author modified the routing table information by adding positive event and negative event. This information indicates that the nodes are malicious or normal node.

Several trust models have been developed for peer-to-peer systems, based on sharing recommendation information to establish reputation. Although in principle these could be applied to routing in MANETs, there are two important problems.

First, there is significant network overhead due to the additional information exchanged. Second, addressing the potential for malicious recommendations requires a trusted third party (or a computationally expensive public-key infrastructure), which goes against the nature of MANETs.

Pirzada et al. proposed a new pragmatic method for establishing trustworthy routes in AODV [10]. The trust models develop with three components: Trust Agent, Reputation Agent and combiner. The Trust agent extracts trust information from the events that are directly experienced by a node. The Reputation agent shares trust information with other nodes in the network. The Combiner calculates the total trust in a node from the information that it receives from the Trust and Reputation agents. Application of the trust agent in AODV uses six categories to derive the events that used to compute the situational trust and subsequently the direct nodes i.e. Acknowledgements (Pa), Packet Precision (Pp), Gratuitous Route Replies (Gr), Blacklists (Bl), Hello Messages (Hm), Destination Unreachable Messages (Du).

Each category will used to define the situational and save into based upon their success or failure rate. To share trust reputation in each node, reputation agent make reputation table that can inform the reputation of nodes inside the network. The application of combiner process the information that is receives from the trust agent and the reputation agent. It will compute the derived and aggregate trust value. Trust value is saved in trust table that running by application trust agent.

Based on this method, Pushpa [11] developed a trust mechanism that node can communicate with others in two security aspect i.e. nodes trust and route trust. To make node and route more trust, the system more complicated because it run three main operations; Node trust calculation, Route trust calculation and Route establishment process. Node trust calculated by using a method proposed by Pirzada, but the data structure from neighbor table has modified. Pushpa modified RREQ and RREP message format to ensure trust mechanism.

In Zhe et al. [8] proposed a security mechanism in AODV routing protocol based on the credence model calculation. This mechanism can react quickly when detecting some malicious behaviors in the network and effectively protect the network from kinds of attacks. The credence mechanism is used to prevent the attack by calculate the communication behaviors by evaluate the routing packet processing and data packet forwarding. Nodes monitor communication behaviors between neighbors, and exchange the information with others to obtain credence values, and store them in the credence table. When the credence mechanism judged a node as an attacker, the security routing protocol will isolate the node attacker from the network. In order to provide secure and reliable data forwarding services, nodes should compare the credence value with his neighbors. In this case, nodes need more space memory to save the credence value of each neighbor.

Griffiths et al. [12] proposed STAODV, a trust model using acknowledgements as the single observable factor for assessing trust. An Acknowledgements offer an effective indication of a node's trustworthiness. An acknowledgement is a means of ensuring that packets which have been sent for forwarding have actually been forwarded. There are a number of ways that this is possible, but passive acknowledgement is the simplest. Passive acknowledgement uses promiscuous mode to monitor the channel, which allows a node to detect any transmitted packets, irrelevant of the actual destination that they are intended for. Using this method a node can ensure that packets it has sent to a neighboring node for forwarding are indeed forwarded. Once a node becomes untrusted it is barred from consideration for packet forwarding by dropping it from the set of neighbors, removing all routes that use it, and sending out a new RREQ to re-establish the removed routes. Similarly, when receiving a RREP the first hop node is checked and if it is untrusted then the reply is disregarded. Thus, only routes where the first hop is trusted are established. Nodes make routing choices based on trust as well as the number of hops, such that the selected next hop gives the shortest trusted path.

Kurosawa et al. [13] proposed an anomaly detection scheme using dynamic training method in which the training data is updated at regular time intervals. When the black attack took place, regardless of the environment the sequence number is increased largely. Also, usually the number of sent out RREQ and the number of received RREP is almost the same. In order to detect this attack, the destination sequence number is taken into account, and evaluates the number of sent out RREQ messages, number of received RREP messages, the average of difference of DstSeq in each time slot between the sequence number of RREP message and the one held in the list.

Rajaram and Palaniswami [14] proposed TCLS, a trust mechanism with an additional data structure called Neighbors' Trust Counter Table (NTT). Trust counter is calculated based on the number of packet that has forwarded through a route. To evaluate the behavior of nodes, the mechanism will compare between the total of success packet forwarded and total accumulation RREQ message in destination. To make communication more secure, in the route reverse process, each node must verify all the packet using digital signature mechanism.

Mistry et al. [15] proposed the improving mechanism for AODV against blackhole attack. Attacker node detect by capturing and comparing the destination sequence number contained in RREP packets with source sequence number. In the original AODV, the greatest one is the most recent routing information.

Islam et.al [16] proposed a possible framework of a link level security protocol (LLSP) to be deployed in a Suburban Ad hoc Network (SAHN). LLSP provides authentication, integrity assurance and encryption for ensuring security at the data link layer. To determine LLSP's practicability, the timing requirement for each authentication process has estimated. The result indicates that LLSP can be a suitable link level security service for an ad hoc network similar to a SAHN. To enhance the security feature of the link layer, the Watchdog module of LLSP monitors channel usage of each neighbor and informs the MAC layer to take necessary steps for misbehaving neighbors. The security services provided by LLSP can be classified into five types: (Type 1) Authenticating a new node, (Type 2) Updating the capability1 (CAP) of a link, (Type 3) Updating the shared key (SHK) of a link, (Type 4) Authenticating received packets and (Type 5) Encrypting payload.

Afzal et al. [17] have presented the design and analysis of a secure on-demand routing protocol, called RSRP which confiscated the problems mentioned in the existing protocols. In addition, RSRP has used a very efficient broadcast authentication mechanism which does not require any clock synchronization and facilitates instant authentication.

Based on the review of secure AODV routing protocol above, we can describe mapping of secure protocol classification based on security method.

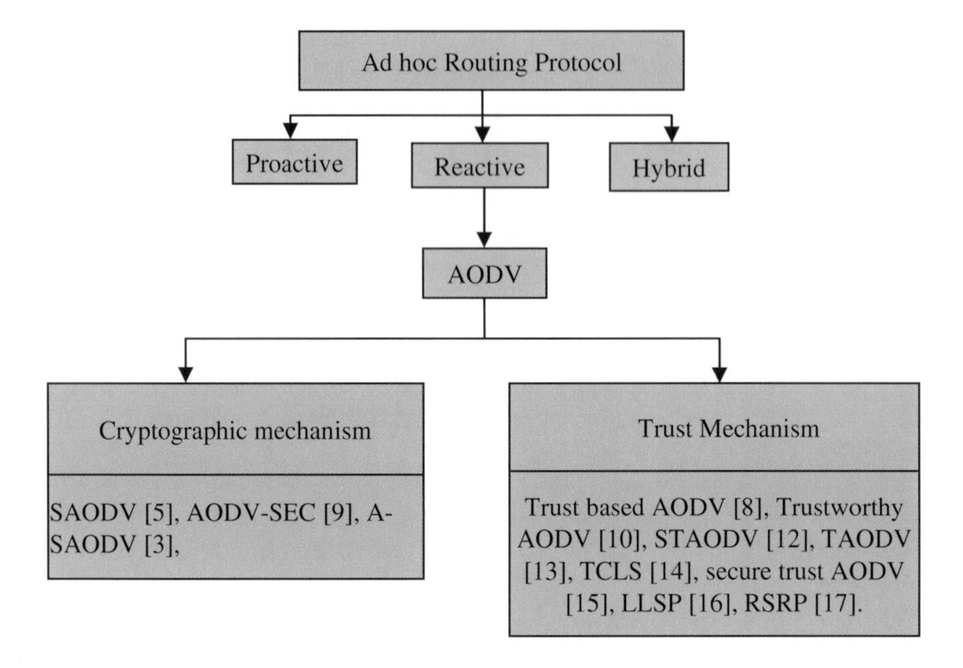

Fig. 1. Mapping Variant AODV Secure Routing Protocol

Trust mechanisms have a several advantages i.e.

1. Trust relation between nodes, no need for a node to request and verify certificates all the time.
2. No need to added any signature or cryptography method in packet message, so the size of messages not big. It can make performance become better.

In this chapter, we choose to improve the security of AODV routing protocol with trust mechanism method to keep the performance of proposed routing protocol.

5 Proposed Trust Mechanism

In this chapter, we proposed a new trust mechanism that has ability to detecting and preventing the potentials attack into a wireless ad hoc network. We have performed some modification AODV routing protocol by adding trust level calculation. The proposed mechanism will detect the attack by trust local calculation and trust global calculation. When a node suspect as an attacker, the security mechanism will isolate the suspect node from the network before communication establish. To perform the trust calculation, each node should collect all the activity information from its neighbor nodes. In order the nodes can hear all the activities of his neighbor, each node will run in promiscuous mode. In standard AODV, promiscuous mode is deactivated. To activated promiscuous mode, we performed minor modification in AODV.

Each node will detect the anomaly in its neighbor node based on the calculation of the activities packet in nodes. Trust calculation performs when the node begin communication process. Each node will hear and calculate the total of received and forwarded RREQ, RREP, and RERR packet.

5.1 Detection Mechanism

A. DOS/DDOS Attack

DOS attack will flood the victim nodes continuously with a useless request and in a big packet size. The victim cannot serve the real request to another node. In our scenario, we assume that when the nodes perform DOS attack, it will send number of request more than number of forwarded. If the neighbor hear and detect that the comparison of number of request and number of forwarded packet is too big, the mechanism will suspect the node as a malicious nodes. Figure 2 describes the DOS attack.

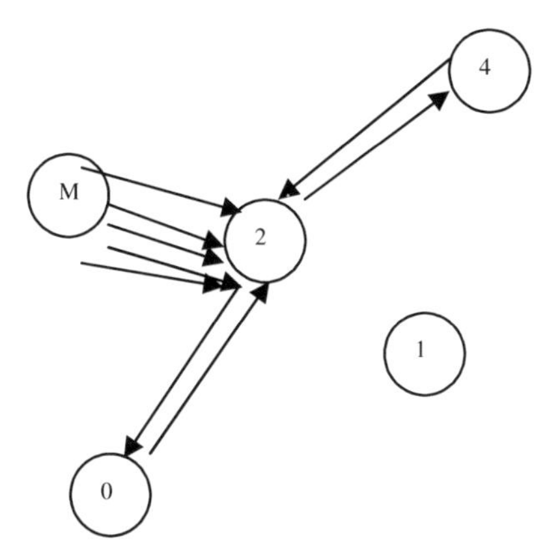

Fig. 2. DOS attack

In figure 2, node M performs DOS attack to node 2 when node 0 establishing the communication with node 4. In this scenario, node 0 and node 2 will hear the request and forwarded packet from node M. All the neighbor of node M will know that the node M is a malicious node, because node M sends more request packet than forwarded packet. Our mechanism will calculate the trust level with formula 1, and 2. If the result calculation indicated node M is a malicious, then node M directly ignored from the communication activities.

B. Backhole Attack

Attacker node will send the fake reply to say to the source node that he has a fresh route or he is a destination. And then source node will establish the communication with malicious nodes. As a consequence, the real destination will never receive the packets because there is no establish communication with the source node. To make source node believe that its reply is a shortest path and a fresh path, malicious node will send the reply packet with a higher destination sequence number. Figure 3 describes the scenario of blackhole attack.

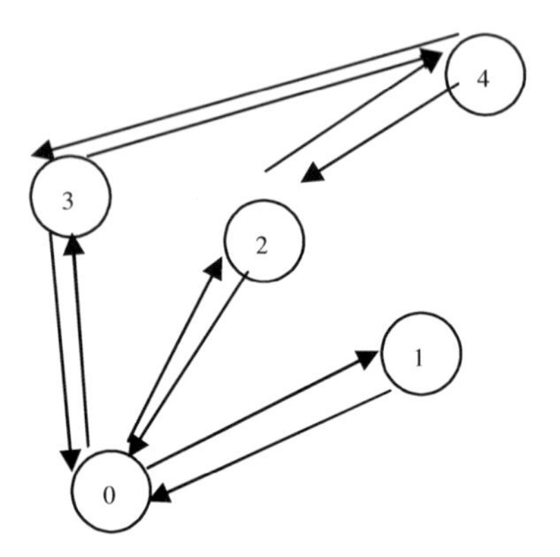

Fig. 3. Blackhole attack scenario

In AODV routing protocol, when the destination node receives a route request (RREQ), it will generate and send route reply (RREP) packet. RREP packet consists of destination packet, source id RREP, sequence number and life time. We use this information to detect blackhole attack. Scenario in figure 3 shows that node 0 wants to establish communication with node 4. During the route discovery process, the malicious node (M) sends reply packet to indicate that he has a fresh route and in order to manipulate the source node that he is the destination. In our trust mechanism, when the source node receive route reply packet from its neighbors, it will compare the reply packets based on their source id. Node 0 will receive RREP from node 2, 3, and M. Source node will check the RREP source id and found that RREP source id from node 2 and node 3 are same, but different RREP source id from node M. Source node will choose communication to node 2 and 3, and suspect that node M is a malicious. Source node will continue the process selecting path based on destination sequence number.

5.2 Trust Calculation

Trust calculation mechanism is an algorithm that will used by each node to compute and to conclude the level of trust of his own neighbor. This calculation will perform each time when the nodes begin to establish connection and sending the packet data.

We divided the trust calculation into two categories i.e. trust global (TG) and trust local (TL).

Trust global (TG) is a trust calculation level based on the total of number packet received compare with the number of forwarded packet in the nodes. The packets are RREQ, RREP, and RERR. And Trust local (TL) is a comparison between total received packet and total forwarded packet from specific nodes. The formula for the trust calculation described as follows, suppose that the topology as figure 4.

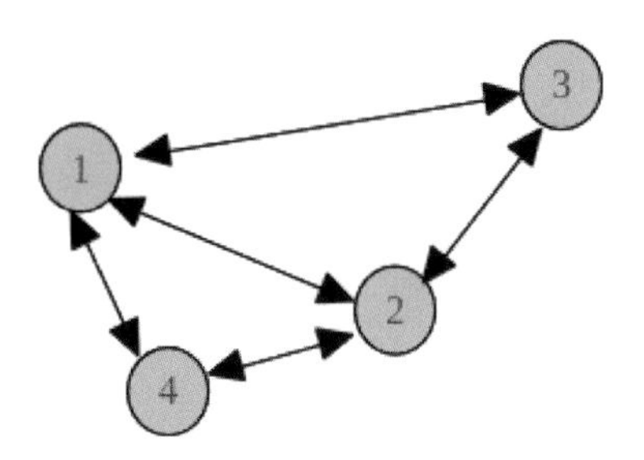

Fig. 4. Description of trust calculation

$$TL_{3-2} = \frac{\sum Packet\ received\ from\ node\ 1\ to\ node\ 2}{\sum Packet\ received\ from\ node\ 1\ forwarded\ by\ node\ 2\ to\ node\ 3};\ where\ TL_{3-2} = 1 \quad (1)$$

$$TG_{3-2} = \frac{\sum packet\ received\ in\ node\ 2}{\sum packet\ forwarded\ in\ node\ 2};where\ TG_{3-2} \leq 1 \quad (2)$$

Where;

TL_{3-2} : Trust local node 2 to node 3.

TG_{3-2} : Trust global node 2 to node 3.

Node 3 wants to calculate the trust level of node 2. First steps, node 3 will calculate TL on node 2. Node 3 will compare total of received packet in node 2 that origin from node 1 with the total of forwarded packet to node 3 that origin from node 1. Next steps, node 3 will calculate TG of node 2 by comparing the total packet received and total packet forward in node 2. Last steps, node 3 will combine the result calculation using formula 4.

The value of trust local (TL) should be equal to 1, because normal nodes will forward the entire received packet. If the nodes only forward some of request or create a new fictitious packet, we suppose that it is a malicious node. For the trust global, the calculation result should be ≤ 1. when the nodes received packet from his neighbor, he will forward it to all intermediate nodes in his communication range. If the nodes not forward the packet to all neighbors, we suppose that he is malicious.

If the calculation result of TL = 1, TL level is true. And if result value of TG ≤ 1, TG level is true. To decide neighbor node is trust or distrust, the nodes will combine the calculation result from TG and TL.

$$Trust\ level = TG\ (and) TL \quad (3)$$

The node is trust when the trust level of TG and TL is true.

The basic algorithm in our trust mechanism shows as follows.

1. The source node broadcasts Route Request (RREQ) message.
2. Once the intermediate node receives the RREQ message, a reverse route towards the upstream node that sends the RREQ message is built.
3. If the node has a fresh route to the destination, it will send Route Reply (RREP) message along the reverse route to the source node, else the RREQ message will be forwarded one by one.
4. Calculates Trust Local and Trust Global,
5. The destination node sends RREP message to the source node through reverse route after it receives RREQ message.
6. Set initial condition for all nodes is trust.
7. Source node will compare the RREP source id, and select a path which sends a same RREP source id.
8. Calculated trust total trust level (TG and TL).
9. If the node is trusted, all nodes on the reverse route update their routing tables, in which a route to the destination node will be built, then the communication and transmission can be done.
10. Else not trust, delete node from the neighbor list, and then select other path.
11. Re-establish new connection.

6 Modification in AODV Routing Protocol

To implement our mechanism, we have modified the original AODV routing protocol. The detail modification describes as follows:

6.1 Activate Promiscuous Mode

a) Make AODV agent a child class of Taping and define the Mac variable in file aodv.h

```
#include <mac.h>
class AODV: public Tap, public Agent {
public:
void tap(const Packet *p);
......
protected:
Mac *mac_;
```

b) Define TCL command "install-tap" and implement AODV::tap( )

```
Int AODV::command(int argc, const char*const* argv) {
......
else if(argc == 3) {
......
else if (strcmp(argv[1], "install-tap") == 0) {
mac_ = (Mac*)TclObject::lookup(argv[2]);
if (mac_ == 0) return TCL_ERROR;
mac_->installTap(this);
return TCL_OK;
}}
return Agent::command(argc, argv);
}
Void AODV::tap(const Packet *p) {
Hear all the activities of neighbor
}
```

c) Modify tcl/lib/ns-mobilenode.tcl

```
Node/MobileNode instproc add-target { agent port } {
$self instvar dmux_ imep_ toraDebug_ mac_
......
# Special processing for AODV
set aodvonly [string first "AODV" [$agent info class]]
if {$aodvonly != -1 } {
$agent if-queue [$self set ifq_(0)] ; # ifq between LL and MAC
$agent install-tap $mac_(0)
......
}
```

6.2 Create Class Counter for Calculate Number of Request, Reply and Forward Packet

a) Added counter class definition in aodv.h

```
int        totrec(Packet *p);
int        totforwd(Packet *p);
int        Totfwd;
int        Totrecvpacket;
int        TotRec[];
int        TotForwd[];
double     calcTrust();
int        trustGlobal[];
```

b) Put the counter class in aodv.cc

```
Int AODV::totrec(Packet *p) {
struct hdr_aodv *ah = HDR_AODV(p);
struct hdr_cmn *ch = HDR_CMN(p);
//for (int i = 0; i <= index; i++) {
  Totrecvpacket=0;
// selecting the type of packet data
 if (ch->ptype() == PT_AODV){
     if ((ah->ah_type==AODVTYPE_RREQ) || (ah->ah_type==AODVTYPE_RREP) ||
(ah->ah_type==AODVTYPE_RERR)) {
           Totrecvpacket++;
           TotRec[index] = Totrecvpacket++;
           fprintf(stdout," Received di Node: %d ",index);
           fprintf(stdout, "= %i \n", TotRec[index] );
              }
   }return TotRec[index];
           //}
}
```

6.3 Create Class for Calculate the Trust Level

```
double AODV::calcTrust(){
  if (TotForwd[index] > 0){
     TG[index]=true;
     trustGlobal[index] =  TotRec[index] / TotForwd[index];
     fprintf(stdout," Calcul Trust di node: %d ",index);
     fprintf(stdout, "= %i \n", trustGlobal[index] );
            if(trustGlobal[index] <= 1)      {
               TG[index]=true;
                                              }
           Else {
              TG[index]=false;               }
 } else {
        fprintf(stdout," nothing forwarded in nodes %d \n ",index);
        }
}
```

6.4 Isolated Unwanted or Malicious Neighbors

To stop forwarding data when the node is detected as a malicious, add this function in class forward and class sendRequest in aodv.cc

```
/ If distrust nodes, ignore communication
 if(TG[index]==false) {
//#ifdef DEBUG
//  fprintf(stderr, "%s: I am Distrust nodes\n", __FUNCTION__);
//#endif // DEBUG
  fprintf(stdout, " I am Distrust nodes %d\n", index);
  Packet::free(p);
  return;
 }
```

6.5 Compare RREP of Source Id

Added this function in class recvReply in aodv.cc

```
Compare rp->rp_src
Choose path with same rq->rq_dst
if ( (rt->rt_seqno < rp->rp_dst_seqno) ||  // newer route
     ((rt->rt_seqno == rp->rp_dst_seqno) &&
      (rt->rt_hops > rp->rp_hop_count)) ) { // shorter or better route

Send packet data.
```

7 Simulation, Result and Analysis

7.1 Parameter Simulation and Topology

Simulation has been conducted using NS-2 version 2.34. In our simulation, 150 mobile nodes move in area of 1250 meters x 1250 meters square for 70 seconds simulation time. We use random waypoint mobility model, and transmission range is 250 meters. In our simulation, the speed are varied from 10 m/s to 50m/s. The data traffic is Constant Bit Rate (CBR). We have performed DOS and blackhole attack to evaluate our proposed protocol by increasing the number of attack. There are 7 nodes at fixed position i.e. node 0, 1, 2, 3, 4, 5, and 6. The other node positions are set randomly. Node 0 will communicate to node 5, and the position of node 1, 2, 3 between node 0 and node 5. During the communication process, DOS will attack node 2. We create the attack scenario from the tcl script. Figure 5 depicted the 7 fixed node positions, and table 2 shows the simulation parameter.

7.2 Performance Metrics

Packet Delivery Ratio (PDR) is the ratio of the number of delivered data packet to the destination. PDR reflects the network processing ability and data transferring ability, and as the main symbols of reliability, integrity, effectiveness and correctness of the protocol.

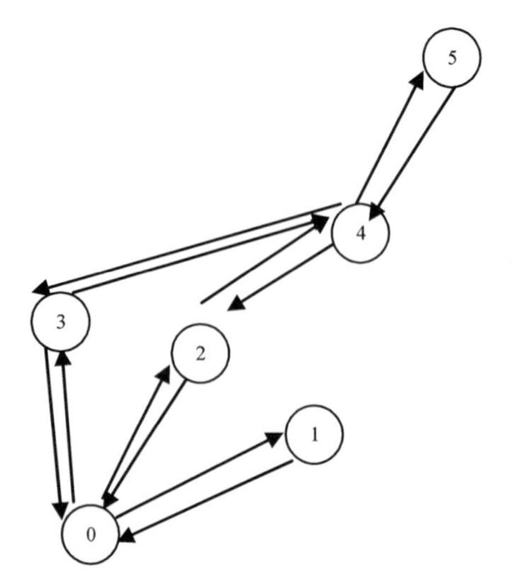

Fig. 5. Simulation scenario

Table 2. Simulation parameter

Parameter	**Value**
Simulation time	70 s
Topology	1250 m x 1250 m
Number of nodes	150
Speed	10,20,30,40,50
Pause time	10 s
Traffic type	CBR
Mobility model	Random way point
Packet size	512 bytes

The formula to calculate packet delivery ratio shows as follows.

$$PDR = \frac{\sum Numbers\ of\ packet\ receive}{\sum Numbers\ of\ packet\ send} * 100\ \% \quad (4)$$

The greater value of packet delivery ratio means the better performance of the protocol.

End-to-end Delay is the average time taken by a data packet to arrive in the destination. It also includes the delay caused by route discovery process and the queue in data packet transmission. Only the data packets that successfully delivered to destinations that counted.

$$End\ to\ end\ delay = \frac{\sum(arrive\ time - send\ time)}{\sum Number\ of\ link\ or\ connection} \quad (5)$$

The lower value of end to end delay means the better performance of the protocol.

Routing overhead is equals to the ratio between the number of routing control packets transferred during the whole simulation process and the number of data packets. It is refer to how many routing control packets are needed for one data packet transmission. It is an important index that compares the performance among different routing protocols; moreover it can evaluate the scalability of routing protocol, the network performance and the energy consumption efficiency under lower bandwidth or congestion.

$$Routing\ overhead = \frac{\sum routing\ packet}{\sum packet\ received} \tag{6}$$

7.3 Result and Analysis

We compare our trust mechanism with another secure protocol such as TCLS [14], LLSP [16], and RSRP [17]. We compare with these protocols due to its used trust mechanism to secure the protocols. We evaluate the protocols under blackhole and DOS attack. We varied the number of attack and the speed of mobile node.

A. Performance Evaluation by Vary the Speed

In the first simulation scenario, we vary the simulation speed as 10, 20,30,40,50 under 5 node attackers included DOS and blackhole attack. After 1 second, the attackers start to attack node 2.

Figure 6 shows the results of end to end delay under attack scenario with varies speed. When DOS attacks perform in the network, the massive RREQ packets will floods the network. It makes the end to end delay of the protocol much higher. Our proposed protocol can find the attacker node directly and ignored the malicious node

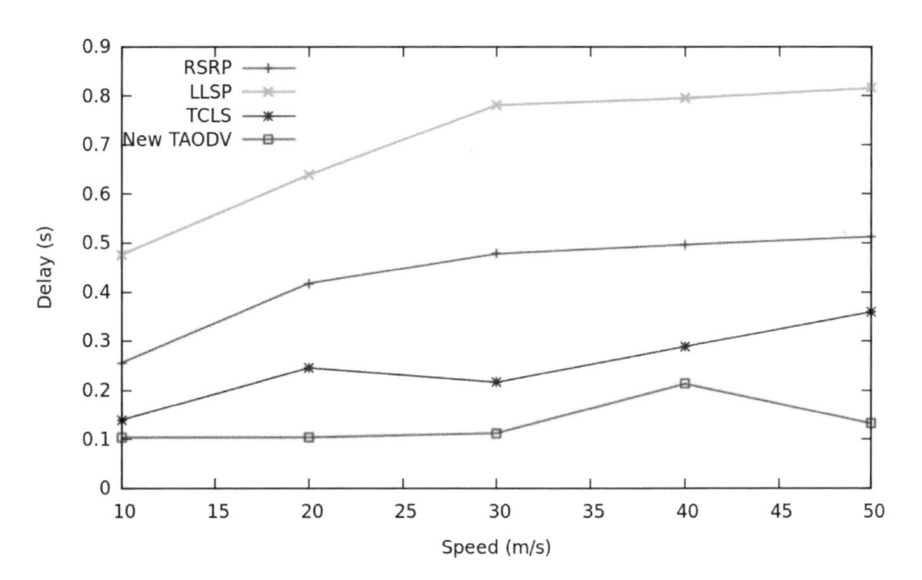

Fig. 6. End to end delay vs speed

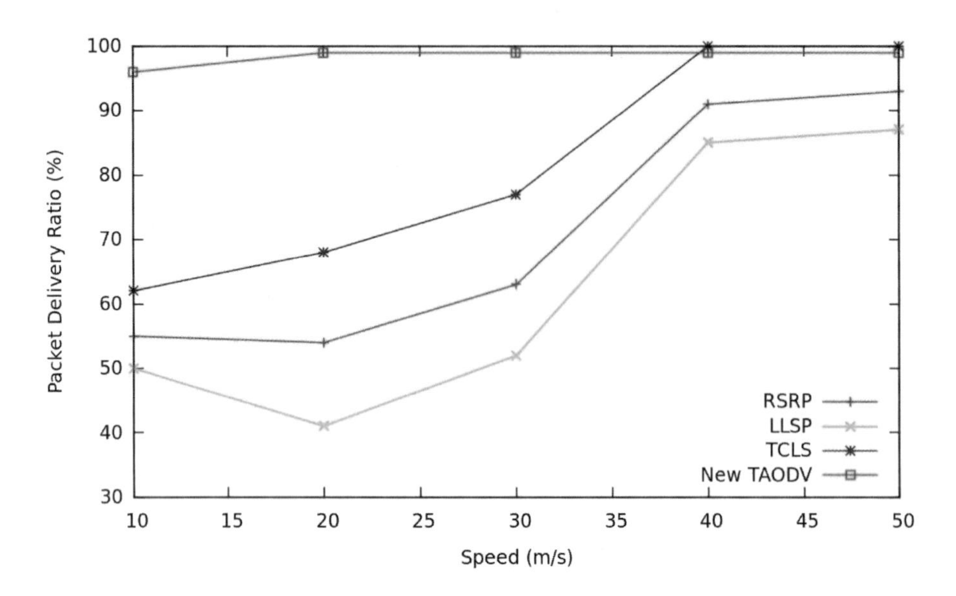

Fig. 7. Packet delivery ratio vs speed

from the network to maintain the delay at the normal level. We can see in figure 6, our proposed protocol outperform TCLS, LLSP and RSRP. When the speed is increased, the delay of communication increases.

Figure 7 depict the packet delivery ratio under attack. The value of packet delivery ratio is between 50% until 65 % for the TCLS, LLSP and RSRP for the low speed scenario. Its means ratio packet successfully delivered to destination is low. In general, the packet delivery ratio becomes higher when the speed of simulation increases. The packet delivery ration value of our proposed protocol is stable between 95% until 100%. Its means the proposed mechanism can guarantee the packet delivery to the destination.

Our proposed protocol directly isolates the attacker and it will stop the attacker to send DOS or fake reply to the victim. Due to the attacker node cannot participate in the network, communication is running as there is no attack in the network. In figure 7, packet delivery ratio of our proposed protocol more stable even in the vary speed condition. It means that the change of speed has no effect on the packet delivery ratio.

Figure 8 shows the performance of secure protocol under attack scenario in term of overhead. The overhead of each protocol become increase when we perform the attack. It means the attacker influence the performance of the routing protocol. The overhead become low when we increase the speed of simulation. Our proposed protocol can mitigate the attack and maintain the routing overhead at the good performance level.

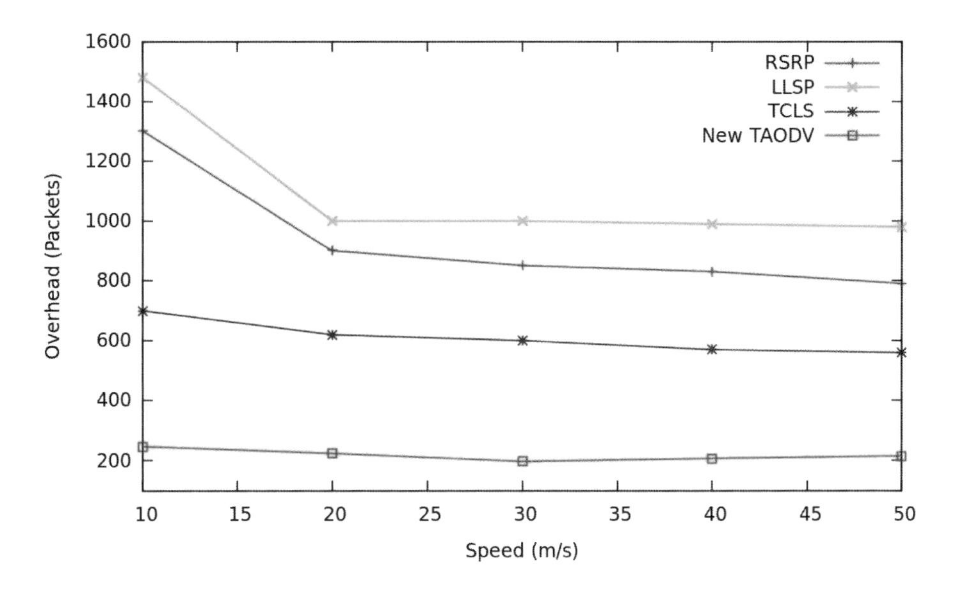

Fig. 8. Overhead vs speed

Our proposed protocol outperform other secure routing protocols (such as LLSP, RSRP, and TCLS) due to the early detection and directly ignore the attacker nodes, simple mechanism without encryption mechanism to secure the packet data, and no need to change the size of packet data.

B. Performance Evaluation by Vary the Number of Attack

In the second simulation scenario, we evaluate our proposed protocol by vary the number of attacks as 5,10,15,20 and 25 under fixed speed as 30 m/s.

Figure 9 shows the effect of number of attacks to the duration of delivered packet from source to destination. As we can see in figure 9, the delay is increase due to the flooding request in the network by DOS attack. Our proposed protocol can keep the delay between 1s until 1.5 s and has a stable delay under various numbers of attacks. It proves that our protocol can mitigate the attacker and then performs the communication normally. Overall, the delay is increase when the number of attack increases.

Figure 10 depicts the packet delivery ratio when the number of node is varied. In general, packet delivery ratio becomes decrease when the number of attack increases. It means, many packets are lost and performance of communication between sources to destination is low. Based on the result, our proposed protocol can maintain the normal work of network at a high packet delivery ratio. The packet delivery ratio value of our proposed protocol is between 96% until 99%. The variation of number of attacks does not give effect to the packet delivery ratio. Due to the proposed protocol can detect and mitigate the attacker early during the route discovery process and perform the communication normally as the network running without attack.

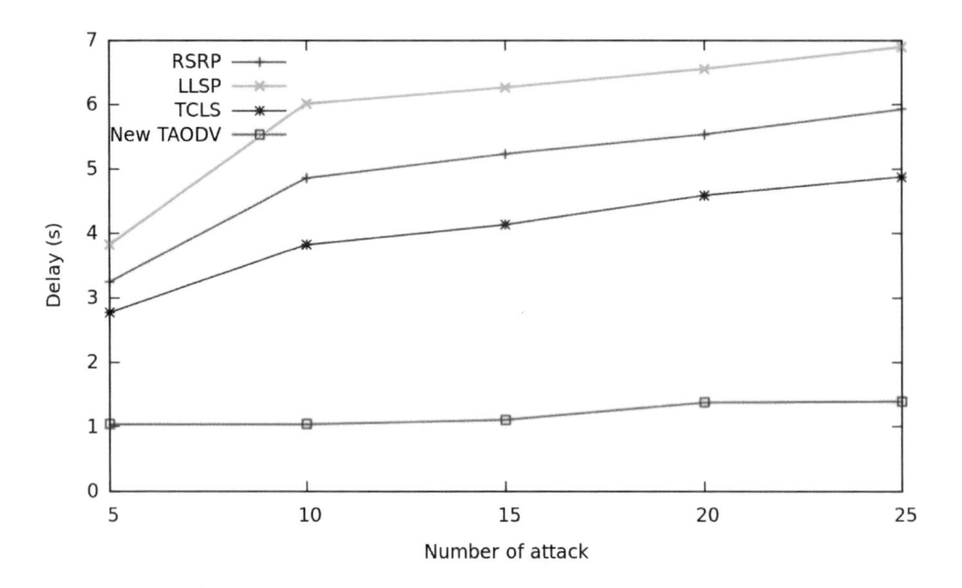

Fig. 9. End to end delay vs Number of attack

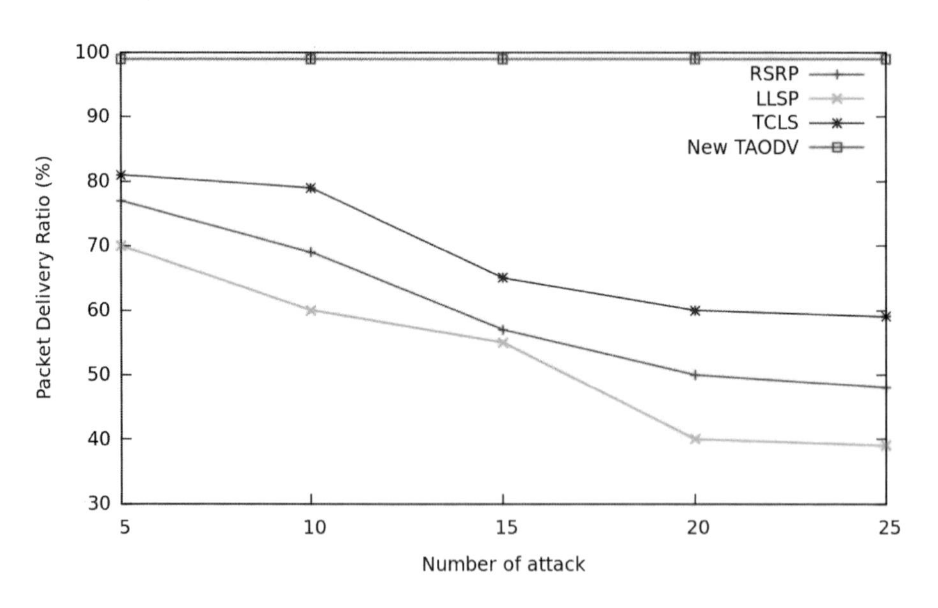

Fig. 10. Packet delivery ratio vs number of attack

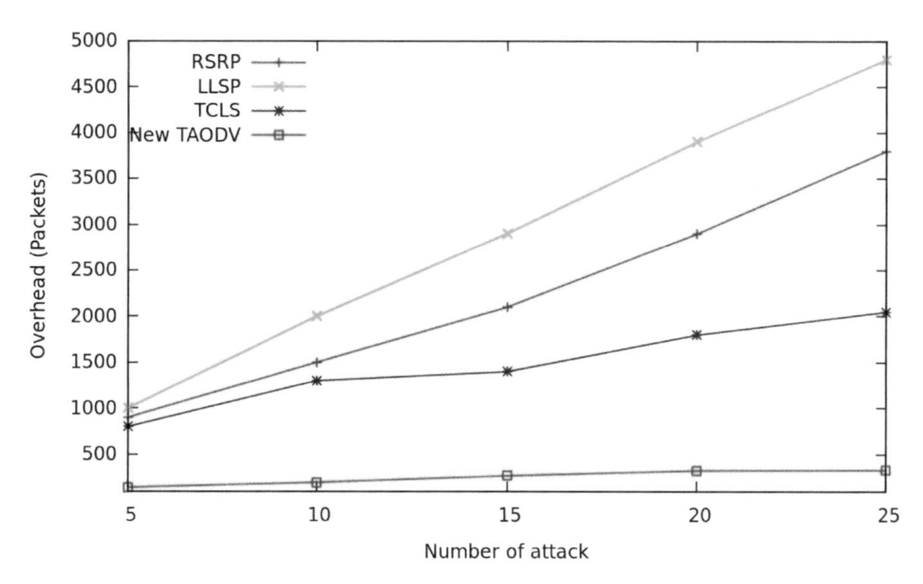

Fig. 11. Routing overhead vs number of attack

Figure 11 depict the performance of secure protocol under varies number of attack in term of routing overhead. In general, routing overhead become increase when the number of attack increases. Based on the result, our proposed protocol has a lower overhead even the number of attack increase. It proves that our propose mechanism can maintain the network communication at low overhead, and mitigate the attacker.

Over all, the performance of our proposed protocol outperform with different speed and different number of attack in term of packet delivery ratio, end to end delay and routing overhead. Due to the proposed protocol can detect the attacker nodes when the node performs route discovery process. The communication is running normally after trust mechanism performed as well as there is no attack inside the network. In addition, our proposed solution does not add any control message to existing AODV neither needs to regenerate any control messages. So, there are minimal chances of rise in Normalized Routing Overhead i.e. in the ratio of number of control packets to data transmissions in a simulation.

Some advantages of our proposed protocol are:

1. No need to perform warning mechanism to whole network. In some mechanism, after calculate the trust level of node, the protocol will inform to all the nodes in the network about the trust status of the node. But in our mechanism, each node has own trust calculation level to its neighbor.
2. No need to save the status of its neighbor, due to the trust calculation performs each time the node doing the communication.

8 Conclusion

We have reviewed some of secure routing protocol based on AODV and explored the security problem in wireless ad hoc network. We also explored the variant of secure routing protocol based on AODV and present the Mapping Variant AODV Secure Routing Protocol. There is two mainstream to enhance the security aspect in AODV routing protocol i.e. cryptographic mechanism and trust based mechanism.

In this chapter, we address the security aspect and proposed a new trust mechanism that has ability to detecting and preventing the attack potentials into a wireless ad hoc network especially for DOS and blackhole attack. Our proposed protocol is design to get the good performance communication if there is attack in the network.

The simulation analysis result prove that the performance of our proposed protocol is better than another secure routing protocol such as LLSP, RSRP and TCLS, in term of packet delivery ratio, end to end delay and overhead especially under DOS and blackhole attack.

In the future work, we will observe and extend the mechanism to be able to detect another type of attack, and apply the bio inspired algorithm to select the shortest and secure path. We also try to apply our mechanism for a real network.

Acknowledgments. This research project is funded by The Directorate General of Higher Education Indonesia and France Embassy.

References

[1] Balakrishna, R., Rajeswar Rao, U., Ramachandra, G.A., Bhagyashekar, M.S.: Trust-based Routing Security in MANETS. International Journal of Computer Science and Information Technolgy 4(3), 547–553 (2010)

[2] Abusalah, L., Khokhar, A., Guizani, M.: A Survey of Secure Mobile Ad Hoc Routing Protocols. IEEE Communications Surveys & Tutorials 10(4), 78–93 (2008)

[3] Cerri, D., Ghioni, A.: Securing AODV: The A-SAODV Secure Routing Prototype. IEEE Communications Magazine 46(2), 120–125 (2008)

[4] Johnson, D.B., Maltz, D.A.: Dynamic Source Routing in Ad hoc Wireless Networks. In: Mobile Computing, pp. 153–181. Kluwer Academic Publishers (1996)

[5] Zapata, M.G.: Secure adhoc on-demand distance vector (S-AODV) Routing. In: Proceeding of ACM Workshop on Wireless Security (WISE), Atlanta (2002)

[6] Yi, S., Naldurg, P., Kravets, R.: A Security-Aware Routing Protocol for Wireless AdHoc Networks. In: ACM Symposium on Mobile Ad Hoc Networking & Computing (ACM Mobihoc 2001), Short paper, Long Beach, CA, USA (October 2001)

[7] Akhlaq, M., Noman Jafri, M., Khan, M.A., Aslam, B.: Addressing Security Concerns of Data Exchange in AODV. Transactions on Engineering, Computing and Technology 16, 29–33 (2006) ISSN 1305-5313

[8] Zhe, L., Jun, L., Dan, L., Ye, L.: A Security Enhanced AODV Routing Protocol. In: Jia, X., Wu, J., He, Y. (eds.) MSN 2005. LNCS, vol. 3794, pp. 298–307. Springer, Heidelberg (2005)

[9] Eichler, S., Roman, C.: Challenges of Secure Routing in MANETs: A Simulative Approach using AODV-SEC. In: 2006 IEEE International Conference on Mobile Adhoc and Sensor Systems (MASS), pp. 481–484 (October 2006)

[10] Pirzada, A.A., Datta, A., McDonald, C.S.: Trustworthy Routing with the AODV Protocol. In: The International Networking and Communications Conference (INCC 2004), pp. 19–24. IEEE Communications Society, Lahore (2004)

[11] Menaka, A., Pushpa, M.E.: Trust Based Secure Routing in AODV Routing Protocol. In: IMSAA 2009 Proceedings of the 3rd IEEE International Conference on Internet Multimedia Services Architecture and Applications. IEEE Press, Piscataway (2009)

[12] Griffiths, N., Jhumka, A., Dawson, A., Myers, R.: A Simple Trust Model for On-Demand Routing in Mobile Ad-Hoc Networks. In: Badica, C., Mangioni, G., Carchiolo, V., Burdescu, D.D. (eds.) IDC 2008. SCI, vol. 162, pp. 105–114. Springer, Heidelberg (2008)

[13] Kurosawa, S., Nakayama, H., Kato, N., Nemoto, Y., Jamalipour, A.: Detecting blackhole attack on aodv-based mobile ad hoc networks by dynamic learning method. International. Journal of Network, Security 5(3), 338–346 (2007)

[14] Rajaram, A., Palaniswami, S.: A trust based cross layer security protocol for ad hoc networks. International Journal of Computer Science And Information Security 6(1) (2009)

[15] Mistry, N., Jinwala, D.C., Zaveri, M.: Improving AODV Protocol against Blackhole Attacks. In: Proceedings of the International Multi Conference of Engineer and Computer Science, vol. 2 (2010)

[16] Islam, A.-M.M., Pose, R., Kopp, C.: A Link Layer Security Protocol for Suburban Ad-Hoc Networks. In: Proceedings of Australian Telecommunication Networks and Applications Conference (December 2004)

[17] Afzal, B., Koh, J.-B., Raza, G.L., Kim, D.-K.: RSRP: A Robust Secure Routing Protocol for Mobile Ad Hoc Networks. In: Proceedings of IEEE Conference on Wireless Communications and Networking, pp. 2313–2318 (April 2008)

Security and Privacy in Vehicular Ad-Hoc Networks: Survey and the Road Ahead

M.A. Razzaque, Ahmad Salehi S., and Seyed M. Cheraghi

Faculty of Computer Science & Information Systems,
University of Technology,
Malaysia
marazzaque@utm.my,
ahmad.salehi.sh@gmail.com,
smcheraghi@yahoo.com

Abstract. Vehicular Ad-hoc Networks (VANETs) can make roads safer, cleaner, and smarter. It can offer a wide range of services, which can be safety and non-safety related. Many safety-related VANETs applications are real-time and mission critical, which would require strict guarantee of security and reliability. Even non-safety related multimedia applications, which will play an important role in the future, will require security support. Lack of such security and privacy in VANETs is one of the key hindrances to the wide spread implementations of it. An insecure and unreliable VANET can be more dangerous than the system without VANET support. So it is essential to make sure that "life-critical safety" information is secure enough to rely on. Securing the VANETs along with appropriate protection of the privacy drivers or vehicle owners is a very challenging task. In this work we summarize the attacks, corresponding security requirements and challenges in VANETs. We also present the most popular generic security policies which are based on prevention as well detection methods. Many VANETs applications require system-wide security support rather than individual layer from the VANETs' protocol stack. In this work we will review the existing works in the perspective of holistic approach of security. Finally, we will provide some possible future directions to achieve system-wide security as well as privacy-friendly security in VANETs.

Keywords: Security, Privacy, VANETs, Roadside Units, Key Management.

1 Introduction

It is now widely accepted by academician and industry that VANETs can significantly improve traffic safety, road efficiency and reduce environmental impact [1]. Studies [2] show that about 60% roadway collisions could be avoided if the driver of the vehicle was provided warning at least one-half second prior to a collision.

S. Khan and A.-S.K. Pathan (Eds.): *Wireless Networks and Security*, SCT, pp. 107–132.
DOI: 10.1007/978-3-642-36169-2_4

VANETs allow vehicles to communicate with each other (V2V) and/or with roadside infrastructure (V2R). Based on these communications VANETs can offer a wide range of services. In a report [3], US Dept. of Transport has already listed more than75 different application scenarios where it can be useful. These can be broadly categorized in two: safety and non-safety related services/applications. Many safety-related ITS applications are real-time and mission critical, which would require strict guarantee of quality of service (QoS), in terms of latency, error rate, and security. For instance, a safety message to prevent a probable accident has to reach concerned vehicles within a fraction of a second (e.g. 100ms [3]) so that the vehicles and their drivers can take necessary actions to prevent the accident. Security is key concern for future VANETs implementations. In VANET a road user will relay on it and does action accordingly whereas on typical systems user takes actions by his/her observation and knowledge. An insecure and unreliable VANET can be more dangerous than the system without it. So, secure VANETs system is more than necessary. Potential security measures could include a method of assuring that the packet/data was generated by a trusted source (neighbor vehicle, sensors, etc.), as well as a method of assuring that the packet/data was not tampered with or altered after it was generated. Any application that involves a financial transaction (such as tolling) requires the capability to perform a secure transaction.

Securing the VANETs along with appropriate protection of the privacy drivers or vehicle owners is a very challenging task. As the applications of VANETs are diverse, their communications and/or system-level security requirements could be diverse too. There are some very good works on VANETs' security and privacy [4-9], which review security related issues attacks, requirements, challenges, and security solutions. But none of these comprehensively covers all of these issues related VANETs' security and privacy except [9]. In [9] security and privacy implementation related issues are missing, precisely communication perspective. In this work we summarize the attacks, corresponding security requirements and challenges in VANETs. We also present the most popular generic security policies which are based on prevention as well detective methods. Many VANETs applications require system-wide security support rather than individual layer from the VANETs' protocol stack. In this work we will review the existing works in the perspective of holistic approach of security. Finally, we will provide some possible future directions to achieve system-wide security as well as privacy-friendly security in VANETs.

The rest of this book chapter is organized as follows. We first present a brief overview of VANETs in section 2. In section 3 we provide an elaboration of the possible adversaries and their possible attacks in VANETs. The security, privacy requirements and major challenging issues faced by VANETs to satisfy the security requirements are described in section 4. This section clearly shows how security and privacy my conflict in VANETs. Section 5 summarizes the generic security mechanisms including some specific attacks based works. We analyze the existing works and provide some future directions in section 6, before we conclude in section 7.

2 Overview of VANET

2.1 What Is VANET?

A modern vehicle can be considered as a network of sensors/actuators on wheels. VANET is a special kind of Mobile Ad-hoc Network (MANET) where vehicles equipped with the technologies are the key constituents. Generally, a VANET differs from MANET in the following aspects:

- Large scale – potentially billion
- Fleeting contact with other vehicles
- Nodes not as constrained in terms of energy, storage and computation.
- Higher mobility
- Privacy requirements

The single most important objective of a VANET is to provide communications between different vehicles on the roads and roads' environments (e.g. roads' condition, weather, traffic, etc.), to improve the driving experience and make driving safer. In doing so, in VANET each vehicle needs to have an OBU (On-Board Units)– communication devices mounted on vehicles and also a WSNs supported roadside unit (RSU) as shown in figure 1. By using OBUs, vehicles can communicate with each other as well as with RSUs. A VANET is a self-organized network that enables communications between vehicles and RSUs, and the RSUs can be connected to a backbone network, so that many other network applications and services, including Internet access, can be provided to the vehicles. So in VANET communications can be Vehicle to Vehicle (V2V)/inter vehicle and/or with roadside infrastructure (V2R) [10]. Figure 1 presents an example VANET, which shows possible communications within a VANET.

To make VANETs intelligent, it integrates multiple ad-hoc networking technologies such as WiFi IEEE 802.11p, WAVE IEEE 1609, WiMAX IEEE 802.16, Bluetooth, IRA, ZigBee for easy, accurate, effective and simple communication between vehicles on dynamic mobility. One of the IEEE1609 (P1609.2) explicitly defines security, secure message formatting, processing, and message exchange. Use of these technologies in VANETs helps in defining safety measures in vehicles, streaming communication between vehicles, infotainment and telematics. VANETs are expected to implement a variety of wireless technologies such as Dedicated Short Range Communications (DSRC), are one or two way short- to medium-range wireless communication channels explicitly designed for automotive use and a corresponding set of protocols and standards. Other candidate wireless technologies are Cellular, Satellite, and WiMAX. VANETs can be envisioned as the most important entity of the Intelligent Transportation Systems (ITS) [1, 3, 10].

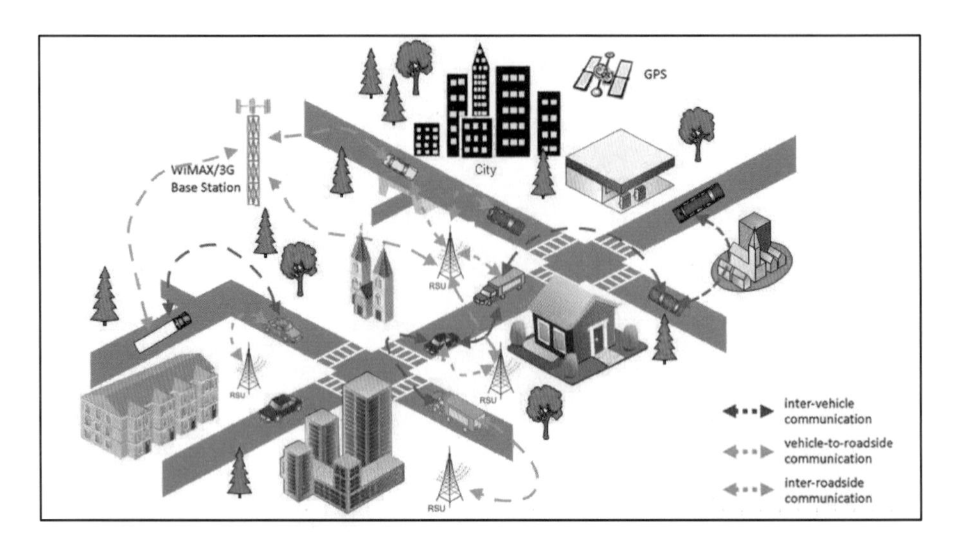

Fig. 1. An example of VANET

2.2 Applications of VANETs

VANETs can offer a wide range of services or applications. In a report [3], US Dept. of Transport has already listed more than75 different application scenarios where VANETs can be useful. These applications can be broadly categorized in two: safety and non-safety related services/applications. Congestion control is one of the non-safety related applications. As shown in Table 1, a little reduction in congestion can contribute very significantly. A snapshot of key applications of VANETs is presented in figure 2.

Table 1. Cost of Congestion in few developed Countries [1]

Country	Congestion Cost (billion$)
USA	200 (890 within 20yrs)
Japan	109
Australia	12.5
UK	35

3 Adversaries and Attacks

Knowing the type and the resources of the adversary can greatly help in determining the scope of the defenses required to secure a VANET. It is really hard to make a precise list of all the possible adversaries in any security system. A realistic analysis

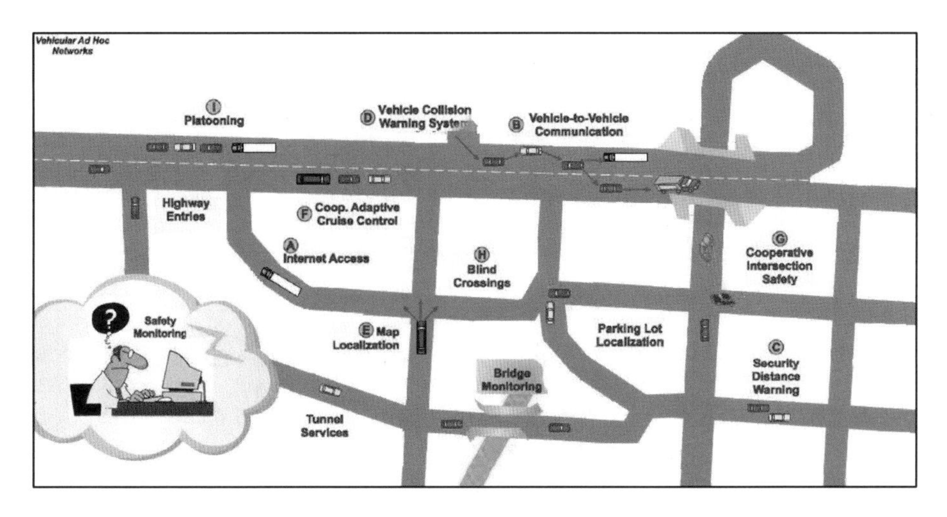

Fig. 2. Possible Applications of VANETs [11]

of the application environment can help in finding the type of typical adversaries. Thus VANETs environment recommends the following categories of adversaries [12, 13]:

Greedy Drivers: It is highly unexpected that all drivers in the system will be trusted to follow the protocols specified by the application. Always there will be some drivers who will attempt to maximize their gains, regardless of the cost to the system. For example, in a congestion control system, a greedy driver might attempt to convince the nearby vehicles that there is considerable congestion ahead, so that they will move to alternate routes and allow the greedy driver a clear path to his/her destination.

Eavesdroppers (Snoops): This type of adversary can includes everyone from a nosy next-door neighbor to a government agency trying to profile drivers. For example, companies may want to track consumers' purchasing habits and use correlated data to alter prices and discounts. Data mining to find pattern over aggregated data may be acceptable, but it can easily conflict with users' privacy concerns if one can extract identifying information about a person.

Pranksters: Like computer and network security, pranksters could be a serious adversary in VANETs. It includes jaded teenagers searching for vulnerabilities and hackers looking for fame via their exploits. For example, a prankster siting by the road can easily create "intelligent collision" by convincing one vehicle to slow down while persuading the vehicle behind it to speed up. The hard real-time response requirement in VANETs potentially leaves it vulnerable to DoS attacks. A prankster could exploit this vulnerability to disable applications or prevent critical information from reaching targeted vehicle.

Industrial Insiders: Inside attackers are very deceptive, and hard defend them. The extent to which VANETs are vulnerable to these depends on other security design decisions. For instance, any mechanic who can update the software on a vehicle can also has the chance to load malicious software. If vehicle makers are in charge of key distribution, then a single rogue employee at one maker could create keys that would be accepted by all other vehicles.

Malicious Attackers: This category of adversary deliberately attempt to cause harm via the applications available on the VANETs. Usually, these attackers have specific targets, and they have access to more resources than the aforementioned attackers. For instance, terrorists might manipulate the warning system to create jam before detonating a bomb. On the other hand criminals might spoof the congestion control application to facilitate getaways. In general, while this class of attackers rarer than those outlined above; their combination of resources and directed malice makes them a serious concern for any security system.

3.1 Attacks

Like other networks attacks in VANETs can be classified into the following categories [12, 13]:

- *Outsider vs. insider attacks:* Outside attacks are defined as attacks from nodes which do not belong to a VANETs; insider attacks occur when legitimate vehicle or node of a VANETs behave in unintended or unauthorized ways.
- *Passive vs. active attacks:* Passive attacks include eavesdropping on or monitoring packets exchanged within a VANET; active attacks involve some modifications of the data steam or the creation of a false stream.
- *Malicious vs. rational:* Usually, a malicious attacker looks for no personal benefits from the attacks, just aims to harm the users or network. Hence, attacker may employ any means disregarding corresponding costs and consequences. On the other hand, a rational attacker looks for personal benefit and hence is more predictable compared to a malicious attacker.
- *Local vs. extended*: An attacker can be limited in scope, even if he controls several entities (vehicles or base stations), which makes him local. An extended attacker controls several entities that are scattered across the network, thus extending his scope. This distinction is especially important in privacy-violating and wormhole attacks that we will describe shortly.

It is also possible to categorize attacks in according to the security requirements in VANETs as:

- *Attacks on secrecy and authentication:* Standard cryptographic techniques can protect the secrecy and authenticity of communication channels from outsider attacks such as eavesdropping, packet replay attacks, and modification or spoofing of packets.

- *Attacks on network availability:* Attacks on availability are often referred to as denial-of-service (DoS) attacks. DoS attacks may target any layer of a sensor network.
- *Stealthy attacks against service integrity:* In a stealthy attack, the goal of the attacker is to make the network accept a false data value.

Being a special implementation of MANETs supported by WSNs (RSUs), a VANET inherits all the known and unknown security weaknesses associated with MANETs and WSNs, and could be subject to many security and privacy threats. In this context we obviously cannot anticipate every possible attack on VANETs; we can enumerate some of the more likely scenarios and ensure that applications are robust against this known set of potential attacks. These attacks can be concerned with the physical security of VANET and messages communicated within it. Here we consider only the attacks against messages rather than vehicles, as the physical security of vehicle electronics is out of the scope of this work. Message related attacks in VANETs can be summarized [13] as below:

Denial of Service (DoS): Like any other networks, it is a very common attack where the attacker can overpower a vehicle's resources or jam the communication channel used by the VANET to bring down the VANET or even cause an accident. This attack is active and malicious in nature. For instance, if a malicious adversary wants to create a massive pileup on the highway, he could provoke an accident and then use a DoS attack to prevent the dissemination of warnings message to other drivers. As shown in figure 3 (iii) jamming can easily cause DoS in VANETs.

Fabrication Attacks: An attacker can initiate a fabrication attack by broadcasting false or bogus information into the network. For example, a greedy driver might behave as an emergency vehicle to speed up his/her own journey. An attacker may also fabricate his/her own information related to his/ her identity, location, or other application-specific parameters. Finding an appropriate defense mechanism against fabrication attacks in VANETs is particularly challenging, as the customary remedy of using strong identities along with cryptographic authentication may conflict with the privacy requirements of drivers or vehicle owners. This generic attack has some variants (may not be mutually exclusive) which are important for VANETs as below [9, 12, 13]:

i. **Bogus Information:** Attackers of this attack are generally insider, rational and active as shown in figure 3(i). They diffuse wrong information in the network to affect the behavior of other drivers (e.g., to divert traffic from a given road and thus free it for themselves).
ii. **Cheating with Sensed Information:** Attackers of this category are insider, rational, active and local who exploit this attack to alter their perceived position, speed, direction, etc. in order to escape liability, notably in the case of an accident.
iii. **Hidden Vehicle:** Here fabrication happens on positioning information. It follows the basic safety messaging protocol described [12], a vehicle broadcasting

warnings will listen for feedback from its neighbors and stop its broadcasts if it realizes that at least one of these neighbors is better positioned for warning other vehicles. As shown in figure 3 (vi), the hidden vehicle attack consists in deceiving vehicle A into believing that the attacker B is better placed for forwarding the warning message, thus leading to silencing A and making it hidden to other vehicles. This ultimately stops the dissemination of the warning message, hence causing a DoS.

iv. **Tunnel:** As in GPS signals disappear in tunnels or underground, an attacker can exploit this temporary loss of positioning information to inject false data once the vehicle leaves the tunnel and before it receives an authentic position update as shown in figure 3 (iv). An area jammed by the attacker may cause the same effects.

v. **Masquerading:** The attacker of this kind actively pretends to be another vehicle by using false identities and can be driven by malicious or rational objectives. Intelligent collision (figure 3 (ii)) is an example of this attack.

Message Suppression Attacks: It is a delicate attack where the attacker may use one or more vehicles to launch a suppression attack by selectively dropping packets from the network. Some popular attacks of MANETs or WSNs such as selective forwarding, black-hole falls under this generic category. For instance, a prankster might suppress congestion avoidance message before selecting an alternate route, thus trapping subsequent vehicles to wait in traffic.

Alteration Attacks: It is an active and inside attack in VANETs that aims to alter existing data. It includes on purpose delaying the transmission of information, replaying earlier transmissions or altering the individual entries within a transmission. For example, if the traffic congestion application requires a vehicle to collect "votes" from other vehicles at the site of the congestion, then an attacker might collect votes while traveling in normal traffic, but alter the locations and timestamps in the votes to make it appear that all of those vehicles were in the same place at the same time, deceitfully indicating a heavily congested highway. A malicious attacker might alter a message alerting vehicles to an obstacle ahead to convince another vehicle that the way is in fact clear.

Tracking: This attack requires ID disclosure of other vehicles. A central monitoring can be used to monitor trajectories of targeted vehicles and use this data for a range of purposes (e.g., the way some car rental companies track their own cars). For the monitoring, the passive attacker can exploit the roadside infrastructure or the vehicles around its target (e.g., by using a virus that infects neighbors of the target and collects the required data). An example of this sort of attack is shown in figure 3(v), where car A is under tracking attack.

Wormhole: In wireless networking attack, this attack consists in tunneling packets between two remote nodes. Similarly, in VANETs, an attacker that controls at least two entities distant from each other and a high speed communication link between them can tunnel packets broadcasted in one location to another, thus disseminating erroneous (but correctly signed) messages in the destination area.

4 Challenges and Security Requirements in VANETs

4.1 Security Requirements

As the applications of VANETs are diverse, their communications and/or system-level security requirements could be diverse too. Potential security measures should include a way of assuring that the packet/data was generated by a trusted source, as well as a way of assuring that the packet/data was not tampered with or altered after it was generated.

VANETs pose some of the most challenging problems in MANETs and WSNs research. In addition, the issue of security in VANETs is particularly challenging due to the unique features of the network, such as high-speed mobility of network nodes or vehicles and the extremely large amount of network entities. It is obvious that any malicious user behavior, such as an alteration and replay attack of the disseminated messages, could be disastrous to other users. So in any situation, it is necessary to make sure that "life-critical safety" information cannot be altered by attackers. A security system needs to be capable of establishing the liability of drivers, while preserving their privacy as much as possible. Considering the aforementioned attacks and suggestion made in other works, VANET security should satisfy the following requirements [3, 12, 13]:

i. **Authentication:** This is the most important requirement in preventing most of the aforementioned attacks in VANETs. Vehicle responses to events should be based on legitimate messages (i.e., generated by legitimate users). Therefore we need to authenticate the OBUs, RSUs and senders of these messages.
ii. **Verification of Data Consistency:** The legality of messages also comprises their consistency with similar ones (those generated in close space and time), as the sender can be legal but the message contains false data. This requirement also known as "plausibility".
iii. **Message Integrity:** Message alteration is very common and crucial attacks in VANETs. We need to maintain the integrity of the message to prevent the alteration attacks.
iv. **Availability:** Attacks like (e.g., DoS by jamming) bring the VANETs down even the considered communication channel is robust. So, availability should be provided by some other means.
v. **Non-repudiation:** Drivers causing accidents should be reliably identified to prove his/her liability. Based on this principle, a sender will not be able to refuse the transmission of a message (it may be key for investigation in determining the correct sequence and content of messages exchanged before the accident).
vi. **Privacy:** People are increasingly cautious of being monitored or tracked. Hence, the privacy of drivers or vehicle owners against unauthorized observers should be protected.
vii. **Traceability and Revocation:** Trace and disable abusing OBUs or RSUs by the authority.

viii. **Real-Time Constraints:** At the very high speeds typical in VANETs, strict time constraints should be respected. This ultimately imposes computation and communication wise efficient schemes.

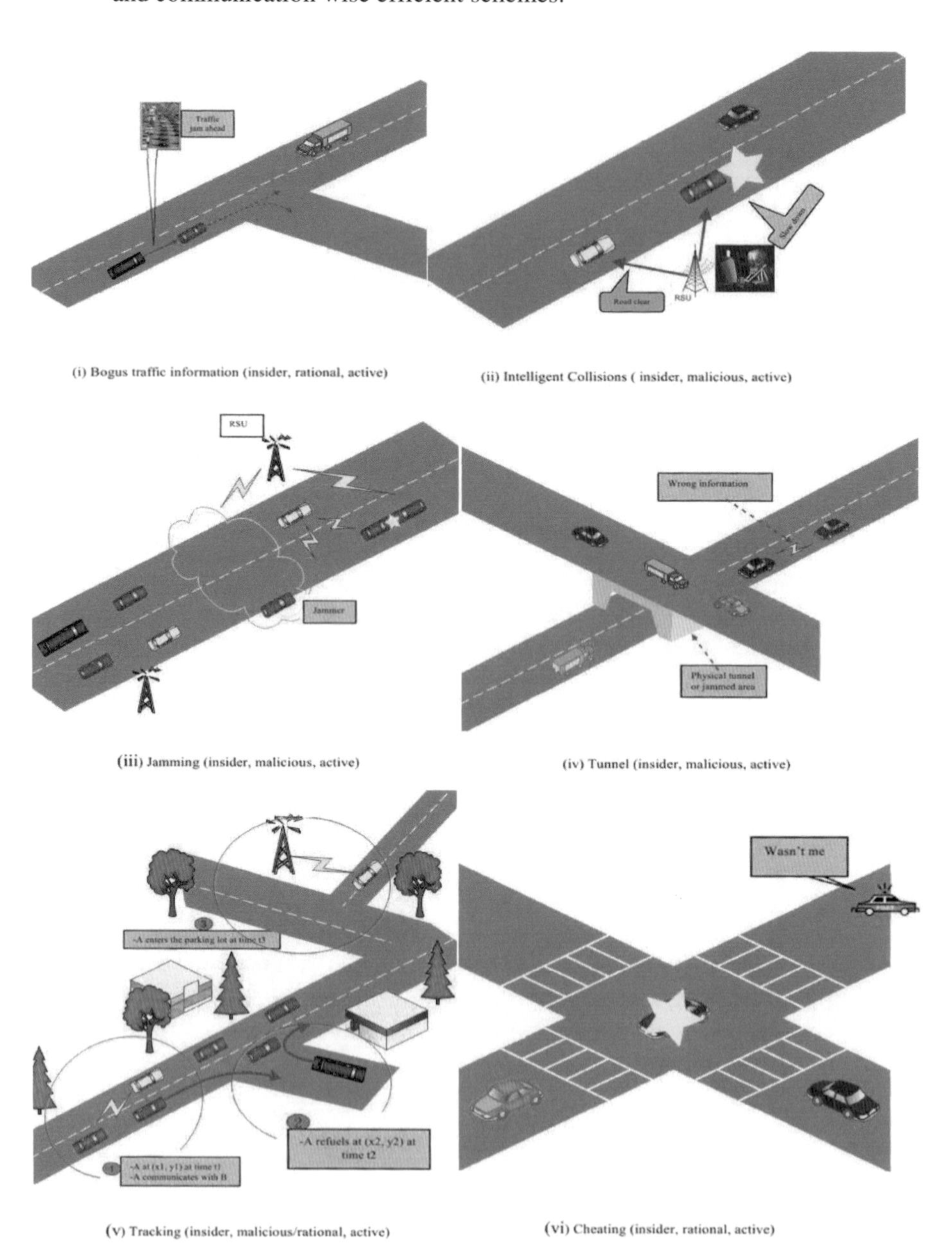

(i) Bogus traffic information (insider, rational, active)

(ii) Intelligent Collisions (insider, malicious, active)

(iii) Jamming (insider, malicious, active)

(iv) Tunnel (insider, malicious, active)

(v) Tracking (insider, malicious/rational, active)

(vi) Cheating (insider, rational, active)

Fig. 3. Few explicit attacks in VANETs [12, 13]

4.2 Challenges

VANETs pose some of the most challenging problems in MANETs and sensor network research. Some of the key challenges [12, 13, etc.] which directly or indirectly related to security of VANETs are summarized below.

Mobility: In general sensor networks often assume a relatively static network, and even MANETs usually assume limited mobility. For vehicular networks, mobility is the norm, and it will be measured in miles, not meters, per hour. This high mobility causes frequent dis-connectivity; hence make the communications highly unreliable which makes security more challenging. The mobility patterns of vehicles on the same road will show strong correlations. Each vehicle will have a frequently shifting set of neighbors, many of whom it has never communicated with before and is unlikely to communicate with again. The short-lived nature of interactions or communications in a VANET will limit the efficacy of reputation-based schemes. For instance, rating other vehicles based on the reliability of their incident reports is unlikely to prove useful; a specific driver is unlikely to receive multiple reports from the same vehicle. Additionally, as two vehicles may only be within communication range for a very short period (e.g. few seconds), we cannot rely on protocols that require significant communication between the sender and receiver.

Privacy vs. Security: Like other IP-based networks (e.g. Internet, MANETs, etc.), it highly desirable to bind each driver or vehicle to a single identity to prevent Sybil or other spoofing attacks. For instance, in the congestion control scheme, it is necessary to prevent one vehicle from claiming to be hundreds in order to create the illusion of a congested road. Authentication is a key security requirement for VANETs that provides valuable forensic evidence and allows us to use external mechanisms, such as traditional law enforcement, to deter or prevent attacks on VANETs. However, drivers or other vehicle users value their privacy and are unlikely to adopt systems that require them to abandon their anonymity. For instance, if we try to prevent spoofing in a way that reveals each vehicle's permanent identity, then we may violate drivers' or users' privacy requirements. So privacy compliant security policies are needed that will require codifying legal, societal and practical considerations. Most countries have widely divergent laws concerning their citizens' right to privacy. As most vehicle makers operate in multinational markets, they will need security solutions that satisfy the most stringent privacy laws, or that can be customized to meet their legal obligations in each market. Authentication schemes must also consider societal expectations of privacy against practical considerations. Vehicles today are not fully anonymous as each vehicle has a publicly displayed license plate that uniquely identifies it and identifies the owner of the car, given access to the appropriate records. Hence, drivers have already sacrificed a portion of their privacy while driving. So, security policies in VANETs should build on these existing compromises instead of encroaching any further upon a driver's right to privacy.

Availability: Number of VANETs applications especially safety-related require real-time, or near real-time, responses and hard real- time guarantees. Other applications may tolerate some margin in their response times; still this requirement is faster than those expected in traditional WSNs or MANETs. However, attempts to meet real-time demands could make applications vulnerable to Denial of Service (DoS) attacks. For instance, in the deceleration application, a delay of even less than a second can render the message meaningless. The problem is further aggravated by the unreliable communications. The current DSRC standard provides an acceptable latency and high data rate; the reliability is still missing [14]. Since vehicles moving in opposite directions will remain within communications range for only a few seconds, opportunities to retry a broadcast will be limited.

Low Tolerance for Errors: Many applications can afford security protocols that rely on probabilistic schemes. However, in VANETs' safety (mission-critical) related applications, even a small probability of error will be unacceptable. Number of vehicles in the world is in billions, even if an application that functions correctly 99.99999999% of the time, the application is still more likely to fail on at least one vehicle than function correctly on all vehicles. So margin of error of any security protocol in VANETs based on deterministic or probabilistic scheme is infinitesimally small. Additionally, for many applications, security must focus on prevention of attacks, rather than detection and recovery. In MANETs it may suffice to detect an attack and alert the user, leaving recovery and clean-up to the humans. However, in many safety-related VANETs applications, detection will be inadequate, as by the time the driver can react, the warning may be too late. So security must focus on preventing attacks in the first place, which requires extensive foresight into the types of attacks likely to occur.

Key Distribution: Key distribution is often a fundamental building block for security protocols. In VANETs, key distribution faces several significant challenges. First, vehicles are manufactured by many different companies, so installing keys at the factory would require coordination and interoperability between manufacturers. If manufacturers are unable or unwilling to agree on standards for key distribution, then we could turn to government-based distribution. Within a country it can hierarchically go to states and then districts that make the coordination complicating. The government can impose standards, but doing so would require significant changes to the current infrastructure for vehicle registration, and thus is unlikely to occur in the near future. However, without a system for key distribution, applications like traffic congestion detection may be vulnerable to spoofing, sybil attacks. A potential approach for secure key distribution would be to empower the Motor Vehicles licensing authority to take the role of a Certificate Authority (CA) and to certify each vehicle's public key. Unfortunately, this approach has number of weaknesses. Moreover, certificate-based key establishment has the danger of violating driver privacy, as the vehicle's identity is revealed during each key establishment. So finding a realistic and privacy friendly key distribution technique is a challenging issue in VANETs.

Cooperation: Successful deployment of VANETs will require cooperation amongst vehicle manufacturers, consumers, and the government, and reconciling their frequently conflicting interests will be challenging. For instance, law-enforcement agencies might quickly adopt a system in which speed-limit signs broadcast the mandated speed and vehicles automatically reported any violations. Understandably, consumers might reject such invasive monitoring, giving vehicle manufacturers little incentive to include such a feature. Equally, consumers might appreciate an application that provides an early warning of a police speed trap. Manufacturers might be keen to meet this demand, but law-enforcement is unlikely to do so.

5 Securing VANETs

Securing VANETs is a very challenging due to the unique features of networks, such as the high-speed mobility of the nodes and the extremely high node density. Moreover, conditional[1] privacy preservation of drivers or vehicle owner crucial information (including the driver's name, license plate, speed, position, and traveling routes along with their relationships) makes it even harder. Thus, it is critical to develop a group of elaborate and carefully designed security mechanisms for achieving security and conditional privacy preservation in a VANET. Up to recently, however, security and privacy issues of VANETs have been given little attention. Lack of such security and privacy concerns have formed the major barrier, preventing many drivers from employing state-of-the-art smart automobile technologies.

In this section we review the existing VANETs security mechanisms. In earlier part of this section we will discuss the generic VANETs security mechanisms and in later part we will discuss on specific mechanism or solutions.

5.1 Generic Security Mechanisms

Like any other networks (e.g. MANETs, WSNs, etc.) security mechanisms in VANETs can be based on prevention (proactive) and detection (reactive) techniques. Considering the criticality nature of VANETs application, preventive security mechanisms are important than the reactive ones. Hence, most of the existing security mechanisms [10, 11, 13, etc.] in VANETs aim to prevent security attacks rather than detect. Even though works on detection techniques in VANETs are very limited, the usefulness of these can be significant in number of situations. For instance, in case of any fabrication attack if prevention mechanism fails, reliable and efficient detection of the fabrication can help drivers in taking the correct action in that situation. Most of the existing security mechanisms directly or indirectly employ cryptography. So, in the first part of this section we briefly introduce the possible key management approaches and keys [12] in VANETs and then we present the three

[1] Conditional means: the authorities should be able to reveal the identities of message senders in case of dispute such as a crime/car accident scene investigation, which can be used in seeking witnesses.

prevention security mechanisms, which are considered to be the most promising candidates to increase security in VANETs. Later part of this section we briefly present reactive based detection techniques.

5.1.1 Key Management

For the cryptographic approaches we need *unique information* about the vehicles which can be an electronic identity called an Electronic License Plate (ELP) [16] issued by a government, or an Electronic Chassis Number (ECN) issued by the vehicle manufacturer. These unique IDs are needed to identify vehicles to the police in case this is required (usually, identities are hidden from the police). Like license plates, the ELP should be changed (i.e., reloaded in the vehicle) when the owner changes or moves, e.g., to a different region or country. But these unique IDs may disclose privacy of the drivers, so anonymous key pairs that can be used to preserve privacy. An anonymous key pair is a public/private key pair that is authenticated by the CA but contains neither information about nor public relationship with the actual ID of the vehicle such as ELP. For the liability purposes this anonymity is conditional. Usually a vehicle will own a set of anonymous keys to prevent tracking. In the following we briefly present the main activities necessary for key management in VANETs.

- *Key bootstrapping and rekeying:* It is an important activity in key management. Like the physical license plate, it should be "installed" in the vehicle using a similar procedure, which means that the governmental transportation authority will preload the ELP at the time of vehicle registration (in the case of the ECN, the manufacturer is responsible for its installation at production time). Anonymous keys are preloaded by the transportation authority or the manufacturer (briefly mentioned in earlier section). As ELPs are unique and fixed, should attach to the vehicle for a long duration, but anonymous key sets have to be periodically renewed after all the keys have been used or their lifetimes have expired.
- *Key certification:* As briefly mentioned in earlier section, governmental transportation authorities or vehicle manufacturer can act as a CA in VANETs but this is not at all an easy process.
- *Key revocation:* Key revocation is necessary to punish the wrong doers in VANETs. One way to do this is to revocate the certificate related to the wrong doer. For instance, the certificates of a detected attacker or malfunctioning device have to be revoked, i.e., it should not be able to use its keys or if it still does, vehicles verifying them should be made aware of their invalidity. The simplest we can do so by distributing CRLs (Certificate Revocation Lists) that contains the most recently revoked certificates; CRLs are provided when infrastructure is available. There are number of ways we can do the revocation. Such as short-lived certificates method proposed in IEEEP1609.2/D2 draft standard [17] automatically revokes keys. It has number of shortcomings which are aimed to solve in RTPD (Revocation Protocol of the Tamper-Proof Device), RCCRL (Revocation protocol using Compressed Certificate Revocation Lists), and DRP (Distributed Revocation Protocol) [18].

- *Anonymous public keys:* There are several types of privacy. As safety messages will not contain any secret data about their senders, vehicle owners will be only concerned about identity and location privacy. Even though anonymous keys do not contain any publicly known relationship to the true identity of the key holders, privacy can still be hijacked by logging the messages containing a given key and thus tracking the sender until discovering his identity (e.g., by associating him with his place of living). Therefore, anonymous keys should be changed in such a way that a pervasive observer cannot track the owner of the keys. But it will require a vehicle to store a large key and certificate set (depending on the key changing frequency). So generation of efficient and reliable anonymous public keys is an open issue.

5.1.2 Prevention Techniques

- **Digital Signature-Based Techniques:** Digital signature is the building block of these security mechanisms, which primarily aim at providing message authenticity. Along with the digital signature, these techniques can exploit cryptography with certification or without certificate [15].
- *Without Certificate:* In this approach cryptographic digital signatures are apply to messages or hashes over messages. Digital message signatures are usually formed by asymmetric cryptography, i.e. by using public-private key cryptography. Messages (or hashes over the respective messages) are signed with the message originators' private keys. This approach can provide three security improvements to communication, namely message authenticity, message integrity protection and non-repudiation. The key advantage of this approach is that that it is simple to realize with small requirements. Mechanisms [shen] based on this approach are widely deployed and well known. However attacks like message forging, DoS, sybil are still possible. Moreover, the approach does not prevent attackers to create fake warning messages.
- *With Certificates:* In order to enhance the above approach, the signatures can be combined with digital certificates provided by a trusted Certificate Authority (CA). The basic notion with certificates is that nodes, which include certificates in their messages, are trusted by other nodes that are able to verify the certificates. The signed messages include a certificate which is cryptographically linked to the public key that belongs to the private key the message issuer uses to sign messages. The advantage of the certificate concept lies in the possibility to exclude external attackers from the system, as well as in the ability to remove malicious or defective nodes. With the support of suitable mechanisms it can also prevent sybil attacks.

  This is the most widely discussed and popular security mechanism in VANETs. Numerous studies and standards exploit certificate-based cryptosystem to support security for VANETs [7, 19-26, 39, etc.]. For instance, authors in [25] propose a vehicular PKI, based on a certificate-based PKC (digital signature) scheme to support security services for message exchange in the VANETs environment. A security architecture based on certificate-based PKC mechanism for VANETs discussed in [20]. However, secure messaging based on digital sign with certificate scheme has a number of limitations, including complexity in certificate verification and management, scalability, performance in a large-scale environment, and timely access to certificate revocation information. Some of the works already acknowledged the shortcomings of using digital sign with certificate–based scheme in VANETs [22]. There has been, however, hardly any discussion on how to improve the scalability of employing certificate-based PKC (digital signature) for VANETs.

- **Proprietary System Design:** This category of security mechanisms aim to exploit non-public (proprietary) protocols or hardware to control the unauthorized access to the networks. In case the protocols or hardware remain undisclosed (or highly expensive), like the certificate approach, this concept prevents non-authorized nodes from participating in the network. Ultimate objective of this concept is to increase the required effort an attacker has to put in order to enter into the system. This scheme does not prevent him from doing so, nor do they prevent any attack from an insider. For example, an attacker is still able to distribute fake warning messages using a vehicle's safety communication system. This approach seems not that promising, as vehicle manufacturers are aiming at the development of a common and open standard for the communication system.

- **Temper Proof Hardware:** In order to complement the aforementioned mechanisms, Tamper Resistant Device (TRD) or Tamper-Proof Device (TPD) hardware is meant to provide secure input to the communication system, by securing the in-vehicle communication system and protecting it from manipulation. Along with the storing secret information, this device will be also responsible for signing outgoing messages. To protect itself of being compromised by attackers, the device should have its own battery, which can be recharged from the vehicle, and clock, which can be securely resynchronized, when passing by a trusted roadside base station. The access to this device should be limited to authorized people. For instance, cryptographic keys can be renewed at the periodic technical checkup of the vehicle. Usually, the TPD contains a set of sensors that can detect hardware tampering and erase (self-destructive) all the stored keys to prevent them from being compromised. This sophisticated feature makes the TPD too sensitive for VANET conditions (e.g. the device can be subject to light shocks because of road imperfections, etc.) as well as too expensive for non-business consumers. A TPM (Trusted Platform Module [27]) that can resist to software attacks but not to sophisticated hardware tampering can be an alternative option to a TPD. These are popular in notebooks and cost only a few tens of dollars. The ultimate notion on of the security hardware will depend mainly on economic and technical factors. A tradeoff between TPD and TPM can be good guide to define it.

5.1.3 Detection Techniques

It is very unlikely that a proactive security measure will be always successful in preventing it concerned attacks. If it fails, then it may lead to a disastrous situation as

VANETs is dealing with life-critical applications. In this situation, detection techniques can help us in avoiding disastrous happenings. For instance, as shown in figure 3(ii) if prevention method fails to detect the possible collision than collision between the cars are must. If we can employ a reliable and efficient detection method which can identify in real-time that the attacker has sent bogus message, then it is possible avoid the collision.

These reactive measures are similar to the intrusion detection of other networks. Both the techniques correlate information which is either already available from normal system operation, or which is introduced additionally. Intrusion detection systems or similar systems for VANETs are still hardly explored (initial publications are [28] and [29]). These systems comprise what is also referred to as plausibility checks, information verification, use of side-channel information or context verification. In VANETs, or more precisely for safety systems in VANETs, reactive security mechanisms have to aim at detecting bogus or fabricated information in warning messages and inconsistencies in the inter-vehicle communication system. To do so, upon the reception of warning messages, nodes assess the validity of the warnings and then process the messages accordingly. If the message content is found to be invalid or bogus, the nodes ignore the message (some systems even try to correct the invalid data) and take action accordingly. Moreover, they may communicate their trusted neighbors to share the experiences.

In detecting security threats in VANETs, along with the common signature based and anomaly based detections we can exploit the contexts of a VANET and its application to detect attacks on it.

- **Signature-Based Detection:** In signature-based detection attacks can be detected by comparing network traffic to known signatures of attacks. As soon as an attack is detected appropriate countermeasures can be initiated. The primary concern of this approach is to realize a mechanism that is capable to detect known attacks on a communication system. The advantages of this detection technique are that it is simple and usually provides reliable detection of known attacks. The frequent updates of the attack signature database, the slow reaction on new attacks and of course the difficulty to define attack signatures are the shortcomings of this detection technique.
- **Anomaly Detection:** This approach is based on a statistical approach that defines normal communication system behavior. Any deviation from that behavior is statistically analyzed and as soon as they reach a defined level, the security system concludes that there is an attack ongoing. The advantage of this detection technique is that it enables the detection of previously unknown attacks without requiring an attack database to be updated. But, there are also some disadvantages. The definition of normal system behavior is pretty complex and anomaly detection is known to produce many false positives.
- **Context Verification:** Context verification is an approach that specifically considers the properties of VANETs and applications in VANETs. The notion is to collect as much information from any information source (e.g. the warning system, data from telemetric monitoring, etc.) available by each vehicle and create an independent view of its current status, its current surrounding (physical)

environment and current or previous neighboring vehicles. In order to do the evaluation of the situation this approach will require to define of rule-sets that determine what is to be expected with which probability in which situation. Situation evaluation mechanisms can be either application independent or application dependent. In application independent case, it can exploit position as well as time related information. On the other hand application context dependent evaluation exploits parameters specific to a certain application.

5.1.4 Standards

In the following we briefly review the IEEE 1609.2 and the Vehicle Safety Communications (VSC) project, which specify methods of securing Wireless Access in Vehicular Environments (WAVE) messages against numerous attacks, such as eavesdropping, alteration, source spoofing, message modification, and replays [30, 31]. This standard is still under revisions.

IEEE 1609.2 and the VSC Project: The IEEE 1609 WAVE communication standards, also known as DSRC protocols, have formed recently to enhance 802.11 to support wireless V2V and V2R communications in VANETs [31]. The IEEE 1609.2 standard addresses the issues of securing WAVE messages against eavesdropping, spoofing, and other attacks. As shown in figure 4, the components of the IEEE 1609.2 security infrastructure are based on industry standards for public key cryptography. It also includes support for elliptic curve cryptography (ECC), WAVE certificate formats, and hybrid encryption methods, in order to provide secure services for WAVE communications. The security infrastructure is also in charge of the administrative functions necessary to support core security functions such as certificate revocation (ongoing work). IEEE 1609.2 yet to define driver identification and privacy protection, and has left a lot of issues open. On the other hand, the VSC project also evaluates the feasibility of supporting vehicle safety related applications through the DSRC standard. It proposes to maintain a list of short-lived anonymous certificates for the purpose of keeping the privacy of drivers. Once the certificates are used, they are discarded. The scheme can provide higher security assurance as the certificates are blindly signed by the certificate authority (CA) in order to deal with any possible insider attack. The CA can abuse its authority and mishandles driver information. A linkage marker is devised for the escrow authorities to associate each blindly signed anonymous certificate with a single vehicle. All compromised and expired vehicles will be revoked by putting certificates belonging to those vehicles into the certificate revocation list (CRL). The main drawback of this scheme is that the CRL may grow quickly and make the real-time validation of certificates impossible. Another shortcoming depends on the fact that for tracing purpose, a unique electronic identity is assigned to each vehicle by which the identity of the vehicle owner can be inspected by the police and authorities in any dispute. Even though this scheme can effectively fulfill the conditional anonymity requirement, it is far from efficient, and suffers in scalability and reliability as the ID management authority has to keep all the anonymous certificates for the vehicles in the administrative region. Once a malicious message is discovered, the authority has to exhaustively search a huge database to find the identity related to the compromised anonymous certificate. To solve these issues, in [6] authors have presented an effective and efficient solution for achieving certificate revocation and conditional privacy preservation.

SeVeCom Project [32]: The SeVeCom project defines a baseline security architecture for VANETs systems. The baseline architecture contains different modules, which addresses different aspects, such as secure communication protocols, privacy protection, and in-vehicle security. The baseline specification provides one instantiation of the baseline architecture, building on well- established mechanisms and cryptographic primitives, thus being easy to implement and to deploy in upcoming VANETs. As shown in figure 5, the security manager is the central part of the SeVeCom system architecture. It instantiates and configures the components of all other security modules and establishes the connection to the cryptographic support module. To deal with different situations, the security manager maintains different policy sets, which can adaptively enable or disable some of the components or adjust their configuration. Even though this architecture is not yet accepted as standard, still it can be exploited as a good guideline for the implementation of security and privacy in VANETs.

5.2 Specific Security Solutions for VANETs

Majority of the existing works address the generic security policies rather than their implementation. Works on the implementation of overall security policies are very limited. Most of the existing works target to detect or prevent very specific attacks, such as sybil [33], DoS [34], etc. Moreover to achieve the full guarantee of security of message intensive VANETs applications, along with the message security in communion security is also necessary. To have a complete secure communication in VANETs, it is necessary have the support from all the layers of communication protocol stack (e.g. TCP/IP) rather than individual layer. Most of the existing research works (e.g. [35, 36, 37, 38, etc.]) on VANETs' security implementation focus on specific layer issues (e.g. routing, link, etc) rather than stack-wide. Moreover, these layer specific works are dominated by secure routing, few on secure MAC for VANETs. In the following we first briefly present few works which addresses some specific attacks and then present secure routing and secure MAC in VANETs respectively.

5.2.1 Specific Attack-Based Solutions

Privacy-Preserving Detection of Abuses of Pseudonyms (P2DAP) [33] explicitly targets sybil attacks in VANETs. It presents a lightweight and scalable protocol to detect Sybil attacks. In this protocol, a malicious user pretending to be multiple (other) vehicles can be detected in a distributed manner through passive overhearing by RSUs. In this scheme, detection of Sybil attacks does not require any vehicle in the network to disclose its identity; hence privacy is preserved at all times. It can detect

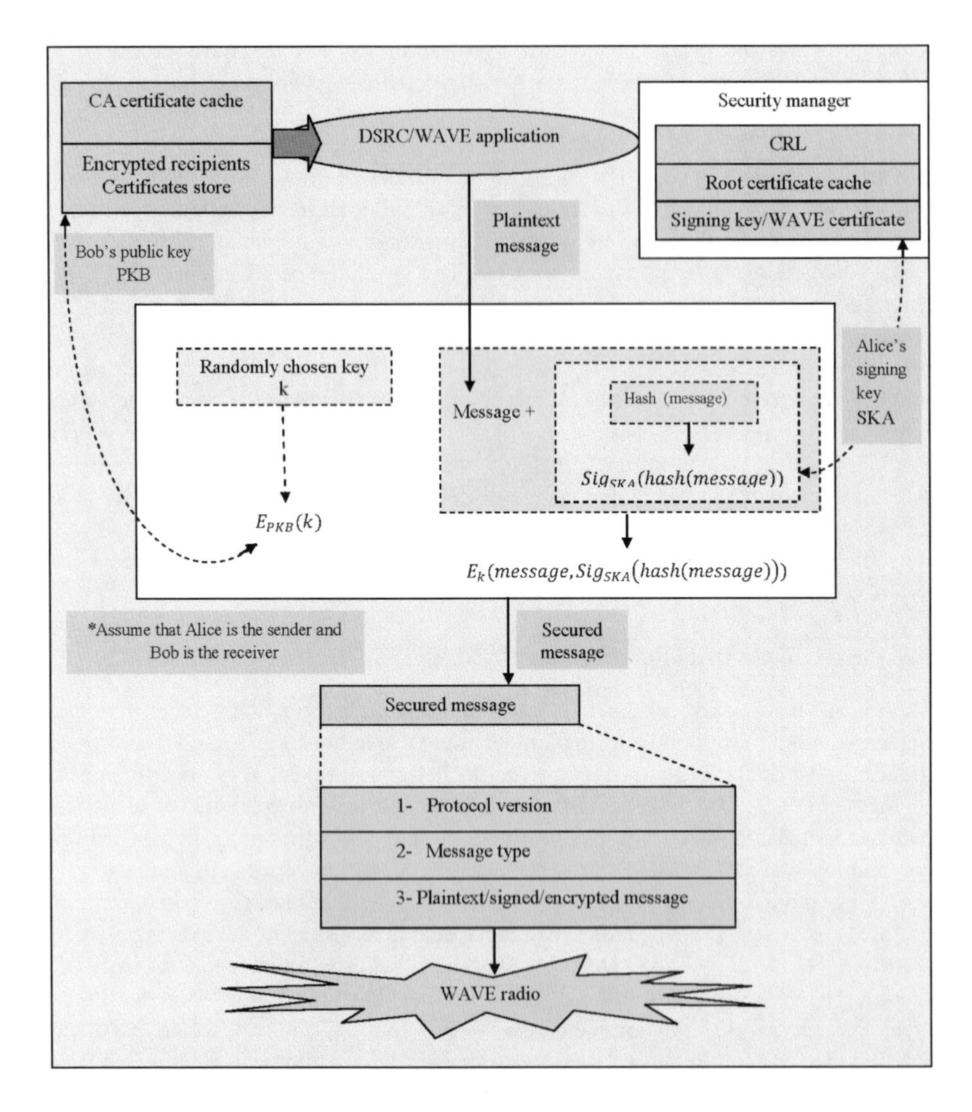

Fig. 4. The IEEE Std 1609.2 security services framework [7]

sybil attacks at low overhead and delay, while preserving privacy of vehicles but it may fail in colluding attacks. Authors in [34] present DoS and Distributed DoS and their severity level in VANET environment. They also introduce a model to secure VANETs from the DoS and DDoS. The solutions are able sorted out DoS but fail to protect privacy, prevent sybil attack, even information cheating.

Misusing VANETs could cause destructive consequences. Authors [38] proposed DRTA (Dynamic Revocation using Threshold Authentication) to punish misbehaving users. This can be employed in both V2V and V2R for anonymous communications.

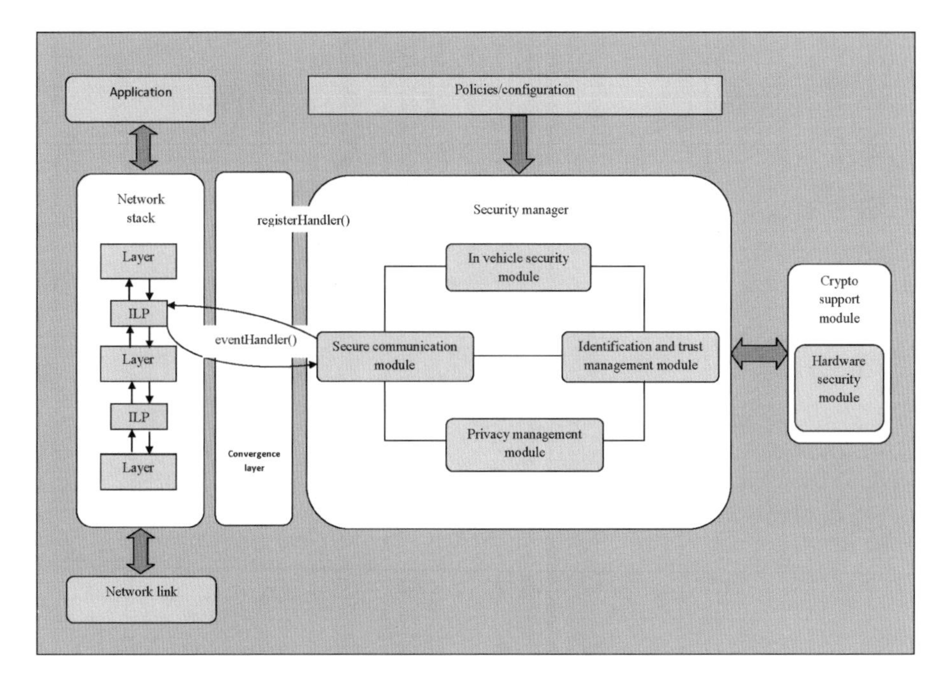

Fig. 5. Baseline architecture: deployment view [8, 32]

It is based on some threshold authentication technique that dynamically revokes a user's credential, while providing the flexibility of whether to reveal the user's identity and tolerating unintentional misbehavior such as hardware malfunctioning. DRTA outperforms its counterparts such as RTPD [12]. Work in [41] proposes protocols, as components of a framework, for the identification and local containment of misbehaving or faulty nodes, and then for their eviction from the system. Results show that the distributed approach to contain nodes and contribute to their eviction is efficiently feasible and achieves a sufficient level of robustness.

5.2.2 Secure Routing

Secure routing is the most important requirement for any secure communication. In VANETs routing can be based on ID of the vehicle or geography of the vehicle. ID methods are for sending data to an individual node, whereas geography methods are for sending data to a group of nodes. ID based routing protocols (e.g. Secure Routing Protocol (SRP) [43], Secure Beaconing [40], etc.) tend to sacrifice privacy frequently but geographic location or position based routing protocols (e.g. PRISM [35], Position-Based Routing [36], etc.) may not. Most secure routing algorithms build on top of insecure routing protocols as no routing protocol was originally built with security in mind.

Table 2, summarizes the secure routing protocols in VANETs. It is clear from the table that most of the secure routing protocols are not privacy compliant. Moreover, privacy compliant secure routing protocols such as PRISM is not secure from sybil attacks.

Table 2. Summary of the secure routing protocols in VANETs

Protocol	Key features	Advantages	Disadvantages
AOD-SEC[42]	-based on ID and a centralized PKI	-based on simple AODV - no impersonation attacks	-based on insecure routing protocol -privacy not protected
SRP[43]	-extension to existing ID-based reactive protocols (AODV, DSR) -assumes secure link between two nodes already exists	-deals with non-colluding malicious nodes. -prevents IP spoofing, ensures privacy.	–route cache poisoning renders efficient algorithms less efficient/ effective. –colluding malicious nodes can "alter" topology
Secure-Beaconing [40]	-ID based, believes most communications are direct. -not all beacons need to be encrypted –tries to strike balance between security and efficiency -omitting Certificates and Certificate Verifications	–saves bandwidth –better data throughput	-no privacy whatsoever. -some messages might be lost. -critical situations mean an exponential load on network
PRISM [35]	-uses AODV to establish path -destination is an area, not a node -uses group signatures on both side -once link is established, create one-time-use secret key between parties -hit and miss approach	-preserves privacy. -avoids creation of pseudonyms(expensive)	-deals with rogue/bad nodes reactively (TTP) -difficult to ensure that DST-AREA value has not been tampered with -sybil attacks are easy
Position-Based Routing [36]	- location table with ID and positions of nodes - location is plausible - end-to-end & hop-by-hop encryption	-two levels of encryption -broadcasts deter wormhole attacks	-caching of location and certificates is a great loss of privacy

5.2.3 Secure MAC

Like secure routing secure MAC protocol is necessary for VANETs. Unfortunately works in this area is very limited [37]. It is important design efficient medium access control (MAC) protocols so that safety related and other application messages can be timely and reliably disseminated through VANETs. In this work authors propose a secure MAC protocol for VANETs, with different message priorities for different types of applications to access DSRC channels. Results show that the proposed MAC protocol can provide secure communications while guarantee the reliability and latency requirements of safety related DSRC applications for VANETs.

6 Analysis of the Existing Mechanisms and Future Directions

It is clear from the discussion of section 4 that, security and privacy may conflict in number of VANET applications. While people or drivers of the vehicles are considering that privacy is their right, on the contrary to have security in certain situations we need to break their privacy. So, a trade-off between privacy and security may be necessary. Moreover, in secure communications precisely in routing it is really hard to find an efficient protocol that can response in real-time as well as maintain the privacy of the drives is very challenging and yet to solve. So scope of further study in this area is highly visible.

Most of the existing security and privacy related works mainly address policies not their implementations. But to make the VANET reality in near future we need to work more on implementation based security solutions which are cost effective and fast responsive. Moreover, certificate-based schemes are still suffering, especially in fixing CAs, real-time certificate verification, certificate revocation, etc. These issues also require further attention from the research community as well Govt. and vehicle manufacturers.

Reactive approach based attack detection techniques has great potential. But these are rarely considered by the VANETs researches. Further research could bring out the potential of these techniques.

For the implementation of comprehensive security and privacy policy, may require the support of all the layers in the protocol stack rather than from single layer support. This protocol stack-wide holistic implementation of security and privacy is missing in the existing works. Further study in this area is a must need. In number of applications, VANETs (e.g. Traffic Signals Violation Warning, Pre-crash Sensing, etc. [3]) exploit V2V and V2R communications. For these application security policy and solution has to take care of both communications which are different in nature. It means any security policy which suits V2V communication might not suit in V2R communication. For example, when a vehicle passes by a RSU; it retrieves fresh environmental data collected by the roadside sensors. After processing, it may interpret the data as a dangerous situation and trigger a safety warning message. In this case if WSNs in RSUs and vehicular communications maintain timeliness (a real-time response requirement for security in VANETs, mentioned section 4) individually (e.g. RSU provides data within 100ms and Vehicle triggers warning within another

100ms, total delay 200ms which is double than the maximum tolerance [3]) than the warning message can be useless and make the situation dangerous. So a combined effort of vehicles and RSUs is needed in guaranteeing overall system-wide security. However, to our best knowledge, there is no published work on the holistic view of security in VANETs. So, further works in this area is an immediate necessity.

Security and efficiency may conflict, especially in WSNs and hard-real time requirements based applications. Even security and QoS in RSUs precisely in WSNs may conflict. So these conflicting issues are needed to be resolved in future.

7 Conclusion

Applications of Vehicular Ad-hoc Networks (VANETs) are very promising and diverse. Majority of the safety-related VANETs applications are real-time and mission critical, which requires strict guarantee of security and reliability. Lack of such security and privacy in VANETs is one of the key difficulties to the wide spread implementations of it. Securing the VANETs along with appropriate protection of the privacy of drivers or vehicle owners is a very challenging task as they conflict with each other in umber of situations. Considering this, in this work we have summarized the attacks, corresponding security requirements and challenges in VANETs. Some of the challenges are not yet tackled at their best level, which require further attention. We have also presented the most popular generic security policies which are based on prevention as well detection methods. Detection-based mechanisms require further attention as they look prospective in VANETs. Many applications in VANETs require stack-wide security support rather than individual layer from the VANETs' protocol stack. In this work we have also discussed the existing works in the perspective of holistic (protocol stack-wide and system-wide) approach of security. These approaches are the concern of our future study.

References

[1] Ezell, S.: Explaining International IT Application Leadership: Intelligent Transportation Systems. The Information Technology & Innovation Foundation (January 2010)

[2] David Wang, C., Thompson, J.P.: Apparatus and method for motion detection and tracking of objects in a region for collision avoidance utilizing a real-time adaptive probabilistic neural network, US. Patent No. 5,613,039 (1997)

[3] US Dept. Transportation, "Vehicle Safety Communications Project Task 3 Final Report" (March 2005), `http://www.its.dot.gov/research_docs/pdf/59vehicle-safety.pdf`

[4] Raya, M., Hubaux, J.-P.: Securing vehicular ad hoc networks. Journal of Computer Security 15(1), 39–68 (2007)

[5] Raya, M., et al.: Securing vehicular communications. IEEE Wireless Communications 13(5), 8–15 (2008)

[6] Papadimitratos, P., et al.: Secure Vehicular Communications: Design and Architecture. IEEE Commun. Mag. (November 2008)

[7] Lin, X., Lu, R., Zhang, C., Zhu, H., Ho, P.H., Shen, X.: Security in vehicular ad hoc networks. IEEE Communications Magazine 46(4), 88–95 (2008)
[8] Kargl, F., et al.: Secure vehicular communication systems: implementation, performance, and research challenges. IEEE Communications Magazine 46(11), 110–118 (2008)
[9] Zeadally, S., Hunt, R., Chen, Y.S., Irwin, A., Hassan, A.: Vehicular ad hoc networks (VANETs): status, results, and challenges. Telecommunication Systems, 1–25 (2010)
[10] Qian, Y., Moayeri, N.: Design of secure and application-oriented VANETs. In: IEEE VTC 2008, pp. 2794–2799 (Spring 2008)
[11] Boukerche, A., Oliveira, H.A.B.F., Nakamura, E.F., Loureiro, A.A.F.: Vehicular ad hoc networks: A new challenge for localization-based systems. Computer Communications 31(12), 2838–2849 (2008)
[12] Raya, M., Hubaux, J.-P.: Securing Vehicular Ad Hoc Networks. J. Computer Security, Special Issue on Secu- rity, Ad Hoc and Sensor Networks 15(1), 39–68 (2007)
[13] Parno, B., Perrig, A.: Challenges in securing vehicular networks. In: Proceedings of the Workshop on Hot Topics in Networks, HotNets-IV (2005)
[14] Yin, J., ElBatt, T., Yeung, G., Ryu, B., Habermas, S., Krishnan, H., Talty, T.: Performance evaluation of safety applications over DSRC vehicular ad hoc networks. In: Proc. of ACM workshop on Vehicular Ad Hoc Networks, VANET (2004)
[15] Leinmuller, T., Schoch, E., Maihofer, C.: Security requirements and solution concepts in vehicular ad hoc networks. In: IEEE 4th Annual Conference on Wireless on Demand Network Systems and Services, pp. 84–91 (2007)
[16] Hubaux, J.-P., Capkun, S., Luo, J.: The security and privacy of smart vehicles. IEEE Security and Privacy Magazine 2(3), 49–55 (2004)
[17] IEEE P1609.2/D2 – Draft Standard for Wireless Access in Vehicular Environments – Security Services for Applications and Management Messages (November 2005)
[18] Jungels, D., Raya, M., Aad, I., Hubaux, J.-P.: Certificate revocation in vehicular ad hoc networks. Technical Report LCA-REPORT-2006-006, EPFL (2006)
[19] Kounga, G., et al.: Proving Reliability of Anonymous Information in VANETs. IEEE Transactions on Vehicular Technology 58(6), 2977–2989 (2009)
[20] Papadimitratos, P., et al.: Architecture for secure and private vehicular communications. In: 7th International Conference on ITS Telecommunications, ITST 2007, pp. 1–6 (2007)
[21] Petit, J.: Analysis of ECDSA Authentication Processing in VANETs. In: 2009 3rd International Conference on New Technologies, Mobility and Security (NTMS), pp. 1–5 (2009)
[22] Plößl, K., Federrath, H.: A privacy aware and efficient security infrastructure for vehicular ad hoc networks. Computer Standards & Interfaces 30(6), 390–397 (2008)
[23] Plößl, K., et al.: Towards a security architecture for vehicular ad hoc networks. In: The First International Conference on Availability, Reliability and Security, ARES 2006, pp. 1–8 (2006)
[24] Rao, A., et al.: Secure V2V Communication With Certificate Revocations. In: 2007 Mobile Networking for Vehicular Environments, pp. 127–132 (2007)
[25] Raya, M., Hubaux, J.-P.: The security of vehicular ad hoc networks. In: Proceedings of the 3rd ACM Workshop on Security of Ad Hoc and Sensor Networks, Alexandria, VA, USA, pp. 11–21. ACM (2005)
[26] Sunnadkal, R., et al.: A Four-Stage Design Approach Towards Securing a Vehicular Ad Hoc Networks Architecture. In: Fifth IEEE International Symposium on Electronic Design, Test and Application, DELTA 2010, pp. 177–182 (2010)
[27] Trusted Platform Module (TPM), https://www.trustedcomputinggroup.org/groups/tpm/

[28] Golle, P., Greene, D., Staddon, J.: Detecting and Correcting Malicious Data in VANETs. In: Proceedings of the First ACM Workshop on Vehicular Ad Hoc Networks (VANET). ACM Press, Philadelphia (2004)
[29] Leinmüller, T., Held, A., Schäfer, G., Wolisz, A.: Intrusion Detection in VANETs. In: Proceedings of 12th IEEE International Conference on Network Protocols (ICNP 2004) Student Poster Session (October 2004)
[30] U.S. Dept. of Transportation, National Highway Traffic Safety Administration, Vehicle Safety Communications Project — Final Report (April 2006), http://www-nrd.nhtsa.dot.gov/pdf/nrd-12/060419-0843/PDFTOC.htm
[31] IEEE Std. 1609.2-2006, IEEE Trial-Use Standard for Wireless Access in Vehicular Environments-Security Services for Applications and Management Messages (2006)
[32] SeVeCom, "Secure Vehicular Communications: Security Architecture and Mechanisms for V2V/V2I, Delivarable 2.1" (2007-2008), http://www.sevecom.org
[33] Zhou, T., Choudhury, R.R., Ning, P., Chakrabarty, K.: P2DAP—Sybil Attacks Detection in Vehicular Ad Hoc Networks. IEEE Journal on Selected Areas in Communications 29(3), 582–594 (2011)
[34] Ahmed Soomro, I., Hasbullah, H., Ab Manan, J.-l.: Denial of Service (DOS) Attack and Its Possible Solution in VANET. In: ICCESSE-WASET 2010 Conference, Tokyo, Japan, May 26-28 (2010)
[35] Karim El Defrawy, M., Tsudik, G.: Privacy-Preserving Location-Based On-Demand Routing in MANETs. IEEE Journal on Selected Areas in Communications 29(10) (December 2011)
[36] Harsch, C., Festag, A., Papadimitratos, P.: Secure Position-Based Routing for VANETs. In: IEEE 66th Vehicular Technology Conference (2007)
[37] Qian, Y., Lu, K., Moayeri, N.: A secure VANET MAC protocol for DSRC applications. In: IEEE GLOBECOM, pp. 1–5 (2008)
[38] Sun, J., Fang, Y.: Defense against misbehavior in anonymous vehicular ad hoc networks. Ad Hoc Networks 7(8), 1515–1525 (2009)
[39] Shen, P.-Y., Liu, V., Tang, M., William, C.: An efficient public key management system: an application in vehicular ad hoc networks. In: Pacific Asia Conference on Information Systems (PACIS), AIS Electronic Library (AISeL), Queensland University of Technology, Brisbane, Qld, p. 175 (2011)
[40] Schoch, E., Kargl, F.: On the Efficiency of Secure Beaconing in VANETs. In: WiSec 2010, March 22-24 (2010)
[41] Raya, M., Papadimitratos, P., Aad, I., Jungels, D., Hubaux, J.-P.: Eviction of Misbehaving and Faulty Nodes in Vehicular Networks. IEEE Journal on Selected Areas in Communications 25(8) (October 2007)
[42] Eichler, S., Dotzer, F., Schwingenschlogl, C., Caro, F.J.F., Eberspaher, J.: Secure routing in a vehicular ad hoc network. In: IEEE 60th Vehicular Technology Conference, pp. 3339–3343 (2004)
[43] Papadimitratos, P., Haas, Z.: Secure Routing for Mobile Ad Hoc Networks. In: SCS Communication Networks and Distributed Systems Modeling and Simulation Conference, pp. 27–31 (January 2002)

Security Issues and Approaches on Wireless M2M Systems

Jorge Granjal, Edmundo Monteiro, and Jorge Sá Silva

University of Coimbra, Portugal
{jgranjal,edmundo,sasilva}@dei.uc.pt

Abstract. Wireless communications will be fundamental in future Machine-to-Machine (M2M) pervasive environments where new applications are expected to employ sensing and actuating devices that are able to autonomously communicate without human intervention. M2M devices using wireless communications are expected to represent fundamental components of a future Internet where applications will allow users to transparently interact with its physical surroundings. The heterogeneity of the characteristics envisioned for M2M devices and applications calls for new approaches regarding how devices communicate wirelessly at the various protocol layers and how security should be designed for such communications. As such devices and communications are expected to support security-critical applications, the security of M2M wireless communications is particularly important.

Since most M2M wireless devices will be seriously constrained in terms of computational capability and energy, security for M2M wireless communications must consider such limitations. This implies that existing security mechanisms may not be appropriate for M2M communications. The particular characteristics and the heterogeneity of the characteristics of M2M devices is currently motivating the design of a plethora of new communication protocols at the various communication layers.

As M2M is a fundamentally recent research area, we currently verify a lack of research contributions that are clearly able to identify the main issues and approaches in targeting security on M2M environments. In this chapter we analyze security for wireless communications considering also protocols in the process of standardization, as such technologies are likely to contribute to future standard communications architecture for wireless M2M systems. We start by addressing the security issues and vulnerabilities related with the usage of wireless M2M communication technologies on applications in various application environments. Such threats to wireless communications are present not only due to the usage of wireless communication in security-threatening environments but also to the inherent constraints of M2M sensing devices. We also discuss ways for strengthening security for wireless communications at the various layers of the communications stack. We also verify that most of the current proposals for M2M wireless communications technologies lack fundamental security assurances and discuss how this major challenge may be targeted by research and standardization work.

The goal of this chapter is twofold, as on the one side we perform a survey on the main security issues of the usage of currently available M2M wireless communication technologies and also discuss the main approaches to introduce

S. Khan and A.-S.K. Pathan (Eds.): *Wireless Networks and Security*, SCT, pp. 133–164.
DOI: 10.1007/978-3-642-36169-2_5

security for such communications, while on the other side we discuss future approaches to security in wireless M2M environments. Various characteristics of such environments will pose challenges and motivate new approaches for security. In fact, many aspects of M2M applications will require a paradigm shift in how security is designed for M2M applications, devices and wireless communications technologies.

Keywords: M2M wireless communications, M2M sensing applications, Cross-layer security, IEEE 802.15.4, 6LoWPAN, CoAP, DTLS, Internet-integrated M2M sensing applications, Security gateway, M2M end-to-end security.

1 Introduction

The Machine-to-machine (M2M) concept refers to applications that are envisioned to be enabled by wireless and wired devices able to communicate autonomously with other sensing or controlling devices, computers or personal appliances. Many types of new sensing applications are envisioned to participate in a world enabled by M2M technologies. One fundamental aspect that differentiates M2M applications and communications from their counterparts in established research areas such as Wireless Sensor Networks (WSNs) is the autonomous nature of M2M devices and communications, as devices in M2M applications are expected to be able to interact autonomously. Interactions between M2M devices will mostly involve wireless communications, and consequently the security of such communications will represent a fundamental enabling factor of most M2M applications. The characteristics envisioned for M2M applications and devices will introduce many challenges for both research and standardization. Security will play a fundamental role in enabling many of the sensing applications currently envisioned for M2M, as such applications will require the processing and transmission of sensitive data between M2M devices, personal appliances and controlling units.

M2M is a concept evolving from a previous connotation with the employment of sensing devices in industrial control environments, towards the support of pervasive and transparent wireless sensing applications. The pervasiveness of public wireless communication networks is expected to provide a major contribution to this paradigm change, as it facilitates unattended communications between sensing devices. Such communications will also provide a major economical motivation for the adoption of new M2M applications, as it may enable efficient and immediate communications between producer and consumers.

Although technologies may be inherited from existing research and standardization work, the design of security mechanisms appropriate for M2M wireless communications is also expected to require new approaches. This will be motivated not only by the unattended nature of M2M wireless communications, but also because M2M applications are expected to encompass a multitude of communication technologies, devices and networks. M2M applications may require the integration of diverse wireless communication technologies such as IEEE 802.11 Wi-Fi, IEEE 802.15.4 Low-Power Personal Area Networks (LOWPANs) or of the

SMS (Short Message Service) service available using GSM networks. Personal and public wireless Wi-Fi networks proliferating worldwide and services such as GSM are expected to provide a major contribution to the integration of M2M sensing applications with the Internet. Short-range communications between constrained devices may also be enabled by new communication standards such as IEEE 802.15.4, particularly considering that many M2M applications may require the usage of constrained and battery-powered sensing or actuating devices. The design of new security technologies will be challenged by this heterogeneity of communication technologies, and M2M applications may require the adoption of mechanisms at different layers of the communications stack.

While a complete communications model or architecture is currently a fuzzy concept, we find it useful to consider a reference communications model for M2M applications, as we illustrate in Figure 1. This model facilitates our discussion on particular wireless communication and security approaches for M2M applications, and its design concentrates particularly on the integration of sensing applications with the Internet. This integration may be supported by mechanisms enabling the interconnection of two separate communication domains, one supporting M2M wireless communications between sensing devices and the other supporting remaining Internet communications. Communications between the two domains may be mediated by a security gateway that implements filtering policies for communications accordingly to the requirements of each application. This entity may also assume other management and security roles and perform as the backend or controlling unit for the M2M sensing application.

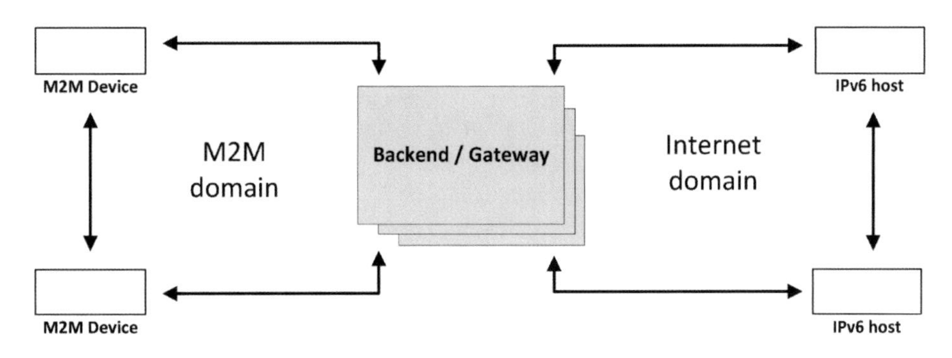

Fig. 1. A reference communications model for Internet-integrated M2M sensing applications

Wireless communications between devices in the M2M domain may take place in an unattended fashion, raising important security concerns such as authentication and trust between devices without previous knowledge of each other. Many applications will also require communications with backend or gateway devices, for example when such devices assume management operations or store sensing data. The gateway entity illustrated in our reference communications model may in practice assume various forms and support diverse management and security mechanisms, depending on the requirements of particular M2M wireless sensing applications. It may support the role of a personal or industrial controlling device or appliance, while also supporting communications with M2M sensing devices and with the Internet.

One important aspect of our model is that in most M2M applications the backend or gateway entity may not present the resource constrains of M2M sensing devices, particularly regarding its processing power, memory and energy availability. This may represent an opportunity to the design of new security solutions, as new security mechanisms may be designed asymmetrically so that the most demanding operations are supported by the gateway. This principle may be applied to mechanisms such as authentication and trust management or intrusion detection, among others. Other important role of the backend or gateway entity may be in the support of protocol conversion mechanisms, similarity to traditional M2M industrial applications requiring protocol translation between technologies from different vendors. In general, the gateway entity may be expected to perform an important role in enabling security mechanisms for M2M applications, since its placement between the M2M and Internet domains enables the usage of security mechanisms for the enforcement of security policies and protection of the M2M domain from attacks originated at the Internet.

This chapter focuses on technologies and solutions that may support the development of pervasive and global M2M applications, in contrast to applications that currently target closed M2M environments. The development of such environments will require the design of open and standardized communications and security mechanisms for the support of Internet-integrated M2M sensing applications. In this context, we distance ourselves from closed and proprietary solutions for M2M communications, also because such solutions reflect an obsolete perception of what M2M wireless communications should enable. In classic industrial M2M environments there are hundreds of automation device protocols and solutions that typically target only very specific market segments and vendor relationships. On the other end, future M2M applications will require open and well-accepted technologies that enable the evolution of the Internet architecture to transparently encompass new pervasive and heterogeneous sensing applications.

2 A Case for Security in M2M Wireless Communications

Most of the envisioned M2M applications can be considered critical in respect to security. Security will be a requirement not only for wireless communications transferring sensitive data between M2M devices but also for the data itself. Given the constrains and the heterogeneity of devices expected to support M2M applications, flexibility will be a desired requirement for new security mechanisms developed to protect M2M wireless communications. Similarly, the protection of data stored on sensing devices may be accomplished by adopting appropriate hardware or software mechanisms designed to guarantee security properties such as privacy and confidentiality of critical data.

The development of new technologies for M2M environments is envisioned to motivate the appearance of new sensing applications in areas as diverse as industrial control, structural monitoring, home automation and entertainment, smart energy, healthcare, vehicular telematics and agricultural monitoring. We must note that this list is certainly not inclusive of all possible application areas, as M2M is itself an evolving concept. We proceed by discussing the usage of M2M application in such areas, with the goal of identifying its main security requirements.

Industrial monitoring and control

The usage of M2M wireless sensing applications in industrial environments will be motivated by benefits such as productivity, energy efficiency and safety. Applications in this domain are expected to target goals such as process monitoring and control, machine surveillance, asset tracking and storage monitoring, among others. The usage of secure and time-bounded data communications between M2M devices will be of fundamental importance in critical deployments where wireless M2M applications are required to replace existing wired devices. On the other end, many of such applications may also benefit from the availability of communications with the Internet or backend devices. Overall, the design of new mechanisms for M2M wireless communications will require efforts from both the security and quality of service domains.

In this application domain the processing of misinterpreted or erroneous information may motivate catastrophic events in critical industrial processes. This implies that fundamental security mechanisms must be in place so that all wireless data communications are protected. Given the criticality of the protected data in industrial environments, algorithms with strong security guarantees will be required to provide confidentiality, integrity and authentication. Authentication of both M2M devices and users of the monitoring and control application will also be a requirement of most applications. The usage of wireless communications in industrial environments raises many challenges, since time-bounded and secure communications between M2M devices will be required in many scenarios. Also relevant is the fact that M2M devices may be battery-powered and employed in environments where the frequent replacement of batteries will not be feasible. Therefore, new security mechanisms must be compatible with such constrains while providing acceptable security levels and lifetime of M2M sensing applications.

Structural monitoring

In the area of structural monitoring M2M applications promise to make safety checks and periodic monitoring of structures such as buildings, tunnels and bridges much more efficient and cheap. The monitored structures may employ mains-powered sensors included in the design-phase or in alternative battery-equipped devices may be added afterwards. These applications will require that all data be verified against its authenticity and integrity, while for particular scenarios confidentiality may also be required. It can be considered that security properties such as confidentiality, authentication and integrity may be achieved by employing algorithms targeting moderate levels of security, at least when compared with more critical applications in areas such as industrial control. As for the remaining application areas, security mechanisms may be employed that are able to adjust its configuration and running parameters to the requirements and characteristics of particular M2M sensing applications. Configuration parameters such as the size of the cryptographic keys or the cryptographic algorithm employed to guarantee specific security properties at the end may dictate the usage of resources on constrained sensing devices and influence significantly the lifetime of the M2M sensing application. In this area quality of service may also be an important property for M2M wireless sensing applications, as such applications may require guarantees in terms of the timely delivery of critical data to central monitoring systems, for example of data related with reports of critical conditions.

Home automation and energy management

Home automation is a promising area for M2M applications, and in this area wireless sensing applications may be designed targeting objectives such as home safety and control, healthcare, smart appliances, entertainment and energy management, among others. Energy management will allow a fine-grained control over the consumption of each home device, and it is likely that market forces will push the smart grid towards the integration of M2M energy applications with the Internet. All communications between M2M devices or with controlling computers via the Internet will require proper security mechanisms, and standard security solutions will also be required for the smart energy area [1]. Other important security requirement of M2M energy applications will be security against tampering of the devices or of the purposes of the applications, given that strong monetary incentives will appear to drive such attacks.

Home automation applications are expected to be very diverse in terms of its requirements on security and reliability. Confidentiality, authentication and integrity may be achieved by employing algorithms with diverse guarantees in terms of security, again favoring the adoption of flexible solutions. For example, a video feed may require authentication and integrity assurances to secure a surveillance application against man-in-the-middle attacks, with confidentiality only being required for particular applications. Also, authentication may be required only for M2M devices or in the other end also for the users of the application. Again, time-bounded data communications may also be required for particular reports or devices. For example, a report on the humidity level or a report from a fire detection sensor may be treated differently in respect to its priority in terms of correct and timely delivery [1].

One important security requirement of M2M applications that will be present particularly in home automation applications is privacy. Users will raise concerns and require guarantees in terms of its privacy before M2M wireless sensing applications are accepted and enter their lives and home environments. One aspect that may also differentiate M2M sensing applications in this area from other areas is that in most situations devices will be mains-powered. This aspect will facilitate the adoption of strongest security and cryptographic mechanisms whenever required, again in the context of a flexible and adaptable set of security mechanisms.

Healthcare

Healthcare M2M sensing applications promise benefits such as the simplification and improvement of patient care and of day-to-day storage and management of materials. Personal networks of sensing devices may be deployed to help in monitoring vital signals of patients or on medicaments have been taken. Tele-assistance applications may help senior citizens that are remotely monitored for their condition and movement around the house. Such environments and applications will require particular attention for the guarantee of security requirements such as confidentiality and integrity of data, and most applications will require strong authentication of all M2M users and devices.

Other important security requirements of M2M applications in the healthcare area are privacy and security against disclosure of sensitive data. Privacy mechanisms will be required to guarantee that personal health information remain private, as patients will be concerned about the disclosure of such information to unauthorized third-parties. Privacy mechanisms may be implemented side-by-side with access control mechanisms to enable control of who may access what information. This will be an important requirement, as for example a medical team may require that specific data will not be available to the patient's family. This may require the design of appropriate role-based access-control mechanisms. Similarly to other application areas, M2M healthcare applications will greatly benefit from the availability of communications with hosts or devices outside of the M2M domain, as this allows for example communications between M2M devices installed at the patient's home and computers at the hospital facilities.

Vehicle telematics

Intelligent transportation systems are expected to adopt M2M applications targeting goals such as traffic flow optimization, reduced road congestion and traffic accident prevention. Vehicular safety systems such as OnStar [2] already provide a subset of the functionalities that M2M vehicle applications may provide in the future. M2M devices that support such applications may be integrated not only with vehicles, but also with traffic signals and roads. Communications supporting M2M vehicle telematics applications will require appropriate confidentiality and integrity assurances, as well as the authentication of all M2M devices and users of the application. As for the previous application areas, the integration of such applications with the Internet may provide various benefits. Internet-integrated applications may allow a user to remotely access information on the status of its vehicle, or allow a vehicle to automatically report driving conditions or accident alerts to law and safety agencies. Communications between vehicles will be required to support applications targeting road traffic optimization and accident prevention.

Vehicular applications will require appropriate mechanisms for the management of identification, trust and privacy for all the communication parties involved. Dynamic identification and trust mechanisms will be required to establish communication relationships between devices and users without previous knowledge of each other, while also guaranteeing privacy and liability in case of accidents. This implies that if on the one side privacy will be required to protect users from the disclosure of personal information, on the other side a driver should not be able to lie in its identification information in order to avoid police identification or speed penalties. Identification mechanisms will be required to guarantee the proper identification of infringing drivers for liability purposes. Thus, new security mechanisms targeting identification, authentication, trust and privacy will play a central role in enabling M2M wireless applications in the vehicle telematics domain.

Agricultural monitoring

The usage of M2M sensing applications for the monitoring of agricultural processes may allow the accurate monitoring of the environmental conditions and consequently greatly improve the productivity and efficiency of agricultural processes. The integration of such M2M applications with the Internet may provide added benefits, such as the remote monitoring of the status of specific crops areas and of information required to optimize productivity of crops and the usage of resources such as water, pesticides and energy [1]. Data obtained from such applications may also be transmitted via the Internet to a system for supply chain management processing purposes.

M2M applications for agricultural monitoring will probably be more concerned with guaranteeing the integrity of wireless data communications, while confidentiality may be required only for very specific applications. Similarly, authentication may be required mainly for M2M devices reporting sensing data. As M2M applications may be envisioned to employ mostly battery-powered devices, the lifetime of such devices (and consequently of the sensing application itself) will be a fundamental design factor. Adaptable security mechanisms will therefore be required that are able to provide acceptable levels of security while not compromising the lifetime of M2M sensing applications.

Overall discussion on the security requirements of M2M wireless sensing applications

Our previous discussion allows us to observe that the heterogeneity of sensing devices and technologies envisioned to enable M2M applications will significantly influence the design of mechanisms to secure M2M wireless communications and applications. As the selection of the most appropriate cryptographic and security mechanisms may be dictated by requirements of particular M2M applications, flexibility and adaptability will be two major aspects of any security architecture adopted for M2M environments. Other factor influencing the selection of particular security mechanisms will be the capabilities of the sensing devices employed and the appropriateness of particular security technologies. Thus, any security architecture for M2M environments should provide mechanisms that enable acceptable compromises between all such aspects. Different mechanisms may also be employed depending on its usage to secure communications between M2M devices or on the other end to secure end-to-end communications between M2M devices and Internet hosts or backend devices. Other than security for M2M wireless communications, mechanisms may also be required to protect sensitive data stored on sensing devices against tampering.

Various classes of M2M sensing applications are envisioned to use constrained sensing devices, in respect to the availability of critical resources such as RAM and ROM memory, computational power and energy. Such limitations will certainly play an important role in the adoption of new security mechanisms, as new proposals must be carefully evaluated as not to compromise the expected lifetime of such devices while providing acceptable security. New security mechanisms must be adopted considering not only the characteristics of the devices, but also that such devices must also support other mechanisms required for the M2M application itself.

As observed, the previously discussed M2M applications will pose new challenges to security, motivated by factors such as the usage of heterogeneous communication technologies and protocols, the necessity of supporting automatic communications between M2M devices without human intervention, limitations on hardware capabilities of M2M devices and expectations from users regarding privacy and liability. Although many lessons and technical solutions are currently available from results in research areas such as Wireless Sensor Networks (WSNs), it is also expected that M2M applications will require new approaches for security. M2M applications and devices are expected to employ a myriad of wired and wireless technologies, such as Ethernet, 3G, GSM, LTE, Bluetooth, IEEE 802.15.4, PLC (Power-Line Communications) and IPv6, among others. Similarly, diverse types of sensing devices are expected to enable such applications, for example smartphones, wireless sensors and actuators, smart appliances and computers. This heterogeneity of technologies and devices will require mechanisms that are able to adapt security to the characteristics of different sensing applications.

As communications may take place in an unattended fashion between M2M devices, the identification and authorization of devices and users will be a fundamental enabling aspect of M2M applications. M2M devices may also be mobile, and in many scenarios a classic trusted third-party authentication scheme or infrastructure might not be available or required. Since many M2M sensing and actuating devices will be seriously limited in terms of computational capability, security technologies must be developed to cope with heterogeneous and constrained classes of devices. Security solutions involving the usage of security infrastructures or the exchange of numerous security-related messages may be totally unfeasible for constrained M2M devices such as passive RFID tags.

Privacy and liability will be fundamental security requirements for most M2M applications. Users will require that systems allow the control of how much personal information is exposed, while on the other end certain applications will require that a certain degree of personal information is guaranteed to be available, namely as we have discussed for liability purposes on vehicular applications. In general, it is possible to observe that the discussed security requirements of M2M applications will require the employment of classic cryptographic solutions side-by-side with new mechanisms designed to cope with the particular characteristics and requirements of M2M applications.

3 Security for Wireless Communications in M2M Environments – A Layered Approach

Although a complete layered communications stack is yet to be defined for M2M environments, we find it is useful to consider a reference stack that facilitates the contextualization and discussion of how security may be applied to wireless communication mechanisms in M2M environments. In Figure 2 we illustrate a reference layered communications stack for M2M wireless communications. This stack inherits many characteristics from the traditional Internet communications stack and also considers the adoption of new cross-layer mechanisms for security and management of M2M applications. Distributed management mechanisms will be required in many applications to enable the control, monitoring and management of M2M wireless sensing devices and applications. Similarly, security requirements may promote the design of cross-layer security mechanisms, as such an approach may promote the efficiency of security operations and of the usage of resources on constrained sensing devices. A cross-layer approach may be appropriate to design security solutions for fundamental mechanisms and services such as key management, trust management, authentication, privacy and intrusion detection, among others. Such security mechanisms are transversal to security and therefore do not relate solely with a particular communications layer or protocol.

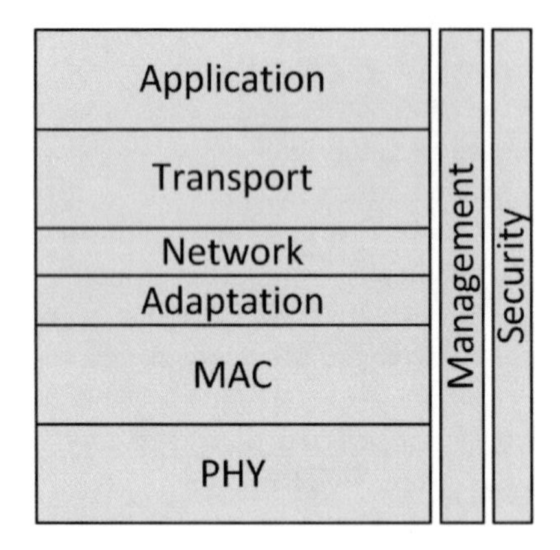

Fig. 2. A reference layered communications stack for M2M sensing applications

It is important to note that the reference communications stack illustrated in Figure 2 is general enough to encompass wireless communications supported by constrained sensing devices and also by more powerful backend or control hosts. This stack therefore identifies a set of protocol layers where communications and security mechanisms may be placed and used as functional components of the M2M application.

Although diverse types of devices are expected to enable M2M applications, our reference communications stack may apply to all such devices. The cross-layer implementation of security mechanisms may provide the optimizations required to enable the usage of more demanding mechanisms on a wide range of device classes. The following discussion considers the various protocol layers of our reference communications stack but, as we verify throughout our discussion, various mechanisms will not clearly belong only to one particular layer, either because they depend on cross-layer interactions or because the constraints of M2M sensing devices demand such a design approach. Our following discussion characterizes the main threats, approaches and future directions in addressing security for M2M wireless communications in the context of the reference stack illustrated in Figure 2. Our goal is in particular to identify possible directions or research and standardization opportunities in targeting the security requirements previously identified for the various envisioned M2M applications.

Security at the Physical (PHY) Layer for M2M Wireless Communications

Our analysis of security for wireless communications on M2M environments considers the most representative technologies designed various classes of wireless networks, namely wireless personal area networks, wireless local area networks and wireless broadband networks. Although other communication technologies may enable M2M applications in the future, we are currently able to identify a set of technologies that are good candidates in playing an important role in the enabling of future M2M environments.

Wireless personal area communications may provide short-range and energy-efficient communications between M2M sensing devices. Technologies such as IEEE 802.15.4 [3] are being designed with this goal in mind and may enable capillary communications between sensing devices in the context of Internet-integrated M2M applications. Such communications may also be supported by Wi-Fi Internet connectivity, as the proliferation of public Wi-Fi networks is expected to greatly contribute to enable many M2M applications. Short-range and Wi-Fi communications may also be complemented by last-mile technologies such as IEEE 802.16 WiMAX [4][5] in providing pervasive broadband communications for sensing devices, irrespective of its placement of deployment.

Wireless M2M connectivity will also be possible using technologies such as GSM (Global System for Mobile Communications) SMS (Short Message Service) text messages or data communications using GPRS (General Packet Radio Services) or EDGE (Enhanced Data rates for GSM Evolution). Newer incarnations of this service as the third generation (3G) UMTS (Universal Mobile Telecommunications Systems) or the fourth generation 4G LTE (Long Term Evolution) advanced standards may also provide pervasive wireless broadband Internet communications to M2M applications. All such communication technologies may enable not only communications between sensing devices but also help in the integration of M2M sensing applications with the Internet.

Several types of threats can be identified against the normal functioning of M2M wireless communications at the physical layer when using the previously discussed communication technologies. Such attacks may be generally classified as passive and active attacks, and may in practice target all wireless communication technologies. Passive attacks normally involve traffic analysis and eavesdropping of wireless communications, while active attacks are generally more disruptive and may involve jamming and scrambling. Jamming attacks consists in the permanent interruption of the communications channel and therefore represents a Denial of Service (DoS) attack against wireless communications. A scrambling attack normally targets more precisely specific frames or parts of a frame and takes place during smaller and well-defined periods of time.

A particularly interesting communications technology likely to contribute to future M2M applications is IEEE 802.15.4 [3]. According to our reference communications model, IEEE 802.15.4 provides mechanisms at the physical (PHY) and MAC protocol layers and targets wireless communications using low-energy devices. IEEE 802.15.4 LoWPANs support communications at 250Kbps between devices separated by a few tens of meters using the 868/915 MHz or 2.4 GHz frequency bands. Despite such moderately low data rates, the goal of IEEE 802.15.4 is to provide communications for sensing applications where constrained sensing devices are employed and therefore where energy is a scarce resource, a factor that is also expected to characterize many M2M wireless sensing applications in the future. Various sensing platforms are currently available that implement IEEE 802.15.4, one of the most used for research and standardization purposes being the TelosB [6].

Several mechanisms have been proposed to obtain security against jamming attacks, as raising the power of the wireless signal or intensifying the bandwidth usage employing transmission techniques such as spread spectrum or frequency hopping. Spread spectrum techniques such as direct-sequence spread-spectrum (DSSS) or frequency-hoping spread-spectrum (FHSS) make it more difficult to illegally monitor the wireless communication signal and in general offer a good solution to target physical layer security. Spread spectrum techniques may provide defense against jamming, eavesdropping and traffic analysis attacks and consequently contribute to the confidentiality of data communications and to the availability of M2M applications.

A possible approach to detect attacks against the physical layer is to design mechanisms to analyze discrepancies in the systems performance [7], although other mechanisms are required in practice to completely avoid such attacks, as they are in general difficult to circumvent using solely the currently available sensing platforms. Future designs of M2M sensing platforms may on the other end promote physical layer security by implementing techniques such as spread spectrum or adaptable communication technologies employing variable modulation schemes or frequency bands, if such design approaches are found to be economically feasible.

Other approach to physical layer security is in the employment of dedicated sensing devices to perform radio spectrum monitoring. While this approach may be viable for M2M static critical applications such as in industrial control, it will not be viable for deployments where unattended M2M devices are used that do not relate with any security infrastructure.

The unattended nature of M2M devices and applications will facilitate attacks against M2M applications and difficult defensive measures against such attacks. Although cryptography will certainly be required, complex M2M deployments may make its usage more complex. Wireless physical security may provide a path towards the implementation of perfect-secrecy data communications between sensing devices, for example by designing cryptographic schemes combined with channel coding techniques. New mechanisms may be designed to ensure perfect-secrecy for communications where an eavesdropper is unable to obtain any useful information [7]. As observed in current proposals for physical layer security, research and standardization may target the development of new mechanisms for secure data communications at the physical layer by exploiting the random nature of the wireless medium to the advantage of security.

As most M2M applications will not support classic approaches such as the usage of a separate secure channel to distribute cryptographic keys required for normal data communications, new security methods will be required to distribute secret cryptographic information in M2M environment, and physical layer security may be an approach for such purpose. Mechanisms may be designed at the physical layer not only to provide confidentiality but also other requirements as location privacy that are important for some M2M applications. Security mechanisms at the physical layer may be employed in the context of mechanisms designed in a cross-layer fashion and consequently complement upper-layers security mechanisms. The usage of cross-layer security solutions implemented from the physical layer up may also improve the resistance of M2M applications and wireless communications against attackers trying to decipher wireless communications.

Current proposals for addressing physical layer security can be of interest in developing future security mechanisms for M2M wireless communications [8]. Location privacy can be implemented at the physical layer by designing information hiding techniques to prevent unauthorized detection of transmission activities, which could be used to obtain the location of the devices. Techniques such as embedding private messages into a background signal or noise process can achieve this goal [8]. Channel fading can have a positive impact on the secrecy capacity of a channel, and techniques can be developed to explore that property of wireless communications.

Various proposals for wireless security at the physical layer involve the usage of multiple antennas. This enables the usage of diversity techniques at the transmitter and receiver that may contribute to the security of wireless communications, for example by designing single-input multiple-output (SIMO) and multiple-input multiple-output (MIMO) communication mechanisms. Randomization of MIMO transmission coefficients [9] has also been proposed to introduce security, and can provide defense against eavesdropping attacks and consequently contribute to the confidentiality of the M2M application. In general, the usage of frequency and time diversity contributes to information security and information hiding, but techniques involving the usage of radios supporting multiple antennas may not be economically viable on the design of future M2M sensing platforms.

The characteristics of the wireless communications channel may also provide a ground for the design of new physical layer security mechanisms. An example can already be found in radio-frequency (RF) fingerprinting techniques [10]. The idea behind radio frequency fingerprinting is to monitor the evolution of the wireless communications signals and alert for the occurrence of any abnormal communications pattern. Detection techniques based on the characteristics of the communications channel may contribute to improve resistance against eavesdropping and masquerading attacks and to the confidentiality and authentication on M2M applications. New fingerprinting techniques may be designed to use the M2M sensors enabling the application or in the other end to employ a dedicated network of sensors. Research may also target the development of new distributed fingerprinting techniques to leverage the capability of detection of abnormal communication patterns. In either case fingerprint techniques may be very resource demanding and therefore its usage with constrained sensing devices must be carefully evaluated.

An alternative to DSSS or FHSS spread spectrum techniques can be found in error correction coding of wireless communications. One example of this technique is the usage of the Advanced Encryption Standard (AES) to generate scrambling sequences to secure code-division multiple access (CDMA) communications [11]. Correction coding and signal scrambling can provide security against eavesdropping and thus contribute to the confidentiality of wireless M2M communications.

Physical layer security may also be implemented employing directional antennas or by injecting artificial noise in the wireless communications. Directional antennas enable nodes in the coverage range of a jammer to still be able to receive information, in contrast to when using omnidirectional antennas. The usage of directional antennas may improve network capacity, avoid physical jamming attacks and enhance data availability. The problem with this approach will of course reside in that the usage of directional antennas will not be viable in most deployment scenarios. On the other end, the main idea behind artificial noise is to achieve security in the condition that the intruder's channel is noisier than the legitimate receiver's channel. This technique may provide security against eavesdropping of M2M wireless communications and in consequence contribute to the confidentiality of such communications. The disadvantages of this technique are its requirements in terms of energy and the fact that detailed knowledge about the wireless communications environment may be required.

Physical layer security for wireless communications is currently a widely open research area, particularly due to the fact that physical layer security depends heavily on extensive knowledge of security theory and on physical layer architecture expertise. Also, as sensing platforms are currently being designed targeting very low production costs and energy-efficient designs, existing physical security proposals are found to be only theoretical. Nevertheless, the proposals and approaches previously discussed allow us to identify possible avenues for further research and standardization work. The usage of spread spectrum communications is interesting in enabling secure wireless communications, but its usefulness for M2M environments is dependent on the design of energy-efficient and inexpensive devices implementing such communication techniques.

A problem with many of the proposed approaches is their dependence on knowledge about the position and communication patterns employed by the transmitter, receiver and attacker devices. Therefore, such proposals may only be applicable in scenarios where pre-deployment of devices is a reality. Other proposals are found to target point-to-point communications, and the redesign of such approaches for multi-user networking communications may be a challenge to research. The evolution of such proposals to target multi-user environments will also require the study and incorporation of fundamental security aspects for M2M environments such as cooperation and trust between devices [8].

Due to hardware complexity, the low-cost implementation of most physical layer security schemes is not possible with current microelectronics technologies. Most current proposals for security of wireless communications rely on cryptographic techniques employed at the upper layers, but cross-layer security will be of paramount importance and physical layer security may be part of a layered approach. The design of hybrid security mechanisms combining classic cryptography with lower layers security is a promising research approach, and the efficient implementation of such mechanisms at the hardware may enable new sensing platforms with the required mechanisms to secure M2M wireless communications.

Security at the Medium Access Control (MAC) Layer for M2M Wireless communications

Various types of threats can be identified against wireless communications at the MAC layer using the wireless M2M communication technologies previously analyzed. Attacks can target the eavesdropping or modification of MAC-layer management or data messages, or the theft of the identity of M2M appliances or devices. Denial of Service (DoS) attacks can also be conducted against constrained M2M devices. The classic prevention of eavesdropping, modification and identity theft attacks is the usage of cryptographic mechanisms to encrypt communications and authenticate the communication parties. The encryption of management and data frames provides confidentiality of wireless communications, while integrity and authentication are similarly enabled by the usage of appropriate message integrity codes or digital signatures.

Security mechanisms supporting cryptographic operations are traditionally implemented at higher layers of the protocol stack, but we observe that recent wireless standards such as IEEE 802.15.4 have pushed such mechanisms to lower layers of the communications stack, due to the necessity of adopting designs that enable efficient hardware implementations. We may thus expect to observe this trend also in future designs of M2M sensing platforms. The availability of security at lower layers may be beneficial in relation to two aspects. One is that it may allow a more efficient detection and response to attacks targeting resources on M2M constrained devices, such as DoS attacks against higher layer protocols or applications. The other is that it may promote the design of energy-efficient hardware security mechanisms, as currently available in platforms implementing IEEE 802.15.4 [3].

IEEE 802.15.4 provides a good example of the adoption of security mechanisms at lower layers in sensing platforms, as it defines security mechanisms at the MAC layer that may be completely implemented at the hardware. IEEE 802.15.4 adopts AES in the Counter and CBC-MAC (combined or CCM) mode to secure data communications. The combined mode allows for the separate support of encryption and authentication of data frames, two operations required for the support of confidentiality, integrity and authentication of wireless communications between sensing devices. As we have previously observed in the context of physical layer security, the availability of efficient cryptographic mechanisms at the MAC layer using IEEE 802.15.4 or other standards can represent an important contribution towards the adoption of efficient cross-layer security mechanisms. Cryptographic operations implemented at the MAC layer can be reused by higher layers to encrypt, decrypt or authenticate data frames. Similarly, cryptographic keys defined at the MAC layer can be of use in the context of cross-layer key management solutions, so that communication protocols at different layers can share the available key material.

Although AES is adopted to implement fundamental security mechanisms at the MAC layer with IEEE 802.15.4, other technologies such as IEEE 802.16 WiMAX [5] provide more complete security solutions at this layer. Of course, these technologies were designed with different purposes in respect to their coverage area and communications bandwidth, but both standards implement security at lower layers. IEEE 802.16 supports not only confidentiality and authentication of data and management frames but also authentication of devices using X.509 public-key certificates. The fact that authentication with certificates is designed in the context of management operations at the MAC layer is interesting, since it differs from traditional public-key security mechanisms typically implemented at higher layers. Future designs of M2M wireless sensing devices can follow this approach and push the implementation of new security mechanisms to lower layers of the communications stack, as this may also promote the usage of energy-efficient hardware implementations of such mechanisms.

Public-key cryptography and authentication at lower communication layers may also make use of ECC cryptography [12] as an effective alternative to X.509-based public-key cryptography. Nevertheless, the usage of public-key cryptography in M2M environments will require research on how M2M applications may be integrated with existing public-key infrastructures. Identification certificates may be manufacturer-issued and factory-installed and previously associated with the particular M2M sensing device, by tying the certificate to a unique identifier of the device, as its Medium Access Control (MAC) address or Extended Universal Identifier (EUI). This would also facilitate the startup up of a new M2M device on an existing secure M2M wireless network.

Another important aspect that may be targeted in tandem with security at the MAC layer is the implementation of Quality of Service (QoS) mechanisms. As we have previously discussed, such mechanisms may be required to guarantee time-bounded data communications for applications in areas where the replacement of wired communications with wireless best-effort counterparts will not be acceptable. This is the case of industrial control applications currently employing wired sensing devices and that are expected to adopt wireless M2M control applications in the future. Not-so-critical applications may also require time-bounded data communications, even if only for the transportation of critical alarm conditions.

The MAC layer may be designed so that time-constraints for data communications or other QoS requirements are guaranteed and built-in from the start, by following a Time Division Multiple Access (TDMA) design scheme. The design of a QoS-aware MAC layer represents an opportunity also for security, as a deterministic approach to communications may provide a ground for mechanisms that guarantee the delivery of critical security management messages. Intrusion detection mechanisms may also benefit from the availability of deterministic wireless communications in order to detect deviations from normal communication patterns. This TDMA design model is already adopted in 802.16 WiMAX [5] and in recent research approaches targeting the design of wireless communication mechanisms for critical industrial environments [13].

It is also relevant to observe that the usage of time-bounded deterministic wireless communication mechanisms may also help in the implementation of security mechanisms on layers higher than the MAC layer. According to the cross-layer approach of our reference communications stack, security mechanisms implemented at the MAC layer may be used by mechanisms implemented at other layers of the communications stack. For example, key management and intrusion detection may be implemented in a cross-layer fashion using information from the MAC layer in conjugation with information from higher layers or specific applications.

One aspect to be addressed in future designs of security at the MAC layer is that of security of management messages. IEEE 802.16 currently presents this problem, as MAC management messages are not encrypted and in some situations not validated. A similar problem is present in IEEE 802.15.4, where acknowledgment messages lack proper authentication [14]. Such limitations open the door to man-in-the-middle (MITM) attacks, replay attacks, passive and active attacks [4], and is therefore a problem to address in M2M challenging environments. Other aspect that may be targeted is how to address DoS attacks against MAC-layer authentication procedures, since lengthy authentication procedures may encourage such attacks. In general, security at the MAC layer should target all possible angles for attacks, including necessarily the protection of management-related messages.

Security at the Network and Adaptation Layers for M2M Wireless Communications

One important aspect of our reference communications model is the usage of an adaptation layer placed between the MAC and network layers. The purpose of the adaptation layer is to enable the usage of existing standard Internet communication technologies in M2M wireless networks. For example, M2M wireless communications employing IEEE 802.15.4 may use data packets of at most 127 bytes available at the MAC layer, in deep contrast with packets measuring a minimum of 1280 bytes using IPv6 for Internet communications. The adaptation layer therefore plays an important role in the transparent integration of M2M sensing applications with the Internet. The reason for the adoption of smaller packets on M2M wireless communications is that standards such as IEEE 802.15.4 or Bluetooth are designed to support wireless networks where communications are error-prone and devices where energy is scarce. Technologies thus adopt lower communications rates and smaller packets in order to decrease the probability of erroneous transmissions and consequently avoid costly retransmissions.

The integration of M2M applications with the Internet calls for the design of mechanisms to enable interoperability between the two communication domains, as our reference model in Figure 1 illustrates. Communication protocols such as IP, UDP or TCP must be adapted for its transparent transport using M2M wireless communication technologies. These adaptation mechanisms are designed for the adaptation layer of our reference communications stack, where interoperability is achieved by providing the necessary adaptation technologies. The smaller packet size available at the MAC layer requires mechanisms to implement header compression and fragmentation of IPv6 packets, and also the simplification of existing communication standards such as IPv6. This approach can also be applied to security, by adapting existing security mechanisms for its usage in the context of the adaptation layer.

Proposals of mechanisms for the adaptation layer are already emanating from working groups such as the 6LoWPAN (IPv6 over Low Power Personal Area Networks) working group of the IETF [15][16]. 6LoWPAN proposes mechanisms for the adaptation layer to enable the transport of IPv6 packets over LoWPANs. The initial focus of 6LoWPAN was understandably on IEEE 802.15.4 wireless communications, but it is nevertheless expected that adaptation mechanisms will be designed for other technologies in a near future. Candidate technologies are Power Line Communication (PLE) or Bluetooth Low Energy (BLE) [17]. Other than the compression of IP headers and options, 6LoWPAN also addresses the design of fundamental networking mechanisms such as neighbor discovery and address auto-configuration, required for a device to activate its presence on an existing network.

The role of adaptation technologies such as 6LoWPAN in the integration of M2M sensing applications with the Internet is expected to be of major importance, as such applications will allow the leverage of existing standard communications mechanisms with new sensing applications. Technologies such as 6LoWPAN enable end-to-end secure communications between M2M devices and Internet hosts using standard communication technologies.

The availability of secure end-to-end communications at the network or higher layer with other sensing devices or with Internet hosts may enable a much richer integration of M2M sensing applications with the Internet. It may also enable new types of sensing applications where smart objects are able to cooperate remotely and securely using Internet communications. Despite the important role of security in the successful integration of 6LoWPAN networks with the Internet, we observe that it has not been properly addressed in 6LoWPAN, as so far only generic recommendations [18] have been produced.

Security in the context of end-to-end wireless communications with IPv6-enabled M2M sensing devices using the 6LoWPAN adaptation layer is currently an open issue, at least from a standardization perspective. Security was in fact not addressed from the start at 6LoWPAN, probably due to the assumption that it should be addressed in other layers of the communications stack. Security mechanisms may be designed and adopted at the 6LoWPAN adaptation layer with the benefit of enabling security for wireless end-to-end communications at the network layer on M2M applications. In this context, the enabling of security at the network layer for 6LoWPAN communications has been previously proposed [19] and, more recently, a secure interconnection model [20] has been presented in the context of which new compressed security headers for 6LoWPAN are theoretically validated [21]. Compressed security headers for 6LoWPAN may play a role equivalent to the Authentication Header (AH) and Encapsulating Security Payload (ESP) headers in the Internet Security Architecture [22].

Other than the design of compressed security headers for the 6LoWPAN adaptation layer, challenging research opportunities remain in the definition of other security mechanisms required to fully enable network-layer security for 6LoWPAN M2M communications, namely regarding mechanisms for key management and management of security associations. While 6LoWPAN security mechanisms belong to the adaptation and network layers of our reference communications model, auxiliary mechanisms such as key management may be implemented as cross-layer security mechanisms. For example, keying material may be shared between the MAC and adaptation layers and consequently a set of cryptographic keys stored on a sensing device can be used with AES/CCM encryption to obtain link-layer security and also with 6LoWPAN security to guarantee network-layer security in the context of the adaptation layer.

As we previously illustrated in Figure 1, a security gateway can provide an important deployment opportunity for the enabling of security mechanisms designed to secure the wireless M2M communications domain from attacks originated at the Internet. Such mechanisms may include security mechanisms designed to secure and filter end-to-end communications at the network layer or in alternative communications from higher layer protocols or applications. Interesting research opportunities reside in the development of mechanisms to filter end-to-end communications using the adaptation and network layers for M2M wireless communications accordingly to predefined security policies. Similarly, the usage of security in the context of 6LoWPAN may require mechanisms to map security headers and security associations, and to manage cryptographic keys in coordination with M2M wireless devices. Communications for higher layers protocols or applications may also benefit from the availability of a security gateway, as similar mechanisms or application proxies may be designed to provide security according to security requirements of particular applications. Other interesting research approach may be to design such security mechanisms asymmetrically, so that the security gateway assumes most of the load of supporting heavy security operations. This design approach would alleviate constrained M2M sensing devices from heavy security or cryptographic operations in deployments where security gateways are devices with more computational power and without energy restrictions.

Security at the Transport Layer for M2M Wireless Communications

Although the design of transport layer protocols may enable reliable wireless communications between wireless sensing devices, the constraints of such devices and the characteristics of the available wireless communication technologies provide challenges at the design of transport layer solutions. Given the scarcity of energy on many sensing devices, care must be taken to design efficient transport layer solutions that implement mechanisms such as congestion control and reliability with minimal overhead and retransmissions [23].

The adoption of transport layer communication mechanisms for sensor networks is currently open to debate, as many consider it goes against the simplicity expected from sensing devices and applications. Nevertheless, sensor networks have evolved to a status where congestion control and reliability mechanisms may be designed and incorporated at individual nodes. We may expect that transport-layer mechanisms may be helpful in M2M environment where reliable communications are required. The requirement for the support of reliable message delivery and congestion control will motivate the design of transport layer mechanisms for M2M wireless networks.

Although it is currently unclear if a single standard transport communications protocol is going to be adopted for sensor networks, it is widely recognized that traditional transport layer protocols such as the Transmission Control Protocol (TCP) are not suitable for networks of constrained wireless sensing devices. This is due mainly to the fact that such networks have characteristics that are different from traditional wired and wireless networks. Several proposals currently exist of transport layer protocols for sensor networks targeting reliable message delivery, congestion control and energy efficiency. All such properties are absent from network-layer or lower layer communications and may therefore be introduced at the transport or application layers. Distributed TCP Caching (DTCP) [24] proposes the optimization of TCP by requiring that intermediate nodes in a communications path cache intermediate segments with the goal of minimizing the usage of end-to-end recovery procedures. Although in general TCP/IP is considered too heavy for most sensing platforms, such optimizations may enable its usage with specific M2M devices that are less resource-constrained. The approach of designing a generic transport layer protocol for wireless communications in sensing applications is also encountered in the Sensor Transmission Control Protocol (STCP) [25] proposal. STCP provides variable reliability, congestion detection and avoidance. Other proposal in the same vein is Adjustable Parallel TCP (AP-TCP) [26], a scheme to control the aggregate throughput of various parallel TCP data flows.

With the Pump Slowly Fetch Quickly (PSFQ) [27] proposal sensing devices that experience the loss of a data segment can recover that segment quickly from neighbor devices. PSFQ implements negative acknowledgments signaling a loss of a segment rather than positive acknowledgments, with the goal of reducing signal overhead. PSFQ is designed for sensing applications requiring reliable delivery of data messages but does not implement any active congestion control scheme. Other proposal is the Reliable Multi-Segment Transport (RMST) [28] protocol, designed to operate on top of direct diffusion communications. RMST adds fragmentation and reassembly of segments and reliable message delivery to such communications, offering the option of end-to-end end or hop-by-hop recovery of lost messages. RMST recovery mechanisms may be combined with MAC-level recovery messages, providing an example of the benefits of cross-layer interactions for the purpose of reliable wireless communications.

Proposals such as DTCP and PSFQ may be difficult to implement due to requirements on buffer space usage on sensing devices. Other problem with such proposals is that they are related with specific traffic patterns or applications, therefore lacking generality for its application in diverse types of M2M applications. A challenging path to research and standardization will therefore be the design of generic and standard transport layer protocols that are able to provide congestion control and ensure reliable delivery of messages for end-to-end communications between M2M devices and between such devices and backend or Internet hosts, considering its usage on Internet-integrated M2M sensing applications.

Current proposals for security with end-to-end communications at the transport layer for M2M wireless sensing environments mostly target the modification of the SSL (Secure Sockets Layer) protocol. SSNAIL [29] provides a light-weighted version of SSL to be supported by Internet hosts and sensing devices. Although it succeeds in demonstrating the usability of SSL in constrained sensing devices, SSNAIL requires the modification of the SSL stack in all communicating devices. Sizzle [30] also provides end-to-end security using SSL, but only for communications between Internet hosts and a security gateway, with communications being translated to a proprietary communications protocol in the wireless domain.

On the one side we observe that SSL can be successfully adopted to secure transport layer communications in M2M wireless environments, provided that applications are designed to explicitly require it and SSL stacks optimized for constrained environments are available, while on the other side we observe that the usage of security mechanisms with fixed configurations may not be the appropriate approach to support security at the transport layer for M2M applications. In fact, security mechanisms should be adaptable to the characteristics and security requirements of particular sensing applications and aspects such as the cryptographic algorithms employed and relevant configuration parameters such as cryptographic key size and frequency of key refreshment deeply influence the lifetime of sensing applications and resource-constrained devices. Security mechanisms should therefore allow the establishment of acceptable compromises between resources required for performing security operations and the security level required for a particular application.

Despite the limitations previously addressed, the design of security at the transport layer using SSL-optimized stacks provides challenging opportunities to research and standardization. As with mechanisms at other layers of the communications stack, SSL security can be designed to benefit from cross-layer interactions. Authentication and key management for sensing devices using SSL can be implemented by cross-layer mechanisms employing cryptographic keys shared also with the MAC, adaptation and application layers. Security mechanisms designed for other communication layers, for example authentication using X.509 certificates in IEEE 802.16, could be reused to support authentication at the transport layer using SSL.

Security at the Application Layer for M2M Wireless Communications

Work on research and standardization of application-layer communication technologies for wireless sensing applications is very recent. The main reference in this area currently is the Constrained Application Protocol (CoAP) [31]. CoAP is currently under active developed at the Constrained RESTful Environments (CoRE) [32] working group of the IETF with the goal of enabling RESTful web communications in applications supported by wireless sensing devices. CoAP is related to HTTP since the purpose of both is to implement the Representational State Transfer (REST) architecture of the web, with CoAP communications being designed to run over UDP on 6LoWPAN-enabled sensing networks. CoAP communications therefore use the 6LoWPAN communication mechanisms available at the adaptation layer, and CoAP will allow accessing resources on CoAP servers running on constrained devices using a subset of HTTP's methods and a similar Universal Resource Identifier (URI) scheme to identify available resources.

The current specification of the CoAP protocol [31] proposes the usage of the Datagram Transport Layer Security (DTLS) [33] protocol to secure wireless communications at the application layer. DLTS adapts the Transport Layer Security (TLS) [34] protocol for connectionless datagram communications using UDP. As TLS is in practice an evolution of SSL, the optimizations designed at DTLS in practice fulfil the same role as the previously discussed proposals to modify SSL for constrained environments. Given its standardized status, DTLS is a strong candidate technology to provide security at the transport layer for M2M 6LoWPAN-enabled wireless applications, and consequently to enable end-to-end secure application layer communications for Internet-integrated M2M sensing applications.

Three security modes are currently proposed for CoAP, in order to enable its usage in deployments with different characteristics in terms of the availability of security infrastructures and of resources on constrained sensing devices. Environments where public-key cryptography is considered unviable or where the deployment of sensing devices facilitates the predefinition of cryptographic keys may use a security mode identified as *PreSharedKey* [31]. In this security mode a device stores the cryptographic keys required to secure communications with other M2M devices or group of devices. In this mode a sensing device stores one key for all its communication partners. The fundamental security properties of confidentiality, integrity and data origin authentication are provided in this mode using AES in the CCM mode, again a design choice motivated by the availability of this cryptographic standard in sensing platforms that implement IEEE 802.15.4.

For M2M sensing applications where the usage of public-key cryptography is viable and desirable, CoAP proposes the usage of the *RawPublicKey* or *Certificates* modes. The two security modes employ ECC public-key cryptography [12] to support authentication of devices and messages and also key agreement. In the *RawPublicKey* mode a sensing device proves its identity using one or various ECC public-keys and uses ECC to perform authentication of other devices. In this mode a device must store the identity and key for each node it communicates with, since a certification chain is not available for authentication purposes. As with the previous security mode,

AES/CCM is used to guarantee confidentiality, integrity and data origin authentication for DTLS communications. During ECC public-key authentication, the client and server agree on a shared premaster secret for DTLS, from which further secrets are derived to secure communications.

In deployment scenarios where a Certification Authority (CA) may be used, trust root anchors are available for certificate validation purposes and CoAP security may employ the *Certificates* security mode. This mode may also use ECC public-key authentication, in this case using identity information extracted from certificates. Since sensing devices may not support ECC encryption, it is currently proposed that this security mode also allows the usage of RSA public keys to perform authentication and pre-shared key agreement. In this alternative scenario confidentiality may also be guaranteed using AES in the CBC mode and integrity using SHA.

The previously described proposed security modes for CoAP lack experimental validation, given that no implementations of CoAP with security currently exist. Other problem is that current sensing platforms such as the TelosB [6] may not possess the required memory or other resources to effectively support the proposed CoAP, since a sensing device would be required to support an operating system such as TinyOS [35] or Contiki [36], together with the required CoAP code plus security with DTLS. Research and experimental validation work will therefore be required to validate the proposed CoAP security modes and to provide guidance information for the evolution of M2M sensing platforms towards the successful adoption of security at the transport and application layers using CoAP.

Other than the integration of security following the same approach as TLS by adopting DTLS to secure CoAP UDP communications, an interesting research approach would be to design security by integrating it at the application protocol itself. CoAP could therefore be extended with the required options to support authentication and key negotiation between CoAP client and server. Although such an approach presents the disadvantage of requiring the design of security for each particular application-layer protocol, it would benefit performance and the implementation of optimized networking stacks. Two other challenges in enabling security at the application layer in the context of Internet-integrated M2M sensing applications appear in guaranteeing end-to-end security between M2M sensing devices and Internet hosts and in the support of secure multicast group communications [37], as we discuss in greater detail in the next section.

Regarding communications at the application layer, we may also analyse how web-scripting languages such as XML or similar technologies may be useful in M2M wireless environments. The interest in such technologies could be in that they may facilitate the exchange of information between M2M devices and the communications between M2M devices and end users or applications. Current proposals for the adaptation of XML to constrained environments are found in BiTXml [38] and M2MXML [39]. The main goal of BiTXml is to standardize how commands and control information is exchanged for each specific target of M2M communication. Similarly, M2MXML seeks the development of an open-standard XML based protocol for M2M communications with a primary focus on simplicity.

Further research opportunities therefore appear in the analysis of how security can target such proposals or similar technologies, again considering its usage in the context of Internet-integrated M2M wireless sensing applications.

4 Cross-Layer Approaches and Future Directions for Security on Internet-Integrated M2M Wireless Sensing Applications

Despite the approaches previously analyzed to target security for communication technologies at particular layers of our reference communications stack, other important security aspects will require either a cross-layer approach or the design of mechanisms targeting security in the context of the integration of M2M wireless sensing applications with the Internet. Such mechanisms may provide network and application security, while requirements will also exist targeting cross-layer issues such as privacy, trust and anonymity, among others. We also realize that although technologies such as 6LoWPAN and CoAP are currently being designed to target the integration of sensing applications with the Internet, security is either lacking or not completely specified for such technologies. Research and standardization challenges will therefore appear regarding the validation of existing security proposals considering its usage with real sensing platforms. Similarly, new security mechanisms will be required for existing M2M communication technologies.

The integration of M2M sensing applications with the Internet will raise new challenges concerning various security aspects and requirements. Internet-integrated sensing applications will require proper authentication of users or devices requesting access to data or resources. At the same time, the identity of users must be protected and security will also be required for wireless communications on M2M environments. In the event of failure of any security mechanism, changes to systems and data should be traceable and reparable, therefore requiring the usage of appropriate audit mechanisms. System availability will also be required, since in most applications it will be necessary to ensure that authorized users or devices can use their access privileges at any time, and therefore critical M2M applications must be resilient to DoS attacks. Other than the classic security requirements that may be guaranteed by the usage of appropriate cryptographic mechanisms, aspects such as trust, privacy, liability and anonymity will require particular attention in the context of Internet-integrated M2M wireless sensing applications.

Given that most of the envisioned M2M applications will either require or benefit from the availability of direct or indirect communications with the Internet, we find it necessary to analyze how such communications may be guaranteed and how security can be addressed in this context. Figure 3 illustrates a reference architecture for the secure integration of M2M sensing applications with the Internet, where we consider the usage of the previously discussed communication and security technologies. We must note that, other than to try to illustrate a closed perception of how M2M sensing applications may be integrated with the Internet, our goal is to contextualize the usage of the previously discussed communications and security technologies to enable secure Internet-integrated M2M wireless applications.

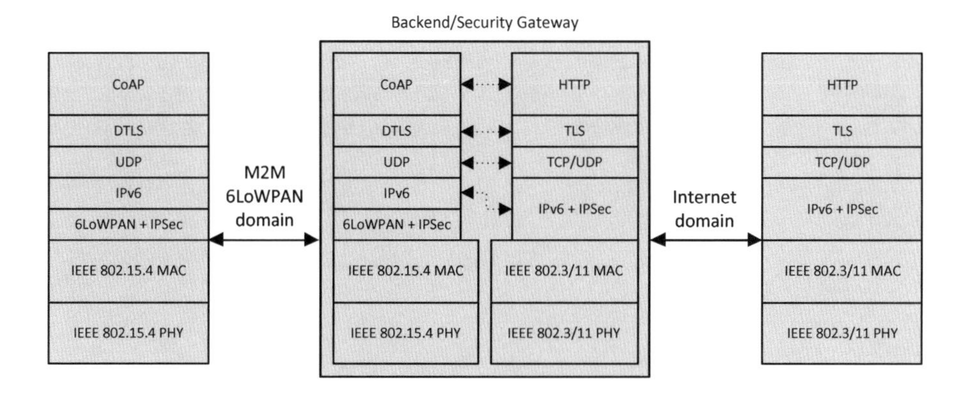

Fig. 3. A reference architecture for end-to-end security in the context of Internet-integrated M2M applications

Our reference architecture in Figure 3 illustrates the usage of a backend or security gateway device in supporting a set of communications and security technologies at the various layers. Such technologies are required for the secure interconnection of the two communication domains previously illustrated in Figure 1, namely the Internet domain and the M2M wireless domain employing 6LoWPAN to support communications at the network and upper layers. This reference architecture again considers the usage of a security gateway in supporting direct or indirect communications between the two domains, while providing good implementation points for specialized security mechanisms. The security gateway is currently a proposed and well-accepted abstraction to guarantee security for Internet-integrated sensing applications, and is proposed in current specifications of technologies such as 6LoWPAN, CoAP [31] or BitXML [38]. In particular, in 6LoWPAN this abstraction materializes itself in a device identified as the 6LoWPAN Border Router (6LBR) in current specifications [15][16]. The architecture illustrated in Figure 3 considers the usage of the communication and security technologies already discussed and facilitate our discussion on the remaining issues that may be addressed in a cross-layer fashion.

Communications between devices on different domains can take place from the network layer up, while for the PHY and MAC layers in Figure 3 we illustrate the usage of IEEE 802.15.4 to support M2M wireless communications. Given it representativeness, 6LoWPAN enables the adaptation layer and therefore provides the required support for network layer communications on the M2M domain. Similarly, UDP and CoAP provide transport and application layer communications, respectively.

Security for network layer communications can be enabled on the Internet domain by using the standardized Internet Security Architecture [22], while on the M2M domain compressed security headers can be designed and adopted at the adaptation layer, as we previously discussed. Until now, there is no solution to ensure complete end-to-end secure communications at the network layer between M2M wireless devices and backend or Internet hosts. In this context, research may target the design of mechanisms to support security header mapping between the two communication domains, meaning that AH and ESP IPSec headers may be mapped to compressed 6LoWPAN security headers on the security gateway. Such mapping mechanisms may be integrated at the networking stack to directly support interfaces with the two communication domains. Other than the mapping of security headers, adaptation and network layer security mechanisms should interface with key management, since end-to-end security must not be broken at the gateway. Possible approaches for this purpose would be to perform tunneling of 6LoWPAN secured packets over IPSec or to protect IPSec and 6LoWPAN security headers using two different cryptographic keys, so that the security gateway and the destination host may separately decrypt and verify the packet.

End-to-end secure communications can also take place at the transport and application layers, using DTLS to secure CoAP communications over UDP on the M2M 6LoWPAN domain. Until now, there is also no solution to implement end-to-end secure communications for web communications between Internet HTTP hosts and M2M CoAP devices. As DTLS introduces modifications to TLS to implement reliable session key negotiation and additional measures to verify exchanges messages, the mapping of DTLS datagrams to TLS sessions is not trivial. Research may therefore address the design of mechanisms to implement a proxy device to perform DTLS-TLS translation at the security gateway. Translation mechanisms must support not only mapping of the DTLS-TLS record layer headers but also of headers from other TLS protocols, namely the handshake, alert and spec protocols. As an HTTP client on the Internet should be able to connect securely to a resource available via CoAP on a constrained M2M device, this proxy must also support HTTP-CoAP protocol translation. Alternative design approaches for such mechanisms [37] would be to implement tunneling of DTLS datagrams over TLS, as in the TLS-DTLS tunnel [40] proposal, or having the sender to encrypt the same packet (CoAP or HTTP) with two separate keys, one for the proxy and the other for the destination device, as in the Integrated Transport Layer Security (ITLS) [41] proposal.

Regarding the usage of CoAP with DTLS security, other issue that will require attention from research is how to address security when multicast communications are employed. Multicast messages may be sent using CoAP to simultaneously manipulate resources in a group of devices, but the problem is that DTLS does not support multicast. Therefore, when multicast communications are used to enable a backend or Internet host to access multiple CoAP resources, no end-to-end security may be provided. A possible approach to address this issue would be to also use a proxy entity to translate an HTTP request to a CoAP multicast request, but this would require a mechanism that allows the proxy to recognize the request as a multicast request, something currently not possible using HTTP. In this scenario the proxy would be required to handle security for all devices belonging to the multicast group. Whatever the solution may be, it would require integration with key management, as group keys may be employed to enable secure communications with the devices belonging to the multicast group. Alternative key management approaches may also be employed, for example by maintaining a separate session key for each device belonging to the multicast group. We also have to note that regarding security for multicast communications, IP Security can provide a better solution, as currently there is a multicast extension for IPSec [42] that may be adapted for M2M environments.

One mechanism that is cornerstone of effective security solutions is key management, as it enables not only the establishment of the initial cryptographic that keys that a host requires to participate in an existing secure network, but also the periodic renegotiation of such keys, a requirement for the maintenance of security. Considering the previously discussed approaches for end-to-end security, the design of appropriate key management solutions in the context of the reference architecture illustrated in Figure 3 is of vital importance. In fact, no standard mechanisms currently exist that are appropriate for this purpose. Such key management solutions must be designed to pair with the end-to-end communication technologies currently being proposed and standardization for M2M wireless environments. One approach would be to simplify the Internet Key Exchange (IKE) [43] protocol adopted for IPSec in order to enable its usage in constrained M2M wireless sensing devices, such as proposed in Lightweight IKEv2 [44]. Other approach may be to design a new key management protocol that is able to adapt to different capabilities of sensing devices, and consequently support different cryptographic primitives according to the sensing platforms deployed to support a particular M2M sensing application [45]. Irrespective of the approach, new key management solutions may be designed so that the security gateway or a similar device supports the most demanding operations. Also, full solutions are required that are able to handle key negotiations for end-to-end security, possibly allowing the security gateway to participate in such key negotiations as a trusted intermediate entity.

Other than the design of security for end-to-end communications, other security issues are expected to motivate a true paradigm shift for its addressing in the context of M2M wireless communications. For many security issues we may also expect that decentralized and distributed approaches will be required, with new security mechanisms implemented in the context of the architecture illustrated in Figure 3. The decentralization of security mechanisms may be required as in many usage scenarios a security infrastructure may not be available. As a consequence, in many situations M2M devices will be unable to derive definitive conclusions about the identity or intents of other entities, and consequently security mechanisms may be designed to implement practical compromises between the enforcement of definitive security controls and the acceptance of controlled risks.

Trust and privacy will be of cornerstone importance in most M2M applications. Distributed and autonomous trust management and verification mechanisms may be required to support autonomous identification and authorization of M2M devices. Two interrelated aspects are privacy and liability, two requirements that many applications will be required to control. In many contexts the user will require the control of how much personal information is exposed to third parties while, on the other end, sensing applications may require that a subset of that information is available if required, for liability purposes. Mechanisms to protect the users privacy may be based on anonymity models, and research may target the design and usage of light weighted formal anonymity models such as k-anonymity [46] on M2M environments. Other approaches may be followed, for example using mechanisms for data transformation and randomization.

The identification and authorization of devices in autonomous M2M environments will also present serious challenges to research and standardization. Authentication mechanisms may be designed to function side-by-side with distributed trust management and identity verification mechanisms. The unattended nature of M2M communications will require that any two devices are able to configure a trust relationship with each other. As authentication is related with identification, M2M applications may require that devices use some form of secure identifier that allows relating the identification of the device or application with secret cryptographic material. Research can consider currently proposed technologies such as IEEE 802.1AR [47] for X.509-based certified secure identifiers as a starting point, or on the other end develop mechanisms to allow M2M devices to use self-generated uncertified secure identifiers. As M2M systems will require that privacy be balanced against disclosure of information, new authentication mechanisms making use of appropriate secure identifiers and incorporating privacy-preserving mechanisms may be required. Three-way autonomous certification technologies can also provide a solution for authentication in the absence of an official certification authority.

Mechanisms will also be required to enable the secure boot up of a device on an existing secure M2M environment. Local state control via a secure boot process may be enforced, enabling the establishment of a trusted environment that provides a hardware security anchor and a root of trust. The Trusted Computing Group (TCG) [48] has proposed autonomous and remote validation models that can be useful in the design of similar mechanisms for M2M environments. Autonomous validation using smart cards may also be a choice in some environments but presents the problem of requiring costly in-field replacements of compromised devices. A possible road for research is also the design of mechanisms for semiautonomous validation of devices. Semiautonomous validation combines local validation with remote validation, allowing a device to validate trust for another device and communicate with a trusted third-party only when required or possible. Therefore, other than the validation of new devices starting up in an existing network, distributed semiautonomous trust verification mechanisms could also be used for distributed trust management purposes.

Engineering and research challenges are also expected to appear in the design of new sensing platforms to support wireless M2M applications in the future. As with existing platforms implementing security mechanisms in the context of IEEE 802.15.4, future designs may employ a security co-processor to enable efficient cryptographic operations. This would enable energy-efficient cryptographic operations such as ECC public-key cryptography. Depending on the application and costs associated with the production of such devices, more complete hardware-based security solutions can also be employed, similarly to existing proposals such as the Trustchip [49] system. Sensing platforms for critical applications may also be required to support secure storage mechanisms to store security-critical data. Such mechanisms may be designed with characteristics similar to the Trusted Platform Module (TPM) proposed by the TCG [48] working group. A module with such characteristics would enable the biding of the device identification to secret cryptographic information with guaranteed security. As cost may prohibit the adoption of such solutions, research may in alternative target the development of software secure storage solutions.

While realizing that security issues other than those discussed in this chapter will require future research and standardization work, one final word goes to the usage of intrusion detection mechanisms on M2M environments. Intrusion detection will be of much relevant in most M2M wireless applications. Distributed intrusion detection mechanisms will be required to enable detection of failing or compromised devices, avoid communications with such devices and disable all previously established trust relationships. For M2M applications supporting autonomous M2M communications, research can target the development of autonomous and cooperative methods allowing the early detection of node compromises [50]. As with other security mechanisms, the security gateway may play an important role in the support of such mechanisms, by gathering and correlation status information from various sources in order to detect intrusions and malfunctions. The most computationally demanding operations related with distributed intrusion detection may be supported by the security gateway, with sensing devices probing and reporting relevant information. This may be a viable approach for environments where sensing devices are deployed in a controlled fashion and intrusions are critical for the application at hand, as with critical industrial M2M wireless applications.

5 Conclusion

Wireless communications will play a fundamental role in the enabling of most of the applications currently envisioned for M2M pervasive environments. Security will certainly be a fundamental enabling factor of most applications, and appropriate mechanisms will be required to protect M2M applications, communications and users.

The design of security for wireless M2M systems may require new approaches due to various characteristics envisioned for M2M applications and devices. One is that users will require guarantees in terms of privacy and liability, given the importance of such aspects for the social acceptance of most of the envisioned M2M applications. Security mechanisms will also be required to support heterogeneous communication technologies, protocols, and devices. Also, other important aspect is that security must cope with autonomous communications between M2M devices. New approaches for security may explore the autonomous and cooperative nature of M2M communications and applications and, given that in many situations a security infrastructure will not be available, decentralized and distributed approaches for security will be required.

Cross-layer design approaches of security mechanisms may guarantee the necessary optimizations to enable the usage of a given security solution with different classes of sensing platforms, and the same applies to the design of security mechanisms in an asymmetric fashion. As previously observed, the current work on the standardization of communications technologies for wireless M2M environments also provides the ground for research and design of new security solutions.

A security architecture adopted for wireless M2M environments must support the integration of M2M sensing applications with the Internet, while also supporting flexible security mechanisms able to enforce acceptable compromises between aspects such as the usage of resources on constrained devices, the enforcement of security requirements and controls, and the acceptance of controlled risks for particular sensing applications. As previously discussed, security mechanisms in the context of this architecture may target on the one side the protection of end-to-end communications with sensing devices at a particular protocol layer, and on the other side the enforcement of fundamental security requirements such as anonymity, privacy and liability.

References

[1] Kranenburg, R., et al.: The Internet of Things. Draft paper Prepared for the 1st Berlin Symposium on Internet and Society, Berlin, Germany (October 2011)

[2] OnStar, http://en.wikipedia.org/wiki/OnStar (accessed September 2012)

[3] Wireless Medium Access Control (MAC) and Physical Layer (PHY) Specifications for Low-Rate Wireless Personal Area Networks (WPANs), IEEE std. 802.15.4 (2006)

[4] An assessment of threats of the Physical and MAC Address Layers in WiMAX/802.16

[5] IEEE Standard 802.16: A Technical Overview of the WirelessMAN Air Interface for Broadband Wireless Access, http://ieee802.org/16/docs/02/C80216-02_05.pdf (accessed September 2012)

[6] TelosB Mote Platform, http://www.xbow.com/pdf/Telos_PR.pdf (accessed September 2012)

[7] Renzo, M., Debbah, M.: Wireless Physical-Layer Security: The Challenges Ahead. In: The 2009 International Conference on Advanced Technologies for Communications, Invited Paper, Haiphong, Vietnam (2009)

[8] Sperandio, C., Flikkema, P.: Wireless Physical-Layer Security via Transmit Precoding Over Dispersive Channels: Optimum Linear Eavesdropping. In: Proceedings of MILCOM 2002, Anaheim, California, USA (2002)

[9] Yi, S., Shih, C.: Physical Layer Security in Wireless Networks: A Tutorial. In: IEEE Wireless Communications (April 2011)

[10] Abbasi-Moghadam, D., Vakili, V., Falahati, A.: Combination of Turbo Coding and Cryptography in Non-Geo Satellite Communication Systems. In: Proceedings of IST 2008, International Symposium on Telecommunications, Tehran, Iran (August 2008)

[11] Li, T., et al.: Physical Layer Built-in Security Analysis and Enhancements of CDMA Systems. In: Proceedings of MILCOM 2005, IEEE Military Communications 2005, New Jeysey, USA (October 2005)

[12] SECG-Elliptic Curve Cryptography-SEC1, http://www.secg.org (accessed September 2012)

[13] Ginseng: Performance control in Wireless Sensor Networks, http://www.zigbee.org/portals/0/documents/events/2011_10_28_wsn/19-GINSENG.pdf (accessed September 2012)

[14] Sastry, N., Wagner, D.: Security Considerations for IEEE 802.15.4 networks. In: Proceedings of the 3rd ACM Workshop on Wireless Security, Philadelphia, USA (October 2004)

[15] Kushalnagar, N., et al.: IPv6 over Low-Power Wireless Personal Area Networks (6LoWPANs): Overview, Assumptions, Problem Statement, and Goals. RFC 4919 (2007)

[16] Hui, J., Thubert, P.: Compression Format for IPv6 Datagrams over IEEE 802.15.4-Based Networks. RFC 6282 (2011)

[17] Nieminen, N., et al.: Transmission of IPv6 Packets over Bluetooth Low Energy (November 2011), http://tools.ietf.org/html/draft-ietf-6lowpan-btle-04

[18] Park, S., et al.: IPv6 over Low Power WPAN Security Analysis (March 2011), http://tools.ietf.org/html/draft-daniel-6lowpan-security-analysis-05

[19] Granjal, J., Silva, R., Monteiro, E., Silva, J.S., Boavida, F.: Why is IPSec a viable option for wireless sensor networks. In: Proceedings of the 5th IEEE International Conference on Mobile Ad Hoc and Sensor Systems (MASS 2008), Atlanta, USA (2008), doi:10.1109/MAHSS.2008.4660130

[20] Granjal, J., Monteiro, E., Silva, J.S.: A secure interconnection model for IPv6 enabled wireless sensor networks. In: Proceedings of IFIP Wireless Days 2010, Venice, Italy (2010), doi:10.1109/WD.2010.5657743

[21] Granjal, J., Monteiro, E., Silva, J.S.: Enabling Network-Layer Security on IPv6 Wireless Sensor Networks. In: Proceedings of IEEE GLOBECOM 2010, Miami, USA (2010), doi:10.1109/GLOCOM.2010.5684293

[22] Kent, S., Seo, K.: Security Architecture for the Internet Protocol. RFC 4301 (2005)

[23] Jones, J., Atiquzzaman, M.: Transport Protocols for Wireless Sensor Networks: State-of-the-Art and Future Directions. International Journal of Distributed Sensor Networks 3, 119–133 (2007), doi:10.1080/15501320601069861

[24] Dunkels A, Alonso J, Voigt T. Making TCP/IP Viable for Wireless Sensor Networks. In: Proceedings of the First European Workshop on Wireless Sensor Networks (EWSN 2004), Berlin, Germany (February 2004)

[25] Iyer, Y., Gandham, S., Venkatesan, S.: STCP: A Generic Transport Layer Protocol for Wireless Sensor Networks. In: Proceedings of the 14th IEEE Intl. Conf. on Computer Communications and Networks (ICCCN 2005), San Diego, California, USA (October 2005)

[26] Yusung, K., Kilnam, C., Lisong, X.: Adjusting the Aggregate Throughput of Parallel TCP flows without Central Coordination. IEICE Transactions on Communications (2010)

[27] Wan, C., Campbell, A., Krishnamurthy, L.: PSFQ: A Reliable Transport Protocol for Wireless Sensor Networks. In: Proceedings of the 1st ACM International Workshop on Wireless Sensor Networks and Applications, Atlanta, USA (2002)

[28] Stann, F., Heidemann, J.: RMST: Reliable Data Transport in Sensor Networks. In: Proceedings of the IEEE International Workshop on Sensor Net Protocols and Applications (SNPA), Anchorage, Alaska, USA (May 2003)

[29] Jung, et al.: SSL-based Lightweight Security of IP-based Wireless Sensor Networks. In: Proceedings of the International Conference on Advanced Information Networking and Applications Workshop (AINA 2009), Bradford, UK (2009)

[30] Gupta, V., et al.: Sizzle: A standards-based end-to-end security architecture for the embedded Internet. In: Proceedings of the Third IEEE International Conference on Pervasive Computing for the Embedded Internet (PERCOM 2005), Hawaii, USA (2005)
[31] Shelby, Z., Hartkle, K., Bormann, C., Frank, B.: Constrained Application Protocol (CoAP) (June 2012), http://www.ietf.org/id/draft-ietf-core-coap-10.txt
[32] Constrained RESTful Environments (core), http://datatracker.ietf.org/wg/core/charter/ (accessed September 2012)
[33] Rescorla, E., Modadugu, N.: Datagram Transport Layer Security, RFC 4347 (2006)
[34] Dierks, T., Rescorla, E.: The Transport Layer Security (TLS) Protocol Version 1.2, RFC 5246 (2008)
[35] TinyOS Operating System, http://www.tinyos.net/ (accessed September 2012)
[36] The Contiki OS, http://www.contiki-os.org/ (accessed September 2012)
[37] Brachmann, M., Garcia-Morchon, O., Kirsche, M.: Security for Practical CoAP Applications: Issues and Solution Approaches. In: GI/ITG KuVS Fachgesprch Sensornetze (FGSN). Universitt Stuttgart (2011)
[38] BiTXml, The ultimate m2m communication protocol, http://www.bitxml.org/ (accessed September 2012)
[39] M2MXML, Open-standard XML based protocol for Machine-To-Machine (M2M) communications, http://m2mxml.sourceforge.net/ (accessed September 2012)
[40] Reardon, J.: Improving Tor using a TCP-over-DTLS Tunnel. Master Thesis, University of Waterloo, Canada (2008)
[41] Kwon, E., Cho, Y., Chai, K.: Integrated Transport Layer Security: End-to-End Security Model between WTLS and TLS. In: Proceedings of the 15th International Conference on Information Networking (ICOIN 2001), Beppu City, Oita, Japan (2001)
[42] Weis, B., Gross, G., Ignjatic, D.: Multicast Extensions to the Security Architecture for the Internet Protocol, RFC 5347 (2008)
[43] Kaufman, C.: Internet Key Exchange (IKEv2) Protocol, RFC 3286 (2005)
[44] Raza, S., Voigt, T., Jutvik, V.: Lightweight IKEv2: A Key Management Solution for both the Compressed IPsec and the IEEE 802.15.4 Security. Position Paper, Workshop on Smart Object Security, Paris, France (March 2012)
[45] Bianchi, G., et al.: Flexible key exchange negotiation for wireless sensor networks. In: Proceedings of the Fifth ACM International Workshop on Wireless Network Testbeds, Experimental Evaluation and Characterization (WiNTECH 2010), Chicago, USA (2010)
[46] Sweeney, L.: k-Anonymity: A Model for Protecting Privacy. International Journal of Uncertainty, Fuzziness & Knowledge-Based Systems 10(5) (October 2002)
[47] IEEE 802.1AR, Standard for Local and Metropolitan Area Networks: Secure Device Identity, http://www.ieee802.org/1/pages/802.1ar.html (accessed September 2012)
[48] Trusted Computing Group, http://www.trustedcomputinggroup.org/ (accessed September 2012)
[49] TrustChip Mobile Device Security, http://www.koolspan.com/trustchip/ (accessed September 2012)
[50] Lu, R., et al.: GRS: The green, reliability, and security of emerging machine to machine communications. IEEE Communications Magazine 49(4) (2011)

Security and Privacy in Wireless Body Area Networks for Health Care Applications

Saeideh Sadat Javadi and M.A. Razzaque

Faculty of Computer Science & Information Systems,
University of Technology,
Malaysia
javadi.saeideh@gmail.com,
marazzaque@utm.my

Abstract. Wireless Body Area Sensor Networks (WBANs) are becoming more and more popular and have shown great potential in real –time monitoring of the human body. With the promise of cost effective, unobtrusive, and unsupervised continuous monitoring, WBANs have attracted a wide range of monitoring applications such as healthcare, sport activity and rehabilitation systems. However, in using the advantage of WBANs, a number of challenging issues should be resolved. Besides open issues in WBANs such as standardization, energy efficiency and Quality of Service (QoS), security and privacy issues are one of the major concerns. Since these wearable systems control life-critical data, they must be secure. Nevertheless, addressing security in these systems faces some difficulties. WBANs inherit most of the well-known security challenges from Wireless Sensor Networks (WSN). However, typical characteristics of WBANs, such as severe resource constraints and harsh environmental conditions, pose additional unique challenges for security and privacy support. In this chapter, we will survey major security and privacy issues and potential attacks in WBANs. In addition, we will explain an unsolved quality of service problem which has great potential to pose a serious security issues in WBANs, and then we discuss a potential future direction.

Keywords: Health Care, Privacy, Quality of Service, Security, Wireless Body Area Networks, Wireless Sensor Networks.

1 Introduction

As wireless devices and sensors are increasingly deployed on people, researchers have begun to focus on Wireless Body Area Networks (WBANs). Applications of wireless body sensor networks include healthcare, entertainment, sport activity and personal assistance, in which sensors collect physiological and activity data from people and their environments. A simple WBAN application scenario has been shown in Figure 1.

S. Khan and A.-S.K. Pathan (Eds.): *Wireless Networks and Security*, SCT, pp. 165–187.
DOI: 10.1007/978-3-642-36169-2_6

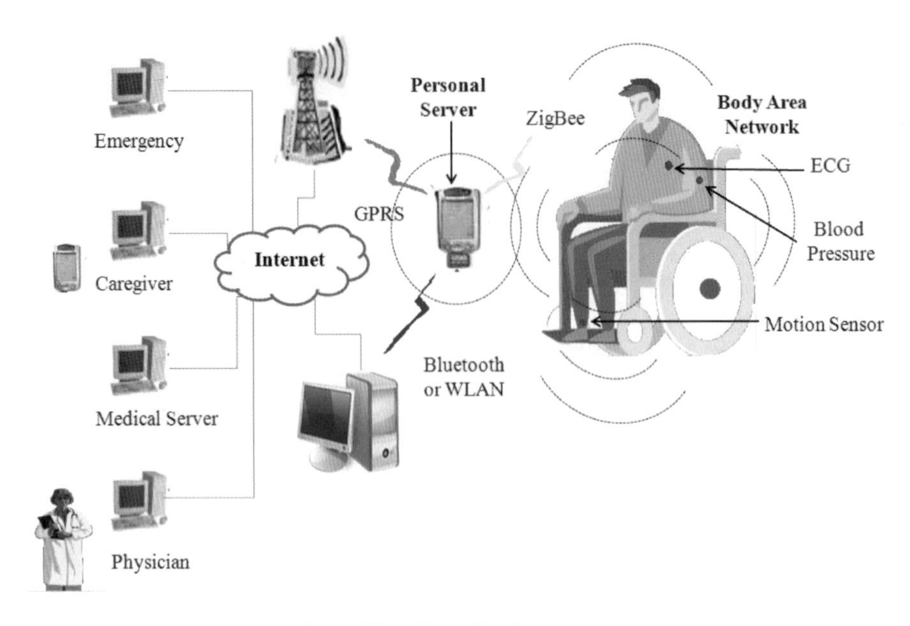

Fig. 1. WBAN application scenario

Recently, WBAN health monitoring systems have attracted researchers' attention. The WBAN is an emerging and promising technology that will change people's healthcare experiences revolutionary [1-3]. The growth of sensor devices in healthcare, medical and biometrics has been increased from 8 percent in 2002 to 46 percent in 2012[4].In compare to traditional healthcare systems, wearable healthcare systems are very cost effective. Automatic monitoring systems release patients from long hospital stays, thus reducing medical labor and infrastructure costs. Reducing length of hospital stay is desirable especially for countries that are short of medical infrastructure and well-trained personnel. Beside the general benefits of WBAN health monitoring systems such as cost effective, unobtrusive and unobtrusive, they provide patients with continuous monitoring of physiological signals, which is very helpful especially for the aging population [5].WBAN enables patients to be monitored continuously, and served quickly by mobile health teams when physiological signals show that is necessary. Continuous monitoring of patients speeds up the patient recovery process, and reduces death rate especially in cardiovascular and diabetic patients. Moreover, the use of WBANs may enable ubiquitous healthcare and could lead to proactive, and even remote, diagnostic of diseases in an early stage. A WBAN may also contain an actuator, which based on measurements and settings, can automatically release medicine or other agents. In addition, WBANs provide health monitoring without interruption of the patient's everyday activities which leads to enhance the quality of life. However, in order to fully utilization of these benefits, some challenging issues such as standardization, social issues, power supply, Quality of Service (QoS) and security and privacy issues should be addressed. Among them, security and privacy issues are very important and need special attention. The transferred and stored data in WBANs play a critical role in medical diagnosis and treatment, thus, it is crucial to ensure the security of these data. Lack of security in WBANs may hamper the wide public acceptance of this technology, and more importantly can cause life-critical events and even death of patients. However, providing a strict and scalable security mechanism to prevent malicious interactions with WBANs is difficult. Open nature of the wireless medium, makes the patient's data prone to being eavesdropped, modified, loss and injected. Moreover, typical channel characteristics in WBANs such as very low Signal-to-Noise-Ratio (SNR) condition and limitation of body sensors in terms of power budget, memory capacity, communication and computational ability make the possibility of security attacks and threads in WBANs more likely than traditional Wireless Sensor Networks (WSNs). In addition, in WBANs, both security and system performance are equally important, therefore, the integration of a high-level security mechanism in such resource-constrained networks is difficult. So far, although there are already several prototype implementations of WBANs that deal with QoS and energy efficiency, studies on data security and privacy issues are few, and existing solutions are far from mature.

The rest of this chapter is organized into 8 sections. Section 2 presents an overview of WBAN and its architecture. The security- privacy requirements and possible attacks in WBANs are explained in section 3 and 4 respectively. Section 5 discusses the major issues and challenges to satisfy WBAN's security requirements. In section 7, we analysis the characteristics of a suitable security scheme for WBANs, and suggest a potential solution. Section 8 discusses the future of WBAN in health monitoring and unknown security and privacy threats that may arise in near future. The final section concludes our work.

2 Overview of WBANs

Wireless body area network is a network, includes low-power, lightweight, small-size, and intelligent sensors that are placed on, in or around the human body ,and used to monitor human's physiological signals and motion for medical, personal entertainment and other applications and purposes. Compared with traditional WLANs, WBANs enable wireless communications in or around a human body by means sophisticated pervasive wireless computing devices. WBAN health monitoring systems include various sensors such as blood pressure, electrocardiograph (ECG), electroencephalography (EEG), electromyography (EMG) and motion sensors. These sensors continuously monitor vital signals and send data to a nearby Personal Server (PS, also known as Network Coordinator (NC)) device. Then, over a Bluetooth/WLAN connection, these data are streamed remotely to a medical application for real time diagnosis, to a medical database for record keeping, or to the corresponding equipment that issues an emergency alert. Generally, the WBANs communications architecture is divided into three levels [6] as shown in Figure 2: level1-intra-BAN communications, which includes communications between body sensors and communications between body sensors and the PS. This level usually references to radio communications of about 2 meters around the human body.Level2-inter-BAN communications, which include communications between the PS and one or more Access Points (APs), and level3- beyond-BAN communications, this level is designed to use in metropolitan areas and involves some components such as medical applications and databases. In order to connect this level to the level 2 (inter-BAN) a gateway device, such as a PDA can be employed.

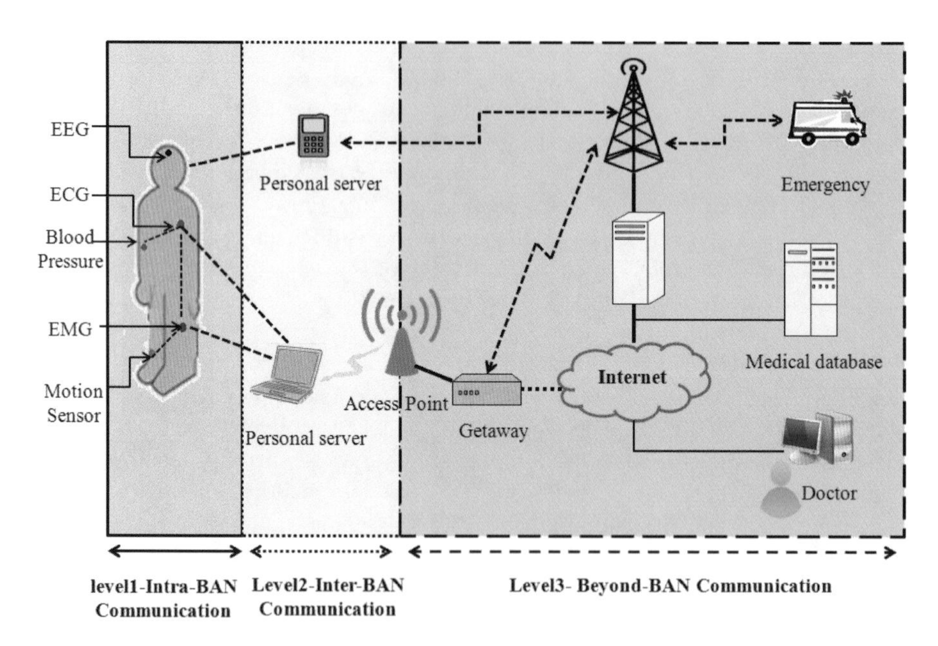

Fig. 2. 3-level WBAN architecture

3 WBAN Security Requirements

Before developing a comprehensive and strong security mechanism for WBANs, it is indispensable to understand the security and privacy requirements of these networks. The security and privacy of data are two essential components for the system security of WBANs. The term of data security means the data is securely stored and transferred, where data privacy means the data can only be accessed and used by the authorized people. In the following subsections, we discuss the major and fundamental security and privacy requirements in WBANs. We categorize the security and privacy requirements into three categories: privacy and data access security requirements, network communication security requirements and data storage security requirements.

3.1 Privacy and Data Access Security Requirements

People usually care deeply about their privacy. There is a risk in public acceptance of a new technology, if the privacy issues associated with it do not be addressed and debated clearly. The health related information is always private and sensitive, without taking care of privacy issues, the WBAN may not be accepted by people widely. In the following subsections, we discuss the major privacy and data access security requirements in WBANs.

3.1.1 Data Confidentiality

Data confidentiality means the transmitted data is strictly protected from leaking and disclosure. WBANs transmit very sensitive and personal information about the patient's health status. Many people would not like their health personal information, such as early stage of pregnancy or details of certain medical conditions be divulged to the public domain [7]. An adversary can monitor the communication between sensors and PS and eavesdrop the transmitted information. The acquired information can be used in many illegal purposes. To protect the user's privacy, all communications over the three levels of WBAN (intra-BAN, inter-BAN and beyond-BAN communications) should be encrypted. Data encryption in traditional WSNs is usually achieved by encrypting the information before sending it by using a secret key shared on a secure communication channel between sender and receiver. In case of intra-BAN communications, considering the scarcity transmission resources of body sensors, the best way for encryption is the use of stream cipher algorithms, because in these kinds of algorithms the size of ciphertext is exactly the same as plaintext, and no extra data needs to be transmitted.

3.1.2 Data Access Control

Data access control is a privacy policy and prevents unauthorized accesses to the patient's data. In WBAN systems, patient's medical data could be accessed by different users and parties such as, doctors, nurses, pharmacies, insurance companies and other supportive staff and agencies. However, if an insurance provider sees patient's medical report, it might discriminate such information against patients by offering health insurance at a high premium [8].therefore at the beyond-BAN layer, a role base access control is required to enforce different access privileges for different users. For example, doctors and nurses can have different access privileges based on their responsibility for patients, or insurance companies might be allowed to see only part of patient records related to reimbursement of medical expenses. An example of role based access control for health care applications is given in [9]. In WBANs, along with the role based access controls applied to beyond-BAN layer's applications, a comprehensive set of control rules is needed at intra-BAN layer. Authors in [10] discuss some rules on patient's privacy at home (intra-BAN layer).For instance, who can decide which sensors should collect the data, or whether patients can completely control how much of the data is sent to the central monitoring station, or they only have a partial control? In this case, guidelines need to be defined explicitly.

3.1.3 Accountability

Accountability is needed for secure data access control in WBANs. When a user abuses his/her privilege to carry out unauthorized actions on patient-related data, he/she should be identified and held accountable. One example is when a user illegally shares a key among unauthorized users. Authors in [11] discuss this problem and then propose a technique to defend against it.

3.1.4 Revocability

Revocability protects the network from compromised nodes/users. If a user or node is detected as malicious or compromised, she/it should be deprived in time from all previously granted permissions [8].

3.1.5 Non-repudiation

Non-Repudiation is a way to guarantee that the sender of a message cannot later deny having sent the message and that the recipient cannot deny having received the message. Generally, non-Repudiation can be obtained through the use of digital signatures.

3.1.6 Policy Requirements

As the sensitive personal health information can be available electronically, the need to have cohesive policies to protect the patient's privacy is raised. Cohesive policies are needed to deal with uncertainties in data ownership, access rights, disclosure, etc. Currently there are different sets of regulations and policies for medical security and privacy in all over the world, because policies and regulations are different from country to country. One example is the American Health Insurance Portability and Accountability Act (HIPAA).The HIPAA Privacy Protection mandates from 2003 were founded for a national standard for health privacy [12].HIPAA is a set of rules to be followed by doctors, hospitals and other health care providers. HIPAA's goal is to make sure that all medical records, medical billing and patient accounts fulfill certain consistent standards with regards to documentation, handling and privacy. However, HIPAA and other existing health policies settings do not make patients sure about their privacy rights, since they only address a minimum set of rules and groundwork. Clear rules should be created that WBAN 's users can rely upon.

3.1.7 Public Awareness

Authors in [7] discuss that an important privacy measure in WBAN systems is to create awareness in the general public. Non-expert people do not understand the technology and its negative impact on their own privacy standards. It is very helpful, if people be educated regarding security and privacy issues and their implications. Moreover, educating people can make them feel more comfortable about the WBAN systems and consequently can help to attain public acceptance of WBANs.

3.2 Network Communication Security Requirements

3.2.1 Data Integrity

Data integrity makes sure that the received information has not been tampered with. Lack of data integrity allows the adversary to change the patient's information before it reaches to the PS. In WBANs, failure to obtain genuine and correct medical data will possibly hold back patients from being treated effectively and even can lead to wrong treatments and disastrous consequences. A data integrity mechanism over transmission time in WBANs can be achieved through Message Authentication Code (MAC). The PS and the body sensors can verify the MAC to ensure that the received data is not modified by an adversary.

3.2.2 Data Authentication

Data authenticity means making sure that the information is sent by the trusted sender. This property is crucial for WBANs because specific actions are launched only if the legitimate nodes requested the action. Absence of this property may lead to situations where an illegitimate entity masquerades as legitimate one and reports false data to the PS or gives wrong instructions to the body sensors possibly causing considerable harm to the host [13]. Therefore, body sensors and the PS need to make sure that the data is sent by a trusted sender and an adversary has not tricked them into accepting false data. To address data authenticity, a calculated Message Authentication Code (MAC) can be applied by using a shared secret key.

3.2.3 Data Freshness

Data freshness guarantees that the received data is fresh. For example, the data frames are in order and not reused to disrupt. Informally, data freshness implies that the data is recent, and it ensures that no adversary replayed old messages. The adversary could capture data over transmission and replay them later by using the old key in order to confuse the PS. There are two types of data freshness: weak freshness and strong freshness. Weak freshness just guarantees ordering of data frames but does not guarantee delay where strong freshness guarantees both delay and frames ordering. WBANs need both weak freshness and strong freshness. Strong freshness in WBANs is required during synchronization, e.g. when a beacon is transmitted by the PS where weak freshness is required by low-duty cycle body sensors such as blood pressure [14].

3.2.4 Localization

For many healthcare applications, it is essential to know about patient's location. Localization identifies the position of target sensor nodes carried by patient in a randomly distributed network. To assign measurement for location, each node has to determine its own position. Absence of smart tracking techniques allows an attacker to send incorrect patient's location by using false signals. A Study of localization techniques in WBANs is given in [15].

3.2.5 Availability

The term of availability means that the patient's information should be always available to the physician even under Denial-of-Service (DoS) attacks. For example, the adversary can capture or disable an ECG sensor which could result to a dangerous situation or even to death. One solution in case of loss of availability is redundancy, which means switch the operation of a disabled node to another available node. Redundancy is necessary especially for those sensor nodes that do vital operations. In case of using redundancy, it is important to consider forward and backward secrecy when a new body sensor is deployed instead of a disabled or captured sensor. Forward secrecy means a sensor should not be able to read future transmitted messages after it leaves the network, while backward secrecy means a new sensor joining the network should not be able to read any previously transmitted messages.

3.3 Data Storage Security Requirements

In the following subsections, we explain the three most important data storage security requirements in WBANs, including data confidentiality, dependability, and integrity. Storage security requirements and related solutions and schemes in WBANs are discussed in detail in [8].

3.3.1 Confidentially

In order to prevent patient-related data from leaking, the data should always be kept confidential. Data confidentially is important in WBANs not only during transmission periods but also during storage periods.

3.3.2 Dynamical Integrity Assurance

In WBANs data integrity is very important because the collected data by the sensors is vital, and modified data could submit patients to dangerous situations. Therefore, data integrity in WBANs should be checked all the time, a node not only should check data integrity over transmitting times but also it should be able to dynamically check and detect modification of stored data in its buffer during storage periods in order to discover potential malicious modification before transmitting the data.

3.3.3 Dependability

Dependability means patient-related data must be readily retrievable in case of individual node failures, sensor compromises or malicious modifications. Dependability is one the critical concerns in WBANs because failure to retrieve correct data may cause life critical events. In order to address dependability, error correcting code techniques can be applied [8]. Although the dependability in WBANs is necessary, so far it has received limited attention.

Cryptographic techniques are one of the main security mechanisms. Many of security requirements described above such as confidentiality, integrity, and authentication can be fulfilled by using cryptographic techniques. Generally, there are two types of cryptographic techniques, symmetric and asymmetric. In symmetric cryptography, sender and receiver use one secret (private) key to encrypt and decrypt the data. Symmetric techniques require that the secret key be known by the party encrypting the data and the party decrypting the data .In asymmetric techniques, sender and receiver use both a public and their own private key. In asymmetric technique, the public key is distributed to anyone, and it is used to encrypt the data which have to be sent .In the receiver side, the encrypted data can only be decoded by the private key. This eliminates the need of having to give someone the secret key (as with symmetric encryption) and risk having it compromised. In usual networks where nodes have enough amounts of processing power and storage space, asymmetric cryptographic techniques are used. However, currently even the simplest version of the asymmetric key exchange techniques involves multiple exponentiations and message exchanges. Also, they assume more energy rather than symmetric techniques [16]. Therefore, since asymmetric techniques based key exchange suffer from heavy transmission overhead and also consume high energy, they are not suitable even for general WSNs [17]. As it will be explained in section 5, body sensors are more limited than generic sensors in terms of power budget and communication ability, so asymmetric techniques that are very expensive in terms of recourse consumption are not suitable for WBANs.

4 WBAN Attacks

WBANs are vulnerable to various types of attacks. Based on the security requirements in WBANs, these attacks can be categorized as:

- Attacks on secrecy and authentication: where an adversary performs eavesdropping, packet replay attacks, or spoofing of packets. One example of eavesdrop attacks in WBANs is activity tracking of users. Based on the patient's recorded data, it might be possible to analyze the activities of patients. This attack is very special to e-Health systems. Authors in [18] discuss a special kind of this attack in WBANs. When a patient is being constantly monitored, it is possible for an attacker to analyze the amount of physical exercise he/she is performing by looking at heart rate and oxygen saturation data. Insurance companies might use this information to limit access to benefits for people with an unhealthy lifestyle. Location tracking of users is another example of these attacks in WBANs. The attacker can eavesdrop the channel and capture the transmitted position signals in order to estimate the real-time patient location and even predict the patient's likely destination. This attack hamper the patients privacy because no one likes his/her location be tracked around the clock. One example of authentication attacks in WBANs is forging of alarms on medical data [18].In this case, attackers can simply create fake messages, which can lead to false system reactions e.g. to unnecessary rescue missions. The secrecy and authenticity of communication channels can be protected by standard cryptographic techniques and Message Authentication Code (MAC).
- Stealthy attacks against service integrity: In this kind of attacks, the attacker attempts to make the network accept a false data value by changing the patient's data before it reaches to the PS. For instance, an attacker can change a high blood pressure value to a normal blood pressure value .This can lead to a disaster event. Integrity attacks can happen during transmission times as well as storing times. Message Authentication Code (MAC) techniques can protect WBANs from these attacks.
- Attacks on network availability: These attacks are referred to as Denial-of-Service (DoS) attacks. DoS attacks attempt to make network resource unavailable to its users and affect the capacity and the performance of a network.

Since WBANs are a kind of wireless sensor networks, they inherit most of DOS attacks from WSN, however, due to the unique characteristics of WBAN, there are some difference between DOS attacks that can take place in WBAN and WSN. In the following subsections, we explain DoS attacks in different layers of Open System Interconnection (OSI) model, from physical to transport layer. In addition, we discuss the applicability of these attacks in WBANs.

4.1 Physical Layer Attacks

4.1.1 Jamming Attack

Jamming is defined as interference with the radio frequencies of the body sensors. In this attack, the adversary tries to prevent, or interfere with the reception of signals at the nodes in the network. In doing so, the attacker sends a continuous random signal on the same frequency used by the body sensors. Affected nodes will not be able to receive messages from other nodes .In this attack, the adversary can use few nodes to emit radio signals in order to disturb the transceivers' operation and block the whole network. However, larger networks are harder to block in their entirety. The key point in successful jamming attacks is SNR [19]. As it will be explained in section 5, since WBANs usually suffer from vary low values of SNR, and also because these kinds of networks are small in size, the likelihood of successful jamming in WBANs is high.

4.1.2 Tampering Attack

In tampering attacks, sensors are physically tampered by an adversary. The adversary may damage a sensor, replace the entire node or a part of its hardware or even electronically interrogate a node to acquire patient's information or shared cryptographic keys. Usually sensor devices have little external security features and hence prone to physical tempering. In a WBAN, the deployed sensors are under surveillance of the person carrying these devices, this means, it is difficult for an attacker to physically access the nodes without this being detected. However, still there is a chance for tampering in WBANs. A good preventive measure against tampering is patient awareness. It could be very helpful to advice patients that only authorized people should be allowed to physically handle the devices [7].

4.2 Data Link Layer Attacks

4.2.1 Collision Attack

Collision attack is synonymous with the jamming attack we just described. In this attack, the attacker listens to the channel, when he/she hears the start of a message, sends out its own signal that interferes with the message. This may cause a frame header corruption, a checksum mismatch, and therefore, the rejection of transmitted packets in the receiver side. This attack is difficult to detect because the only evidence of a collision attack is the reception of incorrect messages. If a frame fails the Cyclic Redundancy Code (CRC) check, the packet is discarded. The countermeasures that can be applied to protect the WBANs from this attack are error correction mechanisms. The same as jamming attack, the likelihood of successful collision attack in WBANs is high.

4.2.2 Unfairness Attack

In unfairness attacks, network performance degrades because medium access control layer priority is generally disrupted according to the application requirements. Use of small frames is a general defense against this attack.

4.2.3 Exhaustion Attack

Exhaustion of battery resources may occur when a self-sacrificing node always keeps the channel busy. In WSN, rate limitation is used to thwart this attack.

4.3 Network Layer Attacks

4.3.1 Selective Forwarding

Selective forwarding occurs when an adversary includes a compromised node in a routing path. When a malicious node receives a packet, it will do nothing and drop it. The malicious node can drop packets both selectively (just for a particular destination) and completely (all packets). Selective forwarding attacks are not applicable to the first communications level (intra-BAN level) of WBAN's architecture, because in intra-BAN communications, the PS is usually in direct communication range of body sensors, hence body sensors can communicate with the PS directory, and they do not require to route packets. Body sensors which have very limited communication range select one nearby node to relay their information to the PS [20]. In WBANs, routing is possible in the second level of communications (inter-BAN level), when multiple APs are deployed to help the body sensors transmit information. This way of interconnection extends the coverage area of a WBAN, and support patient mobility.

4.3.2 Sinkhole Attack

Sinkhole attack is similar to selective forwarding except that it is not a passive attack. In this attack, traffic is attracted towards the compromised or false node. This node drops packets in order to stop packet forwarding. The applicability of this attack in WBANs is the same as selective forwarding attack.

4.3.3 Sybil Attack

In Sybil attacks, a malicious node, called the Sybil node, illegitimately claims multiple false identities by either fabricating new identities or impersonating existing ones. In WSNs, which involve routing, this attack can cause a routing algorithm to calculate two disjoint paths. In WBANs, at the intra-BAN level of communications, this attack can use feigned identities to send false information to the PS [21].

4.3.4 Wormhole Attack

Wormhole attack is carried out using two distant malicious nodes to create a wormhole in the target sensor network. Both malicious nodes have an out of band communication channel. One malicious node is placed near the sensor nodes when the other is placed near the base station. The malicious node, which placed near the sensor nodes, convinces sensors that it has the shortest path to the sink node through the other malicious node, which is placed near the sink node. This creates sinkholes and routing confusions in the target sensor network. Applicability of wormhole attacks in WBANs is the same as selective forwarding and sinkhole attack.

4.3.5 Hello Flood Attack

Many protocols require nodes to broadcast hello packets to announce themselves to their neighbors. When a node receives such hello packets, it may assume that the sender is in its neighbor. In case of hello flood attacks, this assumption may be false. An attacker with a high powered antenna can convince sensors that it is in their neighbor. In addition, the attacker can claim a high quality route and creates a wormhole. Although the creation of wormhole does not affect the intra-BAN communications of WBANs, Hello Flood attack in intra-BAN communications does cause body sensors to reply to the hello packets and therefore, waste their energy.

4.3.6 Spoofing Attack

Spoofing attack targets the routing information exchanged between nodes [22], and attempts to spoof, alter, or replay the information with the intention to complicate the network. For example, an attacker could disturb the network by creating routing loops, generating fake error messages and attracting or repelling network traffic from selected nodes. Applicability of spoofing attack in WBANs is the same as selective forwarding, sinkhole and wormhole attacks.

4.4 Transport Layer Attacks

4.4.1 De-synchronization Attack

De-synchronization attack targets the transport protocols that rely on sequence numbers. The attacker forges some messages with wrong sequence numbers and this leads to infinite retransmissions which waste both energy and bandwidth. WBANs are highly vulnerable against this attack. Since body sensors have a limited power budget, retransmissions could drain sensor's powers quickly and make them unavailable to the network. Authentication can be applied to thwart this attack.

4.4.2 Flooding Attack

Flooding attack is used to exhaust memory resources by sending a large number of connection setup requests. Sine body sensors suffer from low memory space therefore, they are vulnerable against flooding attacks.

In WBANs, The PS is very attractive target for flooding and also for other above mentioned attacks as it is heart of system .In WBANs, the PS is responsible to collects and analyzes all data sent by body sensors, and then transmits them to the remote health applications. If an attacker can make the PS unavailable to the network, he/she can block the whole system. In many cases, the PS is connected to the Internet which allows remote attacks, whereas attackers cannot have direct connectivity to the body sensors. It is essential to provide the PS with high power budget, enough memory space, and strong security mechanisms such as authentication, firewalls, constant monitoring and etc.

Table 1 lists DOS attacks and possible solutions in different layers of OSI model.

Table 1. DOS attacks and possible solutions

Layers	DoS Attacks	Defenses
Physical	Jamming	Spread-spectrum, Priority messages, Lower duty cycle, Region mapping, Mode change
	Tampering	Tamper-proof, Hiding, Patient awareness
Data Link	Collision	Error correction code
	Unfairness	Small frames
	Exhaustion	Rate limitation
Network	Homing	Encryption
	Black holes	Authorization, Monitoring, Redundancy
	Misdirection	Egress filtering, Authorization monitoring
Transport	Flooding	Client Puzzles
	De-synchronization	Authentication

5 Constraints and Challenging Practical Issues

To satisfy the above security and privacy requirements in WBANs, we face several challenging issues. These issues constrain the solution space and need to be considered when designing a security mechanism for WBANs. In the following subsections, we describe these major challenging issues and the constraints.

5.1 Low Power Budget

All sensors are constrained in terms of power budget, but body sensors are more limited in this term. Energy is a crucial resource for body sensors since they use the power to perform all their functions like sensing, computation, and communication.

Replacing this crucial resource in many scenarios is impossible or impractical especially for in-body sensors, which placed inside the human body. So energy limitation is one main consideration to develop WBAN mechanisms and protocols.

5.2 Limited Memory

Memory capacity in body sensors is very limited about few kilobytes. This limitation is because of small size of body sensors. However, the implementation of security mechanism may not need much memory, but keying material is stated in the sensor's memory and takes up most part of the memory.

5.3 Low Computation Capability

Low computation ability in body sensors is caused by both low power budget and limited memory in body sensors. Since the main responsibility of body sensors is communication of the sensed information, therefore, there is very less amount of energy which can be expended on computation processes [13]. Moreover, because of memory limitation in body sensors, they cannot perform heavy computation processes.

5.4 Low Communication Rate

Communication is the most energy consumer function in WBANs. In order to save energy, it is important to minimize the amount of communications in these networks. So, developers have try to minimize the overhead transmissions required by other purposes rather than transforming of actual data.

5.5 Environment Condition of WBANs

Environment characteristics of WBANs pose additional security threads to these networks. Effective bandwidth of WBANs usually degrades due to effect of Radio Frequency (RF) emitting devices such as microwaves around the human body [23]. Furthermore, a number of studies [24-27] prove that the human body presents different adverse fading effects to wireless communication channels that are dependent on body size and posture. In addition, to protect patients against harmful health effects associated with the RF emissions, the Specific Absorption Rate (SAR) in WBANs should be low [28]. SAR is the rate at which the RF energy is absorbed by a body volume or mass. Because of SAR Limitation in WBANs, body sensors must use very low power for transmission. This means that increasing transmission power beyond a certain level in order to reduce transmission losses in WBANs is impossible. Thus, in such environment, very low SNR values are expected. However, interference and noise are generally QoS issues, but they have great potential to pose a serious security issues in WSNs and especially in WBANs. Since WBANs are naturally susceptible to channel fading and interference, and also they suffer from low values of SNR, even introducing a low level of noise into their channel can increase packet loss rates dramatically. Moreover, patient mobility increases the probability of packet loss in WBANs. It is clear that, in such environment, attackers can harm the system by simply presenting a low level of noise into the channel and causing a lot of packet loss. In this scenario, lost packets should be retransmitted. Retransmissions cause the network to waste its bandwidth and sensors to exhaust their power supplies. Moreover, the system will suffer from long delays caused by retransmissions. Retransmission delays have a negative effect on data freshness, which is harmful especially for real-time applications. In some cases such as heart attacks, any delay in receiving the data could lead patients to the death. Therefore, it is easy for attackers to harm WBAN by using the vulnerability of this network to the noise, they even can block the whole system by causing infinite retransmissions.

5.6 Conflict between Security and Safety

A strong access control mechanism should define existing users and regulations and firm guidelines regarding use of data for these users explicitly. Normally, e-Health care scenarios involve only few and limit number of users such as doctors, nurses and supportive staffs. So, a strong access control for WBANs should not include other specific users. However, it should be considered that too strict and inflexible data access control could prevent in time treatment. In some cases, especially in emergency and disasters scenarios disclosure of information to other people (such as mobile health teams) in order to serve the patients is necessary. So, a suitable access control mechanism in WBANs should be flexible enough to accept or compromise users to some extent [7].

5.7 Conflict between Security and Usability

As the operators of WBAN devices are usually non-expert people, therefore, the devices should be simple and easy to use. Moreover, WBAN devices are supposed to be like the plug-and-play devices. Since the setup and control process of the data security mechanisms are patient-related, they shall involve few and intuitive human interactions [8]. However, in case of WBANs security is more important than usability and omitting some manual steps to increase usability is not suggested.

5.8 Lack of Standardization

Each WBAN could include sensors from different manufacturers. Therefore, it is difficult to pre-share any cryptographic materials. In such networks that work with a wide range of devices, it is hard to implement security mechanisms that require the least common settings.

6 IEEE 802.15.4 Security Suits

There are many security mechanisms proposed for generic wireless sensor networks however, just few of them can be applied for WBANs with low-power computation. The IEEE 802.15.4 standard has different security modes that can be improved for WBANs. The IEEE 802.15.4 is a low-power standard designed for low data rate Wireless Personal Area Networks (WPANs).it specifies the physical and media access control layers, focusing on low-cost and low-speed ubiquitous communication between devices. This standard is very close to WBANs because it supports low data rate applications with low cost of power consumption. It is employed by many designers and researchers to develop protocols and mechanisms for WBANs. The IEEE 802.15.4 security suits are classified into null, encryption only (AES-CTR), authentication only (AES-CBC-MAC), and encryption and authentication (AES-CCM) suites. These security suites are shown in Table2 .In the following subsections, we discuss AES-CTR, AES-CBC, and AES-CCM modes.

6.1 AES-CTR

Confidentiality protection in AES-CTR is provided by using Advance Encryption Standard (AES) block cipher with counter mode (CTR) (also known as Integer Counter Mode).In this mode, the cipher text is broken into 16-byte blocks $b_1, b_2, \ldots, b_n$. The sender side, computes the cipher text by $c_i = b_i$ XOR $E_k(x_i)$, where c_i is the encrypted text, b_i is the data block, and $E_k(x_i)$ is the encryption of the counter x_i .The receiver decodes the plaintext by computing $b_i = c_i$ XOR $E_k(x_i)$. The encryption and decryption processes are shown in Figure 3.

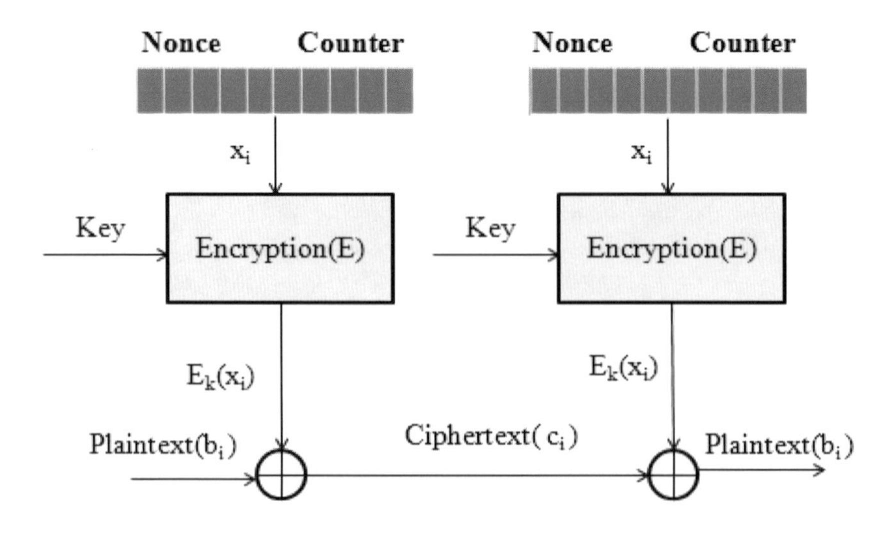

Fig. 3. CRT encryption and decryption

6.2 AES-CBC-MAC

In AES-CBC-MAC, security including authentication and message integrity protection is provided by using a Cipher-Block Chaining Message Authentication Code (CBC-MAC). CBC- MAC specifies that an n block message B = b_1, b_2..., b_n be authenticated among parties who share a secret key (K) for the block cipher (E). The sender can compute either a 4, 8, or 16 byte MAC. The MAC can only be computed by parties with the symmetric key. In this mechanism, the plaintext is XORed with the previous cipher text until the final MAC is created where the cipher texts are generated by $c_i = E_k(b_i \text{ XOR } c_{i-1})$ and plaintexts can be generated by $b_i = D_k(c_i) \text{XOR } c_{i-1}$.The sender appends the plaintext data with the computed MAC. The receiver verifies the integrity by computing its own MAC and comparing it with the received MAC. The receiver accepts the packet if both MACs are equal. Figure 4 shows the block diagram of a CBC-MAC operation.

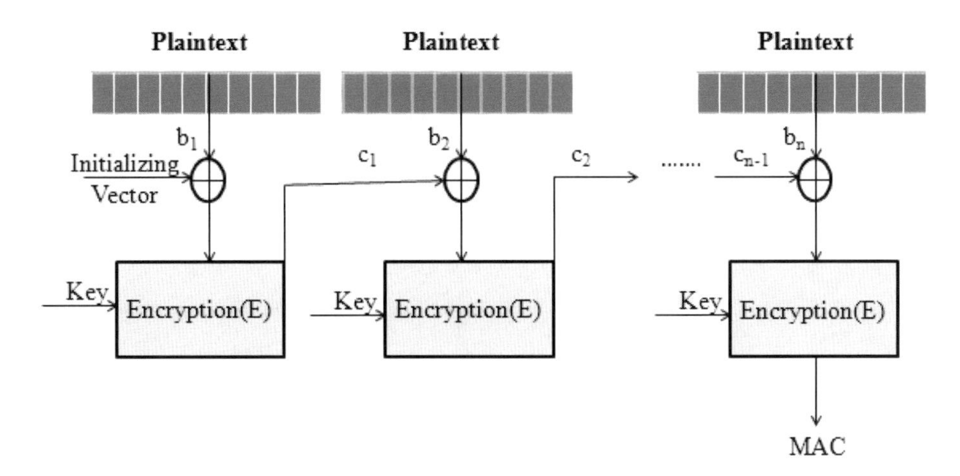

Fig. 4. CBC-MAC operation

Table 2. Security modes in IEEE 802.15.4

Name	Description
Null	No Security
AES-CTR	Encryption only, CTR Mode
AES-CBC-MAC-128	128 bit MAC
AES-CBC-MAC-64	64 bit MAC
AES-CBC-MAC-32	32 bit MAC
AES-CCM-128	Encryption & 128 bit MAC
AES-CCM-64	Encryption & 64 bit MAC
AES-CCM-32	Encryption & 32 bit MAC

6.3 AES-CCM

This security suite addresses a high-level security that includes both data integrity and encryption. It first applies integrity protection over the header and data payload using CBC-MAC mode and then encrypts the data payload and MAC using AES-CTR mode.

7 Discussion and Recommendations

In section 5, we explained limitations of body sensors and other challenging practical issues faced by WBANs. Body sensors are extremely limited in terms of battery power, processing capabilities, memory capacity, buffer size as well as the transmission power. In addition, WBANs usually suffer from low values of SNR and consequently a high level of packet loss, thus they are very vulnerable to the noise. It is clear that, in such networks, traditional and general security approaches are not applicable at all. For these kinds of networks, developers have to look for highly efficient approaches not only because of the resource constraints but also because of application requirements. The cryptographic methods applied in WBANs should be lightweight with fast computation, low storage and low transmission overhead. Otherwise, the power and storage space of the body sensors could be drained quickly. In addition, security mechanisms that can cover major WBAN's security requirements simultaneously are desirable because they can act more efficiency. Moreover, as it is explained in section 5, packet loss is one of the major security concerns in WBANs as it can cause energy and bandwidth waste, long delays and even blocking the whole system. Therefore, a suitable security approach for WBANs should deal with packet loss just fine. Applying an efficient and lightweight error recovery mechanism can be a possible and suitable measure against packet loss in WBANs. In doing so, Network coding seems to be a suitable approach. Network coding is a technique that combines different sets of data at relay nodes in such way that they can be decoded at the destination [29-30]. Network coding was first introduced by Ahlswede et al [29]. A general architecture of network coding is shown in Figure5. This technique uses some intermediate nodes (relay nodes) where each node transmits its data to the destination through these intermediate nodes. Intermediate nodes combine the incoming packets and then transmit a linearly independent of "combination" packet which contains information about all the original (incoming) packets where the size of the "combination" packet is the same as one incoming packet. An overview of network coding is given in [31].

Generally, network coding has been widely known as a suitable approach to improve network performance. It offers several advantages such as delay reduction, transmission energy minimization, and improvement of potential throughput [29]. In addition to all above benefits and, the advantage of light weight operation and no transmission overhead; network coding seems very suitable to satisfy some major QoS and security requirements in WSN and especially in WBANs. Several papers have considered the use of network coding for wireless body area networks.

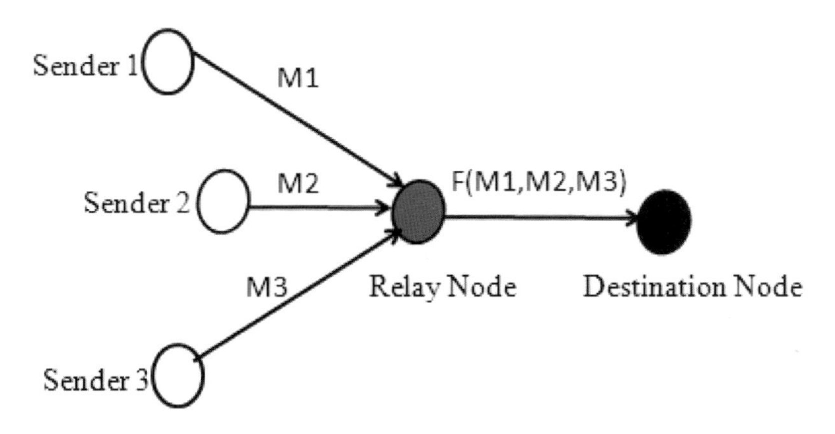

Fig. 5. Network coding architecture

In [32] authors have extended cooperative network coding, from its original configuration (one-to-one) to many-to-many as in multiple-input-multiple-output (MIMO) systems in order to improve the reliability of WBANs in case of node or links failures. Authors in [33] use network coding as an error recovery method in WBANs in which corrupted packets can be recovered at the destination. They show that, the use of network coding provides WBANs with 10 to 100 times better packet loss rate rather than using of a regular redundant sending in order to recover corrupted packets. The simulation result also shows an improvement of packet loss rate by 100 to 10000 times that of the case without any coding and redundancy. Their work proves that network coding can be used as an efficient error recovery mechanism where it can significantly improve network reliability at very low computational and hardware cost. Therefore, applying network coding in WBANs as an error recovery mechanism can reduce packet loss and retransmissions times efficiency, and as a result deals with one of the main security and QoS issues in these networks.

On the other hand, network coding has great potential to be applied as an efficient and lightweight encryption mechanism in WBANs. The nature of coding/mixing operation of network coding can provide a feasible way to thwart the traffic analysis efficiently. In network coding, an unlink ability between incoming packets and outgoing packets can be achieved by mixing the incoming packets at intermediate nodes. However, a simple deployment of network coding cannot address confidentiality since in general deployments of network coding, the mixing operation in intermediate nodes is very simple, and there is a linear dependence between outgoing and incoming packets which can be easily analyzed. Fan et al in [34] apply an efficient network coding based privacy-preserving scheme with a lightweight homomorphic encryption to thwart traffic analysis and flow tracing attacks in multi-hop wireless networks. The proposed scheme offers two significant benefits, packet flow untraced ability and message content confidentiality.

The above mentioned works prove that network coding has great prospective to be used as a light weight and efficient security package in WBANs .We believe that a special implementation of network coding can be applied as a strong and lightweight encryption mechanism in WBANs to address some major security requirements such as confidentiality, integrity, and authentication simultaneously in one package. Employing network coding as an encryption mechanism in WBANs is very efficient since it does not involve any transmission overhead, the only overhead will be the computation overhead in the intermediate nodes to encrypt packets and at the destination to decrypt packets. Moreover, since network coding has a high ability to recover lost and corrupted packets, therefore, it can be used against potential attacks in WBANs such as collision attacks. In addition, network coding can be applied in order to improve the reliability of WBANs in case of node or links failures. Table 3 lists major potentials benefits of using network coding in WBANs.

Table 3. Potentials benefits of using network coding in WBANs

Potential Benefit	Results
Efficient and light weight error recovery mechanism	Packet loss reduction
	Retransmission reduction
	Delay reduction
	Transmission energy minimization
	Increase bandwidth utilization
	Avoid blocking the entire network (to thwart some attacks such as collision)
Addresses confidentiality, integrity, and authentication simultaneously	Reduce transmission overhead
	Transmission energy minimization
	Computation energy minimization
	Increase bandwidth utilization
Reduce the impact of link or node failures	Improve the reliability

8 Open Areas for Research

The role of WBANS in healthcare applications is becoming more and more prominent. As this technology become pervasive, it will be exposed to numerous security and privacy threads. It is better to be ready for such situations before the time comes for it. Currently, WBANs involve homecare and hospital environment scenarios. In homecare and hospital scenarios, body sensors are in direct communication range of the PS, and they do not require to route packets. Therefore presently, we don't need to take care of routing attacks such as selective forwarding, wormhole and sinkhole attacks in the intra-BAN level of WBANs. However, in the near future, with the deployment of mobile and wireless networks, WBANs can play a critical role in treatment of victims in disaster events. In disaster scenarios, body sensors might need to send their data to other devices outside their immediate radio range. Therefore, routing protocols with strong security features will become crucial service for end-to-end communications in the intra-BAN level of WBANs.

The second issue that definitely will become more important in the coming future is lack of cohesive policy sets to protect the patient's privacy. As WBANs become pervasive, more parties such as pharmacies and insurance companies will be involved in the system. Therefore, patient related data will be accessed by more parties, and more attacks on patient privacy are possible. Privacy attacks make people pessimistic about WBANs, and will force major obstacles to growth and development of this technology. We feel that without taking care of current and future privacy issues, WBANs will not be accepted by the public. Strong set of regulations and policies should be formalized and implemented. In new policies sets, all possible future parties and privacy threads associated with them should be considered in which all involved parties feel it difficult to abuse from patient dada.

The next generation of WBANs could benefit from the advantage of cloud computing technology. Combining mobile cloud computing and WBANs is a very new and interested topic [35]. With the support of mobile cloud computing, WBANs can be greatly improved for the deployment of innovative healthcare monitoring applications with richer multimedia contents, more reliable quality of service and more types of convergence services. This combination will involve new security threads that need to be consider.

The growth of WBAN is rapid and fast. Along with current security and privacy issues, new threads will be raised in this area in the near future, in this section we just mentioned to few of them.

9 Conclusion

A WBAN is expected to be a very useful technology with potential to offer a wide range of benefits to patients, medical personnel and society through continuous monitoring and early detection of possible problems. Security is a fundamental feature for the deployment of wireless body area networks. The deployment of WBANs must satisfy the stringent security and privacy requirements. However, the limitations of body sensors and typical characteristics of WBAN's environment make the design of security procedures complicated. The general security approaches are not applicable for WBANs. A suitable security mechanism in WBANs should be lightweight and inexpensive in term of resource consumption. Moreover, we have to keep in our mind that, however, generally noise issues are related to QoS, but in WBANs they can lead to serious security threads. Thus, a suitable security mechanism for WBANs should consider vulnerability of these networks to the noise and apply a powerful and efficient error recovery technique to thwart this weak point. In this chapter, we outlined the main security requirements and attacks in WBANs. We further discussed the major challenging issues for designing security mechanisms in these networks. We also pointed out to network coding .Using network coding in WBANs as a security package is an attractive solution. Network coding has potential to combat packet loss, reduce latency due to retransmissions, avoid single points of failure, and improve the probability of successful recovery of the information at the destination. Moreover, the nature of coding operation of network coding can provide a lightweight encryption mechanism. Therefore, a special implementation of network coding can be applied to WBAN to address its major security requirements and threads efficiency. WBAN is growing fast but so far there is no strong and integrated security framework for this kind of networks. The research in data security and privacy of WBANs is still in its infancy now, more researches and studies in this area are needed.

References

[1] Jovanov, E., et al.: A Wireless Body Area Network of Intelligent Motion Sensors for Computer Assisted Physical Rehabilitation. J. NeuroEng. Rehab. 2(6) (March 2005)
[2] Halperin, D., et al.: Security and Privacy for Implantable Medical Devices. IEEE Pervasive Computing 7, 30–39 (2008)
[3] Lorincz, K., et al.: Sensor Networks for Emergency Response: Challenges and Opportunities. IEEE Pervasive Computing 3, 16–23 (2004)
[4] Kwak, K.S., et al.: A Study on Proposed IEEE 802.15 WBAN MAC Protocols. Presented at the Proceedings of the 9th International Conference on Communications and Information Technologies, Incheon, Korea (2009)
[5] Patel, M., Wang, J.: Applications, challenges, and prospective in emerging body area networking technologies. Wireless Commun. 17, 80–88 (2010)
[6] Chen, M., et al.: Body Area Networks: A Survey. Mob. Netw. Appl. 16, 171–193 (2011)
[7] Ameen, M., et al.: Security and Privacy Issues in Wireless Sensor Networks for Healthcare Applications. J. Med. Syst. 36, 93–101 (2010)
[8] Li, M., et al.: Data security and privacy in wireless body area networks. Wireless Commun. 17, 51–58 (2010)
[9] Evered, M., Bogeholz, S.: A case study in access control requirements for a Health Information System. In: The Second Workshop on Australasian Information Security, Data Mining and Web Intelligence, and Software Internationalisation (2004)
[10] Meingast, M., Roosta, T., Sastry, S.: Security and Privacy Issues with Health Care Information Technology. In: The Proceedings of the 28th IEEE EMBS Annual International Conference (2006)
[11] Yu, S., et al.: Defending against Key Abuse Attacks in KP-ABE Enabled Broadcast Systems. In: 5th International ICST Conference on the Security and Privacy in Communication Networks (2009)
[12] Info Guide to HIPAA Compliance, Implementation and Privacy, http://www.hipaa-101.com (July 31, 2012)
[13] Cherukuri, S., et al.: Biosec: a biometric based approach for securing communication in wireless networks of biosensors implanted in the human body. Presented at the Proceedings of the International Conference on Parallel Processing Workshops (2003)
[14] Saleem, S., et al.: On the Security Issues in Wireless Body Area Networks. International Journal of Digital Content Technology and its Applications 3, 178–184 (2009)
[15] Rehman, O.U., et al.: Performance Study of Localization Techniques in Wireless Body Area Sensor Networks. Presented at the International Symposium on Advances in Ubiquitous, Liverpool,UK (2012)

[16] Carman, D.W., et al.: Constraints and approaches for distributed sensor network security. NAI Labs (2000)
[17] Vijayalakshmi, B.: A Zero- Knowledge authentication for Wireless Sensor Networks based on Congruence. Presented at the Third IEEE International Conference on Advanced Computing (2011)
[18] Kargl, F., et al.: Security, Privacy and Legal Issues in Pervasive eHealth Monitoring Systems. Presented at the 7th International Conference on Mobile Business (2008)
[19] Mpitziopoulos, A., et al.: Jamming in Wireless Sensor Networks. In: Zhang, Y., Kitsos, P. (eds.) Security in RFID and Sensor Networks, pp. 375–397. CRC Press (2009)
[20] Raazi, S.M.K.-U.-R., et al.: BARI+: A Biometric Based Distributed Key Management Approach for Wireless Body Area Networks. Sensors 10, 3911–3933 (2010)
[21] Raazi, U.R., et al.: A Survey on Key Management Strategies for Different Applications of Wireless Sensor Networks. JCSE 4, 23–51 (2010)
[22] Karlof, C.: Secure Routing in Wireless Sensor Networks: Attacks and Countermeasures. Elsevier's AdHoc Networks Journal, Special Issue on Sensor Network Applications and Protocols, 293–315 (2003)
[23] Zhou, G., et al.: BodyQoS: Adaptive and Radio-Agnostic QoS for Body Sensor Networks. Presented at the INFOCOM (2009)
[24] Shah, R.C., et al.: On the performance of Bluetooth and IEEE 802.15.4 radios in a body area network. Presented at the Proceedings of the ICST 3rd International Conference on Body Area Networks (2008)
[25] Natarajan, A., et al.: Investigating network architectures for body sensor networks. In: The 1st ACM SIGMOBILE International Workshop on Systems and Networking Support for Healthcare and Assisted Living Environments (2007)
[26] Roelens, L., et al.: Path loss model for wireless narrowband communication above flat phantom. IEEE Electronics Leters 42 (2006)
[27] Johansson, A.J.: Wave-propagation from medical implants-influence of body shape on radiation pattern. Presented at the in Proceedings of the Second Joint of EMBS/BMES Conference (2002)
[28] IEEE Standard for Safety Levels With Respect to Human Exposure to Radio Frequency Electromagnetic Fields, 3 kHz to 300 GHz, IEEE Standard C95.1 (2005)
[29] Ahlswede, R., et al.: Network information flow. IEEE Transactions on Information Theory 46, 1204–1216 (2002)
[30] Li, S.-Y.R., et al.: Linear network coding. IEEE Transactions on Information Theory 49, 371–381 (2003)
[31] Chou, P.A., Wu, Y.: Network coding for the Internet and wireless networks. IEEE Signal Process. Mag. 24, 77–85 (2007)
[32] Arrobo, G.E., Gitlin, R.D.: Improving the reliability of wireless body area networks. Presented at the Annual International Conference of the IEEE Engineering in Medicine and Biology Society, EMBC (2011)
[33] Marinkovic, S., Popovici, E.: Network Coding for Efficient Error Recovery in Wireless Sensor Networks for Medical Applications. Presented at the Proceedings of the First International Conference on Emerging Network Intelligence (2009)
[34] Fan, Y., et al.: Network Coding Based Privacy Preservation against Traffic Analysis in Multi-Hop Wireless Networks. Trans. Wireless. Comm. 10, 834–843 (2011)
[35] Kurschl, W., Beer, W.: Combining cloud computing and wireless sensor networks. Presented at the Proceedings of the 11th International Conference on Information Integration and Web-based Applications & Services, Kuala Lumpur, Malaysia (2009)

Security and Privacy Issues in Wireless Mesh Networks: A Survey

Jaydip Sen

Innovation Labs, Tata Consultancy Services Ltd. Kolkata, India
jaydip.sen@acm.org,
jaydip.s@gmail.com

Abstract. This chapter presents a detailed survey on various aspects on security and privacy issues in Wireless Mesh Networks. The chapter is written both for the general readers as well as for the experts in the relevant areas. Future research issues and open problems are also mentioned so that the researchers could find appropriate directions to go ahead with their research works after reading the presented materials in this work.

1 Introduction

Wireless mesh networking has emerged as a promising technology to meet the challenges of the next-generation wireless communication networks for providing flexible, adaptive, and reconfigurable architecture and offering cost-effective business solutions to the service providers [1]. The potential applications of *wireless mesh networks* (WMNs) are wide-ranging such as: backhaul connectivity for cellular radio access networks, high-speed *wireless metropolitan area networks* (WMANs), community networking, building automation, *intelligent transportation system* (ITS) networks, defense systems, and city-wide surveillance systems etc [2]. Although several architectures for WMNs have been proposed based on their applications [1], the most generic and widely accepted one is a three tier structure as depicted in Fig. 1. At the bottom tier of this architecture are the *mesh clients* (MCs) which are mobile devices (i.e. users) with limited mobility and having resource constraints in terms of power, memory and computing abilities. At the intermediate tier, a set of *mesh routers* (MRs) or *edge routers* form an interconnected wireless back bone – the *wireless mesh network* (WMN). The MRs are wireless routers which wirelessly connect with each other and provide connectivity to the MCs. At the top tier of the architecture are a group of *gateways* or *Internet gateways* (IGWs). Each IGW is connected with several MRs using wired links or high-speed wireless links. The IGWs are connected to the Internet by wired links. A mesh network thus can provide multi-hop communication paths between the wireless clients (i.e., the MCs), thereby serving as a community network, or can provide multi-hop connectivity between the clients and a gateway router (i.e. an IGW), thereby providing broadband Internet access to the clients. Since deployments of WMNs do not need any wired infrastructures, these networks provide a very cost-effective alternative to the *wireless*

S. Khan and A.-S.K. Pathan (Eds.): *Wireless Networks and Security*, SCT, pp. 189–272.
DOI: 10.1007/978-3-642-36169-2_7

local area networks (WLANs) for the mobile users for the purpose of interconnection and access to the backbone Internet [2]. Wireless technology standards such as IEEE 802.11 (WLAN), IEEE 802.15 (LoWPAN), IEEE 802.16 (mobile WiMAX), IEEE 802.10 are adapted for developing a new wireless standard for mesh networking - IEEE 802.11s.

As WMNs become increasingly popular wireless networking technology for establishing the last-mile connectivity for home networking, community and neighbourhood networking, it is imperative to design efficient and secure communication protocols for these networks. However, the broadcast nature of transmissions in the wireless medium and the dependency on the intermediate nodes for multi-hop communications in such networks lead to several security vulnerabilities. These security loopholes can be exploited by potential external and internal attackers causing a detrimental effect on the network performance and disruption of services. The external attacks are launched by unauthorized users who intrude into the network for eavesdropping on the network packets and replay those packets at a later point of time to gain access to the network resources [3]. On the contrary, the internal attacks are strategized by some legitimate members in the network processing the authenticated credentials for accessing the network services. One example of such an attack is an intermediate node dropping packets which the node is supposed to forward. The internal attacks are more difficult to detect and prevent since the attackers are some members in the network having legitimate access to the network resources. Identifying and defending against these attacks in WMNs, therefore, is a critical requirement in order to provide sustained network services satisfying the quality of services of the user applications [4]. Furthermore, since in a WMN, the traffics from the end users are relayed via multiple wireless MRs, it is possible for these MRs to make a comprehensive traffic analysis for a user, thereby compromising the privacy his/her privacy. Hence, protecting the privacy and defending against privacy attacks on user data are critical requirements for most of the real-world applications in WMNs [5, 6]. Some security and privacy protection protocols for wireless networks are based on the computation and the use of the trust and reputation values of the nodes [7, 8]. However, most of these schemes are primarily designed for deployment in *mobile ad hoc networks* (MANETs) [9, 10], and hence these mechanisms do not fit well into the network architecture and the requirements of the applications in WMNs.

Keeping in mind the critical requirement of security and user privacy in WMNs, this chapter provides a comprehensive overview of various possible attacks on different layers of the communication protocol stack for WMNs and their corresponding defence mechanisms. First, it identifies the security vulnerabilities in the physical, link, network, transport, application layers. Furthermore, various possible attacks on the key management protocols, user authentication and access control protocols, and user privacy preservation protocols are presented. After enumerating various possible attacks, the chapter provides a detailed discussion on various existing security mechanisms and protocols to defend against and wherever

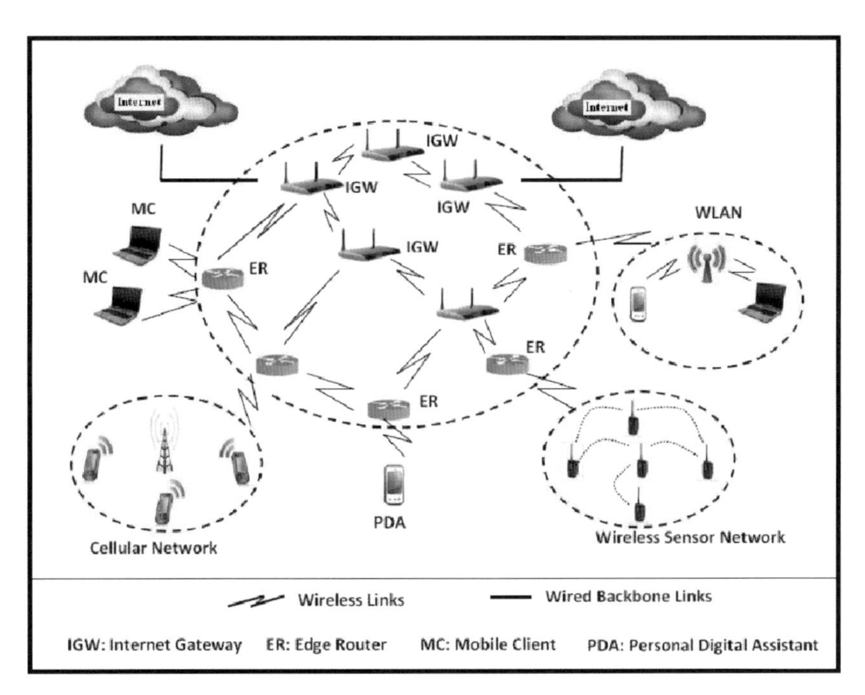

Fig. 1. The three-tier architecture of a wireless mesh network (WMN)

possible prevent the possible attacks. Comparative analyses are also presented on the security schemes with regards to the cryptographic schemes used, key management strategies deployed, use of any trusted third party, computation and communication overhead involved etc. The chapter then presents a brief discussion on various trust management approaches for WMNs since trust and reputation-based schemes are increasingly becoming popular for enforcing security in wireless networks. A number of open problems in security and privacy issues for WMNs are subsequently discussed before the chapter is finally concluded.

The chapter is organized as follows. Section 2 presents various possible attacks on different layers on the communication protocol stack of the WMNs. Section 3 discusses various security mechanisms at different layers for defending the attacks mentioned in Section 2. Section 4 provides a brief discussion on various trust management schemes for enforcing security and collaboration among the nodes in wireless networks with particular focus on WMNs. Section 5 highlights some future research trends on security and privacy issues in WMNs. Finally, Section 6 concludes the chapter.

2 Security Vulnerabilities in WMNs

Different protocols for various layers of WMN communication stack have several vulnerabilities. These vulnerabilities can be exploited by potential attackers to degrade or disrupt the network services. Since many of the protocols assume a pre-existing cooperative relation among the nodes, for successful working of these protocols, the participating nodes need to be honest and well-behaving with no malicious or dishonest intentions. In practice, however, some nodes may behave in a malicious or selfish manner or may be compromised by some other malicious users. The assumption of pre-existing trust relationships among the nodes, and the absence of a central administrator make the protocols at the link, network and transport layers vulnerable to various types of attacks. Furthermore, the application layer protocols can be attacked by worms, viruses, malwares etc. Various possible attacks may also be launched on the protocols used for authentication, key management, and user privacy protection. In this section, we present a comprehensive discussion on various types of attacks in different layers of WMN protocol stack.

2.1 Security Vulnerabilities in the Physical Layer

The physical layer is responsible for frequency selection, carrier frequency generation, signal detection, modulation, and data encryption. As with any radio-based medium, the possibility of a jamming attack in WMNs is high since this attack can be launched without much effort and sophistication. Jamming is a type of attack which interferes with the radio frequencies that the nodes use in a WMN for communication [11]. A jamming source may be powerful enough to disrupt communication in the entire network. Even with less powerful jamming sources, an adversary can potentially disrupt communication in the entire network by strategically distributing the jamming sources. An intermittent jamming source may also prove detrimental as some communications in WMNs may be time-sensitive. Jamming attacks can be more complex to detect if the attacking devices do not obey the MAC layer protocols [12].

2.2 Security Vulnerabilities in the Link Layer

Different types of attacks are possible in the link layer of a WMN. Some of the major attacks at this layer are: passive eavesdropping, jamming, MAC address spoofing, replay, unfairness in allocation, pre-computation and partial matching etc. These attacks are briefly described in this sub-section.

(i) Passive Eavesdropping: The broadcast nature of transmission of the wireless networks makes these networks prone to passive eavesdropping by the external attackers within the transmission range of the communicating nodes. Multi-hop wireless networks like WMNs are also prone to internal eavesdropping by the intermediate hops, wherein a malicious intermediate node may keep the copy of all the data that it forwards without the knowledge of any other nodes in the network. Although passive eavesdropping does not affect the network, functionality directly, it leads to the compromise in data confidentiality and data integrity. Data encryption is generally employed using strong encryption keys to protect the confidentiality and integrity of data.

(ii) Link Layer Jamming Attack: Link layers attacks are more complex compared to blind physical layer jamming attacks. Rather than transmitting random bits constantly, the attacker may transmit regular MAC frame headers (with no payload) on the transmission channel which conforms to the MAC protocol being used in the victim network [13]. Consequently, the legitimate nodes always find the channel busy and back off for a random period of time before sensing the channel again. This leads to the denial-of-service for the legitimate nodes and also enables the jamming node to conserve its energy. In addition to the MAC layer, jamming can also be used to exploit the network and transport layer protocols [14]. Intelligent jamming is not a purely transmit activity. Sophisticated sensors are deployed, which detect and identify victim network activity, with a particular focus on the semantics of higher-layer protocols (e.g., AODV and TCP). Based on the observations of the sensors, the attackers can exploit the predictable timing behavior exhibited by higher-layer protocols and use offline analysis of packet sequences to maximize the potential gain for the jammer. These attacks can be effective even if encryption techniques such as *wired equivalent privacy* (WEP) and *WiFi protocol access* (WPA) have been employed. This is because the sensor that assists the jammer can still monitor the packet size, timing, and sequence to guide the jammer. Because these attacks are based on carefully exploiting protocol patterns and consistencies across size, timing and sequence, preventing them will require modifications to the protocol semantics so that these consistencies are removed wherever possible.
(iii) Intentional Collision of Frames: A collision occurs when two nodes attempt to transmit on the same frequency simultaneously [15]. When frames collide, they are discarded and need to be retransmitted. An adversary may strategically cause collisions in specific packets such as *acknowledgment* (ACK) control messages. A possible result of such collision is the costly exponential back-off. The adversary may simply violate the communication protocol and continuously transmit messages in an attempt to generate collisions. Repeated collisions can also be used by an attacker to cause resource exhaustion. For example a naïve MAC layer implementation may continuously attempt to retransmit the corrupted packets. Unless these retransmissions are detected early, the energy levels of the nodes would be exhausted quickly. An attacker may cause unfairness by intermittently using the MAC layer attacks. In this case, the adversary causes degradation of real-time applications running on other nodes by intermittently disrupting their frame transmissions.
(iv) MAC Spoofing Attack: MAC addresses have long been used as the singularly unique layer-2 network identifiers in both wired and wireless LANs. MAC addresses which are globally unique have often been used as an authentication factor or as a unique identifier for granting varying levels of network privileges to a user. This is particularly common in 802.11 WiFi networks. However, the MAC protocol in 802.11 standard and the network interface cards do not provide any safeguards against a potential attacker from modifying the source MAC address in its transmitted frames. On the contrary, there is often full support in the form of drivers from the manufacturers to change the MAC address in the transmitted frames. Modifying the

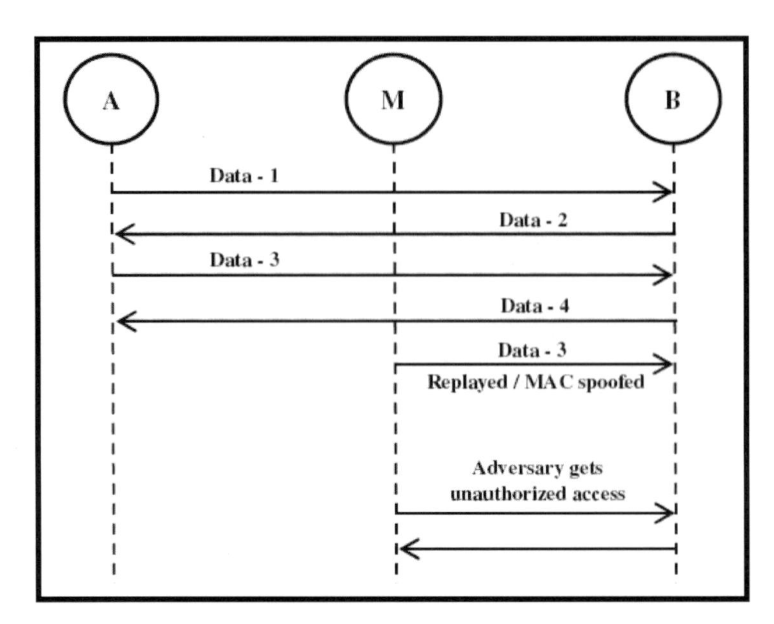

Fig. 2. Illustration of MAC spoofing and replay attacks launched by the malicious node *M*

MAC addresses in transmitted frames is referred to as *MAC spoofing*, and it can be used by attackers in a variety of ways. MAC spoofing enables the attacker to evade *intrusion detection systems* (IDSs) in the networks. Further, the network administrators often use MAC addresses in access control lists. For example, only registered MAC addresses are allowed to connect to the access points. An attacker can easily eavesdrop on the network to determine the MAC addresses of legitimate devices. This enables the attacker to masquerade as a legitimate user and gain access to the network. An attacker can even inject a large number of bogus frames into the network to deplete the resources (in particular, bandwidth and energy), which may lead to denial of services for the legitimate nodes.

(v) Replay Attack: The replay attack, often known as the *man-in-the-middle* attack [16], can be launched by external as well as internal nodes. As shown in Fig. 2, an external malicious node (*M*) can eavesdrop on the broadcast communication between two nodes *A* and *B*. It can then replay the (eavesdropped) messages later to gain access to the network resources. Generally, the authentication information is replayed where the attacker *M* deceives a node (node *B* in Fig. 2) to believe that the attacker is a legitimate node (node *A* in Fig. 2). On a similar note, the malicious node *M*, which is an intermediate hop between two nodes *A* and *B*, can keep a copy of all relayed data. It can then retransmit this data later to gain an unauthorized access to the network resources.

(vi) Pre-computation and Partial Matching Attack: Unlike the above-mentioned attacks, where MAC protocol vulnerabilities are exploited, these attacks exploit the vulnerabilities in the security mechanisms that are employed to secure the MAC layer of the network. Pre-computation and partial matching attacks exploit the cryptographic primitives that are used at the MAC layer for secure communication. In a pre-computation attack or *time memory trade-off attack* (TMTO), the attacker computes a large amount of information (key, plaintext, and respective ciphertext) and stores that information before launching the attack. When the actual transmission starts, the attacker uses the pre-computed information to speed up the cryptanalysis process. TMTO attacks are highly effective against a large number of cryptographic solutions. On the other hand, in a partial matching attack, the attacker has access to some (cipher text, plaintext) pairs, which in turn decreases the encryption key strength, and improves the chances of success of the brute force mechanisms. Partial matching attacks exploit the weak implementations of encryption algorithms. For example, in the IEEE 802.11 standard for MAC layer security in wireless networks, the MAC address fields in the MAC header are used in the *message integrity code* (MIC). The MAC header is transmitted as plaintext while the MIC field is transmitted in the encrypted form. Partial knowledge of the plaintext (MAC address) and the cipher text (MIC) makes IEEE 802.11i vulnerable to partial matching attacks.

DoS attacks may also be launched by exploiting the security mechanisms. For example, the IEEE 802.11i standard for MAC layer security in wireless networks is prone to the sensor hijacking attack and the man-in-the-middle attack, exploiting the vulnerabilities in IEEE 802.1X, and DoS attack, exploiting vulnerabilities in the four-way handshake procedure in IEEEE 802.11i.

2.3 Security Vulnerabilities in the Network Layer

The attacks on the network layer can be broadly divided into two types: *control packets attacks* and *data packets attacks*. Furthermore, both these attacks could be either active or passive in nature [17]. Control packets attacks generally target the routing functionality of the network layer. The objective of the attacker is to make routes unavailable or force the network to choose sub-optimal routes. On the other hand, the data packet attacks affect the packet forwarding functionality of the network. The objective of the attacker is to cause the denial of service for the legitimate user by making user data undeliverable or injecting malicious data into the network. We first consider the network layer control packets attacks, and then the network layer data packets attacks.

(i) Attacks on the Control Packets: *Rushing* attacks that target the on-demand routing protocols (e.g., AODV), were among the first exposed attacks identified by Hu et al. [18] on the network layer of multi-hop wireless networks. Rushing attacks exploit the route discovery mechanism of on-demand routing protocols. In these protocols, the node requiring a route to the destination floods the *route_request* (RREQ) message, which is identified by a sequence number. To limit the flooding, each node only forwards the first message that it receives and drops remaining messages with the same sequence number. The protocol specify a specific amount of delay between receiving the RREQ message by a particular node and forwarding it, to avoid collusion of these messages. The malicious node launching the rushing attack forwards the RREQ message to the target node before any other intermediate node from the source to destination. This can easily be achieved by ignoring the specified delay. Consequently, the route from the source to the destination includes the malicious node as an intermediate hop, which can then drop the packets of the flow resulting in data plane DoS attack.

Hu et al. identified the *wormhole* attack that has a similar objective as that of the rushing attack but it uses a different strategy [19]. During a wormhole attack, two or more malicious nodes collude together by establishing a tunnel using an efficient communication medium (i.e., wired connection or high-speed wireless connection etc.), as shown in Fig. 3. During the route discovery phase of the on-demand routing protocols, the RREQ messages are forwarded between the malicious nodes using the established tunnel. Therefore, the first RREQ message that reaches the destination node is the one forwarded by the malicious nodes. Consequently, the malicious nodes are added in the path from the source to the destination. Once the malicious nodes are included in the routing path, the malicious nodes either drop all the packets, resulting in complete denial of service, or drop the packets selectively to avoid detection.

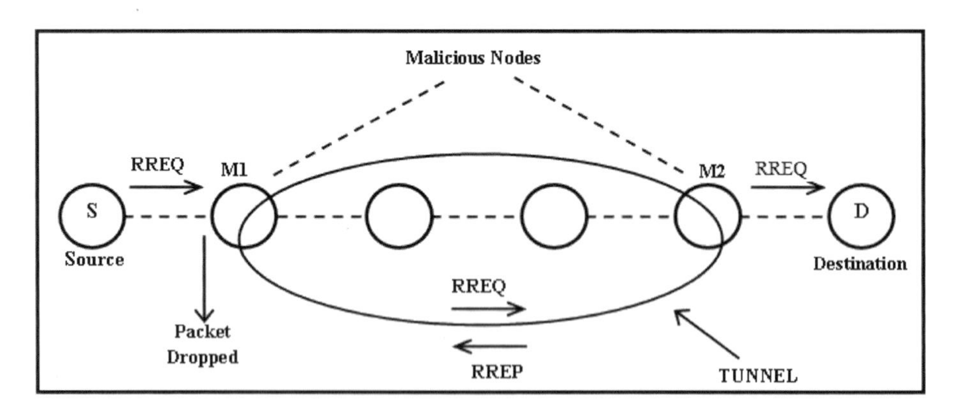

Fig. 3. Illustration of wormhole attack launched by nodes *M1* and *M2*

A *blackhole* attack (or *sinkhole* attack) [20] is another attack that leads to denial of service in WMNs. It also exploits the route discovery mechanism of on-demand routing protocols. In a blackhole attack, the malicious node always replies positively to a RREQ, although it may not have a valid route to the destination. Because the malicious node does not check its routing entries, it will always be the first to reply to the RREQ message. Therefore, almost all the traffic within the neighborhood of the malicious node will be directed towards the malicious node, which may drop all the packets, causing a denial of service. Fig. 4 shows the effect of a blackhole attack in the neighborhood of the malicious node where the traffic is directed towards the malicious node. A more complex form of the attack is the cooperative blackhole attack where multiple nodes collude together, resulting in complete disruption of routing and packet forwarding functionality of the network. Ramaswamy et al. have proposed a scheme for prevention of cooperative blackhole attack in which multiple blackhole nodes cooperate to launch a packet dropping attack in a wireless ad hoc network [21].

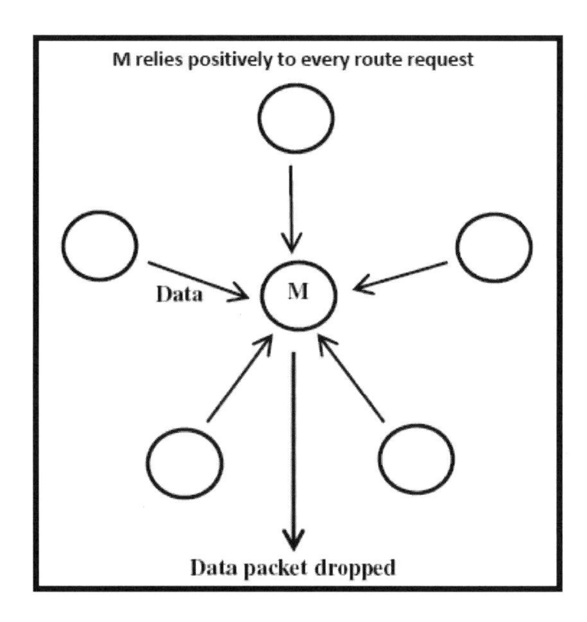

Fig. 4. Illustration of blackhole attack launched by node *M*

A *grayhole* attack is a variant of the blackhole attack. In a blackhole attack, the malicious node drops all the traffic that it is supposed to forward. This may lead to possible detection of the malicious node. In a grayhole attack, the adversary avoids the detection by dropping the packets selectively. A grayhole does not lead to complete denial of service, but it may go undetected for a longer duration of time. This is because the malicious packet dropping may be considered congestion in the network, which also leads to selective packet loss. Sen et al. have proposed a cooperative detection scheme for grayhole attack in a wireless ad hoc network [22].

A *Sybil* attack is the form of attack where a malicious node creates multiple identities in the network, each appearing as a legitimate node [23]. A Sybil attack was first exposed in distributed computing applications where the redundancy in the system was exploited by creating multiple identities and controlling the considerable system resources. In the networking scenario, a number of services like packet forwarding, routing, and collaborative security mechanisms can be disrupted by the adversary using a Sybil attack. Following form of the attack affects the network layer of WMNs, which are supposed to take advantage of the path diversity in the network to increase the available bandwidth and reliability. If the malicious node creates multiple identities in the network, the legitimate nodes, assuming these identities to be distinct network nodes, will add these identities in the list of distinct paths available to a particular destination. When the packets are forwarded to these fake nodes, the malicious node that created the identities processes these packets. Consequently, all the distinct routing paths will pass through the malicious node. The malicious node may then launch any of the above-mentioned attacks. Even if no other attack is launched, the advantage of path diversity is diminished, resulting in degraded performance.

In addition to the above-mentioned attacks, the network layer of WMNs are also prone to various types of attack such as: *route request (RREQ) flooding attack*, *route reply (RREP) loop attack*, *route re-direction attack*, *false route fabrication attack*, *network partitioning* attack etc. *RREQ flooding* is one of the simplest attacks that a malicious node can launch. An attacker tries to flood the entire network with the RREQ message. As a consequence, this causes a large number of unnecessary broadcast communications resulting in energy drains and bandwidth wastage in the network. A *routing loop* is a path that goes through the same nodes over and over again. As a result, this kind of attack will deplete the resources of every node in the loop.

Fig. 5 describes two instances where *route re-direction attack* has been launched by a malicious node *M*. In case *A*, the malicious node *M* tries to initiate the attack by modifying the mutable fields in the routing messages. These mutable fields include hop count, sequence numbers and other metric-related fields. The malicious node *M* could divert the traffic through itself by advertising a route to the destination with a larger *destination sequence number* (DSN) than the one it received from the destination. In case *B*, route re-direction attack may be launched by modifying the metric field in the AODV routing message, which is the hop-count field in this case. The malicious node *M* simply modifies the hop count field to zero in order to claim that it has a shorter path to the destination.

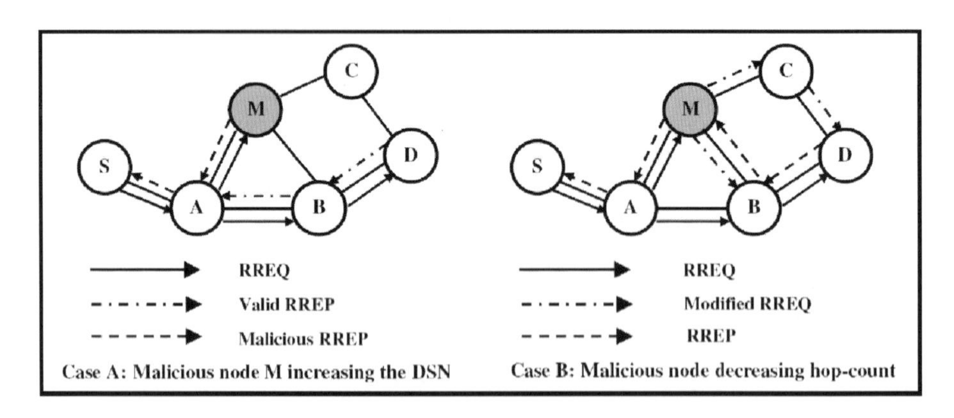

Fig. 5. Illustration of the route re-direction attack

An adversary may fabricate false routing messages in order to disrupt routing in the network. For example, a malicious node may fabricate a *route error* (RERR) message in the AODV protocol. This may result in the upstream nodes re-initiating the route request to the unreachable destination so as to discover and establish alternative routes to them leading to energy and bandwidth wastage in the network. In a network partitioning attack, the malicious nodes collude together to disrupt the routing tables in such a way that the network is divided into non-connected partitions, resulting in denial of service for a certain network portion. Routing loop attacks affect the packet-forwarding capability of the network where the packets keep circulating in loop until they reach the maximum hop count, at which stage the packets are simply dropped.

(ii) Attacks on the Data Packets: The attacks on the data packets are primarily launched by selfish and malicious (i.e., compromised) nodes in the network and in the network and lead to performance degradation or denial of service of the legitimate user data traffic. The simplest of the data plane attacks is *passive eavesdropping*. Eavesdropping is a MAC layer attack. Selfish behavior of the participating WMN nodes is a major security issue because the WMN nodes are dependent on each other for data forwarding. The intermediate-hop selfish nodes may not perform the packet-forwarding functionality as per the protocol. The selfish node may drop all the data packets, resulting in complete denial of service, or it may drop the data packets selectively or randomly. It is hard to distinguish between such a selfish behavior and the link failure or network congestion. On the other hand, malicious intermediate-hop nodes may inject junk packets into the network. Considerable network resources (i.e., bandwidth and packet processing time) may be consumed to forward the junk packets, which may lead to denial of service for legitimate user traffic. The malicious nodes may also inject the maliciously crafted control packets, which may lead to the disruption of routing functionality. The control plane attacks are dependent on such maliciously crafted control packets. The malicious and selfish behaviors of nodes in WMNs have been studied in [24, 25]. The multi-hop wireless networks such as *mobile ad hoc networks* (MANETs), *wireless sensor networks* (WSNs), and *wireless mesh networks* (WMN) have many common security vulnerabilities in the network layer. Detailed discussions on various attacks on the network layer and their defense mechanisms for WSNs and WMNs can be found in [26] and [4] respectively.

(iii) Attacks on Multicast Routing Protocols: Multicast routing protocols deliver data from a source node to multiple destinations which are organized in a multicast group. Since many of the applications that use multicast services in a WMN have high-throughput requirements, and hop-count does not serve as a good metric for maximizing throughput, some protocols [27, 28] focus on maximizing path throughput, where paths are selected based on metrics that are dependent on the wireless link qualities. In these protocols, nodes periodically send probes to their neighbors to measure the quality of the links from their neighbors. Selection of the best path for maximizing throughput is done based on collaboration of nodes. An aggressive strategy for the best path selection assuming a perfect collaboration among all participating nodes provides an easy opportunity to a malicious node to manipulate the link metrics to its own advantage. In other words, an attacker may suitably adjust the link metrics so that it gets selected on the best routing path for a source-destination pair. In this way, it draws more traffic towards itself. However, since its intention is to disrupt network communication, it starts dropping packets which can lead to a possible network partitioning or can help the malicious node to carry out a traffic analysis on the network. Roy et al. have proposed a secure multicast routing protocol on a tree-based architecture of a WMN using hop-count as the metric for path selection [29]. In Section 3.3.11, we have discussed various attacks on the multicast routing protocols for wireless networks.

2.4 Security Vulnerabilities in the Transport Layer

The attacks that can be launched on the transport layer of a WMN are: (i) *SYN flooding attack*, (ii) *de-synchronization attack*, and (iii) *session hijacking attack.*

SYN flooding attacks are easy to launch on a transport layer protocol like TCP. TCP requires state information to be maintained at both ends of a connection between two nodes, which makes the protocol vulnerable to memory exhaustion through flooding. An attacker may repeatedly make new connection request until the resources required by each connection are exhausted or reach a maximum limit. In either case, further legitimate requests will be ignored. One variant of such DoS attacks is the SYN flooding attack, in which an attacker creates a large number of half-open TCP connections with a target node without completing any of these requests. In the TCP protocol, two nodes have to successfully complete a *three-way handshake* mechanism before a session can be established between the pair of nodes. As shown in Fig. 6, in the first message, the node initiating the communication sends a SYN packet to the receiver node along with a sequence number. The receiver node sends a SYN/ACK message containing a sequence number and an acknowledgment sequence number to the initiator node. The initiator node then completes the handshake process by sending an ACK message containing an acknowledgment number. An attacker can exploit this protocol by sending a large number of SYN packets to a target node and spoofing the return address of the SYN packets. The SYN/ACK packets are sent back by the target node to the spoofed return address. The target node also waits for the final ACK message from the attacker keeping the half-open data structure open in its memory. When the number of such half-open connections becomes too high to create an overflow in the table which stores these data structures in the target victim node, the victim node will not be able to accept any further connections requests even from any legitimate nodes in the network, causing disruption in the network services.

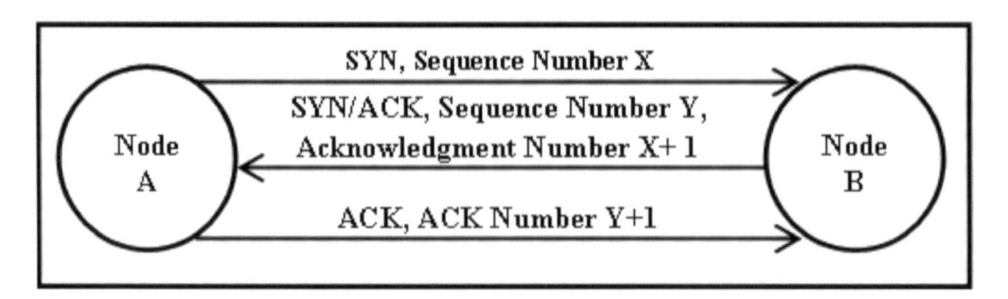

Fig. 6. Three-way handshake protocol for establishing a TCP session

Session hijacking attacks exploits the vulnerability of the transport protocols (e.g., TCP) that do not provide any security check during an on-going session. All security mechanisms are applied only during the session establishment time. In a TCP session hijacking attack, an attacker spoofs the IP address of a victim node, correctly determines the current sequence number that is expected to be generated at the victim node, and then performs a DoS attack on the victim node.

A *de-synchronization attack* refers to the disruption of an existing connection [15]. An attacker may, for example, repeatedly spoof messages to an end host causing the host to request the retransmission of missed frames. If timed correctly, an attacker may degrade or even prevent the ability of the end hosts to successfully exchange data causing them instead to waste energy attempting to recover from errors which never really exist. Wu et al. have illustrated the de-synchronization attack that leads to *TCP ACK storm problem* [30]. In this attack, an attacker injects false data in an ongoing session between two nodes by launching a session hijacking attack. The false injected data is received by one of the nodes in the communicating pair and on receipt of the data, the node sends an ACK to the other node. Since the node at the other end was not expecting the sequence number of this ACK packet, the node tries to re-synchronize the session with its communicating peer. This cycle goes on indefinitely as the ACK packets travelling back and forth in the network causes an ACK storm.

2.5 Security Vulnerabilities in the Application Layer

At the application layer, a compromise requires a full knowledge of the communicating applications (e.g., application layer formatting for traffic snooping) as well as compromising all the lower layers. The impact of such attacks can be extremely detrimental. For example, a *flooding attack* can affect the availability of the victim node as well as a large portion of the network. *Snooping attack* at the application layer can affect the integrity of the messages being communicated. However, snooping attack at the application has very low likelihood of success due to various defense mechanisms which are usually in place for protecting the lower layers. The use of encryption and authentication schemes at the higher layers also ensures that integrity of the messages is protected. The attacks in the application layer are mainly due to the viruses, malwares and worms or the repudiation attacks launched by insider nodes [30].

Mobile viruses and worms contain malicious codes which spread or replicate rapidly in a network and in the hosts and launch various types of attacks such as memory exhaustion, information leakage, phishing etc. Some types of Internet worms can scan the IP address of the nodes in a network and then send probe packets to critical UDP and TCP ports which are found in the port scanning process. The worms then attack the hosts using some application.

Repudiation attacks launched in the application layer cannot be detected or prevented by deploying firewalls at the network layer or by end-to-end encryption of traffic at the transport layer. An attacker getting an access to the information in network or in a host by sophisticated techniques can repudiate having conducting such an activity. Detection of such attacks needs sophisticated intrusion detection systems at multiple layers.

2.6 Security Vulnerabilities in the Authentication Protocols

Several vulnerabilities exist in different authentication protocols used in WMNs. Notable among these attacks are: (i) unauthorized access, (ii) spoofing attack, (iii) *denial of service* (DoS) attack, and (iv) compromised or forged MRs.

In *unauthorized access* attack, a user who is not authorized to access a resource gets access to the network services by masquerading a legitimate user. The masquerader gains all the privileges of the legitimate node. Once an attacker is successful in launching such an attack, it becomes extremely difficult for a security mechanism to detect the attacker. *Spoofing* is the act of forging a legitimate MAC or IP address of a node. IP spoofing is quite common in multi-hop communication networks like WMNs. In IP spoofing attack, an adversary inserts a false source address or puts the address of a legitimate node on the packets forwarded by it. Using such a spoofed address, the malicious attacker can intercept a termination request and hijack a session. In MAC address spoofing, the attacker modifies the MAC address in the transmitted frames originating from a legitimate node. MAC address spoofing enables attackers to evade *intrusion detection systems* (IDSs) that may be place in different nodes in a WMN. In DoS attacks, a malicious attacker sends a flood of packets to an MR, thereby making a buffer overflow in the router (i.e. in an MR). In one variant of such an attack, a malicious node can send false termination messages on behalf of a legitimate MC, thereby preventing a legitimate user from accessing network services.

In compromised or forged MR attack, an attacker is able to compromise one or more MRs in a network by physical tampering or logical break-in. The adversary may also introduce rogue MRs to launch various types of attacks in a WMN. The fake or compromised MRs may be used to attack the wireless link, thereby implementing attacks such as: passive eavesdropping, jamming, relay and false message injection, traffic analysis etc. The attacker may also advertise itself as a genuine MR by forging duplicate beacons procured by eavesdropping on legitimate MRs in the network. When an MC receives these beacon messages, it assumes that it is within the radio coverage of a genuine MR, and initiates a registration procedure. The false MR now can extract the secret credentials of the MC and can launch a spoofing attack on the network. This attack is possible in protocols which require an MC to be authenticated by an MR and not the vice versa [31].

2.7 Security Vulnerabilities in the Key Management Mechanisms

Since the robustness and security of the cryptographic protocols used in WMNs are dependent on the strength of the keys used, key management is a very critical security function in WMNs. The functions of a key management protocol include: key generation, storage, distribution, updating, revocation and providing certificate services to the legitimate nodes in the network. Sophisticated attacks may be launched by malicious attackers to get access to the keys stored in a node or during the transit of the key from the key issuing server to the nodes in a WMN. For example, any key exchange protocol based on the *Diffie-Hellman* (DH) key exchange protocol [32] is vulnerable to the *man-in-the-middle* attack [33]. The key management protocols which are based on issuing of certificates to the network nodes by a trusted key distribution server or by a trusted third party are all vulnerable to DoS attacks.

2.8 Security Vulnerabilities in the User Privacy Protection Mechanisms

Protection of user privacy is an important issue in wireless network communication. However, ensuring privacy of the users is difficult to achieve even if the messages in the network are protected, as there are no security solutions or mechanisms which can guarantee that data is not revealed by the authorized parties themselves [34]. Communication privacy cannot be assured with message encryption since the attackers can still observe who is communicating with whom as well as the frequency and duration of each communication session. In addition, unauthorized parties can get access to the location information about the positions of different MCs by observing their communication and traffic patterns. Hence, there is a need to ensure location privacy in WMNs as well. In Section 3.8, we will see how privacy can be protected with respect to message contents, data traffic and location information.

Table 1 presents a summary of various types of vulnerabilities in different layers of the communication protocol stack of a WMN and their possible defense mechanisms. The details of the different defense mechanisms are discussed in Section 3.

Table 1. Summary of different attacks on WMN protocol stack and their countermeasures

Layer	Attacks	Defense Mechanisms
Physical	Jamming	Spread-spectrum, priority messages, lower duty cycle, region mapping, mode change
MAC	Collision	Error-correction code
	Exhaustion	Rate limitation
	Unfairness	Small frames
Network	Spoofed routing information & selective forwarding	Egress filtering, authentication, monitoring
	Sinkhole	Redundancy checking
	Sybil	Authentication, monitoring, redundancy
	Wormhole	Authentication, probing
	Hello Flood	Authentication, packet leashes by using geographic and temporal information
	Ack. Flooding	Authentication, bi-directional link authentication verification
Transport	SYN Flooding De-synchronization	Client puzzles, SSL-TLS authentication, EAP
Application	Logic errors Buffer overflow	Application authentication Trusted computing, Antivirus
Privacy	Traffic analysis, Attack on data privacy and location privacy	Homomorphic encryption, Onion routing, schemes based on traffic entropy computation, group signature based anonymity schemes, use of pseudonyms.

3 Security Mechanisms against Various Attacks in WMNs

In this section, we present a detailed discussion on the various security mechanisms for defending the attacks that we mentioned in the mentioned in Section 2. We provide description of various defense techniques at each layer of the protocol stack - physical, link, network, transport and application. In addition, some secure authentication mechanisms, user privacy protection schemes, and key management protocols are also discussed.

3.1 Security Mechanisms for the Physical Layer

The jamming attack at the physical layer can be defended by employing different spread-spectrum technologies such as frequency hopping and code spreading [15]. In *frequency hopping spread spectrum* (FHSS) [35], signals are transmitted by rapidly switching a carrier signal among many frequency channels using a pseudo-random sequence which is known to both the transmitter and the receiver. Since it will be impossible for an attacker to predict the frequency selection sequence a priori, it will be difficult for him/her to jam the frequency being used at a given point of time. The interference is also minimized as the signal is spread over multiple frequencies.

In *direct sequence spread spectrum* (DSSS), each data bit in the original signal is represented by multiple bits in the transmitted signal using a spreading code. The spreading code spreads the signal over a wider frequency band which is directly in proportion to the number of bits being used. The receiver can use the spreading code with the signal to recover the original data.

Both FHSS and DSSS prohibit an attacker to intercept the radio signals. In order to successfully eavesdrop on the signal, the attacker must know the frequency band, the spreading code, and the modulation techniques being used. Spread spectrum technology also reduces the chance of interference from other radios and electromagnetic signals.

3.2 Security Mechanisms for the Link Layer

Use of error-correcting codes is a common strategy for defending against *frame collision attack* [15]. However, these codes also add additional processing and communication overhead. Although it is reasonably easier to detect any malicious collision of frames, no comprehensive defense mechanism against such an attack is known to us today.

A strategy for defending against *energy exhaustion attack* is to apply a *rate limiting MAC admission control* mechanism. This will allow the network to ignore the requests that intend to exhaust the energy of a battery driven *mesh client* (MC) node. Use of *time division multiplexing* (TDM) can be another effective strategy in which each node is allotted a time-slot for transmission of its packets [15]. However, this mechanism is vulnerable to the frame collision attack, even when it can ensure that there is no possibility of an indefinite postponement of packet transmission in the back-off algorithm in the MAC layer.

The effect of unfairness caused by a malicious attacker can be partially eliminated by using small frames. Use of smaller packets reduces the time for the attacker to capture the channel making it harder for the attacker to launch an attack [15]. However, this technique often reduces the throughput in the network due to more control overhead. In addition, it is susceptible to further unfairness as the attacker may try to retransmit quickly instead of waiting for a random interval of time.

Various other security mechanisms [36, 37] have been proposed for multi-hop wireless networks that can be applied to WMNs possibly with slight modifications. All of these schemes are based on *data confidentiality service*, *data and header integrity services*, and *robust key management service* provided by the underlying cryptographic mechanisms. The data confidentiality service provides protection against the *passive eavesdropping attack*. Although, an eavesdropper can still intercept the encrypted message, he/she cannot decrypt it for extracting any information from the message since he/she does not have any access to the encryption key. The data and header integrity services provide protection against MAC spoofing attacks. The integrity verification algorithm at the receiver node will be able to detect any message with spoofed MAC address since the message will fail integrity verification test. Replay attacks in multi-hop wireless networks can be avoided by using per-packet authentication and integrity verification [36]. These approaches are based on using a fresh key for each packet which is synchronously computed by the sender and the receiver before the packet is sent by the sender node. Any replayed packet which is encrypted by an outdated key fails the integrity check at the receiver node due to key mismatch and automatically gets discarded. Use of a fresh key for each message also protects the data from pre-computation and partial matching attacks. Since the pre-computed information needs to be applied on every message in order to decrypt it, an attack becomes extremely costly [17].

In the following sub-sections, we discuss some of the existing security mechanisms for the link and the *medium access control* (MAC) layer of WMNs.

3.2.1 Application of Synchronous Dynamic Encryption System in Mobile Wireless Domains

Soliman and Omari propose a stream-cipher cryptosystem named *synchronous dynamic encryption system* (SDES) for wireless environment that is based on permutation vector generation [36]. The proposed light-weight cryptographic scheme has a high level of security. Specifically, the protocol is robust against (i) key compromise, (ii) biased bytes analysis (an attack, in which the attacker can analyze the byte distribution in the transmitted data to derive the key in a key-stream in a stream cipher), (iii) integrity violation. The number of key exchanges between the supplicants (SUP), the *access points* (AP) and the *authentication server* (AS) is kept at the minimum in order to reduce the communication overhead and the possible vulnerability during the key exchange process. The SUPs and the APs are always kept synchronized with the AS with respect to their shared encryption keys in such a way that it is impossible for a malicious intruder to get synchronized with the AS with the dynamically changing shared secret key. The node registration process is simple and it is carried out only once during the initial registration of the node with the AS.

For ensuring security, use of two types of shared keys is proposed: (i) *secret authentication keys* (SAK) and (ii) *secret session keys* (SSK). The AS generates and transmits the initial SAK to each SUP and AP. For all subsequent mutual authentication processes with the AS, each SUP and AP uses its shared SAK. Once an SUP is initially authenticated by the AS, the AS forwards the SUP's SAK to the AP with which the SUP is associated. This reduces the delay in the authentication process. The SSK is generated per-session basis between the APs and the SUPs. The validity of an SSK is only during the session for which it is generated. For communication between two APs, the generation and distribution of the SSK is done by the AS. However, for secure communication between two SUPs, the AP associated with the source SUP generates and distributes the SSK to each SUPs. Both the keys (SAK and SSK) are used in the process of shuffling the *permutation vectors* (PVs) during the encryption process.

Since the protocol uses stream ciphers, the encryption and decryption processes are fairly simple and light-weight. For encryption, the source node carries out an XOR operation between the plaintext data and the corresponding PV to produce the ciphertext, and sends the ciphertext to the receiver node. The receiver node performs the decryption process by XORing the ciphertext with the same PV (generated at the receiver node). For the next cycle of encryption/decryption process, both the nodes synchronously generate a new PV based on their shared SAK and SSK.

Since the keys SAK and SSK serve as the seeds for generation of the stream of PV, the security of the protocol depends on the way these keys are generated and managed. The authors have proposed three modes for the generation of SAK/SSK, each mode providing different levels of security and involving different computing overhead. The three modes of operations are: (i) static shared keys, (ii) stream of shared keys, and (iii) dynamic stream of shared keys. In the first mode, the secret keys at both the communicating nodes are not changed. This makes the scheme vulnerable to cryptanalysis and successful key compromise attack. Since the permutation vectors may lead to the same stream of keys in successive cycles, it is easy to launch known plaintext-ciphertext pair attack. While this mode provides a very low level of security, it is computationally efficient since no key management is required. In the second mode, the shared keys are dynamically generated and changed after each encryption/decryption cycle. This makes the protocol secure against the known plaintext-ciphertext pair attack since it is not possible to make an easy cryptanalysis on the cipher. In addition, this mode is also secure against biased byte analysis. The additional overhead is also very low since it involves only an extra addition operation. However, in case of multiple simultaneous sessions between two nodes, due to use of the same key streams for all the sessions, breaking of one session will break all the sessions. This mode, therefore, fails to provide independent security to multiple simultaneous sessions. In the third mode, which provides the highest level of security, the data being transmitted is also used in the key generation process. Since the key generation process involves the data transmitted in the session, different sets of shared keys are generated for multiple simultaneous sessions, thereby eliminating the security loophole of the second mode. Another advantage of this approach is that data integrity guarantees that keys are not compromised during the transit. If the cipher is manipulated during the transit, it would break the synchronization of the shared keys at the two nodes. The additional overhead in this mode is due to two extra addition operations. The authors have provided detailed simulation results demonstrating the performance of the protocol.

3.2.2 A Threshold and Identity-Based Key Management and Authentication Scheme

Deng et al. [38] propose a distributed key management and authentication approach in multi-hop wireless ad hoc network using the concepts of *identity-based authentication* [39, 40] and *threshold secret sharing* [41]. The scheme proposed by the authors follows a self-organized approach that does not assume any *a priori* trust association between the nodes or any centralized trusted entity in the network. This is in contrast to the traditional PKI-based authentication for key distribution and management, wherein a trusted server is deployed to generate, distribute and manage the keys.

The scheme assumes that each node in the network has an IP address or an identity, which is unique and remains unchanged throughout the lifetime of the node in the network. Each node discovers the identities of its one-hop neighbor by a neighbor discovery mechanism. The key generation process has two phases: (i) distributed key generation and (ii) identity-based authentication. The key generation phase is responsible for distributing the master key and the public/private key pair to each node in a distributed manner. The generated private keys are used for authentication. Authentication is realized by identity-based mechanism.

In the threshold cryptography-based solution proposed by the authors, the network has a public/private key pair, which is called the *master key*. The master key is used for key generation. The master public key (say, PK) is generated by the key generator and it is known to all the nodes in the network. The master private key (say, SK) is shared among the nodes in a threshold cryptographic manner. While no node can reconstruct the master private key (secret key) alone, any k nodes among the total n nodes in the network can jointly reconstruct the key. It is, however, infeasible even for k -1 nodes to reconstruct the key by colluding among themselves. At the time of joining the networks, a node needs to acquire its private key corresponding to its identity by requesting the *private key generation* (PKG) service from at least k neighbor nodes. The identity of the node is used as its public key. The authors have proposed the computation of the public key as $QID = H(ID \parallel Expire\_time)$, where $H(\,)$ is a hash function, *ID* stands for the identity of the node, and the *Expire_time* is a time stamp expressing the time of validity of the public key. When the public key of a node expires, the node contacts at least k neighbors and presents its identity and requests for the PKG services. In the proposed scheme, since all the nodes have the master private key, any of them can act as the PKG node for any other node. Each of the k PKG service nodes generates a secret share of the new private key and sends the same to the requesting node. In this way, any group of k nodes can act as the PKG nodes for rest of the nodes such that a potential adversary who is able to compromise less than k nodes cannot get access to a node's private key. The private key generation process is depicted in Fig. 7.

The scheme uses each node's identity as its public key. Since the identity of a node can be much shorter than a 1024 bit RSA public key, less communication and storage overhead is incurred in transmitting and storing the keys. The communication overhead incurred by the scheme is mainly due to the key generation process. In the network bootstrapping time, all the n nodes have to participate in the generation of master key pair which induces large delay in set up. In addition, each node needs to broadcast a key generation request to its k neighbors at the time of joining the network. In response, each PKG service node has to send its share of the generated private key. All these messages involve appreciable communication overhead. However, a trade-off can be made between the level of security and communication overhead in the scheme. A lower value of k will reduce the communication overhead while providing a lower level of security (since fewer nodes need to be compromised by an adversary to get access to the private key of a node). For higher level of security requirement, a larger value of k should be chosen. The authors have experimentally shown how the master key generation time varies with the size of the network and the effect of the value of the parameter k on PKG service time and the ratio of successful PKG service.

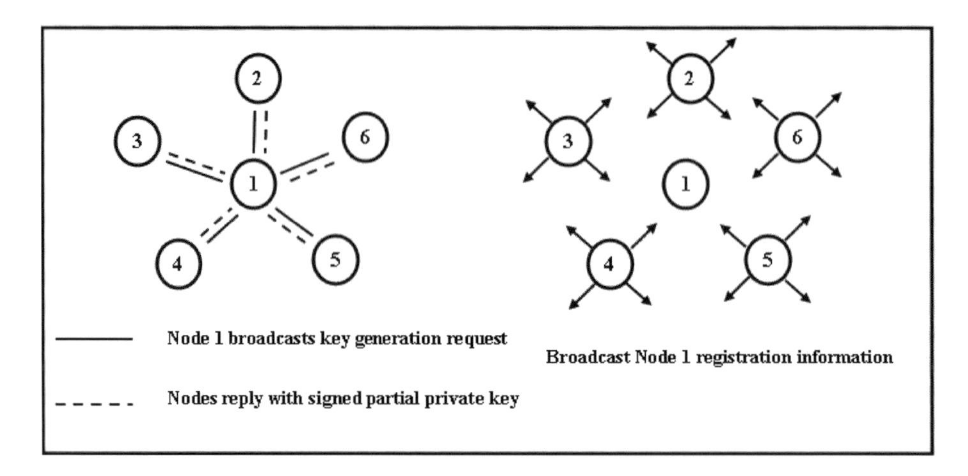

Fig. 7. The private key generation process of a node in Deng et al.'s scheme [38]

3.2.3 Wireless Intrusion Detection and Response Mechanisms

Lim et al. [42] propose an intrusion detection system for wireless networks that consists of a number of devices deployed throughout the network. Each device is placed near an *access point* (AP) and all such devices are connected to a standard wired network to allow for remote management of the networked system. The intrusion detection system works at different levels. At the basic level, the system tracks the MAC address of the network adapter. If the MAC address is not found in the whitelist, or if it is found in the blacklist, then an alert is flagged. This is known as *MAC address filtering*.

The authors have also proposed to detect passive intruders using the IEEE 802.11b *request to send* (RTS) and *clear to send* (CTS) frames. The RTS frames are normally used to check whether the transmission medium is clear and to reserve a time slot for transmission of data. The CTS frames are used for acknowledging the RTS frames. The relationships between these frames may be used to detect presence of intruders in a network. If an active Wardriver is detected, RTS messages are sent to that MAC address. If the intruder is passively eavesdropping on the network, the card will respond with a CTS message, thereby revealing its presence. Stateful monitoring of packets in the network provides further detection of intrusions. Arrival of unexpected packets like unsolicited random responses might indicate a possible probing by an intruder.

In the proposed system, several detection devices are deployed that are connected to a central server so that it is possible to determine the exact position of an attacker or a rogue access point by *triangulation*. The position information may help in determining whether the source is a valid user with a possibly unregistered MAC address or a real intruder outside the premises. The central server may be augmented with additional authentication mechanisms such as *remote authentication dial-in user service* (RADIUS) authentication to actually identify whether a valid interface card is really being used by its assigned user or by some unauthorized person.

For intrusion response, the authors have suggested techniques like *address resolution protocol* (ARP) poisoning and disassociation-reassociation on the intruder. Since DoS attacks against the intruder will have an adverse impact on the overall network performance, a possible alternative is to send specially designed malformed frames targeted to the intruder. These frames may cause crashing on the intruder's computer. However, these intrusion response mechanisms are computationally expensive and their use will surely have an adverse impact on the network services.

3.2.4 MobiSEC: A Security Architecture for Wireless Mesh Networks

Martignon et al. have presented a security architecture – MobiSEC – that provides access control in a WMN [43]. In this scheme, for authentication and key agreement between a node (an MC or an MR) with a *mesh access point* (MAP), a two-step approach is proposed. As shown in Fig. 8, in the first step, the new node (MC or MR) authenticates to the nearest MAP using 802.11i protocol [44]. In the second phase, the node uses a protocol based on *transport layer security* (TLS) and a certificate issued by a *certificate authority* (CA) with the AAA server to additionally authenticate as router and obtain the keying material required for this role in the WMN. For key distribution, use of two protocols is proposed – *server driven* and *client driven*. In the server driven protocol, each MR contacts a key distribution server for getting a key list. In the client driven protocol, the MRs obtains a seed from the server and a hash function type to generate the cryptographic keys as done in a hash chain method. Both the protocols need a mutual authentication based on certificate exchanges between the MRs and the key distribution server. MobiSEC supports mobility for both mesh clients and mesh routers. The client mobility is ensured since 802.11i protocol has client mobility support and MobiSEC is based on 802.11i authentication. The mobility of the routers in the backbone network is ensured by having all the routers using the same keying materials from the key server. Since all the routers in the backbone use the same key for authentication, router mobility in the backbone does not need any re-authentication process.

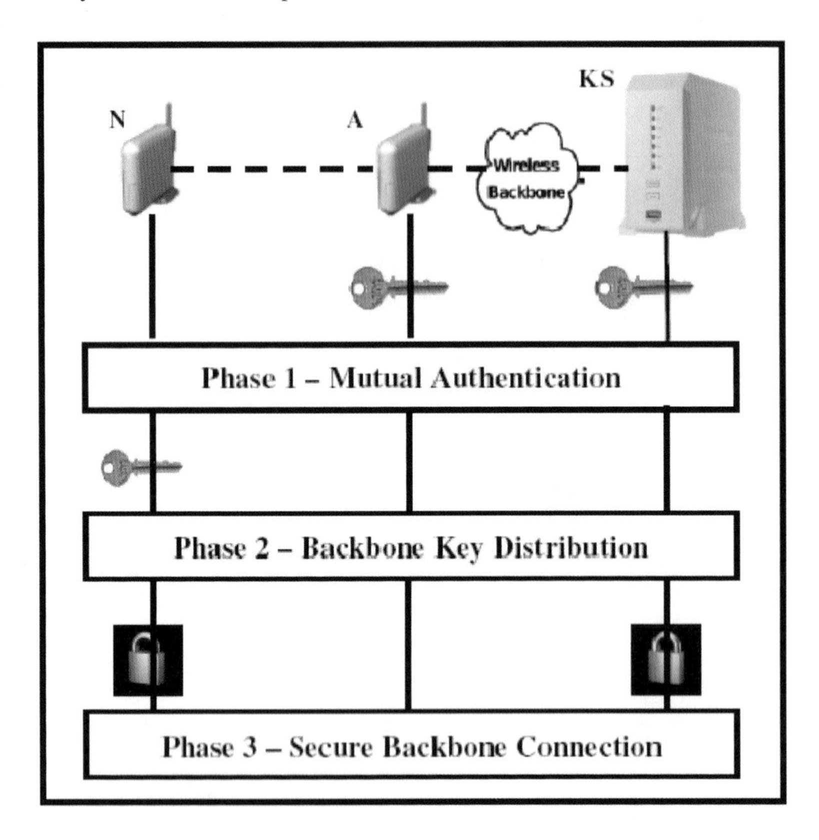

Fig. 8. Different phases of the connection process performed by a new mesh router *N* in MobiSEC

The server driven protocol for key distribution is a reactive process for delivering the keys from the key distribution server to the mesh routers. This key is used for protecting the integrity and confidentiality of the traffic exchanged in the backbone during a specific interval. The protocol ensures that all the routers in the backbone have the same key for encryption and integrity protection of the messages transmitted in the mesh backbone network. In the client driven protocol, for key distribution, the key distribution server provides only a seed and a function type that should be used to compute the sequence of keys used by the mesh routers. The generation of the sequence of keys is similar to a hash chain computation, in which the computation of the next key is based on the output of a hash function to which the input was the key used in the previous round.

MobiSEC addresses access control issues including authentication and key establishment for the mesh clients and mesh routers in a WMN. However, the architecture does not explicitly addresses issues like message confidentiality, message integrity, and protection against replay attacks. In particular, the proposal only supports communications between the mesh clients and the mesh access points and between a pair of mesh routers [45]. In addition, use of a network-wide key for the protection of all messages in the mesh backbone is another issue which may lead to a complete breakdown of the security in backbone if a single mesh router is compromised. In addition, an attacker who is in possession of the backbone key can insert bogus traffic into the network thereby causing congestion and denial of service attack. Furthermore, the use of the mesh access point as the authenticator, implicitly assumes that key distribution server will transfer the keying material to the MAP during the authentication process. However, the mesh access point and the key server do have any shared secret for establishing a secure communication session between them, and only way to transfer the key material is to encrypt it using the key in the mesh backbone. If the backbone key is used for transferring the key from the key distribution server to the mesh access point, any malicious mesh router which is in neighborhood of the mesh access point will be able to capture the key. In spite of several security loopholes, MobiSEC provides a simple architecture for handling access control and mobility management issues in a WMN.

3.2.5 Other Security Mechanisms for MAC Layer Misbehavior Detection in WMNs

Identifying various possible misbehaviors in the MAC layer and designing detection mechanisms for them has been a subject of extensive research in WLANs and ad hoc networks [46-48]. Some mechanisms for MAC layer misbehavior detection and their defense for WMNs have also been proposed [49-51].

Kyasanur and Vaidya have argued that the distributed contention resolution mechanism used in the MAC layer of IEEE 802.11 protocol is susceptible to abuse by a selfish node that does not adhere to the protocol and obtains an unfair share of the channel bandwidth [47]. To identify and penalize such selfish node, the authors have proposed a modification to 802.11 protocol. In the proposed modification, instead of the sender node selecting the random backoff time to initialize the backoff counter, the receiver node selects the backoff value and sends it in the *clear to send* (CTS) and ACK packets to the sender. The sender node uses this value of backoff in its next transmission to the receiver node. A receiver node can identify whether a sender node has deviated from the assigned backoff time by observing the number of idle slots between consecutive transmissions from the sender. If the observed number of idle slots is less than the assigned backoff, then there is a probability that the sender has deviated from the assigned backoff. The magnitudes of the observed deviations over a small number of packets transmissions are used to infer sender misbehavior with a high probability. If the sender node deviates from the assigned value, it will be assigned high backoff values in the next round to compensate for this deviation. However, this mechanism will be ineffective in case of a possible collusion between the sender and the receiver nodes or if the receiver node itself is a misbehaving node. Cardenas et al. have addressed the issue of preventing a possible colluding sender-receiver pair by ensuring randomness in the MAC protocol [52].

Konorski and Kurant have proposed a protocol called *R-hash* to prevent MAC layer misbehavior [53]. In the proposition, the winner of a contention is determined using a public hash function to the feedback each station gets from the contention. This confuses a potential misbehaving station is such a way that no modification of the probability distribution of transmission delay should be beneficial to these station.

Raya et al. have shown how a greedy user in a hotspot can substantially increase his/her share of bandwidth in the shared wireless medium by slightly modifying the driver of the network adapter of the wireless node [54]. A software system - DOMINO (Detection Of greedy behavior in the MAC layer of IEEE 802.11 public NetwOrks) - is designed that can detect and identify greedy stations without needing any modifications in the standard-compliant access points.

A proposition based on game theory for handling misbehavior in the MAC is been presented by Cagalj et al. [55]. The optimum strategy for each node has been derived in terms of controlling the cannel access probability by adjusting the contention window, so that the equilibrium point is reached in the overall network. The authors have also derived conditions under which the Nash equilibrium of the network is Pareto optimal for each node in the network as well, when some of the nodes in the network are misbehaving.

Radosavac et al. have proposed a *minimax* robust MAC layer misbehavior detection framework, with the goal of having the optimum performance of the network under the worst-case attack scenario [46]. The network performance is measured using the required number of observations to arrive at a reliable decision. The framework not only captures the presence of an uncertainty in the attacks but also pays more attention to the attacks that are most significant in terms of their adverse impact on the network performance. It also considers scenarios in which an intelligent attacker launches an adaptive attack so that its detection becomes difficult.

Naveed and Kanhere have studied attacks on dynamic channel assignment in 802.11-based WMNs, in which a compromised mesh node manipulates control messages of the channel assignment protocol in such a way that the mesh links are forced to use heavily congested channels [51].

Table 2 presents a summary of the aforementioned MAC layer security schemes.

3.3 Security Mechanisms for the Network Layer

A large number of schemes exist in the literature dealing with the issue of securing the network layer of WMNs [56-62]. In this section, we provide an overview of various security mechanisms in the network layer. A detailed discussion on these schemes can be found in [4].

As mentioned in Section 2.3, the attacks on the network layer can be either on the *route establishment* process or on the *data delivery* process, or on both. The protocols Ariadne [56] and SRP [63] intend to secure on-demand source routing protocols by using hop-by-hop authentication approach to prevent malicious packet manipulations in the route discovery process. On the other hand, SAODV [64], SEAD [57], and ARAN [58] use one-way hash chains to secure the propagation of hop counts in on-demand distance vector routing protocols. Papadimitratos and Hass have proposed a secure link state routing protocol that ensures correctness of the link state updates by using digital signatures and one-way hash chains [65]. To ensure correct data delivery, Marti et al. have presented two mechanisms - *watchdog* and *pathrater-* that can detect adversarial nodes by monitoring the packet forwarding behaviours of the nodes in a neighbourhood

[59]. SMT [60] and Ariadne [56] use multi-hop routing to prevent malicious nodes from selectively dropping data packets. Sen et al. have proposed a co-operative detection scheme for identifying malicious packet dropping nodes in an ad hoc network that is robust in presence of Byzantine failure of nodes [66]. ODSBR protocol [61, 62] provides resilience to colluding Byzantine attacks by detecting malicious links based on end-to-end acknowledgment-based feedback technique. HWMP protocol [67, 68] allows two *mesh points* (MPs) to communicate using peer-to-peer paths. This model is primarily used if nodes experience a changing environment and no root MP is configured. While the proactive tree building mode is an efficient choice for nodes in a fixed network topology, HWMP does not address security issues and is vulnerable to a numerous attacks such as RREQ flooding attack, RREP routing loop attack, route re-direction attack, fabrication attack, tunnelling attack and so on [69]. LHAP [70] is a lightweight transparent authentication protocol for wireless ad hoc networks. It uses TESLA [71] to maintain the trust relationship among nodes.

In contrast to secure unicast routing, work studying security problems specific to multicast routing in wireless networks is particularly scarce. Two notable propositions on the secure multicast routing in wireless networks are [29] and [72]. Roy et al. propose an authentication framework [29] that prevents outsider attacks in a tree-based multicast protocol - MAODV [73]. Curtmola and Nita-Rotaru have presented a protocol named "BSMR" that addresses insider attacks in tree-based multicast protocols in wireless mesh networks [72].

Table 2. Summary of some link and MAC layer defense mechanisms for WMN communication

Protocol	Salient Features
SDES [36]	It is a stream cipher-based cryptosystem for wireless networks that uses permutation vectors. The supplicants and the access points are always synchronized with the authentication server with respect to their shared keys so that it is impossible for an intruder to dynamically change the key and launch an attack. Use of stream ciphers makes the encryption and decryption processes fairly simple and light-weight. Two types of shared keys are used: (i) secret authentication keys (SAKs) and (ii) secret session keys (SSKs). Both these keys are used in the process of shuffling the permutation vectors during the encryption process. The protocol is robust against key compromise, biased bytes analysis, and integrity violation attacks.
Threshold and identity-based key management [38]	This authentication and key management scheme uses the concepts of identity-based authentication and threshold secret sharing. It assumes that each node has an IP address which is unique and remains unchanged throughout the lifetime of the network. The key generation process has two phases: (i) distributed key generation and (ii) identity-based authentication. In the key generation phase the master key and the public/private key pair are distributed to each node. The generated private key is used for authentication which is based on identity-based cryptography. The scheme is highly secure due to the deployment of a threshold authentication mechanism.

Table 2. (*continued*)

Wireless intrusion detection and response system [42]	The scheme proposes a wireless intrusion detection system (IDS) that consists of a number detection devices deployed in strategic points in a network. The IDS works at different level. At the basic level, it carries out a MAC address filtering if it cannot find the MAC address of a device in the white-list. For intrusion response, the system uses ARP poisoning and a disassociation-reassociation strategy with the suspected node. However, the proposed intrusion response mechanisms are computationally expensive and their invocation may adversely affect network performance.
MobiSEC [43]	It is an efficient scheme for secure authentication and access control in WMNs. It proposes a two-step approach for authentication of an MC with its MR. In the first step, the MC authenticates to the nearest MR. In the second phase, the MC uses a protocol that is based on the transport layer security and uses a certificate issued by a CA with the AAA server to additionally authenticate as a router. The key distribution may be server driven or client driven. In the server driven, each MR contacts a key distribution server for getting the key list, while in the client driven protocol, the MR obtains a seed from the server and a hash function to generate the key. The mobility of the MRs in the backbone is facilitated by having each router using the same key for authentication. The protocol addresses access control issues including authentication and key establishment. However, it does not address issues like message confidentiality, message integrity, and protection against replay attacks.
R-hash [53]	The scheme intends to prevent MAC layer misbehavior of nodes by using a hash function-based mechanism. The winner of a contention for accessing the wireless channel is determined by using a public hash function to the feedback that each station gets from the contention. This strategy effectively confuses a potential misbehaving station so that no possible modification can be made on the probability distribution of transmission delay for the contending stations.
Game theory-based minimax framework [46]	The goal of this game-theoretic proposition is to have a robust MAC layer misbehavior detection for optimizing the network performance under the worst-case attack scenario. It captures the presence of an uncertainty in the attacks and pays more attention to the attacks that are most significant in terms of their adverse impact on the network. The framework also considers adaptive strategy followed by sophisticated attackers which are very difficult to detect.

A key point to note is that all of the above-mentioned secure protocols for unicast or multicast routing use only some basic routing metrics such as hop-count or latency. None of them consider routing protocols that incorporate high-throughput metrics, which are critical for achieving high performance in wireless networks. On the contrary, many of them even have to remove important performance optimizations in the existing protocols in order to prevent security attacks. There are also a few studies on secure QoS routing in wireless networks [74, 75]. However, these schemes are based on strong assumptions, such as existence of symmetric links, correct trust evaluation on nodes, ability to correctly determine link metrics even in an attack scenario etc. In addition, none of them consider attacks on the data delivery phase. Dong has proposed a scheme that considers both high performance and security as goals in multicast routing and deals with attacks on both path establishment and data delivery phases [76].

As mentioned in Section 2.3, wireless networks are also subject to attacks such as rushing attacks and wormhole attacks. Defences against these attacks have been extensively studied in [77-80]. RAP [18] prevents the rushing attack by waiting for several flood requests and then randomly selecting one to forward, rather than always forwarding only the first one. Techniques to defend against wormhole attacks include *packet leashes* [77] which restrict the maximum transmission distance by using time or location information, *Truelink* [79] which uses MAC level acknowledgments to infer whether a link exists or not between two nodes, and the use of directional antennas for detecting wormhole nodes [80].

In the following sub-sections, we provide brief discussions on some of the existing well-known secure routing protocols for WMNs. For more details on several such protocols, readers may refer to [4].

3.3.1 Authenticated Routing for Ad Hoc Networks (ARAN)

Authenticated routing for ad hoc networks (ARAN) is an on demand routing protocol that provides authentication of route discovery, route setup, and route path maintenance using cryptographic certificates [58]. It can detect and protect against malicious attackers without requiring any pre-deployed network infrastructure. However, it assumes a small amount of prior security coordination among the nodes. A trusted certificate server is used whose public key is assumed to be known to all nodes. On joining the network, each node receives a certificate issued by the trusted server. The certificate received by a node contains the IP address of the node, the public key of the node, the timestamp of creation of the certificate and the time at which the certificate would expire. A node uses its certificate for authenticating itself during the routing process. At the time of route discovery, a node broadcasts a signed *route discovery packet* (RDP). The RDP includes the IP address of the destination node, the certificate of the source node, a *nonce*, and a timestamp. The RDP is signed by the private key of the source node. Each node in the route discovery path validates the signature of the previous node, removes the certificate and the signature of the previous node, and records the IP address of the previous node. The node then signs the original contents of the packet, appends its own certificate and forwards the message after signing it with its private key. When the RDP reaches the intended destination node, the node creates a *route reply packet* (REP) and unicasts it back along the reverse path. The REP includes an identifier of the packet type, the IP address of the source, its certificate, the nonce, and the associated timestamp that was initially sent by the source node. On receiving the REP, the source node verifies the signature of the destination node, and the nonce. An *error message* (ERR) is generated if the timestamp or nonce does not match the requirements or if the certificate fails in the authenticity validation process. ARAN is a secure protocol that can prevent a number of attacks such as unauthorized participation of nodes, spoofed route signaling, spurious routing messages, alteration of routing packets, manipulation of the TTL values in the packets, and replay attacks. However, it is vulnerable to DoS attacks which are launched by flooding the network with bogus control packets. Since signature verification for each packet is required, the attacker can force a node to discard some of the control packets if the node cannot verify the signatures at the rate which is equal to or greater than the rate at which the attacker is injecting the bogus control packets.

3.3.2 Secure Efficient Ad Hoc Distance Vector (SEAD) Routing Protocol

The *secure efficient ad hoc distance vector* (SEAD) [57] is a secure and proactive ad hoc routing protocol based on the *destination-sequenced distance vector* (DSDV) routing protocol [81]. The protocol deploys a one-way hash function for computing the hash chain elements which are used to authenticate the sequence numbers and the metrics of the update messages of the routing tables. The protocol ensures a mutual authentication between a source and a destination pair. The source of each routing table update message is also authenticated so as to prevent creation of any possible routing loop by an attacker which may try to launch an impersonation attack. Although the hash chains are useful for authenticating the metric and the sequence number, they are not sufficient for defending against a malicious node which can advertise the same distance and sequence number that the node has received. To defend against such malicious nodes, *hash tree chains* are used in conjunction with *packet leashes* [77], in which the address of the authenticator is tied with the address of the sender node. This prevents an attacker from replaying to an authenticator that it hears in its neighborhood. The protocol uses TESLA TIK [71] for shared key distribution among each pair of nodes in the network. SEAD can defend against routing loop attack if the loop does not contain more than one attacker. The protocol is simple and easy to implement by making a slight modifications to the DSDV protocol. The use of one-way hash chain for authentication reduces the computational complexity. The main drawback of the protocol, however, is the requirement of a trusted entity for distribution and maintenance of the verification element of each node. The trusted entity can also be a single-point-of-failure in the protocol operation.

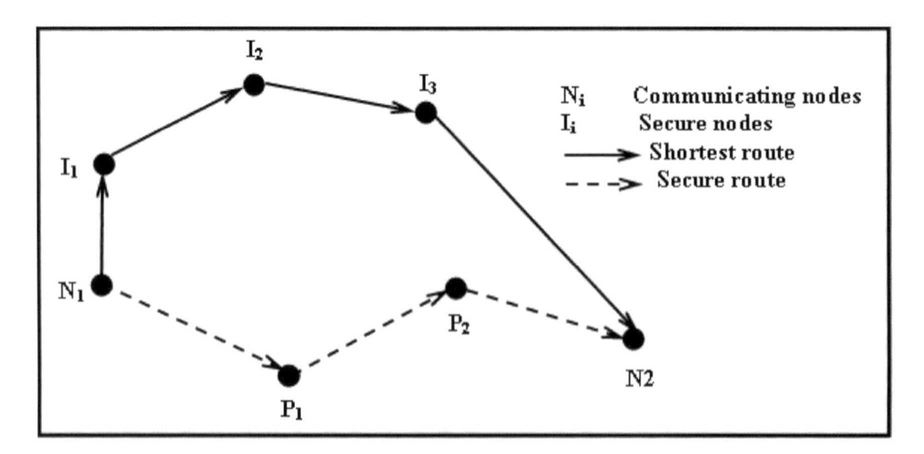

Fig. 9. Illustration of the use of trust metrics of nodes in SAR protocol

3.3.3 Security-Aware Ad Hoc Routing (SAR) Protocol

The *security-aware ad hoc routing* (SAR) protocol uses security as one of the key metrics in the route discovery and maintenance operations, and provides a framework for enforcing and measuring the attributes of the security metric [82]. Unlike traditional routing protocols which utilize distance (measured by the hop-counts), location, power and other metrics for routing path determination, SAR uses security attributes (such as trust values and trust relationships among nodes) in order to define a routing metric. SAR extends on-demand ad hoc routing protocols such as AODV [83] or DSR [84] in order to incorporate the security metric into the route discovery messages. The protocol ensures that an intermediate node that receives an RREQ packet can process or forward it only if the node can provide the required security or has the required authorization and trust level. If the node cannot provide the required security, the RREQ packet is dropped. If an end-to-end path with the required security attributes can be found, a suitably modified RREP message is sent from an intermediate node or the destination node. The security metric of SAR can be specified by a hierarchy of trust among the nodes. In order to define the trust levels, a key distribution or secret sharing mechanism is utilized in which the nodes belonging to a particular trust level share a key among them. Since the nodes of different security levels do not share any key, they cannot decrypt or process routing packets. SAR allows an application to choose its required level of security. However, the protocol needs different keys for different levels of security. Hence, with the increase in the number of security levels to be maintained, the number of keys to be managed also increases leading to an increase in storage and computational overheads.

Fig. 9 illustrates how trust metric is used in SAR. As shown in Fig. 9, the packets from the source node N_1 have two paths to travel to the destination node N_2. The shorter among these two paths, however, passes through nodes P_1 and P_2, whose trust levels are low. Hence, the protocol chooses a longer but secure path that passes through the trusted nodes I_1, I_2, and I_3.

3.3.4 Secure Ad Hoc On-Demand Distance Vector (SAODV) Routing Protocol

The *secure ad hoc on-demand distance vector* (SAODV) routing protocol [64] is a secure extension of the AODV protocol [83]. The main objective of SAODV is to ensure integrity, authentication, and non-repudiation of the messages used in the AODV protocol. SAODV uses two mechanisms to secure routing messages: (i) digital signatures to authenticate the non-mutable fields of the messages, and (ii) hash chains to secure the hop count field which is the only mutable information in the packets. Since the protocol uses asymmetric cryptography for digital signatures, a key management mechanism is needed for enabling a node to acquire and verify the public key of other nodes in the network. SAODV uses the following additional fields in a routing packet header: (i) the *hash function* field identifies the one-way hash function used for securing the hop-count information, (ii) *max hop count* is a counter that specifies the maximum number of nodes a packet is allowed to go through, (iii) *top hash* field is the result of the application of the hash function on the max hop count times to a randomly generated number, and (iv) *hash* field is the random number used for routing. Each time a node sends an RREQ or an RREP message, it generates a random number and sets the value of the *max hop count* field same as the *time to live* (TTL) field in the IP header. The node then sets the *hash* field with the random number and also sets the *identifier* field of the hash function. Finally, the node computes the *top hash* by hashing the random number *max hop count* times. The protocol enables the receiver node to verify the hop count of each message by applying the hash function (*maximum hop count – hop count*) times to the value in the hash field. If the computed hash value and the value in the top hash field match, the hop count is successfully verified. Each time an RREQ message is re-broadcasted or an RREP is forwarded, the node has to apply the hash function to the hash field. Digital signatures are used to sign every field except the *hop count* and the *hash field*. Although the use of hash function and digital signature makes the scheme secure, the intermediate nodes cannot reply to an RREQ message if they have a fresh route to the destination node in their caches. In order to overcome this problem, the authors propose two solutions. The first solution does not allow the intermediate nodes to respond to a RREQ message and make then simply forward the RREQ message, since they cannot sign the message on behalf of the destination node. The second solution involves addition of a signature that can be used by intermediate nodes to reply to an RREQ by the node that originally created the RREQ. The *route error* (RERRs) messages are secured using digital signatures. A node that generates or forwards an RERR message, signs the whole message (except the destination sequence number) using its shared key with its neighbor node. Since the destination node does not authenticate the destination sequence number, a node should not update the destination sequence numbers of the entries in its routing table based on the RERR messages. The performance characteristics of SAODV are similar to those of the AODV protocol. However, the communication overhead in SAODV increases very rapidly with increase in mobility of the nodes due to the use of expensive asymmetric cryptographic operations.

3.3.5 Secure Routing Protocol (SRP)

The *secure routing protocol* (SRP) [63] is a secure extension that can be applied to many of the existing routing protocols especially to the DSR protocol [84]. The protocol requires the existence of a *security association* (SA) between a source-destination pair. This security association is utilized to establish a shared secret key between the two nodes. The protocol appends a header to each routing packet. The source node sends an RREQ with a *query sequence* (QSEQ) number which is used by the destination node to check whether the RREQ is outdated or valid, a random *query identifier* (QID) that identifies the specific request, and the output of a keyed hash function. The input to the function is the IP header, the header of the base protocol, and the shared secret key between the pair of nodes. The RREQ message generated by the source node is protected by a *message authentication code* (MAC) computed using the shared key between the source-destination pair. The RRQEs are broadcast to all the neighbors of the source node. Each neighbor that receives the RREQ for the first time appends its identifier to the RREQ and further broadcasts it in the network. All nodes maintain a priority ranking of its neighbors based on the rate at which the queries are generated from them. Higher priorities are assigned to nodes which generate queries at lower rates. The destination node checks the validity of the query and verifies its integrity and authenticity by computing and matching the keyed hash value. If the query is found to be valid and if it passes the integrity and authentication verification tests, the destination node generates a number of replies (RREPs) using different routes. This protects against attacks from malicious nodes that may attempt to modify the RREPs. An RREP includes the entire path from the source to the destination, the *query sequence* (QSEQ) number, and the *query identification* (QID) number. The integrity and authenticity of an RREP message is done using message authentication code in the same manner as in case of an RREQ message. Route maintenance is done using route error messages. The route error messages are source-routed along the path which is reported to be broken by an intermediate node. When the notified node receives a route error packet, it compares the route followed by the packet with the prefix of the corresponding route as reported in the route error packet. However, this approach has a security loophole since a fabricated route error attack can be easily launched by a malicious node. SRP is a light-weight protocol that can be easily implemented on a base routing protocol. However, as mentioned earlier, it cannot prevent unauthorized modifications of routes by malicious nodes.

3.3.6 ARIADNE: A Secure On-Demand Routing Protocol for Ad Hoc Networks

Ariadne [56] is a secure on-demand routing protocol that is an extension of the *dynamic source routing* (DSR) protocol [84]. In contrast to the SEAD protocol [57] which is based on hop-by-hop authentication and message integrity, Ariadne assumes an end-to-end security approach. The protocol assumes the existence of a shared secret key between a pair of nodes and uses a *message authentication code* (MAC) for authenticating messages using this secret key. In fact, Ariadne proposes three schemes for authentication of messages: (i) authentication between two nodes using their shared secret key, (ii) shared secrets between communicating nodes combined with broadcast authentication using TESLA [71, 85], and (iii) digital signatures. In TESLA, a sender node generates a one-way key chain and defines a schedule based on which the keys are disclosed in the reverse order of their generation [71, 85]. This makes time synchronization a critical requirement for Ariadne. In the route discovery phase, the source node sends an RREQ message that includes the IP address of the source node, an ID that identifies the current route discovery process, a TESLA time interval for indicating the expected arrival time of the request to the destination, a hash chain that includes the address of the source node, the destination node address, the ID of the destination, and two empty lists – a *node list* and a *MAC list*. A neighbor, node on receiving the RREQ message, first checks the validity of the TESLA time interval so that the time interval is not too far in the future and its corresponding keys are not disclosed yet. A request with an invalid time interval is dropped by the neighbor nodes. If the time interval is valid, then the neighbor node inserts its address in the node list, replaces the hash chain with a new one that contains the address of the neighbor nodes along with the addresses of the nodes in the previous hash chain, and appends a *message authentication code* (MAC) of the entire packet to the MAC list. The MAC is computed using the TESLA key that corresponds to the time interval of the RREQ message. The neighbor node then broadcasts the RREQ message further in the network. The destination node buffers the RREQ and checks for its validity. An RREQ is considered to be valid if the keys with respect to the specified time interval have not yet been disclosed, and if the included hash chain can be verified. If the RREQ message is found to be valid, the destination node generates and broadcasts an RREP message. An RREP message contains all the fields of an RREQ message. In addition, it also contains a *target MAC* field and an empty *key list*. The target MAC field is filled in using the computed MAC of the preceding fields of the RREP message and the key that the destination shares with the initiator node. The RREP message is forwarded back to the initiator along the reverse path included in the node list as specified by the DSR protocol. An intermediate node, on receiving the RREP message, waits until the specified time interval allows it to disclose its key. On expiry of the specified time interval, the intermediate node discloses the key and appends the RREP to the key list and forwards the message to the next node. Upon receiving an RREP message, the initiator node verifies the validity of each key in the key list, checks the authenticity of the target MAC, and each MAC in the MAC list. The route maintenance in Ariadne is done in a similar manner as in DSR protocol. A node forwarding a packet to the next hop along the source route returns an RERR message to the packet's original sender if it is unable to deliver the packet to the next hop after a limited number of retransmission attempts. The most critical requirement for the operation of the Ariadne protocol is the existence of a clock synchronization mechanism. The base Ariadne protocol is vulnerable to wormhole attack. Hu et al. have proposed a security solution to defend against the wormhole attack using a mechanism called *packet leashes* [77].

3.3.7 Security Enhanced AODV Protocol

Li et al. have proposed a *security enhanced AODV* (SEAODV) routing protocol [69] that employs Bloom's key pre-distribution scheme [86] to compute *pair-wise transit key* (PTK) through the flooding of enhanced hello message. The protocol uses the established PTK to distribute the *group transit key* (GTK). The PTKs and GTKs are used for authenticating unicast and broadcast routing messages respectively. A unique PTK is shared between each pair of nodes, while the GTK is shared secretly between a node and all of its one-hop neighbors. A *message authentication code* (MAC) is attached as the extension to the original AODV routing message to guarantee the authenticity and integrity of the messages in a hop-by-hop manner. In order to ensure hop-by-hop authentication, each node must verify the incoming messages from its one-hop neighbors before re-broadcasting or unicasting the messages. The route discovery process in SEAODV is similar to that in AODV except for a minor difference. In SEAODV, an MAC extension is appended to the AODV routing packet. The MAC is computed based on the GTK of the node that broadcasts an RREQ message in its neighborhood. A neighbor node, on receiving the RREQ message, computes the MAC of the received message using the GTK. If the computed MAC matches with the received one, the received RREQ is considered to be authentic. The neighbor node then updates the hop-count of the RREQ message and its routing table. Further, it sets up a reverse path back to the source node by recording the node from which it has received the RREQ message. Finally, the node computes a message authentication code of the updated RREQ using the GTK and appends the MAC to the RREQ before re-broadcasting the RREQ. The destination node on receiving an RREQ generates an RREP message and unicasts it back to the source node along the reverse path. Since the RREP message is authenticated at each hop using the PTKs, an adversary can no way re-direct the traffic to some other route. A node generates a route error (RERR) message if it receives a packet for which it does not have an active route in its routing table, or the node possibly detects a broken link for the next hop of an active route. Although SEAODV is a secure extension of the AODV protocol, it is vulnerable to *RREQ flooding attack*. However, since the protocol provides authentication for RREQs from nodes that are in the list of active one-hop neighbors, such an attack would be detected very quite early before it can cause a serious damage in network communication.

3.3.8 Secure Link State Routing Protocol (SLSP)

The *secure link state routing protocol* (SLSP) [65] is a secure proactive routing protocol for multi-hop wireless networks like MANET and WMNs. Its major goal is to enable a secure topology discovery and distribution of link state information across a wireless network. The critical requirement of SLSP protocol is the existence of an asymmetric key pair for every network interfaces of a node. The participating nodes in the network are identified by the IP addresses of their respective network interfaces. The key management is done by a group of nodes or by the use of *threshold cryptography* [41, 87]. The operation of SLSP can be logically divided into three parts: (i) public key distribution and management, (ii) neighbor discovery, and (iii) link state updates. The nodes broadcast their public key certificates within their zone using *public key distribution* (PKD) packets. The nodes verify the subsequent packets from the source node by matching its signed PKD packet. The link state information is also broadcasted periodically using *neighbor lookup protocol* (NLP) [65]. The NLP protocol uses signed *HELLO* messages which include the sender's MAC address and the IP address for the current network interface. NLP can inform SLSP about any suspicious observations (e.g. two different IP addresses having the same MAC address, or a node claiming the MAC address of the current node etc.) by generating notification messages. SLSP discards suspicious packets for which it has received a notification message. The hop count information in a packet is authenticated using hash chains. The *link state update* (LSU) packets are identified by the IP address of the initiating node [65]. The hash chains are authenticated using a digitally signed part of the LSU message. When a node receives an LSU it verifies the attached signature using a public key that it received earlier in the public key distribution phase of the protocol. To protect against DoS attacks, the nodes maintain a priority ranking of each neighboring node based on the rate of out-bound traffic. Nodes with lower rates of LSU generation are assigned higher priorities. This prevents a possible attack by a malicious node that may attempt to flood the network with spurious control packets, since the node will be always assigned a very low priority due its high rate of traffic generation. SLSP protocol provides security in the neighbor discovery process and uses NLP to identify spoofing attack by detecting discrepancies between the IP and the MAC addresses. However, the protocol is vulnerable to colluding malicious nodes that fabricate spurious links between themselves and flood this information in their neighborhood. Further, due to the use of asymmetric key cryptography, the protocol involves higher computational overhead.

3.3.9 Secure Optimized Link State Routing (SOLSR) Protocol

Secure optimized link state routing (SOLSR) protocol [88] is a secure extension of the base *optimized link state routing* (OLSR) protocol [89]. OLSR is a proactive link state routing protocol that employs an optimized flooding mechanism for diffusing link-state information. The optimization in OLSR is achieved by the use of *multi point relays* (MPRs). Fig.10 illustrates how the use of MPRs drastically reduces the overhead of control message communication.

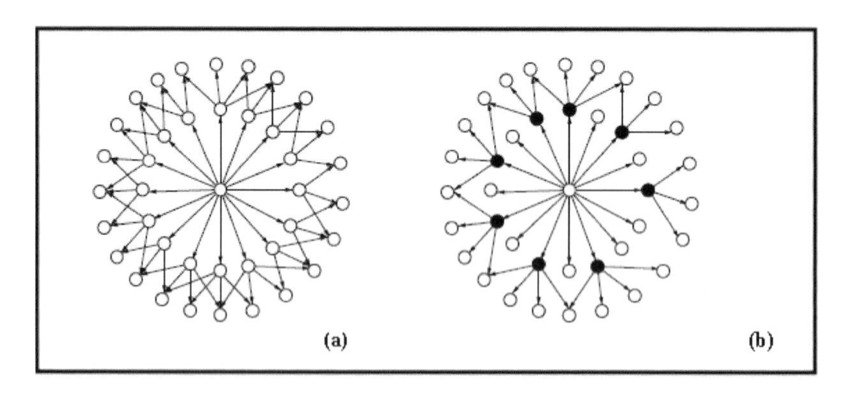

Fig. 10. OLSR: (a) Each2-hop neighbor broadcasts. (b) Only MPRs transmit the broadcast [87].

In OLSR, each node selects MPRs from among its neighbors in such a way that a message emitted by a node and further forwarded by the MPR nodes will be received by all nodes which are two-hops away from the source. Each node maintains its *MPR selector set*. On receiving an OLSR control message, a node consults its MPR selector set for deciding if the message is to be retransmitted. If the last hop of the control message is an *MPR selector*, then the message is to be retransmitted; otherwise, it is not retransmitted. If a message is to be broadcasted network-wide, it is sufficient to send it to a subset of the neighbors of the source node. This subset consists of the MPR set of the source node. In this way, OLSR optimizes message communication in a multi-hop wireless network. However, the OLSR protocol has a number of security vulnerabilities which can be exploited by a malicious node to launch attacks such as: (i) incorrect control traffic generation, (ii) incorrect *HELLO* message generation by identity spoofing or link spoofing, (iii) incorrect *topology control* (TC) message generation by identity spoofing or link spoofing, and (iv) incorrect control traffic relaying. The SOLSR protocol defends against such attacks by providing authentication for the OLSR signalling packets. The protocol uses *message authentication codes* (MACs) in every hop to ensure integrity and authentication of the routing messages. Every message is also time-stamped in order to ensure the freshness of the message. To prevent false message injection by malicious nodes, a signature is generated by the source node of each control message and the signature is appended with the control message. The receiver node checks the authenticity of the signature and the integrity of the message. Depending on the level of security desired, either an asymmetric key cryptographic method or a shared secret key cryptographic method is used for signature generation and verification and message integrity checking. The time stamps in the control messages are used to defend against replay attack. For each message that is generated by a source node, a unique timestamp is included in the message. If the difference between the time at which a message is received by a receiver and the timestamp of generation of the message at the source node does not exceed a threshold value than the message is considered to be fresh and it is accepted by the receiver provided it passes the authentication and integrity verification. There are various approaches for timestamp generation: (i) synchronous, real-time timestamps, (ii) non-volatile timestamps, and (iii) timestamps obtained using a challenge-response mechanism [88]. SOLSR involves more communication overhead than the base OLSR protocol. However, the computational overhead may be reduced by the use of symmetric key cryptography for message authentication and integrity verification purposes. The protocol is ideally suited to networks with low mobility like the WMNs. However, with a large network, it exhibits a scalability problem in its performance.

3.3.10 Hybrid Wireless Mesh Protocol (HWMP)

Bahr has proposed a *hybrid wireless mesh protocol* (HWMP) [90]. It is the default routing protocol for IEEE 802.11s WLAN mesh networking. Every IEEE 802.11s compliant device is able to use this protocol for selecting routing paths. HWMP has both reactive and proactive routing capabilities. It is based on the adaptation of AODV routing protocol [83] into a novel protocol called *radio-metric AODV* (RM-AODV) [91]. Unlike the AODV protocol that works on the network layer using the IP addresses, RM-AODV works on the MAC layer using the MAC addresses. RM-AODV uses a radio-aware metric for routing that helps in path selection. A *mesh portal* (a mesh point that has a connection to a wired network and acts as a bridge between the mesh network and the wired network) is configured to periodically broadcast mesh portal announcements to set up a tree with the mesh portal acting as the root of the tree. The created and maintained tree allows proactive routing with the mesh portal acting as the destination node. The proactive extension of HWMP uses the same distance vector routing strategy as RM-AODV and utilizes the routing control messages of RM-AODV for routing purpose. HWMP uses destination sequence numbers for detecting expired and outdated routing information. Routing packets with newer sequence numbers are always considered for routing and the packets with older sequence numbers are discarded. All routing table entries have specified validity time. The lifetime associated with a routing path is reset every time data frames are transmitted over that path.

The reactive components of HWMP uses a route discovery process which is similar to that used in the AODV [83] and the DSR [84] protocols. A source mesh points that needs to discover a path towards a destination mesh point broadcasts a route request (RREQ) packet. The destination mesh point or an intermediate mesh point that has a fresh route information to the destination node replies with a unicast route reply (RREP) message. However, the route discovery process in HWMP is adapted to the requirements of the IEEE 802.11s path selection protocol, and hence the MAC addresses of the nodes are used in routing and radio-aware links metrics are used for determining the optimal route path. The protocol uses the *airtime link metric* as defined by IEEE 802.11s standard [92] for this purpose.

HWMP has a proactive routing component as well. In deployment scenarios (for instance in a wireless mesh network that provides access to the Internet), large proportion of the traffic in a mesh network are destined for only one or a few mesh points. Since a proactive routing strategy to the mesh portal will be more efficient for such scenarios [90], the mesh portals are configured to periodically broadcast mesh portal announcements through wireless mesh network. A tree with the mesh portal as the root is constructed and a distance vector-based routing strategy as used in RM-AODV is adopted. The messages of RM-AODV are gainfully utilized in proactive routing.

The use of the proactive extension of RM-AODV and the reactive component of HWMP can be configured in the mesh portal node. This implies that the proactive component is optional in a mesh portal. For operation of the proactive component, a mesh portal is to be configured so that it can periodically broadcast mesh portal announcements. This triggers a root selection and routing tree construction process for the operation of the proactive routing protocol.

3.3.11 Byzantine-Resilient Secure Multicast Routing (BSMR) Protocol

In multicast routing, data is delivered from a source node to multiple destination nodes which belong to a multicast group. Multicast routing protocols for wireless multi-hop networks use various approaches such as flooding, gossiping, geographical positions and are based on various communication structures such as meshes or trees. Designing a secure multicast routing protocol for wireless networks is more difficult than designing a unicast routing protocol due to several unique challenges that multicast communications bring in [72]. Curtmola and Nita-Rotaru have proposed a secure multicast routing protocol, named BSMR, that is resilient against Byzantine attacks [72]. The authors have first identified various possible attacks on multicast routing such as: Byzantine behavior of malicious nodes either alone or in collusion, which may lead to packet dropping, false packet injection, modification or replaying of packets etc at the network layer, intentional collision of frames at the MAC layer, and jamming at the physical layer. Further, in a multicast routing protocol, an adversary can attack the control messages for route discovery, route setup, and tree construction and management etc, and the data packets. In addition to attacks such as false route advertisement, generation of malicious route error messages may lead to network or multicast tree partitioning. Attacks on data packets include eavesdropping, modification, replay, false data injection, selective packet forwarding etc. Many of these attacks such as selective packet forwarding and DoS attacks cannot be prevented by use of cryptographic mechanisms only.

In the BSMR protocol [72], multicast data is communicated from the source to the members of a multicast group even if there are Byzantine attackers in the network as long as the multicast group members can be reached from the source node using paths that do not contain any adversarial node. An authentication mechanism is used that ensures that only authenticated nodes are allowed to perform certain critical operations such as joining in the multicast tree using valid group certificates. BSMR is also robust against a possible attack by a malicious node that may attempt to prevent a legitimate node from establishing a route to the multicast tree by flooding spurious route request or route reply messages. Selective packet forwarding attack is mitigated by using a reliability metric that detects adversarial behavior. The metric uses a list of link weights. A link with higher weight has lower reliability. Each node maintains its list corresponding to the weights of its links. This list is appended in each route request sent by the node so that the adversarial links are always avoided due to their higher weights when a new route to the tree is established. The reliability of a link is determined by the throughput of the link, and the nodes dynamically update their weight lists based on the link reliabilities. The authentication framework involves the use of a *tree token* by each of the authenticated members in the multicast tree. The tree token is periodically refreshed and distributed by the multicast group leader. The tree token is encrypted using the *pair-wise shared keys* established between each pair of neighbor nodes in the multicast tree. To allow any node in the network to verify whether the tree token possessed by a tree node is really a valid one, the group leader periodically broadcasts a *tree token authenticator*. The tree token authenticator can be expressed as *f (tree token)*, where *f* is a *collision-resistant one-way trap door function*. Any node can check the authenticity of a given tree token by applying the function *f* on it and checking the result with the value received from the tree token authenticator.

In order to prevent a node from falsely claiming that it is at a smaller hop distance from the group leader node than actually it is, the authors have proposed a technique based on *one-way hash chains*. The last element of the hash chain is referred to as the *hop count anchor*, which is periodically broadcasted in the network by the group leader thereby preventing a node to make any false claim about its distance from the group leader.

For joining a multicast group, a node needs to make a route discovery to the multicast tree. To prevent any possible attack, all route discovery messages are authenticated using the public key corresponding to the group certificate. All tree nodes use tree token to prove their membership in the current multicast group. For joining a multicast group, the requesting node first broadcasts an RREQ message that includes the node identifier, its weight list, the multicast group identifier, the last known group sequence number, and a request sequence number. The RREQ is flooded in the network till it reaches a tree node that has a group sequence number which is either greater than or equal to the group sequence number in the RREQ message. On receiving the RREQ message, the tree node initiates a response. The RREP message includes the node identifier, its recorded group sequence number, the requester's identifier, a response sequence number, the group identifier, and the weight list from the RREQ. To prove its current tree node status, the node also includes the current token encrypted with the requester's public key in the RREP message. The RREP is also flooded in the network till it reaches the requester node. The BSMR protocol uses a robust multicast tree maintenance strategy which is activated on occurrence of events such as pruning, link breaks, and network tree partitioning. The pruning messages are authenticated using the pair-wise keys shared between the tree neighbors. Even if a malicious node that has a sub-tree under it prunes itself, the legitimate nodes in the sub-tree will be able to reconnect to the tree using a procedure proposed in the protocol.

3.3.12 Secure On Demand Multicast Routing Protocol (SODMRP)

Dong et al. have proposed a secure version (SODMRP) [93] of the *on demand multicast routing protocol* (ODMRP) [94]. Before discussing the salient features of SODMRP, we first provide a brief overview of ODMRP.

ODMRP is an on-demand multicast routing protocol for multi-hop wireless networks. The protocol uses a mesh of nodes that constitutes a multicast group. Nodes are added to multicast groups using a route selection and activation protocol. The source node periodically reconstructs the mesh by flooding a JOIN QUERY message in the network so that membership information and the routing information are updated regularly. The interval between two successive mesh constructions is known as a *round*. JOIN QUERY messages are flooded in the network using a *basic flood suppression mechanism* which only allows the processing of the first received copy of a flooded message. When a JOIN QUERY message reaches a receiver node, the latter activates the path from itself to the source node by constructing a JOIN REPLY message and then broadcasting it. The JOIN REPLY message contains entries for each multicast group it wants to join. Each entry has a next hop field which is filled with the corresponding upstream node. When an intermediate node receives a JOIN REPLY message, it checks whether it is on the path to the source or not by verifying if the next hop field of any of the entries in the message matches with its own identifier. If the node finds that it lies on a path to the source, it makes itself a part of the mesh (the FORWARDING GROUP), and creates a new JOIN REPLY message using the matched entries. The node then broadcasts the JOIN REPLY message. As the JOIN REPLY messages reach the source node, the multicast receivers become connected to the source through a mesh of nodes (the FORWARDING GROUP) which guarantees delivery of multicast data. As long as a node is in the FORWARDING GROUP, it rebroadcasts any non-duplicate multicast data packets that it receives from its neighbors. To leave a multicast group, the receiver nodes just do not reply to the JOIN QUERY messages. They are not required to explicitly send any messages for this purpose. The participation of a node in the FORWARDING GROUP expires if its forwarding-node status is not updated in each time interval.

For enhancing the throughput of the ODMRP protocol Dong et al. first propose a high throughput algorithm called ODMRT-HT [93]. The fundamental differences between ODMRP and ODMRP-HT are: (i) unlike ODMRP which chooses links with minimum delay for routing, ODMRP-HT selects routes based on link quality metrics for achieving high throughput, and (ii) ODMRP-HT uses a *weighted flood suppression mechanism* to flood JOIN QUERY messages instead of a *basic flood suppression mechanism* [93]. Each node measures the link quality of each of the links with its neighbors based on a probing mechanism. The source node floods the JOIN QUERY message periodically which contains a route cost field based on the cumulative costs of the links of the route on which the message has travelled. When a node receives a JOIN QUERY message, it updates the route cost field by adding the metric of the last link over which the message has travelled. JOIN QUERY messages are flooded using a weighted flood suppression mechanism. In this approach, a node processes duplicate messages received over a fixed interval of time and rebroadcasts flood messages that advertise a better metric as indicated in the route cost field in the messages. Each node also records the node in the upstream path to the source node from which it has received the best link quality metric in the JOIN QUERY message. The receiver node, as in case of ODMRP, constructs a JOIN REPLY packet which is forwarded towards the source node through the *best path* as determined by the metric. The nodes on this best path are chosen as the members of the FORWARDING GROUP.

Dong et al. have identified various metric manipulation attacks that may be launched on ODMRP-HT protocol [93]. These attacks have been broadly categorized into two groups: (i) *local metric manipulation* (LMM) and (ii) *global manipulation* (GMM) [93]. Both these attacks types are Byzantine in nature since they may be launched by legitimate member nodes in the network which possess the necessary credentials. In the LMM attack, a malicious node intentionally increases the quality of its adjacent links and thereby creates a false perception among its neighbor about the link qualities. These falsely advertised good quality links have higher chances of being chosen by the neighbors and in this way the malicious node gets included on the selected routes. The GMM attack, on the other hand, involves a malicious node that arbitrarily changes the cumulative value of the route metric in a flooded packet before rebroadcasting it. In this way, the malicious node is able to not only manipulate its own contribution to the path metric in terms of its advertised link quality, but it can also adjust the contributions of the previous nodes on the routing path. Both these attacks are *epidemic* in nature and can have a detrimental effect on the performance of the throughput of the multicast routing.

To defend against the LMM and GMM attacks, Dong et al. have proposed the SODMRP protocol [93]. SODMRP uses an authentication framework which ensures that each node in the mesh network has a public-private key pair. In addition, each node possesses a *client certificate* that binds its public key to its one unique identity. Every packet is authenticated so that it is not possible for an outsider to inject any spurious packet in the network. For detection of attacks, two reactive approaches have been proposed: (i) a *measurement-based attack detection* protocol, and (ii) an *accusation-based reaction* protocol. The measurement-based attack detection strategy is based on the ability of the honest nodes in the network to detect discrepancy between the *expected packet delivery ratio* (ePDR) and the *perceived packet delivery ratio* (pPDR). The ePDR of a route is estimated from the value of the metric of the route, while the pPDR of a route can be determined by measuring the throughput along the route. Both the FORWARDING GROUP members and the received nodes monitor the pPDR along their routes. An alert is raised if the deference between the ePDR and pPDR exceeds a threshold value. In the accusation-based reaction, a node on detecting malicious behavior of another node, accuses the suspected node and floods the network with an ACCUSATION message. The ACCUSATION message contains the identity of the accuser as well as that of the accused node. The metrics advertised by the accused node are ignored and the accused node is not considered for inclusion in any subsequent FORWARDING GROUP selection. To prevent any possible *bad-mouthing attack*, a node is not allowed to issue any further accusation before the expiry of its previously made accusation. Any possible *metric poisoning effect* caused due to a metric manipulation attack is prevented by refreshing the metrics in the network immediately after an attack is detected. This is achieved by automatic and periodic broadcasting of JOIN QUERY messages.

SODMRP can defend against metric manipulation attack in wireless mesh networks and ensures high throughput in multicast communications. Since it uses asymmetric key cryptography, the computational overhead on the nodes and communication overhead in the network are higher which can be justified for applications which need high security and sustained high throughput.

Table 3 presents a summary of some of the aforementioned security schemes in the network layer of a WMN communication protocol stack.

Table 3. Summary of some network layer security schemes for WMNs

Protocol	Salient Features
ARIADNE [56]	It is an on-demand routing protocol that assumes clock synchronization and the existence of a shared secret between each pair of nodes. It also assume an authentic TESLA key for each node in the network and an authentic route discovery chain element for each node for which this node will forward RREQs. TESLA keys are distributed to the participating nodes via an online key distribution center. Freedom from routing loop is guaranteed. Routing metric is the routing path length. In routing, shortest path identification is not done. Intermediate nodes are not allowed to reply to RREQs. It is resistant to: replay, DoS, routing table poisoning attacks. It is vulnerable to: location disclosure, black hole, wormhole attacks.
SRP [63]	It is an on-demand routing protocol that assumes the existence of a security association between each source and destination node. Malicious nodes are assumed not to collude. Freedom from routing loops guaranteed. Path length is the routing metric. The shortest path identification is not done. Intermediate nodes are allowed to optionally reply to RREQs. It is resistant to: replay, DoS, routing table poisoning attacks. It is vulnerable to: location disclosure, black hole, wormhole attacks.
SAODV [64]	It uses an on-demand routing approach that assumes the presence of an online key management scheme for association and verification of the public keys. Freedom from routing loops is guaranteed. Routing metric is the routing path length. It does not identify the shortest path in routing. Intermediate nodes are allowed to optionally reply to RREQs. It is resistant to replay, routing table poisoning attacks; and vulnerable to location disclosure, black hole, wormhole, DoS attacks.
SEAD [57]	It follows a table-driven (reactive) routing approach and assumes the existence of a clock synchronization, or a shared secret between each pair of nodes. Freedom from routing loop is guaranteed. Routing metric is the path length. It does not identify the shortest path in routing. Intermediate nodes are not allowed to reply to RREQs. It is resistant to replay, DoS, routing table poisoning attacks; vulnerable to location disclosure, blackhole, wormhole attacks.

Table 3. (*continued*)

ARAN [58]	It is an on-demand routing protocol that requires the presence of an online certification authority. Each node knows the public key of the CA a priori. Freedom from loops is guaranteed. Selection of the shortest path in routing is not mandatory. Intermediate nodes are not allowed to reply to RREQs. It is resistant to replay, routing table poisoning attacks; vulnerable to location disclosure, black hole, wormhole, DoS attacks.
SMT [60]	It is an on-demand routing protocol that assumes an initial trust between source and destination using public key cryptography. It also assumes a shared finite field for purposes of data dispersion in pre-computed set of columns. Freedom from routing loop is guaranteed. Selection of the shortest path in routing is not mandatory. Intermediate nodes may optionally reply to RREQs. It is resistant to replay, routing table poisoning attacks; vulnerable to location disclosure, black hole, wormhole, DoS attacks.
SAR [82]	This on-demand routing protocol assumes the existence of a key distribution or secret sharing mechanism. Freedom from routing loops not guaranteed-depends on the selected security requirements. Shortest routing path selection is not possible. Intermediate nodes are not allowed to reply to RREQs. It is resistant to replay, routing table poisoning attacks; vulnerable to location disclosure, black hole, wormhole, DoS attacks.
SEAODV [69]	It is an on-demand routing approach that assumes the presence of an online key management scheme for the association and verification of the public keys. Freedom from routing loops is guaranteed. Routing metric is the routing path length. It does not identify the shortest path in routing. Intermediate nodes are allowed to optionally reply to RREQs. It is resistant to replay, routing table poisoning attacks; vulnerable to location disclosure, blackhole, wormhole, DoS attacks.
SLSP [65]	It is a table-driven (proactive) protocol and assumes that the nodes have their public keys certified by a trusted third party (TTP). Malicious nodes are assumed not to collude. Freedom from loop is guaranteed. Routing metric is the routing path length. It does not involve any shortest path identification. Intermediate nodes are not allowed to reply to RREQs. It is resistant to replay, DoS, routing table poisoning attacks; vulnerable to: location disclosure, black hole, wormhole attacks.
SOLSR [88]	It is a table-driven (proactive) link state routing protocol and assumes a loose clock synchronization for time-stamping the messages. A key distribution center is also assumed to be present to manage the public keys or generation of the secret keys for message authentication, integrity and other security-related operations. Freedom from routing loop is guaranteed. Routing metric is the routing path length. It does not involve any shortest path identification. Intermediate nodes are not allowed to reply to the RREQs. It is resistant to replay attack, routing table poisoning attack, incorrect control traffic generation, incorrect HELLO message generation by identity spoofing or link spoofing, incorrect topology control (TC) message generation, incorrect control traffic relaying attacks; vulnerable to blackhole, wormhole, DoS and location disclosure attacks.

Table 3. (*continued*)

SODMRP [93]	It is an on-demand multicast routing protocol and assumes a public key cryptographic framework in place. It also assumes that each node possesses a client certificate that binds its public key to its unique identity. Freedom from routing loop is guaranteed. It ensures high throughput in multicast communications to support rich user experience. It is resistant to local metric manipulation (LMM) and global metric manipulation (GMM) attacks; vulnerable to minor bad mouthing (false accusation) by malicious nodes.
TESLA [71, 85]	It is a source authentication scheme for multicast communication which is based on loose time synchronization between the sender and the receivers followed by a delayed release of the authentication key by the sender. The sender attaches a MAC to each packet using the key which initially is known to the sender only; the receiver buffers it without being able to authenticate the packet. A short while late, the sender discloses the key and the receiver is then able to authenticate the packet. A single MAC per packet is able to ensure source authentication if the receiver has a synchronized clock which is ahead in time as that of the sender. TESLA is light-weight, scalable and can be used in the network or in the application layer. For security, the sender and the receiver must have a loose time synchronization in which the receiver needs to know an upper bound on its deviation from the sender' clock although precise time synchronization is not required. It is resistant to DoS attack on the sender if indirect time synchronization is used. However, it is prone to DoS attack on the sender if direct time synchronization is used. A powerful buffer overflow attack on the receivers and DoS attacks on the authentication key chains are also possible.
Packet Leashes [77]	It is a mechanism to defend against the wormhole attack in which an attacker records a packet at one location in a network and tunnels the data to another location and replays the packet there. The attacker can perform the attack even if cryptographic services providing confidentiality, authentication and integrity protection are available in the network. The scheme assumes that each node can obtain an authenticated key from any other node. A trusted entity is also assumed that signs the public-key certificates for each node. Two types of leashes are distinguished-geographical leashes and temporal leashes. In temporal leashes, an extremely precise clock synchronization mechanism is assumed to be present. In the geographical, the scheme assumes geographical location information and loosely synchronized clocks. Geographical leashes are less efficient, since they require broadcast authentication. The scheme is particularly designed to defend against wormhole attack to which most of the wireless routing protocols are vulnerable.
HWMP [67,68]	It is a hybrid routing protocol that has both reactive and proactive routing capabilities. It assumes the existence of a mesh portal that is configured to periodically broadcast beacons. Routing metric is the routing path length. The intermediate nodes can optionally respond to the RREQs. The route discovery process is adapted to the IEEE 80.11s path selection protocol in which MAC addresses of the nodes are used in routing. Optimal routing path is determined based on radio-aware link metrics. It is resistant to replay, routing table poisoning attacks; vulnerable to location disclosure, blackhole, wormhole, DoS attacks.

Table 3. (*continued*)

Secure MAODV [29]	It is a secure multicast on-demand routing protocol in which each node (multicast group members as well as non-members) possesses a pair of public/private keys and a certificate signed by a CA. The certificate binds the public key of a node to its IP address. A group member has a group membership certificate which binds the group member's public key and IP address with the IP address of the multicast group. Each node in a multicast tree establishes pairwise shared keys with its neighbor. The multicast group leader digitally signs the group HELLO packets to prevent spoofing. Tree key credentials are used for distinguishing between tree nodes and other nodes. The hop counts are authenticated using one-way has chains. It is resistant to outsider attacks and insider attacks such as attacks on route discovery process by a non-tree or a tree node, attacks on link activation, attacks on multicast tree maintenance – on the tree pruning process, on the link repair process, and on the partition merge process. It is vulnerable to DDoS attack.
ODSBR [61, 62]	It is an on-demand (reactive) routing protocol that assumes bi-directional communication links and requires pairwise shared keys among the nodes which are established on-demand. The services of a public key infrastructure (PKI) are assumed to be available for key distribution, management and revocation. The protocol establishes a reliability metric based on the behavioral analysis of the nodes and selects the best path based on it. The metric is represented by a list of link weights. It is resistant to Byzantine attacks by insider nodes, such as creation of routing loops, routing packets via non-optimal paths, selectively dropping packets etc; vulnerable to DoS and wormhole attacks.
BSMR [72]	It is a multicast routing protocol that assumes a public key infrastructure to enable public key cryptographic operations. For multicast communication, a multicast tree is constructed. For joining a multicast group a new node makes a route discovery to the multicast tree by broadcasting RREQ messages. Tree maintenance algorithms are invoked on occurrence of events such as pruning, link breaks, and network tree partitioning. It is resistant to Byzantine attacks by insider nodes, packet dropping, false packet injection, modification or replaying of packets, false route advertisement, generation of malicious route error message that leads to network or multicast tree partitioning, intentional collision of frames at the MAC layer and jamming at the physical layer. However, it is vulnerable to DDoS attacks.

3.4 Security Mechanisms in the Transport Layer

Secure socket layer (SSL) [95], *transport layer security* (TLS) [95] and *private communications transport* (PCT) [95] protocols are usually used for securing the transport layer in wireless networks including the WMNs. SSL/TLS uses asymmetric key cryptographic techniques to ensure secure communication sessions. It can also help in protecting against masquerading attack, man-in-the middle attack, rollback attack, replay attack and buffer overflow attack.

For securing the transport layer in WMNs, an upper layer authentication protocol - *extensible authentication protocol encapsulating transport layer security* (EAP-TLS) protocol - is proposed by Aboba and Simon [96]. Although EAP-TLS offers mutual authentication between a *mesh router* (MR) and a *mesh client* (MC) or between a pair of MCs, it introduces high latency in WMNs because each terminal acts as an authenticator for its previous neighbor before the authentication request reaches an *authentication server* (AS). Furthermore, for nodes with high mobility, frequent re-authentications due to handoffs can have a very adverse impact on the quality of service of the applications. As a result, variants of EAP-TLS have been proposed to adapt IEEE 802.1X authentication model for multi-hop WMNs [3].

3.5 Security Mechanisms in the Application Layer

The most usual ways of securing the application layers is the use of firewalls and *intrusion detection systems* (IDS). In a wireless network with a firewall installed, the firewall provides easy controls for achieving access control, user authentication, packet filtering, and logging and accounting services etc. Application-level firewalls provide protection against various attacks such as detection of malwares, spywares etc. However, the access control policies in a firewall are static in nature which makes a firewall unable to detect a new attack based on an anomaly-detection technique.

For detecting more sophisticated novel attacks, *intrusion detection systems* (IDSs) are used as the second line of defense along with firewalls. Interested readers may refer to [97, 98] for a comprehensive discussion on IDSs in wireless networks. An architecture of a cooperative, distributed IDS based on clustering of nodes for a multi-hop wireless network (CWIDS) that can detect attacks at multiple layers (including the application layer) is presented in [99]. An agent-based IDS for the network layer is proposed that can be adapted to large-scale distributed WMNs [100].

Table 4 presents a list of vulnerabilities in different layers of the protocol stack of WMNs and the corresponding security schemes for defending these attacks.

Table 4. Summary of various attacks and their defense mechanisms for WMN communications

Attack	Targeted layer in the protocol stack	Protocols
Jamming	Physical and MAC layers	Frequency hopping spread spectrum (FHSS) [35], Direct sequence spread spectrum (DSSS) [35]
Wormhole	Network layer	Packet Leashes [77]
Blackhole	Network layer	SAR [82]
Grayhole	Network layer	GRAYSEC [22], SAR [82]
Sybil	Network layer	SYIBSEC [23]
Selective packet dropping	Network layer	SMT [60], ARIADNE [56], Sen [7], Sen[9], Sen [66]
Rushing	Network layer	ARAN [58], SAR [82], SEAD [57], ARIADNE [56], SAODV [64] , SRP [63], SEAODV [69]
Byzantine	Network layer	ODSBR [61, 62]
Resource depletion	Network layer	SEAD [57]
Information disclosure	Network layer	SMT [60]
Location disclosure	Network layer	SRP [63]
Routing table modification	Network layer	ARAN [58], SAR [82], SRP [63], SEAD [57], ARIADNE [56], SAODV [64], SEAODV [69]
Multicast routing metrics manipulation attack	Network layer	SODMRP [93]

Table 4. (*continued*)

Byzantine multicast routing insider attack	Network layer	BSMR [72]
SYN flooding	Transport layer	SSL [95], TLS [95], EAP-TLS
Session hijacking	Transport layer	SSL [95], TLS [95], PCT [95], EAP-TLS[96]
Repudiation	Application layer	ARAN [58], CWIDS [99]
Denial of service	Multi-layer	SRP [63], SEAD [57], ARIADNE [56]
Impersonation	Multi-layer	ARAN [58], SEAD [57], SEAODV [69]

3.6 Secure Authentication Mechanisms

Robust authentication and authorization mechanisms provide adequate safeguards against fraudulent access by unauthorized users in WMNs. Authentication ensures that an MC and the corresponding MR can mutually validate their credentials with each other before the MC is allowed to access the network services. Since secure authentication is a critical requirement for real-world deployments of WMNs, an extensive work has been done by researchers on this topic. In the following, we present a brief discussion on some of the secure authentication mechanisms and then provide a detailed discussion on four such propositions.

Mishra and Arbaugh propose a standard mechanism for client authentication and access control to guarantee a high-level of flexibility and transparency to all users in a wireless network [16]. The users can access the mesh network without requiring any change in their devices and softwares. However, client mobility can pose severe problems to the security architecture, especially when real-time traffic is transmitted. To cope with this problem, proactive key distribution approach is proposed [101,102].

Providing security in the backbone network for WMNs is another important challenge. Mesh networks typically employ resource constrained mobile clients. It is sometimes difficult to protect these devices against removal, tampering, or replication attacks. If a device can be remotely managed, a distant hacking into the device would work perfectly [103]. Accordingly, several research works have been done to investigate the use of cryptographic techniques to achieve secure communication in WMNs. Cheikhrouhou et al. have proposed a security architecture [104] that is suitable for multi-hop WMNs employing PANA (Protocol for carrying Authentication for Network Access) [105]. In the scheme proposed by the authors, the wireless clients are authenticated on production of the cryptographic credentials necessary to create an encrypted tunnel with the remote access router to which they are associated. Even though such framework protects the confidentiality of the information exchanged, it cannot prevent adversaries to perform active attacks against the network itself. For instance, a malicious adversary can replicate, modify and forge the topology information exchanged among mesh devices, in order to launch a denial of service attack. Moreover, PANA necessitates the existence of IP addresses in all the mesh nodes. This poses a serious constraint on deployment of this protocol.

Prasad et al. have presented a lightweight *authentication, authorization and accounting* (AAA) infrastructure for providing continuous, on-demand, end-to-end security in heterogeneous networks including WMNs [106]. The notion of a security manager is used by deploying an AAA broker. The broker acts as a settlement agent, providing security and a central point of contact for many service providers.

Lee et al. propose a distributed authentication scheme for minimizing authentication delay in a wireless network [107]. In this scheme, multiple trusted nodes are distributed over a WMN which act on the behalf of an authentication server. Deployment of multiple authentication servers makes management of the network easy, and it also involves less storage overhead in the MRs. However, the performance of the scheme will degrade when multiple MCs send out their authentication requests, since the number of trusted nodes acting as the authentication server is limited compared to the number of access routers.

The authentication schemes for MANETs can also be adapted in WMNs. Sen and Subramanyam have proposed and evaluated the performance of a distributed certificate authority based on threshold cryptography [108]. The scheme is an extension of the MOCA protocol [109] in which a collection of nodes selected on several parameters acts as the certificate authority and provides an attack resilient and robust certificate distribution and verification service.

In the following sub-sections, we provide a brief discussion on a few authentication schemes for WMNs. For a more comprehensive discussion on this topic, interested readers may refer to [3].

3.6.1 ARSA: An Attack-Resilient Security Architecture for Multihop WMNs

Zhang and Fang have proposed an attack-resilient architecture for large-scale WMNs that deploys three categories of network entities: (i) brokers, (ii) users, and (iii) network operators [110]. In this architecture, each WMN domain is assumed to be operated by an operator, and it consists of a certain number of mesh routers. The mesh routers are assumed to be powerful in computing and communication capabilities. Hence, the packets transmitted by a mesh router reach their intended mesh client nodes (in the coverage area of the mesh router) in a single hop. On the other hand, the communication from the mesh clients to their mesh router may be multi-hop in nature because of the limited computing and transmission power of the mesh clients. Each user (i.e., mesh client) acquires a *universal pass* form the network operator which allows it to get ubiquitous access to the network. Multiple network operators need to have bilateral *service level agreements* (SLAs) between them. Instead, each network operator only needs to have an agreement with one or more brokers. The number of brokers is far less than the number of network operators since the WMN is very large in scale. For authentication purpose, the mesh clients need to locally communicate with its serving WMN domain without requiring any communication with the corresponding broker. This approach reduces authentication delay and signaling overhead in the authentication process. In addition, ARSA also provides efficient mechanisms for mutual authentication between any pair of node belonging to the same WMN domain. The scheme is not only efficient but also has been shown to be secure against various kinds of attacks such as: attack on location privacy, DoS attacks, bogus beacon flooding attack, bandwidth exhaustion attack etc [110].

ARSA assumes multiple trust domains in the mesh network, each domain being managed by a broker or a network operator. For accessing network services, each mesh client has to first register with at least one broker. Upon successful registration, the broker issues an *electronic universal pass* to the client. Each network operator also needs to establish trust relationship with one or more brokers. A network operator allows network access to a mesh client which has a valid universal pass issued by a broker with which the network operator has pre-established trust relationship. The passes are the important components in the authentication process in ARSA. In order to minimize the bandwidth and signaling overhead in the authentication process, the size of the passes is made as small as possible utilizing the concept of *identity-based cryptography* (IBC). Although the concept of IBC was first introduced by Shamir [39], a fully functional IBC scheme was not established till Boneh and Franklin applied *Weil pairing* to construct a *bilinear map* [40]. Use of IBC in ARSA requires the presence of one network entity in each trust domain. This entity, known as the *domain administrator*, performs some essential *trust domain initialization* activities [110].

ARSA uses three types of passes: (i) *router passes* which are issued by a network operator to its mesh routers, (ii) *client passes* which are provided by a broker to its registered clients, and (iii) *temporary client passes* which are issued by a network operator to the mesh clients present in its domain. ARSA utilizes router passes and client passes to realize authentication and key agreement between a mesh router and a mesh client and also between a pair of mesh clients. The authentication may be either inter-domain or intra-domain. In case of inert-domain authentication, a mesh client migrates from one WMN domain to another. In intra-domain scenario, a mesh client changes its association from its current mesh router to a new mesh router in the same domain. While inter-domain authentication is a more expensive operation than authentication in intra-domain, it occurs less frequently. ARSA provides a very efficient way for client-client authentication by using temporary client credentials for clients which are associated with the same mesh domain. In summary, the protocol provides an efficient, secure and ubiquitous network access to the users of a WMN and is ideally suited for large-scale networks with limited client mobility.

3.6.2 AKES: An Efficient Authenticated Key Establishment Scheme for WMNs

He et al. have proposed a distributed *authenticated key establishment scheme* (AKES) that is based on *hierarchical multi-variable symmetric functions* (HMSF) [111]. In the proposed scheme, MCs and MRs can mutually authenticate each other and establish pair-wise communication keys without the need of any interaction with a central authentication server. This leads to reduced communication overhead and delay in the authentication process.

The WMN architecture assumed in the scheme is same as that used in ARSA scheme discussed in Section 3.6.1. The mesh network is divided into a number of *domains*. Each *mesh client* (MC) registers itself in its *home domain*. Each domain is managed by an *Internet service provider* (ISP) which relies on an *authentication authorization and accounting* (AAA) server for managing the entities in its domain.

There are a few *Internet gateways* (IGWs) and *mesh routers* (MRs) and a large number of *mesh clients* (MCs) in a domain managed by an ISP. The scheme assumes pre-existing security frameworks for communications between AAAs and IGWS, between MRs and IGWs and between the MRs themselves. The goal of the scheme is to secure the communication sessions between the MCs and between the MRs and the MCs. In designing the proposed scheme, the authors have utilized the concept of *polynomial-based key generation* concept proposed by Blundo et al. [112]. Using the polynomial-based key distribution scheme an authorized server (e.g., AAA server) distributes a small piece of information (e.g., coefficients of polynomials) among a group of users in such a way that each user can compute a shared key with every other user in the group by only exchanging their IDs. These shared keys are used for pair-wise authentication and encryption of messages among the users. The proposed authentication scheme extends the concept of polynomial-based key generation so that it can be applied for *asymmetric mutual authentication* and key establishment. The need for asymmetric mutual authentication arises since the MRs and MCs in a WMN are unequal entities. Using the symmetric polynomial and asymmetric function, the authors have designed a "*hierarchical multi-variable symmetric function based authenticated key establishment scheme*" [111] that finally enables generating the MR-MC keys, MC-MC keys, and pair-wise session keys.

The robustness of the pair-wise keys is dependent on selection of the pair-wise master key generation functions [111]. In fact, it has been proved that for a polynomial of degree t, the polynomial-based key distribution scheme will be t-secure [113]. A t-secure key distribution scheme is robust against collusion attacks by a maximum of t nodes. Therefore, the robustness of the keys can be increased by increasing the value of t in the polynomial used for key generation. Since MCs authenticate locally with the MRs, spoofing attacks and DoS attacks in the wireless backbone involving the IGWs and AAA servers are very difficult to launch. The computation overhead of the scheme depends on the function used in the key generation.

3.6.3 SLAB: A Secure Localized Authentication and Billing Scheme for WMNs

Zhu et al. have proposed a *secure localized authentication and billing* (SLAB) scheme for wireless mesh networks to address security requirements and performance efficiency in terms of reducing inter-domain handoff authentication latency, and computation load on the roaming broker [114].

Most of the security solutions for WMNs are based on *authentication, authorization and accounting* (AAA) architecture [115], in which the authentication request from a *mobile user* (MU) is sent through the *serving mesh access point* (sMAP) and the *mesh gateway* (MGW) to a centralized authentication server (e.g., RADIUS server) that can grant access to the MU after verifying the authenticity of the MU. Since such lengthy authentication processes involve unacceptable delay in real-time applications, faster solutions are in demand. One approach that is followed for fast *inter-domain authentication* (authentication of an MU when it is roaming in a domain other than its home domain) is to have a pre-established trust among different *wireless Internet service providers* (WISPs) by deploying a centralized *roaming broker* (RB) trusted by all the WISPs [116]. In this approach, the foreign WISP, in whose domain the MU is currently roaming, forwards the current AAA session of the MU to the home WISP of the MU for authorization via the RB. For enhanced security, the RB may be configured not only as a *trusted third party* (TTP), but it also serves as a centralized *certificate authority* (CA) that issues public key certificates to the WISPs and MUs. This enables faster trust establishment among the WISPs or between a WISP and MUs since it only requires verification of the *public key certificates* (PKCs) issued by the RB [110, 117].

In designing the proposed scheme, the authors have observed that from the point of view of the WISPs, an inter-domain handoff can be considered as an inter-WISP payment, while from the point of view of the MUs, an MU can roam into another WISP domain only if it has enough credits remaining with it. From this perspective, a WISP can issue a digital signature based on PKI which is equivalent to a digital currency for inter-domain roaming payment with another WISP. This transaction does not involve any intervention of the RB. Moreover, this digital signature can also be taken as an authentication credential of the MU to which it is issued. The authors have called such digital signature as *D-coin* (digital coin).

The authors have also observed that if the *mesh access points* (MAPs) are pre-loaded with necessary cryptographic mechanisms, some important security-related operations - e.g., roaming/handoff authentication and billing -- can be performed in a localized manner with much better scalability and efficiency, thereby solving the scalability problem with respect to a centralized RB. However, this localized approach to authentication leads to security issues. The MAPs are low cost devices and susceptible to easy compromise [103]. An attacker can retrieve the cryptographic keys from a compromised MAP and launch some serious attacks such as *coin fraud attacks* (arbitrary issue of D-coins to an illegal MU or accepting D-coin from an MU and not providing service against it) [114].

SLAB exploits the advantages provided by localized authentication approach while providing adequate security protections against the vulnerabilities associated with the MAPs. To thwart coin fraud and overcharging attacks [114], a *local voting* strategy and *threshold digital signature* mechanism are adopted in designing the authentication scheme [118]. Local voting strategy enforces a requirement that the issued D-coin is not only endorsed by the *serving MAP* (sMAP) but also by the *neighboring MAPs* (nMAPs) to avoid a possible single-point-of-compromise at the sMAP. A *local user accounting profile* (LUAP) for the MU is maintained both at the sMAP and the nMAP for recording each network access and roaming operation performed by the MU. Since LUAPs for each MU are maintained, the on-line billing can be done easily. SLAB also enables inter-domain handoff authentication and billing to be done in a peer-to-peer manner without any intervention of the RB when an MU performs an inter-domain handoff. The RD is involved only during the "clearance phase" in which a WISP submits its collected d-coin issued by the other WISP for receiving payment against the D-coin. This operation is performed off-line and does not put much load on the RD. However, to further reduce the load on the RD during the off-line clearance phase, SLAB exploits the use of short and aggregate digital signature to minimize the overhead due to the verification and storage of the D-coin [119].

In summary, SLAB has five phases in its operation: (i) signing key distribution, (ii) secure session maintenance and LUAP generation phase, (iii) localized LUAP transfer during intra-domain handoff phase, (iv) D-coin issuance and inter-domain handoff authentication phase, and (v) clearance phase.

In the signing key distribution phase, for issuing a D-coin on behalf of the MGW, each of the MAPs first obtains its own share (i.e., partition) of the signing key using the threshold digital signature technique. In the secure session maintenance and LUAP generation phase, the sMAP of an MU collaborates with some of the nMAPs to generate and maintain the LUAP of the MU in order to track the spending information of the MU. In order to ensure the authenticity of the LUAP information exchange over a secure session, the MUs are mandatorily required to submit non-repudiation proof of the previous spending information so that the session consistency is maintained. The localized LUAP transfer during intra-domain handoff phase is involved in ensuring that every new nMAP of the MU can obtain a copy of the MU's authentic LUAP. During an intra-domain handoff, an MU roams within a common WISP resulting in a switch of the sMAP and the corresponding nMAPs. SLAP reduces signaling overhead during intra-domain handoff by invoking a localized LUAP transfer algorithm based on a local voting strategy. The algorithm accepts an LUAP as a valid one only if more than k valid LUAP copies from the nMAPs are found to be consistent. In the D-coin issuance and inter-domain phase, the MU roams in domains managed by different WISPs. Inter-domain handoff involves mutual authentication between the MU and the target WISP and the inter-domain WISP payment-related issues. SLAB handles inter-WISP authentication and billing using D-coin. In the clearance phase, the RB handles the inter-WISP payments in an efficient manner. An event-driven clearance procedure is used in which D-coin is regarded as an event. Since the processing is done in a batch mode, the D-coin can be only submitted to the RB when a given size of D-coin has been gathered or a specified time interval has elapsed. The batch processing enables the RB to verify the gathered D-coin and perform aggregate signature [119] simultaneously so that transmission and verification cost is minimized.

3.6.4 LHAP: A Lightweight Hop-by-Hop Authentication for Ad Hoc Networks

Zhu et al. have proposed a *light-weight hop-by-hop access protocol* (LHAP) for authenticating data packets and preventing resource consumption attacks [70, 120]. The protocol uses a light-weight hop-by-hop authentication approach in which intermediate nodes authenticate the data packets they receive before forwarding them to the next hop node. LHAP employs one-way hash chains [121] for traffic authentication and the TESLA [71, 85, 122] protocol for bootstrapping and maintaining trust between the nodes.

A one-way hash chain is a chain of keys generated through repeated application of a one-way hash function on a random number. For example, if a node needs to generate a key chain of size N, it first chooses a key - $K(N)$ - randomly. The node, then, successively computes the remaining keys in reverse order so that $K(N-1) = F(K(N))$, $K(N-2) = F(K(N-1))$… till it get $K(0) = F(K(1))$. To use one-way hash chain for the purpose of authentication, a sender node first signs the last value in the chain (i.e., $K(0)$) with its private key so that another node that has the knowledge of its public key can verify the signature and the authenticity of $K(0)$. The sender, then, discloses the successive keys in the chain in the reverse order in which they were generated. The receiver can verify the authenticity of $K(j)$ by checking $K(j-1) = F(K(j))$.

LHAP uses the TESLA protocol [71, 85, 122]. TESLA is a broadcast authentication scheme that uses a one-way hash chain and an approach of delayed key disclosure. After bootstrapping an authentic key derived from a one-way hash chain, the sender digitally signs it and sends it to the receivers. TESLA computes the *message authentication codes* (MACs) in the subsequent broadcast authentications but discloses the key at the receivers with a delay. In the basic scheme of TESLA, the sender node uses a key K from its hash chain as the MAC key to compute a MAC over packet $P(i)$, and then appends the MAC to $P(i)$. The key K is disclosed in the next packet $P(i+1)$, which allows the receiver nodes to verify the authenticity of the key K and hence the MAC of $P(i)$. If both K and the MAC are verified to be correct, and if the packet $P(i)$ is guaranteed to be received before the packet $P(i+1)$ was sent, the receiver nodes conclude that the packet $P(i)$ is authentic. A critical requirement of TESLA is a receiver's ability to determine the sending time of each packet. This requirement is met by periodic key disclosure and loose time synchronization [85].

LHAP assumes that each node in the network has a public key certificate signed by a trusted *certificate authority* (CA), and the public key of the CA is known to all the nodes in the network. A loose time synchronization scheme is also assumed to be available so that the TESLA protocol can be used.

Since LHAP authenticates all traffic packets at each node on the route from the source to the destination, it is mandatory that the authentication protocol should be lightweight. In the LHAP data packet authentication, each source node generates a one-way hash chain of keys which are used by its immediate neighbor nodes. The keys generated from this one-way hash chain are termed as "TRAFFIC KEYs". Each neighbor of a node obtains an authentic key in this TRAFFIC KEY chain when it first establishes a trust relationship with the node. A node transmitting a packet will append a new TRAFFIC KEY to the packet. All the neighboring nodes that receive this packet can verify the its authenticity by checking the validity of the attached TRAFFIC KEY.

One naïve way of bootstrapping trust among the node is to exchange TRAFFIC KEYs using public key encryption in which each node signs its most recently released TRAFFIC KEY and sends it to each of its neighbors. However, this approach lacks scalability in a large-scale dense wireless network. To solve this problem, LHAP uses TESLA to minimize the number of signature operations. EACH node only uses digital signatures to bootstrap a TESLA key chain, and the TESLA keys are used to generate subsequent authentic TRAFFIC keys. To maintain the trust relationship, each node broadcasts its latest TRAFFIC KEY at periodic intervals. The broadcast TRAFFIC KEYs are authenticated using the TESLA keys of the node. After the updated TRAFFIC KEY from a node is received, all of its neighbors drop any packets they receive which are authenticated by the older TRAFFIC key. This periodic broadcast of KEYUPDATE messages ensures that the protocol is secure against any possible replay attack.

Security analysis of LHAP shows that the protocol is robust against various outsider and insider attacks [70, 120]. Among the outsider attacks, the protocol is resistant to attacks such as: impersonation attack, wormhole attack, hidden terminal attack etc. LHAP is also resistant to insider attacks such as: insider-clone attack. However, it cannot defend against multiple insider attacks launched by multiple insider compromised nodes.

3.6.5 A Localized Two-Factor Authentication Scheme for WMNs

Lin et al. have proposed a two-factor localized authentication scheme for inter-domain handover and mobility management in IEEE 802.11 standard compliant WMNs [123]. The localized authentication scheme is based on Rabin cryptosystem [124]. Rabin cryptosystem has asymmetric computational overheads – while the encryption and signature verification operations are very fast, the decryption and signature generation operations are computationally intensive [125].This asymmetric property of Rabin cryptography makes it particularly suitable for scenarios like inter-domain handover in wireless networks where the *access points* (APs) have high computational power and the *mobile stations* (MSs) have limited resources.

The authentication scheme is designed for a standard wireless network depicted in Fig.11 that consists of four types of entities: (i) the *mobile units* (MUs), (ii) a *trusted third party* (TTP), (iii) the *wireless Internet service providers* (WISPs), and (iv) the *hotspots*. As shown in Fig.11, several hotspots may be under the operation of a single WISP, and these hotspots may not necessarily be adjacent to each other. Each hotspot is assumed to have one AP. The use of the terms AP and hotspot can be interchangeable. Before an MU can access the network services, it has to first subscribe to the TTP. The TTP and each of the WISPs have mutual agreement such that an MU after subscribing to the TTP can access the hotpot operated by the corresponding WISP. The TTP also acts as the *certificate authority* (CA) that issues certificates to the MUs and WISPs. The issued certificates are essentially the public keys of the MUs and the WISPs which are digitally signed by the TTP. After successful registration, each MU is also issued a *smart card* that contains the authentication credentials of the MU. The smart card serves as an electronic pass for the MU for roaming across different WISPs.

The authentication protocol has three phases: (i) the *initialization* phase, (ii) *the login and mutual authentication* phase, and (iii) the *handoff* phase.

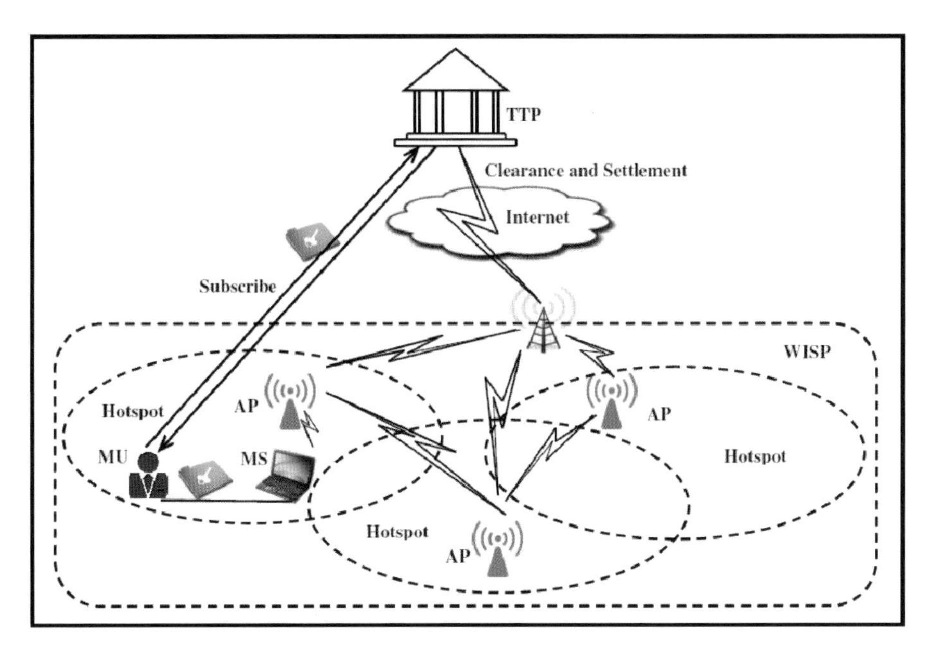

Fig. 11. WMN architecture and different network entities in the local authentication protocol [122]

In the initialization phase, various operations are performed by different entities in the network. The TTP generates its public-private key pair and publishes its public key while keeping the private key secret. Each WISP sets up a mutual agreement with the TTP such that a registered and authenticated MU can log in and access the network services provided by the WISP. For this purpose, each WISP sends a chosen identity of it to the TTP and also generates its private key. The TTP checks the validity and uniqueness of the identity. If the validation process passes correctly, TTP signs the public key of the WISP and also sets up a linkage between the public key and the identity of the WISP. The TTP sends this signed public key of the WISP and the linkage of the public key with the identity of the WISP to the concerned WISP. After the initialization of the WISP is complete, its starts the initialization process for the hotspots operated by it. Each hotspot generates its own private key and its *service set identifier* (SSID). The hotspot sends its SSID to its home WISP for verification and keeps its private key secret. The home WISP checks for uniqueness of the SSID. If the SSID provided by the hotspot is found to be unique, the home WISP create a signature on the public key of the hotspot and also sets up a linkage between the public key of the hotspot and its SSID. The home WISP sends the signed public key of the hotspot and the linkage of the public key and the SSID to the concerned hotspot thereby completing the initialization of the hotspot.

The next step of the initialization process involves registration of the MUs. For this purpose each MU generates a unique identity of it and sends the identity to the TTP through a secure channel. After checking for the uniqueness of the identity of the MU, the TTP provides a smart card to the MU which can be used as an electronic pass by the MU.

To provide enhanced security, the scheme uses two-factor authentication for a roaming MU. In order to defend against scenarios where the confidential security credentials of an MU might have been leaked, the second factor using a smart card is adopted. If an attacker has to successfully impersonate a legitimate user, the attacker must not only know the authentication credentials of the user, but also capture or replicate the corresponding smart card. Two-factor authentication provides increased robustness to the localized authentication approach.

In the login and mutual authentication phase, the MU and the hotspot (i.e., the AP) mutually authenticate each other using a robust challenge and response mechanism [123].

The handoff procedure is performed between an MU and APs when the MU moves from the current AP to an adjacent AP. In order to minimize the delay in handoff, the proposed localized authentication scheme adopts an approach in which the session key is cached in the current network domain. When an MU sends a handoff request to an AP with which the MU shares a valid shared session key, an authenticated symmetric key agreement protocol is invoked instead of a full authentication procedure that involves asymmetric key cryptographic operations. This makes the handoff operation fast while compromising on the security in the authentication process.

The security analysis of the localized authentication scheme has shown that it is resistant to replay attack, impersonation attack, password guessing attack, and attack on the privacy of the users (i.e. MUs).

Table 5 provides a brief summary of a few authentication mechanisms for WMNs.

3.7 Security Mechanisms against Attacks on Key Management Protocols

Several key management protocols for wireless networks exist in the literature. Due to the computational resource constraints in the mobile hand-held devices, schemes which involve high computational overhead for security-related operations are not particularly suitable for wireless networks. Again, the Diffie-Hellman key exchange protocol [32] and many of its variations do not fit to the requirement of an online *certificate authority* (CA) that computes the shared keys between the nodes on-demand in a WMN. In this section, we provide a brief overview of a few key management protocols in wireless networks which can be adapted to WMNs. For a detailed discussion on these schemes, the readers may refer to [177].

Capkun et al. have proposed a *public key infrastructure* (PKI)-based key management approach for wireless multi-hop networks [126]. The proposed scheme does not use any trusted third party or a CA for key distribution and management purposes. Instead, it follows an approach of using self-signed certificates as in *pretty good privacy* (PGP) [127]. Based on their individual experiences, the users (i.e., nodes) issue public key certificates to each other. Each node maintains a repository of certificates that it has issued to others and that have been issued to it by other nodes. The certificate revocation process can be explicit or implicit. In explicit revocation process, the issuer of the certificate explicitly informs other nodes about the certificates being revoked. In the implicit revocation approach, the nodes use expiry time for each certificate. The certificates which have not been renewed before the expiry of their time of validity are assumed to be revoked. In this scheme, if node

A wants to communicate to node B, the two nodes first merge their certificate repositories before node A tries to find a public key certificate chain that has a path from node A to node B. The chain is constructed progressively in such a way that the first certificate can be verified using node A's public key, and each of the subsequent certificates can be verified using the key included in the previous certificate in the chain. The last certificate must include the public key of the node B so that the verification process is successfully completed. The authors have introduced trust-based confidence metrics which are included in each certificate so as to defend against any possible attacks by malicious nodes that may issue false certificates.

Table 5. Summary of some authentication schemes for WMN communication

Protocol	Salient Features
ARSA [110]	This efficient authentication architecture assumes a WMN to be consisting of multiple trust domains, each domain being managed by a broker. Each client registers with a broker which issues an electronic universal pass to the client. The pass serves as the authenticated credential. ARSA supports inter-domain and intra-domain mobility of the clients efficiently by providing efficient and ubiquitous network access to the clients. It is resistant to attacks on location privacy of the users, DoS attacks, bogus beacon flooding attack and bandwidth exhaustion attack.
AKES [111]	It is a distributed authenticated key establishment scheme for WMNs. The WMN consists of a number of domains each having an AAA server for managing the entities. A hierarchical multi-variable symmetric function-based authenticated key establishment scheme is used for generating the MR-MC keys, MC-MC keys, and the pair-wise session keys. The robustness of the pair-wise keys generated depends on the selection of the pair-wise master key generation functions. Due to the localized authentication used by the MCs with the MRs, spoofing attacks and DoS attacks in the wireless backbone network is very difficult to launch on this scheme. The computational overhead depends on the polynomial function used in the key generation.
SLAB [114]	This authentication scheme exploits the advantages provided by localized authentication approach while providing adequate security protections against the vulnerabilities associated with the mesh access points (MAPs). The scheme works in five phases: (i) distribution of keys, (ii) secure session maintenance and local user accounting profile (LUAP) generation phase, (iii) localized LUAP transfer during intra-domain handoff phase, (iv) D-coin issuance and inter-domain handoff authentication phase, and (v) clearance phase. SLAB provides a strong authentication framework while supporting user mobility by reducing the delay in handoff.

Table 5. (*continued*)

LHAP [70, 120]	It is a light-weight hop-by-hop access control and authentication protocol for data packets that is resistant to resource consumption attack. It uses one-way hash chains for data traffic authentication and the TESLA protocol for bootstrapping and maintaining trust between the nodes. It assumes that each node has a public key certificate signed by a trusted CA, and the public key of the CA is known to all the nodes. A loose time synchronization scheme is also assumed to be available to support TESLA. The protocol is robust against various outsider and insider attacks such as: impersonation, wormhole, hidden terminal, insider-clone attacks. However, it is vulnerable to cooperative insider attacks launched by multiple insider compromised nodes.
Localized two-Factor Authentication [123]	This authentication scheme supports inter-domain handover and mobility in IEEE 802.11 compliant WMNs. It utilizes the asymmetric property of Rabin cryptosystem in mobility management issues where APs have high computational power and the MSs are resource-constrained. For providing enhanced security, it uses two-factor authentication for roaming mobile users. The scheme is resistant to attacks such as: replay, impersonation, password guessing and attack on the privacy of the users.

Based on the concept of *self-certified key* (SCK) [128], Li and Garcia-Luna-Aceves have proposed a protocol named NIKAP for facilitating key agreement processes in WMNs [129]. In the NIKAP protocol, pair-wise keys are computed between two nodes in a non-interactive manner. The services of a CA are needed only during the initial phase of the network bootstrapping. The non-interactive progression capability of NIKAP makes it particularly suitable for WMNs in which a pair of nodes needs to establish pair-wise shared keys without any negotiation over an insecure channel.

Zapata has proposed a key management system named *simple ad hoc key management* (SAKM) that allows the nodes of an ad hoc network to use asymmetric key cryptography for providing security and robustness to multi-hop routing used in such networks [130].

Key distribution protocols using ID-based cryptography [40] or the combination of threshold and ID-based cryptography [131] have the same advantage as SCK because IDs are used to obtain the corresponding public keys of nodes, instead of using a certificate to bind the ID and its public key. However, online CA services must exist for such protocols to work. The requirement of an online CA service is the drawback of an ID-based cryptosystem and a threshold cryptography-based system [132].

An approach to combine threshold secret sharing and probabilistic key sharing is proposed by Zhu et al. [133]. In this scheme, a source node splits its shared secret key with the destination node into several parts and sends the parts to the destination node in such way that the destination node can reconstruct the full key from the individual parts with a high probability. However, in addition to having high overhead of communication (due to transmission of several parts of the key) and intensive computation overhead (in splitting and recovering the shared secret key), the protocol fails in its operation if at least a threshold minimum number of parts do not reach the destination node.

In group key agreement protocols [134, 135], usually a shared key is distributed among all the members of a multicast group. While these protocols have low storage complexity compared to the schemes that use pair-wise keys for each node pair in a network, the cost of re-keying operations in these protocols is very high because of leaving and joining of nodes in the multicast group. Moreover, if a group key is compromised, it will lead to a complete collapse of security in the group as opposed to a possible compromise of a pair-wise secret key which will affect only the concerned node pair.

In order to increase the efficiency and robustness in the certificate construction process by combining the secret shares in a threshold cryptography-based system, Joshi et al. propose to allocate each designated distributed CA node more than one share of the CA's secret [136].

Signature aggregation [137] mechanism aggregates all the certificates in a chain of certificates into a single short signature so as to reduce the size of the certificate chain. The resultant smaller certificate incurs less communication overhead and consumes less bandwidth. The basic principle of signature aggregation is based on the idea that if N distinct messages are signed by N distinct signing authorities (i.e. nodes), then it is possible for a trusted node in the network to aggregate all these N signatures into a single signature in a secure way such that a verifier node at the receiver side can securely verify that each user indeed signed its own message.

3.8 Security Mechanisms against Attacks on Privacy Protection Schemes

The issue of user privacy in WMNs has attracted serious attention from the research community. Several propositions have been made in the literature for protecting privacy of user data, location and other sensitive information in wireless communication networks.

Wu et al. propose a light-weight privacy preserving solution to achieve a well-maintained balance between network performance and traffic privacy preservation [5]. At the center of the scheme is an information-theoretic metric called *traffic entropy*, which quantifies the amount of information required to describe the traffic pattern and to characterize the performance of traffic privacy preservation. The authors have also presented a penalty-based shortest path routing algorithm that maximally preserves the traffic privacy by minimizing the mutual information of traffic entropy observed at each individual relaying node while controlling the possible degradation of network within an acceptable region. Extensive simulation results demonstrate the effectiveness of the solution and its resilience to a possible collusion between two malicious nodes. However, one of the major problems of the solution is that the algorithm is evaluated in a single-radio, single channel WMN.

The performance of the algorithm in multiple radios, multiple channels scenario will be a really questionable issue. Moreover, the solution has a scalability problem.

Wu and Li have proposed a mechanism with the objective of hiding an active node that connects to a gateway router, where the active mesh node has to be anonymous [6]. A communication protocol is designed to protect the node's privacy using both cryptography and redundancy. The protocol uses the concept of *onion routing* [138]. A mobile user who requires anonymous communication sends a request to an *onion router* (OR). The OR acts as a proxy to the mobile user and constructs an onion route consisting of other ORs using the public keys of the routers. The onion is constructed in such a way that the inner most part is the message for the intended destination and the message is wrapped by encrypting it using the public keys of the ORs in the route. The mechanism protects the routing information from insider and outsider attack. However, it has a high computation and communication overhead.

Sen proposes a reputation-based trust management system has been combined with a privacy preservation scheme for designing a secure and efficient searching protocol for unstructured and de-centralized peer-to-peer networks [139]. The scheme utilizes network topology adaptation by constructing an overlay of trusted peers where the nodes in a neighborhood are selected based on their trust ratings. The overlays of trusted peers are utilized to achieve a user-preserving searching protocol.

To control the misuse of personal information and to prohibit disclosure of personal data, different types of information hiding mechanisms like anonymity, data masking can be implemented in WMN applications. The following approaches can be useful in information hiding, depending on what is needed to be protected:

- *Anonymity*: This is concerned with hiding the identity of the sender or receiver of the message or both of them. In fact, hiding the identity of both the sender and the receiver of the message can assure communication privacy. Thus, the possible attackers who may monitor the messages being communicated could not know who is communicating with whom.
- *Confidentiality*: It is concerned with hiding the transferred messages by using suitable data encryption algorithms. Instead of hiding the identity of the sender and the receiver of a message, the message itself is hidden in this approach.
- *Use of pseudonyms*: This is concerned with replacing the identity of the sender and the receiver of the message by pseudonyms which function as identifiers. The pseudonyms can be used as a reference to the communicating parties without infringing on their privacy. This ensures that the users in the WMNs cannot be traced or identified by malicious adversaries. However, it is important to ensure that there is no indirect way by which the adversaries can link the pseudonyms with their corresponding real world entities.

In the following sub-sections, we provide a brief discussion on some well-known scheme for privacy protection in WMNs.

3.8.1 Traffic Privacy Preservation Using Penalty-Based Multipath Routing

Privacy is a critical issue in the context of WMN-based Internet access by users in their residences where users' traffic is forwarded via multiple mesh routers [5].

Since the inbound and outbound traffic of a residence can be easily observed by the mesh routers residing at the neighboring locations, there is no privacy protection of the users. In order to provide protection to traffic privacy of the users in a community mesh network, Wu et al., propose a privacy-preservation scheme which is resistant to any traffic analysis of the users' sensitive personal communications [5]. The scheme exploits the intrinsic redundancies in WMNs by routing traffic from a source node to a destination node through *multiple paths*. The traffic is split in a random manner both in spatial and temporal domains so that the intermediate nodes can only have limited knowledge of the traffic pattern. This *traffic concealment* mechanism makes traffic analysis an impossible proposition. More formally, in this scheme, the data packets are routed in such a way that the statistical distributions of the traffic data as observed in each intermediate node on the routing path are *independent* of the actual traffic data between the source-destination pair. An information theoretic metric -*traffic entropy*- is defined that quantitatively identifies the amount of information required for profiling a traffic pattern. Based on the computation of traffic entropy, a penalty-based routing is proposed that minimizes the mutual information of traffic entropy observed at each intermediate node on the multiple routing paths.

The scheme assumes a three-tier architecture of WMN, in which the client devices communicate with stationary mesh routers. Multiple mesh routers communicate with each other to form a wireless backbone and a group of mesh routers are connected to an Internet gateway by wired links (or high-speed wireless links). Each mesh node and the gateway node have a public-private key pair. The gateway maintains a directory of certified public keys of all mesh nodes. Each mesh node knows the public key of the gateway. Before the start of each session, a pair of mesh nodes first establishes a secret key (derived using the public keys of the node pairs and the public key of the gateway). The shared secret key is used to encrypt all messages in that session. When an IP packet is sent from a source node (s) in the Internet to a client (d) in the mesh network, the packet is encrypted at the gateway using the shared key between the gateway (g) and the client (d). If i is the mesh router of the client d, to route the encrypted packet to destination node d , the gateway g prefixes to the packet the source route from the gateway g to the mesh router i. The encapsulated packet is then forwarded by relaying routers in the WMN. Since the source address and all higher layer information including the port numbers are encrypted, the intermediate routers cannot access any information about the source with which the client is communicating. However, this encryption scheme can only provide data confidentiality and it is easy to make traffic analysis on the scheme. Since the source route is transmitted in clear text form in an encapsulated packet, the mesh routers can observe the traffic pattern of a mesh client node. To address this problem, traffic from the gateway to the mesh routers is routed through multiple paths in such a way that for each individual packet the chosen route provides traffic information to routers in the route that is independent of the overall traffic.

The penalty-based routing algorithm for traffic privacy preservation executes in three phases: (i) *path pool generation*, (ii) *candidate path selection*, and (iii) *individual packet routing*.

In the *path pool generation phase*, a large number of paths from the gateway node (g) to the destination node (x) are identified. The paths are identified in such a way that the majority of them are disjoint paths, i.e., they do not have any common nodes. The set of these paths is denoted as S_{paths}. Each node is assigned a penalty weight. The weight of a link is the average of the penalty weights at its two end nodes. The weight (i.e. cost) of a path is the sum of the penalty weights of all the edges that constitute the path. The path pool generation algorithm works iteratively. At the beginning of the first iteration, each node is assigned a penalty weight of 1. The *Dijkstra's shortest path algorithm* [140] is utilized to identify the shortest path between the gateway (g) and the destination node (x). Once, the first shortest path is identified, the penalty weights associated with each nodes on this route are increased arbitrarily so that these nodes do not become potential candidates in the next round of execution of the algorithm. The algorithm is executed iteratively to find the optimum paths in successive iterations. The iteration continues till the set S_{paths} has acquired sufficient number of candidate paths.

In the *candidate path selection phase*, a subset $S_{selected}$ is chosen from the set S_{paths}. The elements of $S_{selected}$ are chosen randomly from the set $S_{selected}$. After a path is selected in the set $S_{selected}$, to reduce its probability of getting selected again in the next round, a suitable factor is employed. The use of this probability factor prevents selection of multiple identical paths in the set $S_{selected}$.

In the packet routing phase, for each packet, one path from $S_{selected}$ is randomly chosen and the packet is routed through that path. Every time a particular path is chosen for routing a packet, the value of a counter corresponding to that chosen path is increased by one. If the value of this counter for a path reaches a threshold value, then the entire $S_{selected}$ set is assumed to have expired and a new $S_{selected}$ is chosen by invoking the *candidate path selection phase* once again. Since each packet is routed through a randomly selected path, and the candidate paths are mostly disjoint, the probability that the packets are routed through the same path will be negligibly small. The algorithm seeks to achieve a tradeoff between *routing efficiency* (i.e. routing through the shortest path between a source-destination pair) and *traffic pattern privacy* (routing through disjoint multipaths).

The authors have also shown that the proposed mechanism is robust against collusion attack by two neighboring nodes which exchange their knowledge (*colluded traffic mutual information*) about the same destination. However, the traffic splitting increases delay in communication and hence this mechanism lacks scalability. It may not, therefore, be suitable for large-scale WMN deployment scenarios delivering real-time applications.

3.8.2 PEACE: Privacy-Enhanced Yet Accountable Security Framework

Ren et al. have proposed a novel privacy and security scheme named PEACE (Privacy-Enhanced yet Accountable security framework) for WMNs [141]. The authors have assumed a three-tier architecture of a WMN under the control of a *network operator* (NO). At the top level, there are gateways connected to the public Internet, at the next level there are *access points* (APs) and *mesh routers* (MRs) which are connected to the Internet gateways by either high-speed wireless links or high-speed wired link. The MRs and APs are also interconnected among them by high-speed wireless links forming a WMN. At the lowest level of the architecture, there are wireless mobile devices which are connected to the MRs and APs by lower-speed wireless links. The security associations between the gateways and the MRs and APs are assumed to be already in place. The goal of PEACE is to secure the communication between the mobile devices with the MRs and APs. In view of this, the security objectives of PEACE are to achieve the following: (i) authentication and key management between the user devices and the routers, (ii) mutual authentication and key agreements among the users (i.e., mobile devices), (iii) user privacy protection, (iv) user accountability, and (v) membership maintenance.

To protect user privacy, one essential requirement is that the user information should be well protected from network communications against an adversary and even a network operator (NO). To satisfy this requirements, PEACE aims to achieve the following: (i) no communication sessions should reveal any information related to the identity of a user except that the user is a legitimate member in the network, (ii) no adversary and NO should be able to link two different communication sessions to a particular specific user in the network, and (iii) a given communication session under the audit by an NO can only be linked to the attributes of the user without disclosing his/her full identity, (iv) each user is assumed to belong to a user group and each user group is supposed to be managed by a group manager. The full identity of a user can be identified only by the joint effort from the user group manager and the NO and not by any of them individually.

To address the aforementioned security and privacy requirements, PEACE proposes a trust and key management model. The high-level trust model includes four network entities: (i) the network operator, (ii) user group managers, (iii) user groups, and (iv) a *trusted third party* (TTP). Each user group consists of a set of users according to certain aspects of their *non-essential attributes*. Each group has one group manager who is delegated with the responsibility of adding or removing users from that group. The manger of a group knows the essential and non-essential attributes of each of the members in his group since these information is shared by the users to the manager at time of their joining the network. Each group manager subscribes to the network operator on behalf of its group members. The network operator allocates a set of *group secret keys* to the user group after successful registration of the group. Each group secret key is divided into two parts before they are distributed by the network operator. One part of the key is sent to the corresponding group manager and the other part is communicated to the TTP. To access network services, each user sends two requests: one request to its group manager and the other to the TTP for recovering the complete group secret key. On receiving the key parts, the user sends signed acknowledgments to the group manager and the TTP. This principle of key management which is based on *separation of powers* enables PEACE to achieve a high level of robustness in its security and privacy protection strength. From the network operator's point of view, access security is guaranteed since each user has to gather a valid complete group secret key and generate a valid access credential based on that key. User privacy is highly enhanced since the user identity information and the corresponding key information are divided among three autonomous network entities – the group manager, the TTP and the network operator. Although the network operator has the access to the complete user secret key, it does not have any information about how these keys are mapped to the essential attributes of the users. On the other hand, the group manager and the TTP both know the essential attributes of the users, but none of them have complete information about the secret key. Unless two among the three entities (network operator, TTP and group manager) collude, PEACE makes it impossible for any of them to have access to a user's essential attributes from the access credentials submitted by the user. However, the accountability of a user is fully ensured since in case of any possible incident, a legal authority can collect information from the network operator, the group manager, and the TTP to precisely identify the user.

Since the conventional blind signatures and ring signature schemes can only provide irrevocable anonymity, and PEACE requires revocable anonymity to guarantee user accountability, the authors have developed a variation of the short group signature scheme proposed in [142] to meet the requirement. PEACE also provides a robust mechanism for dynamic addition and deletion of users and mesh routers to and from the network. A comprehensive security analysis has shown that the scheme is resistant to bogus data injection attacks, data phishing attacks and DoS attacks [141].

3.8.3 SAT: A Security Architecture for Achieving Anonymity and Traceability

Sun et al. present a security architecture named "SAT" [143, 144]. The system uses ticket-based protocols for resolving the conflicting security requirements of unconditional anonymity for honest users and traceability of misbehaving users in a WMN. By using the tickets, self-generated pseudonyms, and the hierarchical identity-based cryptography, the architecture has been demonstrated to achieve the desired security objectives while maintaining the performance efficiency at a high level. The system uses a blind signature technique from payment systems [145-148] to achieve anonymity by delinking user identities from their activities. The pseudonym technique also renders user location information unexposed. The pseudonym generation mechanism does not rely on a central authority unlike the *broker* in ARSA [110], the *domain authority* in [149], the *transportation authority* or the *manufacturer* in [150], and the *trusted authority* in [151], who can derive the user's identity from his/her pseudonyms and illegally trace on an honest user. However, the system is not intended for achieving routing anonymity. *Hierarchical identity-based cryptography* (HIBC) for inter-domain authentication is adopted to avoid domain parameter certification in order to ensure anonymous access control.

Fig. 12 depicts the architecture of a WMN on which the SAT scheme is implemented. The scheme assumes a wireless mesh backbone consisting of the *mesh routers* (MRs) and the *gateways* (GWs) which are connected by wireless links. The MRs and GWs serve as the access points for individual WMN domains such as hospital, campus etc. It is assumed that each domain is managed by a *trusted authority* (TA). Each TA is connected to its GWs by high-speed links. The TAs and the GWs are powerful computing devices and these devices are physically protected from any possible attacks. The scheme assumes the existence of a mechanism that distributes pre-shared keys and establishes secure communication channels between the TAs, the GWs, and the MRs at the backbone. SAT, however, focuses on the authentication and the key establishment problems during the network access of the MCs. It adopts the *hierarchical ID-based signature scheme* (HIDS) [152] for inter-domain authentication of clients which are affiliated to a home TA and are currently visiting a foreign TA. For trust domain initialization, the authors have proposed the use of hierarchical *identity based cryptography* (IBC) [152]. SAT uses a *ticket-based security architecture* for client authentication, data integrity, and confidentiality in communications both in the home domain and in inter-domain. The ticket-based architecture is responsible for ticket issuance, ticket deposit, fraud detection, and ticket revocation protocols. When a client joins the network, the TA issues a ticket to it. The client needs to reveal its real ID to the TA at this time so that TA can verify the authenticity of the client. The client employs some blinding techniques to transform the ticket so that it cannot be linked to a specific execution of the ticket generation algorithm while the verifiability of the ticket is maintained. This ensures that the TA cannot link the issued ticket to the real identity of the client. The authors have proposed use of a *partially restrictive blind signature scheme* [153, 154] in the ticket generation algorithm. The TA publishes the domain parameters to be used by the clients in its trust domain following the standard IBC domain initialization process. After obtaining a ticket, a client may deposit the ticket anytime for accessing network services before the ticket expires. The ticket is to be deposited only once to the first encountered gateway that provides the network access. The gateway (also called the *deposit gateway*) verifies the signature and the integrity of the deposited ticket and generates a signature on the client's pseudonym along with the ID of the *deposit gateway* (DGW). This signature is required so that the other access points in the trust domain can determine whether and where to forward the client's access request, in case the deposited ticket is to be further used from other access points. The client is not allowed to change its pseudonym, since the DGW will refuse network access to the client if the current pseudonym does not match with the one that was initially recorded with the ticket. The real ID of a client has to be revealed to the home TA at the time the ticket is issued due to the requirement of client authorization. However, if the privacy of the client is to be protected, this ID is hidden from the access point. Unless the access point colludes with the home TA, user identity privacy cannot be compromised. For obtaining new tickets, the client simply deposits the old ticket using the ticket deposit protocol and gets the new ticket. The request for ticket is sent to the home TA in encrypted form.

The ticket revocation protocol is used for two purposes: (i) revocation of new tickets and (ii) revocation of deposited tickets. The client may have some unused tickets which have not been deposited before. For revoking these tickets, the client sends a signed and time-stamped request for revocation to a GW. The GW authenticates the client and records the ticket serial number as revoked. For revoking already deposited tickets, the client sends an authenticated revocation request to the

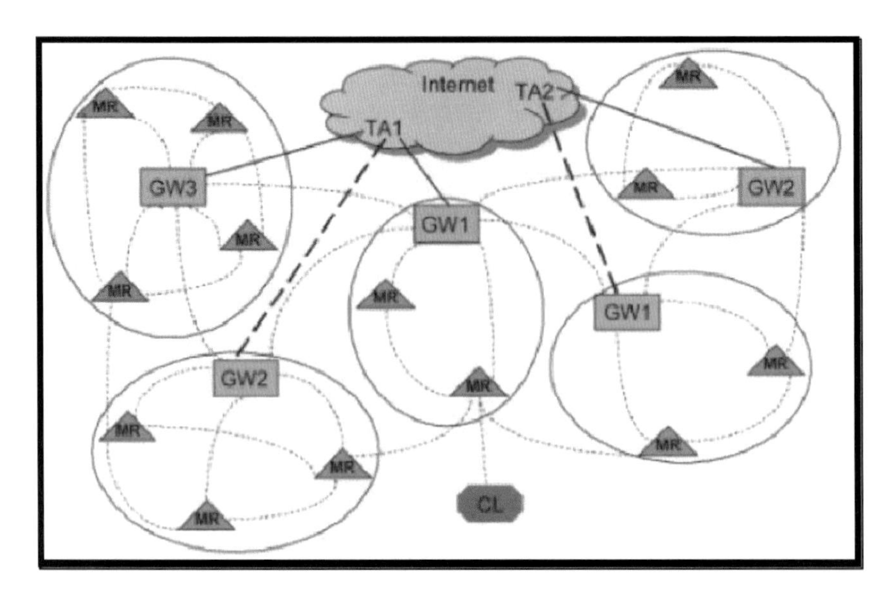

Fig. 12. The WMN architecture used in SAT scheme [144]

DGW. The DGW verifies the authenticity of the client and marks the associated ticket as revoked. A GW after successfully executing a revocation request, reports the revocation details to the home TA. The home TA updates and distributes the revocation list for all GWs in the trust domain.

When a client attempts to access network services from a foreign (visiting) trust domain, two possible scenarios of authentication for the client may arise. In the first case, the client does not have any available ticket for accessing the network from the foreign domain. In this case, a foreign *mesh router* (MR) forwards the client's new ticket request to the home domain of the client. In the second scenario, the client has available new tickets issued by its home TA. In this case also the foreign mesh router forwards the tickets to the home domain for verifying the authenticity of the available tickets. In SAT, the clients use a robust and efficient way for generating their pseudonyms. The method of self-generation of pseudonym is similar to that proposed by Rahman et al. in [155]. Self-generation of pseudonyms by clients incurs very low overhead and hence, the clients can use it frequently to regenerate their pseudonyms. This increases the strength of the anonymity and privacy protection.

3.8.4 An Efficient and User Privacy Preserving Routing Protocol for WMNs

Sen has proposed an efficient and secure routing protocol for WMNs for handling stringent *quality of service* (QoS) requirements of real-time applications while ensuring protection of user privacy [156]. The scheme assumes a three-tier architecture of WMNs in which the top level contains the *Internet gateways* (IGWs) which are connected to the wired Internet forming the backbone of the WMN. At the next level of the architecture are the wireless mesh routers which are connected to the IGWs and among each other forming a wireless mesh. The mesh clients (i.e., the wireless mobile user devices) are at the lowest level of the architecture which can connect to the MRs using wireless networking for accessing the Internet services. The author has identified several challenges in designing an efficient routing protocol such as: (i) accurate measurement of link reliabilities, (ii) robust estimation of the end-to-end delay in routing, (iii) reducing control overhead to maximize throughput in the network. To address these challenges the proposed protocol has incorporated several salient features such as: (i) designing a robust estimate for reliability of each routing path for a source-destination pair, (ii) exploiting the network topological information for efficient route discovery, (iii) developing a reliable model for estimating the end-to-end delay along a routing path, (iv) use of *multi-point relay* (MPR) nodes to reduce the control overhead in routing, (v) accurately estimating the available bandwidth in a wireless link and along a wireless path for a source-destination pair, (vi) routing through a path in the fixed network through wired IGWs, (vii) identification of selfish nodes, (viii) protecting the user privacy. In the following, these features are described very briefly. A more detailed discussion can be found in [156].

In the protocol, every node computes the reliability of each of its out-bound wireless links. For computing the reliability of a link, the number of control packets a node receives during a given time interval is used as the base parameter. An *exponentially weighted moving average* (EWMA) approach is use to update the link reliability estimate for a link [156]. Each node maintains and periodically updates a *link reliability table* after computing the reliability metric for each of its out-bound wireless links. The reliability for an end-to-end routing path is computed based on the average of the reliabilities of all the links constituting the path.

The protocol also exploits the knowledge of network topology by using a strategy of selective flooding of control messages, and hence avoids broadcast of control messages. If a source-destination pair is under the control of the same MR, the flooding of the control message are made only within the part of the network served by the MR. To further reduce the overhead of control messages, the nodes accept broadcast control messages from only those neighbors which have link reliabilities greater than a threshold minimum value. The paths with less reliability values are not discovered and hence not considered for routing.

For achieving an accurate estimation in end-to-end delay over a routing path, the protocol uses *probe packets* during the route discovery phase. In this approach, when a source node receives the RREP packet from a destination in response to an RREQ message, it constructs a table and stores all the RREP packets that have been received along with the routes through which the RREP messages have arrived. The source node then sends some probe packets to the destination through the (reverse) path along which the first RREP message arrived from the destination. This ensures that the probe packets are sent along the path which is likely to induce the minimum end-to-end delay. The number of probe packets is kept limited to $2H$ for a path of H hops to make a tradeoff between the control overhead and the measurement accuracy. The destination node computes the average delay experienced by all the probe packets it has received, and sends the computed value to the source node piggybacking it on an RREP message. If the computed average delay is less than the maximum delay that can be tolerated to satisfy the application QoS, the source node selects the route.

Otherwise, the source tries with the next path in its table till it finds a path that satisfies the delay tolerance.

For further reducing the control overhead, the protocol selects MPR nodes as in *optimized link sate routing* (OLSR) protocol, and routes the control packets through these nodes. In addition to the computation of path reliability and use of MPR nodes to minimize the overhead due to flooding, the protocol uses a robust estimation of the available bandwidth of the wireless paths used in routing by computing the packet loss due to congestion in the wireless links [156].

The efficiency in routing is further improved by occasional routing of packets through the fixed wired network backbone. Every MC knows its hop distance from its IGW. This hop count information is included in the RREQ messages from the MC. When the destination MC receives the RREQ, since it also knows its hop distance from its gateway, it checks which path is better from routing – the path through the WMN or the path through the fixed network. If the destination node finds that the better route is through the fixed network, it sends the RREP message through wired backbone network. Therefore, in such situations, the forward route is established between the source and the destination through the wired network, while the reverse route is setup through the WMN. Since the nodes in the forward and the reverse routes are on node-disjoint paths, they do not contend for accessing the wireless medium resulting in an improved throughput and reduced end-to-end delay. For handling any possible selfish behavior of nodes (a selfish node utilizes network resources for routing its own packet but avoids forwarding packets for other nodes to conserve its energy), the protocol employs a simple mechanism to discourage selfish behavior and encourage cooperation among nodes. Each node forwards routing packets to only to those neighbors which have link reliability greater than a threshold value. Since the link reliabilities for the selfish nodes are all zero (the selfish nodes do not forward any packets), these nodes never get any opportunity to participate in routing. Hence, packets originating from these nodes are always dropped. The link reliability metric serves the dual purpose of enhancing the reliability and enforcing cooperation among the nodes.

The protocol also ensures user anonymity and privacy protection by using a *novel* approach that extends the ring signature authentication scheme [157]. A security analysis of the protocol has shown that the protocol ensures: (i) *user anonymity*, (ii) *anonymous authentication of a user with an authentication server (AS) and also anonymous authentication among a pair of users*, (iii) *forward secrecy* (forward secrecy of a scheme refers to its ability to defend against leaking of its keys of the previous sessions when an attacker is able to catch hold of the key of a particular session). Since forward secrecy is ensured, the protocol is resistant to any replay attack [156].

Table 6 presents a summary of some privacy protection schemes in WMN communication.

3.9 Cross-Layer Security Mechanisms

Kidston et al. have proposed a cross-layer architecture for security in multi-hop wireless networks [158]. The authors argue that by utilizing the metrics from the security services at one layer, such as authentication systems and *intrusion detection systems* (IDS), robustness in the security services being provided at other layers can be increased. For example, application layer security services such as secure authentication and IDSs can provide real-time attack profiles into an integrated cross-layer security service. These profiles may be used by the lower layers to improve their detection efficiencies since the lower layers don't have to compute any security metrics at all. While the computational complexity in a node increases for using this cross-layer security services, the communication overhead in the network is reduced to a large extent. In the proposed cross-layer architecture, the authors have used a *publish-subscribe* system that acts a vertical interface between all the layers of the communication protocol stack [158]. This interface is used as a real-time messaging framework by the layers for communicating critical security metrics. Other layers subscribe to the metrics. An independent cross-layer service may also be designed that will exclusively subscribe to the security metrics. Each protocol stack focuses only on its core security functionality while subscribing to the value-added metrics published by other layers. The per-layer information is combined to build a consolidated security picture that can assist the operator of the network in designing an automated response system for responding to any possible attack.

Table 6. Summary of some privacy preservation schemes in WMN communication

Protocol	Salient Features
Penalty-based multipath routing [5]	This mechanism aims to achieve a balance between network performance and traffic privacy in a WMN. It utilizes a novel concept called traffic entropy that quantifies the amount of information required to describe the traffic pattern and to characterize the performance of traffic privacy preservation. The scheme uses a penalty-based shortest path routing algorithm that maximally preserves the traffic privacy by minimizing the mutual information of traffic entropy observed at intermediate node while ensuring that network performance is maintained at high level. The scheme is effective in preserving traffic privacy and resilient against colluding malicious nodes. However, it has a scalability problem.
PEACE [141]	It is a privacy architecture for WMNs that assumes three-tiers with the gateways connected to the Internet at the topmost level, MRs at the second level and user mobile devices at the lowest level. The security associations between the gateways and the MRs are assumed to be in place. The protocol has the following objectives: (i) ensure authentication and robust key management between the user devices and the MRs, (ii) provide mutual authentication and key distribution among the mobile user devices, (iii) protect the privacy of the users, (iv) ensure user accountability in the system, and (v) maintain the node membership in the system. The protocol has a robust trust and key management system to satisfy its security and privacy requirements. In addition to providing revocable anonymity, the protocol is resistant to bogus data injection attack, data phishing attacks and DoS attacks.

Table 6. (*continued*)

SAT [143, 144]	The protocol uses tickets, self-generated pseudonyms, and a hierarchical identity-based cryptographic structure to achieve the conflicting tradeoff between security and user privacy while maintaining the performance efficiency at a high level. The pseudonym generation technique does not rely on a central authority and provides user location privacy while avoiding any single point of failure in the system architecture. The protocol provides anonymous access control to the users but cannot ensure routing anonymity. The ticket-based security architecture provides client authentication, data integrity, and confidentiality in communications both in the home domain and in inter-domain. A blinding technique is used by a client to transform the issued ticket by the TA (trusted authority) so that the TA cannot link the issued ticket to the real identity of the client.
Privacy preserving routing [156]	It is a secure routing protocol that can handle stringent quality of service (QoS) requirements of real-time applications while ensuring protection of user privacy. Various modules of the protocol perform the following functions: (i) accurate measurements of wireless link reliabilities, (ii) robust estimation of end-to-end delay in routing, (iii) reducing the control packet overhead and maximizing the throughput in the network, (iv) use of multi-point relay (MPR) nodes to reduce control overhead, (v) accurate estimation of the available bandwidth in the wireless links on the routing path for a source-destination pair, (vi) identification of selfish nodes, (vii) protection of user privacy. User anonymity and privacy is ensured by a novel scheme that extends the ring signature authentication scheme.
Privacy in P2P networks [139]	This protocol is designed to combat the problem of unauthentic file downloads as well as to improve the search scalability and efficiency in an unstructured peer-to-peer network while protecting the privacy of the users. The adaptive trust-aware protocol is based on the construction of an overlay of trusted peers where neighbors are selected based on their trust ratings and content similarities. The semantic communities of peers also enable the protocol for form a neighborhood of trust that is utilized to protect user privacy.
Anonymity using onion routing [6]	The objective of this mechanism is to hide an active node that connects to a gateway router for accessing the Internet, where the active mesh nodes are to be kept anonymous. The communication protocol uses the concept of onion routing (OR) to protect the privacy of the node. The mobile user who requires anonymous communication sends a request to the onion router which constructs an onion route consisting of the public keys of the routers. The onion is constructed in such a way that the inner most part is the core message which is wrapped by encrypting it using the public keys of the ORs on the route. The mechanism, however, has a high computing and communication overhead.

4 Trust Management in WMNs

The use of trust and reputation-based frameworks for enforcing security and collaboration among nodes in wireless multi-hop networks is a very popular approach. A trust-based scheme can protect against attacks that are sometime beyond the capabilities of cryptographic mechanisms. For example, issues like judging the quality and reliability of wireless links for establishing routing paths in high-throughput multicast routing, robust key management, reliable packet forwarding over multiple hops etc. can be elegantly addressed by utilizing the services of a trust-based framework. Several trust- and reputation-based frameworks for mobile ad hoc and wireless sensor networks exist in the literature [8]. Most of these schemes can be easily adapted to WMNs.

Pirzada and McDonald propose an approach to building trust among the nodes in an ad hoc network that can be deployed in a WMN which does not have a centralized trusted entity [159]. In this scheme, the nodes passively monitor the packets received and forwarded by their neighbors. The receiving and forwarding of packets are considered as events. The events are assigned different weights based on the applications and the forwarding behaviors of the nodes. The weights reflect the significance of the concerned event with respect to the associated applications. The trust values associated with all the events of a node are combined to arrive at an aggregate trust metric for the node. The computed trust values of a pair of nodes are used for deriving the trust value of the link connecting those nodes. The links which have higher trusted values are assigned smaller weights and a shortest-path algorithm is utilized to find the most trusted path (i.e., the path with the minimum weight) between a pair of nodes.

Yen et al. propose a security solution for ad hoc networks that is based on a trust framework to ensure data protection, secure routing, and message authentication [160]. In this scheme, logical and computational trust analysis and evaluation methodologies are applied in each node. Each node then evaluates the trust metric for each of its peers using factors such as experience statistics, data value, intrusion detection results, and recommendations from its neighbors leading to a robust trust management framework.

Sen et al. have proposed a trust establishment mechanism among nodes in a MANET [161]. The proposed framework is a probabilistic solution based on a distributed trust model. A *secret dealer* is introduced during the network bootstrapping phase to initiate the trust establishment process. Subsequently, the nodes use a self-organized certificate distribution and management procedure based on threshold cryptography. The mechanism has been shown to be very effective in a large-scale dynamic ad hoc network.

In a multi-operator WMN deployment, the major issue is the lack of trust between the heterogeneous network entities belonging to different service providers and network operators. In such scenarios, reputation-based systems are useful for establishing and sustaining trust between mobile users and different network operators and service providers. Ben Salem et al. have addressed the issue of interoperability for trust building in a wireless network [162]. The authors argue that since the home network of a service provider can be the home network for some mobile nodes and a foreign network for some other mobile nodes, the home network of a service provider cannot always be trusted by all mobile nodes. The proposed reputation-based system allows mobile nodes to evaluate the trustworthiness of a service provider based on the latter's behavior. At the same time, the system also allows a service provider to authorize mobile users to access its network services based on their trust levels with the service provider.

Jarrett and Ward have presented a trusted computing-based routing protocol-*trusted computing ad hoc on-demand distance vector* (TCAODV) [163] – that extends the base *ad hoc on-demand distance vector* (AODV) [83] routing protocol. In TCAODV, each node has its public key certified by a *certificate authority* (CA) and the public key certificate of each node is stored within a trusted root. The source node broadcasts its public key certificate with its *HELLO* message. Each of the neighbor nodes, on receiving the certificate, first verifies the authenticity of the certificate and then stores it as the broadcaster's public key. The RREQ packet sent by each node is signed using the integrity metric from the routing module of the sender. Each neighbor, on receiving the RREQ message, verifies the signature using the previously stored broadcaster's key and determines whether the provided integrity measurements are trustworthy. The intermediate nodes on the path from the source to the destination strip off the signature in the RREQ message and put their own signatures and integrity measurements. A *per-route symmetric encryption key* is also established for allowing only the trusted nodes along the route to use the route. All traffic sent along this route is encrypted using this symmetric key.

Sen presents a reputation- and trust-based security framework for *mobile ad hoc networks* (MANETs) that can detect packet dropping attacks by malicious insider nodes [164]. The mechanism uses a trust model that computes the reputation metrics for all nodes in the network based on their packet forwarding statistics. The proposition includes a robust trust metric computation, propagation and update process which involves low computation and communication overheads. Moreover, it can be easily adapted to WMNs which do not have any centralized point of administration or a *certificate authority* (CA) for key management or a deployed *trusted third party* (TTP).

Tang and Wu propose a trust-delegation-based *efficient mobile authentication scheme* (EMAS) that uses the elliptic curve discrete logarithm problem [165]. Lee et al. have presented a distributed authentication model for reducing the authentication delay by deploying multiple trusted nodes which serve the role of an *authentication server* in a WMN [107]. Zhang et al. propose an *attack resilient security architecture* (ARSA) for WMNs that aims at providing secure roaming in multi-domain WMNs using a *user-broker-operator trust model* [110]. Lin et al. have presented an authentication scheme for WMNs that utilizes the services of a *trusted third party* (TTP) which also acts as a trusted *certificate authority* (CA) for issuing certificates to service providers and mobile users [123].

For establishing a trust relationship among different *wireless Internet service providers* (WISPs) in a multi-operator WMN, an approach of deploying a centralized *roaming broker* (RB) has been proposed by Leu et al. [116]. The RB is trusted by all the participating WISPs. In this approach, when a *mobile user* (MU) roams into the domain of a foreign network, the foreign WISP forwards the AAA session of the MU to its home WISP for authorization verification via the RB. In the scheme proposed by Long et al., the RB not only acts as TTP, but also as a certificate authority that issues public key certificates to the WISPs and the MUs [117]. The trust relationships among the WISPs or between a WISP and MUs can be easily established by validating the public key certificates by the RB. The foreign WISP reports the accounting information of the roaming MU to its home WISP at the completion of the session. Although, the deployment of an RB effectively reduces handoff authentication delay, unfortunately, it becomes a point of performance bottleneck and single-point-of-failure. Zhu et al. have presented a digital signature-based inter-domain roaming and billing architecture wherein a WISP not only accepts a valid public key certificate issued by another WISP but also accepts the digital signature issued by that WISP as a payment technique [166]. Hence, inter-domain authentication and billing are performed simultaneously by verifying and validating a digital signature. The scheme assumes that every *access point* (AP) is enough trustworthy is authorized to issue digital signatures on behalf of its WISP. However, this assumption may not be realistic in large-scale distributed WMNs in which some of the *mesh access points* (MAP) may be easily compromised.

Several other trust and reputation-based schemes are available for wireless multi-hop networks such as: Watchdog and Pathrater [59], CONFIDANT [167], CORE [168], RFSN [169], DRBTS [170], OCEAN [171]. The interested readers may refer to [8] for further details on these schemes.

5 Security in WMNs – Open Problems and Future Research Challenges

WMNs have become the focus of research in wireless networks in the recent years owing to their great promise in realizing numerous next-generation wireless services. Driven by the demand for rich and high-speed content access with stringent QoS requirements, recent research on WMNs has focused on developing high performance communication protocols, while the security and privacy issues of these protocols have received relatively less attention. However, given the wireless and multi-hop nature of communication in WMNs, these networks are vulnerable to a wide variety of attacks at all layers of the communication protocol stack. Although, the researchers have made substantial contributions in the areas of security and privacy in WMNs, there are still many challenges that remain to be addressed. Some of these challenges and a few future research directions in security for WMNs are briefly discussed in the following.

Existing authentication schemes for WMNs under high mobility conditions involve large delay and latency that sometimes adversely affect the QoS of the applications and network services. Efficient, lightweight and robust authentication protocols for MRs and MCs with high mobility need to be designed that minimize authentication and handoff delay. For this purpose, scalable key management systems are also required. Similarly, for reliable routing in wireless environment that involves high rate of packet drops, energy-aware, efficient multipath routing schemes are also in demand for high throughput applications.

For strategic deployments of WMNs, robust hop integrity verification protocols need to be designed. In addition, due to hardware/software compatibility and efficiency considerations, it may be a challenging issue to design a strategic plan for deploying the hop integrity verification protocols in the nodes in a WMN. While the deployment of an integrity verification protocol in each MR may be a very secure strategy, from efficiency point of view this may not be an optimal one. An adaptable strategy that can be varied based on the requirements of the applications will be ideal.

Existing security schemes for deployment in hybrid WMNs are also inadequate. In hybrid WMNs, the networks are designed to be integrated with other types of networks, such as wired networks and cellular wireless networks. Such networks are vulnerable to a wide range of attacks which are not present in the individual network components. For instance, a mesh network for wireless Internet access can be targeted with DoS attacks launched from the Internet [172]. The scarcity of bandwidth resource on the wireless network makes the attack more difficult to defend and the existing defense mechanisms for DoS attacks will be ineffective in this scenario. On the other hand, hybrid networks also present additional resources and opportunities for defending against attacks. For example, for defending attacks on a WMN connected to a wired network, it is possible to leverage the high bandwidth, low latency wired links, and powerful computers in the wired network for handling security issues in the wireless networks.

Authentication and key management in WMNs with highly mobile MCs and MRs pose particularly difficult challenges. Owing to the very limited coverage of IEEE 802.11-based MRs (e.g., 100 meters), high mobility users migrate very fast from the coverage area of an MR to another. It is not acceptable for a user to authenticate and negotiate the key with each MR. Novel solutions possibly by using group keys are needed for this purpose. A very interesting research direction in this regard is to investigate whether an approach of *signature aggregation* [137] can be utilized for developing an efficient key management scheme for WMNs. This can be extremely useful for group key establishment leading to reduction in the overhead incurred in the group key creation and re-keying process.

Use of novel approaches like *dynamic topology-aware adaptations* and *dynamic network coding* for securing multicast messages in WMNs are very interesting emerging trends [172]. The essential principle of dynamic topology adaptation is to improve network performance by dynamically adjusting the protocol structures based on the variations of the wireless links. On the other hand, dynamic network coding exploits the broadcast nature of the wireless medium and makes use of the common occurrence of packet overhearing in wireless networks with the ultimate goal to improve network performance. Ahlswede et al. have proposed the use of network coding for secure routing in wireless networks [173]. However, the existing network coding systems are vulnerable to a wide range of attacks besides the most well-known *packet pollution attacks* [174]. Many of the weaknesses of the existing systems lie in their single focus on performance optimizations. A more balanced approach, which can provide improved security guarantees, is crucial for the actual adoption of network coding in real-world applications. A future direction of research is to uncover the security implications of different design and optimization techniques, and explore balanced system designs with network coding that achieve appropriate tradeoffs between security and performance that is suitable for different applications' requirements.

Cross-layer security approach is another possible future direction of research for wireless networks. While cross-layer protocols for wireless networks have been widely researched in the literature [175], their applications in security domain have found very limited attention. As mentioned in Section 3.9, cross-layer approach to security could lead to efficient and proactive intrusion and anomaly detection in some mission-critical wireless network deployments.

For user privacy preservation, optimized deployments of emerging techniques like *fully homomorphic encryption* [176] in WMNs can be another very interesting direction of research.

6 Conclusion

WMNs have become an important focus area of research in the recent years owing to their great potentials in realizing numerous next-generation wireless services with stringent QoS guarantees and with high mobility support for the users. Driven by the increasing demand for rich, high-speed and bandwidth intensive content access, recent research has focused on developing high performance communication protocols for such networks, while issues like security, privacy, access control, intrusion detection, secure authentication etc. have taken the back seat. However, given the inherent vulnerabilities of the wireless medium due to its broadcast nature and multi-hop communications in WMNs, these networks are subject to a wide range of threats. This chapter has made a comprehensive presentation on the various attacks on different layers of the communication protocol stack of WMNs. While highlighting various vulnerabilities in the physical, link, network, transport and application layers, this chapter has also focused its attention on how attacks can be launched on authentication, privacy and key management protocols on WMNs. After identifying various security threats, the chapter has presented a comprehensive state of the art survey on various defense mechanisms for defending those attacks. Some of these defense mechanisms are also compared with respect to their different approaches towards security and their performance efficiencies. Finally, some of the emerging trends in research and future research issues related to security and privacy in WMNs are presented.

References

1. Akyildiz, I.F., Wang, X., Wang, W.: Wireless mesh networks: a survey. Computer Networks 47(4), 445–487 (2005)
2. Franklin, A.A., Murthy, C.S.R.: An introduction to wireless mesh networks. In: Zhang, Y., et al. (eds.) Security in Wireless Mesh Networks, pp. 3–44. CRC Press, USA (2007)
3. Sen, J.: Secure and privacy-preserving authentication protocols for wireless mesh networks. In: Sen, J. (ed.) Applied Cryptography and Network Security, pp. 3–34. INTECH, Croatia (2012)

4. Sen, J.: Secure routing in wireless mesh networks. In: Funabiki, N. (ed.) Wireless Mesh Networks, pp. 237–280. INTECH, Croatia (2011)
5. Wu, T., Xue, Y., Cui, Y.: Preserving traffic privacy in wireless mesh networks. In: Proceedings of the International Symposium on a World of Wireless, Mobile and Multimedia Networks (WoWMoM 2006), Buffalo-Niagara Falls, NY, USA, pp. 459–461 (June 2006)
6. Wu, X., Li, N.: Achieving privacy in mesh networks. In: Proceedings of the 4th ACM Workshop on Security of Ad Hoc and Sensor Networks (SASN 2006), Alexandria, VA, USA, pp. 13–22 (October 2006)
7. Sen, J.: A Distributed Trust and Reputation Framework for Mobile Ad Hoc Networks. In: Meghanathan, N., Boumerdassi, S., Chaki, N., Nagamalai, D. (eds.) CNSA 2010. CCIS, vol. 89, pp. 538–547. Springer, Heidelberg (2010)
8. Sen, J.: Reputation- and trust-based systems for wireless self-organizing networks. In: Pathan, A.-S.K. (ed.) Security of Self-Organizing Networks: MANET, WSN, WMN, VANET, pp. 91–122. Aurbach Publications, CRC Press, USA (2010)
9. Sen, J., Chowdhury, P.R., Sengupta, I.: A distributed trust mechanism for mobile ad hoc networks. In: Proceedings of the International Symposium on Ad Hoc and Ubiquitous Computing (ISAHUC 2006), Surathkal, Mangalore, India, pp. 62–67 (December 2006)
10. Sen, J.: A robust and efficient node authentication protocol for mobile ad hoc networks. In: Proceedings of the 2nd International Conference on Computational Intelligence, Modelling and Simulation (CIMSiM 2010), Bali, Indonesia, pp. 476–481 (September 2010)
11. Shi, E., Perrig, A.: Designing secure sensor networks. IEEE Wireless Communication Magazine 11(6), 38–43 (2004)
12. Xu, W., Trappe, W., Zhang, Y., Wood, T.: The feasibility of launching and detecting jamming attacks in wireless networks. In: Proceedings of the 6th ACM International Symposium on Mobile Ad Hoc Networking and Computing (MobiHoc 2005), Urbana-Champaign, IL, USA, pp. 46–47. ACM Press (May 2005)
13. Law, Y., Palaniswami, M., Hoesel, L.V., Doumen, J., Hartel, P., Havinga, P.: Energy-efficient link-layer jamming attacks against wireless sensor network MAC protocols. ACM Transactions on Sensor Networks (TOSN) 5(1), article no. 6 (February 2009)
14. Brown, T.X., James, J.E., Sethi, A.: Jamming and sensing of encrypted wireless ad hoc networks. In: Proceedings of the 7th ACM International Symposium on Mobile Ad Hoc Networking and Computing (MobiHoc 2006), Florence, Italy, pp. 120–130 (May 2006)
15. Wood, A.D., Stankovic, J.A.: Denial of service in sensor networks. IEEE Computer 35(10), 54–62 (2002)
16. Mishra, A., Arbaugh, W.A.: An initial security analysis of the IEEE 802.1X standard. Technical Report CS-TR-4328, Computer Science Department, University of Maryland, USA (2002)
17. Naveed, A., Kanhere, S.S., Jha, S.K.: Attacks and security mechanisms. In: Zhang, Y., et al. (eds.) Security in Wireless Mesh Networks, pp. 111–144. Auerbach Publications, CRC Press, USA (2008)
18. Hu, Y.-C., Perrig, A., Johnson, D.B.: Rushing attacks and defense in wireless ad hoc network routing protocols. In: Proceedings of the ACM Workshop on Wireless Security (WiSe 2003) in conjunction with ACM MobiCom 2003, San Diego, CL, USA, pp. 30–40. ACM Press (September 2003)

19. Hu, Y.-C., Perrig, A., Johnson, D.B.: Packet leashes: a defense against wormhole attacks in wireless ad hoc networks. In: Proceedings of the 22nd IEEE Joint Conference of IEEE Computer and Communications Societies (INFOCOM 2003), San Francisco, USA, pp. 1976–1986. IEEE Press (March-April 2003)
20. Al-Shurman, M., Yoo, S.-M., Park, S.: Black hole attack in mobile ad hoc networks. In: Proceedings of the 42nd Annual Southeast Regional Conference (ACM-SE), Huntsville, Alabama, USA, pp. 96–97 (April 2004)
21. Ramaswamy, S., Fu, H., Sreekantaradhya, M., Dixon, J., Nygard, K.E.: Prevention of cooperative black hole attacks in wireless ad hoc networks. In: Proceedings of the International Conference on Wireless Networks (ICWN 2003), Las Vegas, Nevada, USA, pp. 570–575. CSREA Press (June 2003)
22. Sen, J., Chandra, M.G., Harihara, S.G., Reddy, H., Balamuralidhar, P.: A mechanism for detection of grayhole attack in mobile ad hoc networks. In: Proceedings of the 6th IEEE International Conference on Information, Communications, and Signal Processing (ICICS 2007), Singapore (2007)
23. Newsome, J., Shi, E., Song, D., Perrig, A.: The Sybil attack in sensor networks: analysis and defenses. In: Proceedings of the 3rd International Symposium on Information Processing in Sensor Networks (IPSN 2004), Berkeley, CA, USA, pp. 259–268. ACM Press (April 2004)
24. Zhong, S., Li, L.E., Liu, Y.G., Yang, Y.R.: On designing incentive-compatible routing and forwarding protocols in wireless ad-hoc networks: an integrated approach using game theoretical and cryptographic techniques. Wireless Networks 13(6), 799–816 (2007)
25. Ben Salem, N., Buttyan, L., Hubaux, J.-P., Jacobson, M.: A charging and rewarding scheme for packet forwarding in multi-hop cellular networks. In: Proceedings of the 4th ACM International Symposium on Mobile Ad Hoc Networking and Computing (MobiHoc 2003), Annapolis, MD, USA, pp. 13–24. ACM Press (June 2003)
26. Sen, J.: Routing security issues in wireless sensor networks: attacks and defenses. In: Tan, Y.K., et al. (eds.) Sustainable Wireless Sensor Networks, pp. 279–309. INTECH Publishers, Croatia (2010)
27. Roy, S., Koutsonikolas, D., Das, S., Hu, C.: High-throughput multicast routing metrics in wireless mesh networks. In: Proceedings of the 26th IEEE International Conference on Distributed Computing Systems (ICDCS 2006), Lisbon, Portugal, p. 48. IEEE Computer Society Press (July 2006)
28. Chen, A., Lee, D., Chandrasekaran, G., Sinha, P.: HIMAC: high throughput MAC layer multicasting in wireless networks. In: Proceedings of the 3rd IEEE International Conference on Mobile Adhoc and Sensor Systems (MASS 2006), Vancouver, British Columbia, Canada, pp. 41–50 (October 2006)
29. Roy, S., Addada, V.G., Sethia, S., Jajodia, S.: Securing MAODV: attacks and countermeasures. In: Proceedings of the 2nd Annual IEEE Conference on Sensor and Ad Hoc Communications and Networks (SECON 2005), Santa Clara, CL, USA, pp. 521–532 (September 2005)
30. Wu, B., Chen, J., Wu, J., Cardei, M.: A Survey on Attacks and Countermeasures in Mobile Ad Hoc Networks. In: Xiao, Y., Shen, X., Du, D.-Z. (eds.) Wireless Network Security. SCT, pp. 103–135. Springer, Heidelberg (2006)
31. He, B., Xie, B., Zhao, D., Reddy, R.: Secure access control and authentication in wireless mesh networks. In: Pathan, A.-S.K. (ed.) Security of Self-Organizing Networks: MANET, WSN, WMN, WANET, pp. 545–569. Auerbach Publications, CRC Press, USA (2010)

32. Diffie, W., Hellman, M.: New directions in cryptography. IEEE Transactions on Information Theory 22, 644–654 (1976)
33. Raymond, J.-F., Stiglic, A.: Security issues in the Diffie-Hellman key agreement protocol. IEEE Transactions on Information Theory, 1–17 (2000)
34. Moustafa, H.: Providing authentication, trust, and privacy in wireless mesh networks. In: Zhang, Y., et al. (eds.) Security in Wireless Mesh Networks, pp. 261–295. CRC Press, USA (2007)
35. Stallings, W.: Wireless Communication and Networks, 2nd edn. Person Education (2009)
36. Soliman, H.S., Omari, M.: Application of synchronous dynamic encryption system in mobile wireless domains. In: Proceedings of the 1st ACM International Workshop on Quality of Service and Security in Wireless and Mobile Networks (Q2SWinet 2005), Montreal, Quebec, Canada, pp. 24–30. ACM Press (2005)
37. Ren, K., Lou, W., Zhang, Y.: LEDS: providing location-aware end-to-end data security in wireless sensor networks. IEEE Transactions on Mobile Computing 7(5), 585–598 (2008)
38. Deng, H., Mukherjee, A., Agrawal, D.P.: Threshold and identity-based key management and authentication for wireless ad hoc networks. In: Proceedings of International Conference on Information Technology: Coding and Computing (ITCC 2004), vol. 1, pp. 107–111. IEEE Computer Society Press (April 2004)
39. Shamir, A.: Identity-Based Cryptosystems and Signature Schemes. In: Blakely, G.R., Chaum, D. (eds.) CRYPTO 1984. LNCS, vol. 196, pp. 47–53. Springer, Heidelberg (1985)
40. Boneh, D., Franklin, M.: Identity-Based Encryption from the Weil Pairing. In: Kilian, J. (ed.) CRYPTO 2001. LNCS, vol. 2139, pp. 213–229. Springer, Heidelberg (2001)
41. Shamir, A.: How to share a secret. Communications of the ACM (1979)
42. Lim, Y.-X., Yer, T.S., Levine, J., Owen, H.L.: Wireless intrusion detection and response. In: Proceedings of Information Assurance Workshop, IEEE Systems, Man and Cybernetics Society, West Point, NY, USA, pp. 68–75 (June 2003)
43. Martignon, F., Paris, S., Capone, A.: MobiSEC: a novel security architecture for wireless mesh networks. In: Proceedings of the 4th ACM Symposium on QoS and Security for Wireless and Mobile Networks (Q2SWinet 2008), Vancouver, Canada, pp. 35–42 (2008)
44. IEEE Standard 802.11i Medium Access Control (MAC) security enhancements, amendment 6. IEEE Computer Society (2004)
45. Egners, A., Meyer, U.: Wireless mesh network security: state of affairs. In: Proceedings of the 35th Annual IEEE Conference on Local Computer Networks (LCN 2010), Denver, Colorado, USA, pp. 997–1004 (2010)
46. Radosavac, D.S., Baras, J.S., Koutsopoulos, I.: A framework for MAC protocol misbehaviour detection in wireless networks. In: Proceedings of the 4th ACM Workshop on Wireless Security (WiSe 2005), pp. 33–42. ACM Press, New York (2005)
47. Kyasanur, P., Vaidya, N.: Detection and handling of MAC layer misbehaviour in wireless networks. In: Proceedings of the International Conference on Dependable Systems and Networks (DSN 2003), San Francisco, CA, USA, pp. 173–182 (June 2003)
48. Konorski, J.: Multiple Access in Ad-Hoc Wireless LANs with Noncooperative Stations. In: Gregori, E., Conti, M., Campbell, A.T., Omidyar, G., Zukerman, M. (eds.) NETWORKING 2002. LNCS, vol. 2345, pp. 1141–1146. Springer, Heidelberg (2002)
49. Li, H., Xu, M., Li, Y.: Selfish MAC Layer Misbehavior Detection Model for the IEEE 802.11-Based Wireless Mesh Networks. In: Xu, M., Zhan, Y.-W., Cao, J., Liu, Y. (eds.) APPT 2007. LNCS, vol. 4847, pp. 382–391. Springer, Heidelberg (2007)

50. Santhanam, L., Nandiraju, D., Nandiraju, N., Agrawal, D.P.: Active cache-based defense against DoS attacks in wireless mesh networks. In: Proceedings of the 2nd International Symposium on Wireless Pervasive Computing (ISWPC 2007), San Juan, PR, USA, pp. 419–424 (February 2007)
51. Naveed, A., Kanhere, S.: Security vulnerabilities in channel assignment of multi-radio multi-channel wireless mesh networks. In: Proceedings of the 49th Annual IEEE Global Telecommunications Conference (GLOBECOM 2006), San Francisco, USA, pp. 1–5 (November-December 2006)
52. Cardenas, A.A., Radosavac, S., Baras, J.S.: Detection and prevention of MAC layer misbehavior in ad hoc networks. In: Proceedings of the 2nd ACM Workshop on Security of Ad Hoc and Sensor Networks (SASN 2004), Washington, DC, USA, pp. 17–24 (October 2004)
53. Konorski, J., Kurant, M.: Application of a hash function to discourage MAC-layer misbehaviour in wireless LANS. Journal of Telecommunications and Information Technology 2, 38–46 (2004)
54. Raya, M., Hubaux, J.-P., Aad, I.: DOMINO: a system to detect greedy behaviour in IEEE 802.11 hotspots. In: Proceedings of the 2nd International Conference on Mobile Systems, Applications, and Services (MobiSys 2004), Boston, Massachusetts, USA, pp. 84–97. ACM Press (June 2004)
55. Cagalj, M., Ganeriwal, S., Aad, I., Hubaux, J.-P.: On cheating in CSMA/CA ad hoc networks. Technical Report IC/2004/27, EPFL-DI-ICA (March 2004)
56. Hu, Y.-C., Perrig, A., Johnson, D.: Ariadne: a secure on-demand routing protocol for ad hoc networks. In: Proceedings of ACM Annual International Conference on Mobile Computing (MobiCom 2002), Atlanta, GA, USA, pp. 21–38 (September 2002)
57. Hu, Y.-C., Johnson, D.B., Perrig, A.: SEAD: secure efficient distance vector routing for mobile wireless ad hoc networks. In: Proceedings of the 4th IEEE Workshop on Mobile Computing Systems and Applications (WMCSA 2002), Callicoon, NY, USA, pp. 3–13 (June 2002)
58. Sanzgiri, K., Dahill, B., Levine, B.N., Shields, C., Belding-Royer, E.M.: A secure routing protocol for ad hoc networks. In: Proceedings of the 10th IEEE International Conference on Network Protocols (ICNP 2002), Paris, France, pp. 78–87 (November 2002)
59. Marti, S., Guili, T., Lai, K., Baker, M.: Mitigating routing misbehavior in mobile ad hoc networks. In: Proceedings of the 6th ACM Annual International Conference on Mobile Computing and Networking (MobiCom 2000), Boston, Massachusetts, USA, pp. 255–265 (2000)
60. Papadimitratos, P., Haas, Z.J.: Secure data transmission in mobile ad hoc networks. In: Proceedings of the 2nd ACM Workshop on Wireless Security (WiSe 2003), San Diego, CA, USA, pp. 41–50 (September 2003)
61. Awerbuch, B., Holmer, D., Nita-Rotaru, C., Rubens, H.: An on-demand secure routing protocol resilient to Byzantine failure. In: Proceedings of the 1st ACM Workshop on Wireless Security (WiSe 2002), Atlanta, GA, USA, pp. 21–30. ACM Press (September 2002)
62. Awerbuch, B., Curtmola, R., Holmer, D., Nita-Rotaru, C., Rubens, H.: On the survivability of routing protocols in ad hoc wireless networks. In: Proceedings of the 1st International Conference on Security and Privacy for Emerging Areas in Communications Networks (SecureComm 2005), Athens, Greece, pp. 327–338 (September 2005)
63. Papadimitratos, P., Haas, Z.J.: Secure routing for mobile ad hoc networks. In: Proceedings of the SCS Communication Networks and Distributed Systems Modelling and Simulation Conference (CNDS 2002), San Antonio, TX, USA, pp. 27–31 (January 2002)

64. Zapata, M.G., Asokan, N.: Securing ad hoc routing protocols. In: Proceedings of the 1st ACM Workshop on Wireless Security (WiSe 2002), Atlanta, GA, USA, pp. 1–10. ACM Press (September 2002)
65. Papadimitratos, P., Hass, Z.J.: Secure link state routing for mobile ad hoc networks. In: Proceedings of the Symposium on Applications and the Internet Workshops (SAINT 203 Workshops), Washington DC, USA, pp. 379–383 (2003)
66. Sen, J., Chandra, M.G., Balamuralidhar, P., Harihara, S.G., Reddy, H.: A distributed protocol for detection of packet dropping attack in mobile ad hoc networks. In: Proceedings of the International Conference on Telecommunications and Malaysian International Conference on Communications (ICT-MICC 2007), Penang, Malaysia (May 2007)
67. Bahr, M.: Proposed routing for IEEE 802.11s WLAN mesh networks. In: Proceedings of the 2nd Annual International Wireless Internet Conference (WICON), Boston, MA, USA, pp. 133–144 (2006)
68. Bahr, M.: Update on the hybrid wireless mesh protocol 80.11s. In: Proceedings of the 4th IEEE International Conference on Mobile Ad Hoc and Sensor Systems (MASS 2007), Pisa, Italy, pp. 1–6 (October 2007)
69. Li, C., Wang, Z., Yang, C.: Secure routing for wireless mesh networks. International Journal of Network Security 13(2), 109–120 (2011)
70. Zhu, S., Xu, S., Setia, S., Jajodia, S.: LHAP: a lightweight hop-by-hop authentication protocol for ad-hoc networks. In: Proceedings of the 23rd IEEE International Conference on Distributed Systems (ICDCS 2003): Workshop on Mobile and Wireless Network, Providence, Rhode Island, USA, pp. 749–755 (2003)
71. Perrig, A., Canetti, R., Tygar, J.D., Song, D.: Efficient and secure source authentication for multicast. In: Proceedings of the Network and Distributed System Security Symposium (NDSS 2001), San Diego, CL, USA, pp. 35–46 (February 2001)
72. Curtmola, R., Nita-Rotaru, C.: BSMR: Byzantine-resilient secure multicast routing in multi-hop wireless networks. IEEE Transactions on Mobile Computing 8(4), 445–459 (2009)
73. Belding-Royer, E.M., Perkins, C.E.: Multicast Ad-hoc On-demand Distance Vector (MAODV) Routing, Internet Draft (July 2000)
74. Papadimitratos, P., Haas, Z.J.: Secure route discovery for QoS-aware routing in ad hoc networks. In: Proceedings of IEEE Sarnoff Symposium, Princeton, NJ, USA, pp. 176–179 (April 2005)
75. Zhu, T., Yu, M.: A dynamic secure QoS routing protocol for wireless ad hoc networks. In: Proceedings of IEEE Sarnoff Symposium, Princeton, NJ, USA, pp. 1–4 (March 2006)
76. Dong, J.: Secure and Robust Communication in Wireless Mesh Networks. PhD Thesis, Purdue University, West Lafayette, Indiana USA (2009)
77. Hu, Y.-C., Perrig, A., Johnson, D.B.: Packet leashes: a defense against wormhole attacks in wireless ad hoc networks. In: Proceedings of the 22nd Annual Joint Conference of the IEEE Computer and Communications Societies (IEEE INFOCOM 2003), San Francisco, CL, USA, vol. 3, pp. 1976–1986 (March-April 2003)
78. Hu, Y.-C., Perrig, A., Johnson, D.B.: Ariadne: a secure on-demand routing protocol for ad hoc networks. In: Proceedings of the 8th ACM Annual International Conference on Mobile Computing (MobiCom 2002), Atlanta, GA, USA, pp. 21–38. ACM Press (September 2002)
79. Eriksson, J., Krishnamurthy, S.V., Faloutsos, M.: Truelink: a practical countermeasure to the wormhole attack in wireless networks. In: Proceedings of the 14th IEEE International Conference on Network Protocols (ICNP 2006), Santa Barbara, CL, USA, pp. 75–84 (November 2006)

80. Hu, L., Evans, D.: Using directional antennas to prevent wormhole attacks. In: Proceedings of the 11th Network and Distributed System Security Symposium (NDSS 2003), San Diego, CL, USA, pp. 131–141 (February 2004)
81. Perkins, C.E., Bhagwat, P.: Highly dynamic destination-sequenced distance-vector routing (DSDV) for mobile computers. In: Proceedings of the ACM Conference on Communications Architectures, Protocols and Applications (SIGCOMM 1994), London, UK, pp. 234–244. ACM Press (August-September 1994)
82. Yi, S., Naldurg, P., Kravets, R.: Security-aware ad hoc routing for wireless networks. In: Proceedings of the ACM Symposium on Mobile Ad Hoc Networking and Computing (MobiHoc 2001), Long Beach, CL, USA, pp. 299–302 (October 2001)
83. Perkins, C., Belding-Royer, E., Das, S.: Ad hoc on-demand distance vector (AODV) routing. IETF RFC 3561 (2003)
84. Johnson, D.B.: The dynamic source routing protocol (DSR) for mobile ad hoc networks for IPv4. IETF RFC 4728 (2007)
85. Perrig, A., Canneti, R., Tygar, J.D., Song, D.: The TESLA broadcast authentication protocol. RSA CryptoBytes 5(2), 1–13 (2002)
86. Blom, R.: An Optimal Class of Symmetric Key Generation Systems. In: Beth, T., Cot, N., Ingemarsson, I. (eds.) EUROCRYPT 1984. LNCS, vol. 209, pp. 335–338. Springer, Heidelberg (1985)
87. Desmedt, Y.G.: Threshold cryptography. European Transactions on Telecommunications 5(4), 449–457 (1994)
88. Adjih, C., Clausen, T., Jacquet, P., Laouiti, A., Muhlethaler, P., Raffo, D.: Securing the OLSR protocol. In: Proceedings of the 2nd IFIP Annual Mediterranean Ad Hoc Networking Workshop (Med-Hoc-Net 2003), Mahdia, Tunisia, pp. 25–27 (June 2003)
89. Clausen, T., Jacquet, P.: Optimized link state routing (OLSR) protocol. RFC 3626 (2003)
90. Bahr, M.: Proposed routing for IEEE 802.11s WLAN mesh networks. In: Proceedings of the 2nd Annual International Wireless Internet Conference (WICON 2006), Article No. 5, Boston, MA, USA (2006)
91. Abraham, S., et al.: Simple efficient extensible mesh (SEE-mesh) proposal overview. IEEE P802.11 Wireless LANs, Document IEEE 802.11-05/0562r0 (July 2005)
92. IEEE P802.11s draft amendment to standard IEEE 802.11: ESS mesh networking. IEEE (2006)
93. Dong, J., Curtmola, R., Nita-Rotaru, C.: Secure high-throughput multicast routing in wireless mesh networks. Technical Report TR#08-014, Computer Science Department, Purdue University, Indiana, USA (2008)
94. Lee, S.J., Su, W., Gerla, M.: On-demand multicast routing protocol in multihop wireless mobile networks. ACM Mobile Networks and Applications 7, 441–453 (2002)
95. Stallings, W.: Cryptography and Network Security, 4th edn. Pearson Education (2010)
96. Aboba, B., Simon, D.: PPP EAP TLS authentication protocol. RFC 2716 (1999)
97. Chen, T.M., Kuo, G.-S., Li, Z.-P., Zhu, G.-M.: Intrusion detection in wireless mesh networks. In: Zhang, Y., et al. (eds.) Security in Wireless Mesh Networks, pp. 146–169. Aurbach Publications, CRC Press, USA (2008)
98. Mishra, A., Nadkarni, K., Patcha, A.: Intrusion detection in wireless ad hoc networks. IEEE Wireless Communications 11, 48–60 (2004)
99. Sen, J.: An intrusion detection architecture for clustered wireless ad hoc networks. In: Proceedings of the 2nd IEEE International Conference on Intelligence Communication Systems and Networks (CICSyN 2010), Liverpool, UK, pp. 202–207 (July 2010)

100. Sen, J., Sengupta, I.: Autonomous Agent Based Distributed Fault-Tolerant Intrusion Detection System. In: Chakraborty, G. (ed.) ICDCIT 2005. LNCS, vol. 3816, pp. 125–131. Springer, Heidelberg (2005)
101. Kassab, M., Belghith, A., Bonnin, J.-M., Sassi, S.: Fast pre-authentication based on proactive key distribution for 802.11 infrastructure networks. In: Proceedings of the 1st ACM Workshop on Wireless Multimedia Networking and Performance Modeling (WMuNeP 2005), Montreal, Canada, pp. 46–53 (2005)
102. Prasad, A.R., Wang, H.: Roaming key based fast handover in WLANs. In: Proceedings of IEEE Wireless Communications and Networking Conference (WCNC 2003), New Orleans, Louisiana, USA, vol. 3, pp. 1570–1576 (2005)
103. Ben Salem, N., Hubaux, J.-P.: Securing wireless mesh networks. IEEE Wireless Communication 13(2), 50–55 (2006)
104. Cheikhrouhou, O., Maknavicius, M., Chaouchi, H.: Security architecture in a multi-hop mesh network. In: Proceedings of the 5th Conference on Security and Network Architectures (SAR 2006), Seignosse, France (June 2006)
105. Parthasarathy, M.: Protocol for Carrying Authentication and Network Access (PANA) Threat Analysis and Security Requirements. RFC 4016 (March 2005)
106. Prasad, N.R., Alam, M., Ruggieri, M.: Light-weight AAA infrastructure for mobility support across heterogeneous networks. Wireless Personal Communications 29(3-4), 205–219 (2004)
107. Lee, I., Lee, J., Arbaugh, W., Kim, D.: Dynamic Distributed Authentication Scheme for Wireless LAN-Based Mesh Networks. In: Vazão, T., Freire, M.M., Chong, I. (eds.) ICOIN 2007. LNCS, vol. 5200, pp. 649–658. Springer, Heidelberg (2008)
108. Sen, J., Subramanyam, H.: An Efficient Certificate Authority for Ad Hoc Networks. In: Janowski, T., Mohanty, H. (eds.) ICDCIT 2007. LNCS, vol. 4882, pp. 97–109. Springer, Heidelberg (2007)
109. Yi, S., Kravets, R.: MOCA: mobile certificate authority for wireless ad hoc networks. In: Proceedings of the 2nd Annual PKI Research Workshop Program (PKI 2003), Gaithersburg, Maryland, pp. 52–64 (April 2003)
110. Zhang, Y., Fang, Y.: ARSA: an attack-resilient security architecture for multihop wireless mesh networks. IEEE Journal of Selected Areas in Communication 24(10), 1916–1928 (2006)
111. He, B., Joshi, S., Agrawal, D.P., Sun, D.: An efficient authenticated key establishment scheme for wireless mesh networks. In: Proceedings of IEEE Global Telecommunications Conference (GLOBECOM 2010), Miami, Florida, USA, pp. 1–5 (2010)
112. Blundo, C., De Santis, A., Herzberg, A., Kutten, S., Vaccaro, U., Yung, M.: Perfectly-Secure Key Distribution for Dynamic Conferences. In: Brickell, E.F. (ed.) CRYPTO 1992. LNCS, vol. 740, pp. 471–486. Springer, Heidelberg (1993)
113. Gupta, A., Mukherjee, A., Xie, B., Agrawal, D.P.: Decentralized key generation scheme for cellular-based heterogeneous wireless ad hoc networks. Journal of Parallel and Distributed Computing 67(9), 981–991 (2007)
114. Zhu, H., Lin, X., Lu, R., Ho, P.-H., Shen, X.: SLAB: A secure localized authentication and billing scheme for wireless mesh networks. IEEE Transactions on Wireless Communications 7(10), 3858–3868 (2008)
115. de Laat, C., Gross, G., Gommans, L.: Generic AAA architecture. IETF, RFC 2903 (2000)
116. Leu, J., Lai, R., Lin, H., Shih, W.: Running cellular/PWLAN services: practical considerations for cellular/PWLAN architecture supporting interoperator roaming. IEEE Communications Magazine 44(2), 73–84 (2006)

117. Long, M., Wu, C.H., Irwin, J.D.: Localised authentication for inter-network roaming across wireless LANs. IEEE Proceedings on Communication 151(5), 496–500 (2004)
118. Cao, Z., Zhu, H., Lu, R.: Provably secure robust threshold partial blind signature. Science in China Series e 35(12), 1254–1265 (2005)
119. Boneh, D., Lynn, B., Shacham, H.: Short signature from the Weil pairing. Journal of Cryptology 17(4), 297–319 (2004)
120. Zhu, S., Xu, S., Setia, S., Jajodia, S.: LHAP: a lightweight network access control protocol for ad hoc networks. Ad Hoc Networks 4(5), 567–585 (2006)
121. Lamport, L.: Password authentication with insecure communication. Communications of the ACM 24(11), 770–772 (1981)
122. Perrig, A., Canetti, R., Tygar, J., Song, D.X.: Efficient authentication and signing of multicast streams over lossy channels. In: Proceedings of the IEEE Symposium on Security and Privacy (SP 2000), Berkeley, California, USA, pp. 56–73 (May 2000)
123. Lin, X., Ling, X., Zhu, H., Ho, P.-H., Shen, X.S.: A novel localised authentication scheme in IEEE 802.11 based wireless mesh networks. International Journal of Security and Networks 3(2), 122–132 (2008)
124. Rabin, M.O.: Digitized Signatures and Public-key Functions as Intractable as Factorization. Technical Report, Massachusetts Institute of Technology, Cambridge, MA, USA (1979)
125. Gaubatz, G., Kaps, J.-P., Sunar, B.: Public Key Cryptography in Sensor Networks—Revisited. In: Castelluccia, C., Hartenstein, H., Paar, C., Westhoff, D. (eds.) ESAS 2004. LNCS, vol. 3313, pp. 2–18. Springer, Heidelberg (2005)
126. Capkun, S., Buttyan, L., Hubaux, J.-P.: Self-organized public-key management for mobile ad hoc networks. IEEE Transactions on Mobile Computing 2(1), 52–64 (2003)
127. Zimmerman, P.: The Official PGP User's Guide. MIT Press (1995)
128. Petersen, H., Horster, P.: Self-certified keys-concepts, and applications. In: Proceedings of the 3rd International Conference on Communications and Multimedia Security, Athens, Greece, pp. 102–116 (September 1997)
129. Li, Z., Garcia-Luna-Aceves, J.J.: Non-interactive key establishment in wireless mesh networks. In: Zhang, Y., et al. (eds.) Security in Wireless Mesh Networks, pp. 297–321. Aurbach Publication, CRC Press, USA (2008)
130. Zapata, M.G.: Key management in wireless mesh networks. In: Zhang, Y., et al. (eds.) Security in Wireless Mesh Networks, pp. 323–346. Aurbach Publication, CRC Press, USA (2008)
131. Khalili, A., Katz, J., Arbaugh, W.A.: Towards secure key distribution in truly ad-hoc networks. In: Proceedings of the IEEE Workshop on Security and Assurance in Ad Hoc Networks in conjunction with the IEEE International Symposium on Applications and the Internet (SAINT 2003), Orlando, Florida, USA, pp. 342–346. IEEE Computer Society Press (January 2003)
132. Zhou, L., Haas, Z.J.: Securing ad hoc networks. IEEE Networks, Special Issue on Network Security 13(6), 24–30 (1999)
133. Zhu, S., Xu, S., Setia, S., Jajodia, S.: Establishing pairwise keys for secure communication in ad hoc networks: a probabilistic approach. In: Proceedings of the 11th IEEE International Conference on Network Protocols (ICNP), Atlanta, Georgia, USA, pp. 326–335 (November 2003)
134. Amir, Y., Kim, Y., Nita-Rotaru, C., Tzudik, G.: On the performance of group key agreement protocols. In: Proceeding of the 22nd IEEE International Conference on Distributed Computing Systems (ICDCS 2002), Vienna, Austria, pp. 463–464 (July 2002)

135. Chan, H., Perrig, A., Song, D.: Random key predistribution schemes for sensor networks. In: Proceedings of the IEEE Symposium on Security and Privacy, Berkeley, CA, USA, pp. 197–213 (May 2003)
136. Joshi, D., Namuduri, K., Pendse, R.: Secure, redundant and fully distributed key management scheme for mobile ad hoc networks: an analysis. EURASIP Journal on Wireless Communications and Networking 2005(4), 579–589 (2005)
137. Boneh, D., Gentry, C., Shacham, H., Lynn, B.: Aggregate and Verifiably Encrypted Signatures from Bilinear Maps. In: Biham, E. (ed.) EUROCRYPT 2003. LNCS, vol. 2656, pp. 416–432. Springer, Heidelberg (2003)
138. Reed, M., Syverson, P., Goldschlag, D.D.: Anonymous connections and onion routing. IEEE Journal on Selected Areas in Communications 16(4), 482–494 (1998)
139. Sen, J.: Secure and Privacy-Aware Searching in Peer-to-Peer Networks. In: Garcia-Alfaro, J., Navarro-Arribas, G., Cuppens-Boulahia, N., de Capitani di Vimercati, S. (eds.) DPM 2011 and SETOP 2011. LNCS, vol. 7122, pp. 72–89. Springer, Heidelberg (2012)
140. Dijkstra, E.W.: A note on two problems in connexion with graphs. Numerische Mathematik 1, 269–271 (1959)
141. Ren, K., Yu, S., Lou, W., Zhang, Y.: PEACE: a novel privacy-enhanced yet accountable security framework for metropolitan wireless mesh networks. IEEE Transactions on Parallel and Distributed Systems 21(2), 203–215 (2010)
142. Boneh, D., Shacham, H.: Group signatures with verifier-local revocation. In: Proceedings of the 11th ACM Conference on Computer and Communication Security (CCS), Washington DC, USA, pp. 168–177 (October 2004)
143. Sun, J., Zhang, C., Fang, Y.: A security architecture achieving anonymity and traceability in wireless mesh networks. In: Proceedings of the 27th IEEE International Conference on Computer Communications (IEEE INFOCOM 2008), pp. 1687–1695 (April 2008)
144. Sun, J., Zhang, C., Zhang, Y., Fang, Y.: SAT: A security architecture achieving anonymity and traceability in wireless mesh networks. IEEE Transactions on Dependable and Secure Computing 8(2), 295–307 (2011)
145. Brands, S.: Untraceable Off-Line Cash in Wallets with Observers. In: Stinson, D.R. (ed.) CRYPTO 1993. LNCS, vol. 773, pp. 302–318. Springer, Heidelberg (1994)
146. Wei, K., Chen, Y.R., Smith, A.J., Vo, B.: WhoPay: a scalable and anonymous payment system for peer-to-peer environments. In: Proceedings of the 26th IEEE International Conference on Distributed Computing Systems (ICDCS 2006), Lisbon, Portugal, p. 13 (July 2006)
147. Figueiredo, D., Shapiro, J., Towsley, D.: Incentives to promote availability in peer-to-peer anonymity systems. In: Proceedings of the 13th IEEE International Conference on Network Protocols (ICNP 2005), Boston, MA, USA, pp. 110–121. IEEE Computer Society Press (November 2005)
148. Chaum, D.: Blind signatures for untraceable payments. In: Proceedings of the Annual International Cryptology Conference (CRYPTO 1982). Advances in Cryptology, pp. 199–203. Plenum Press, New York (1983)
149. Ateniese, G., Herzberg, A., Krawczyk, H., Tsudik, G.: Untraceable mobility or how to travel incognito. Computer Networks 31(8), 871–884 (1999)
150. Raya, M., Hubaux, J.-P.: Securing vehicular ad hoc networks. Journal of Computer Security, Special Issue on Security of Ad Hoc and Sensor Networks 15(1), 39–68 (2007)
151. Zhang, Y., Liu, W., Lou, W., Fang, Y.: MASK: anonymous on-demand routing in mobile ad hoc networks. IEEE Transactions on Wireless Communications 5(9), 2376–2385 (2006)

152. Gentry, C., Silverberg, A.: Hierarchical ID-Based Cryptography. In: Zheng, Y. (ed.) ASIACRYPT 2002. LNCS, vol. 2501, pp. 548–566. Springer, Heidelberg (2002)
153. Chen, X., Zhang, F., Mu, Y., Susilo, W.: Efficient Provably Secure Restrictive Partially Blind Signatures from Bilinear Pairings. In: Di Crescenzo, G., Rubin, A. (eds.) FC 2006. LNCS, vol. 4107, pp. 251–265. Springer, Heidelberg (2006)
154. Chen, X., Zhang, F., Liu, S.: ID-based restrictive partially blind signatures and applications. Journal of Systems and Software 80(2), 164–171 (2007)
155. Rahman, S.M. M., Inomata, A., Okamoto, T., Mambo, M., Okamoto, E.: Anonymous Secure Communication in Wireless Mobile Ad-Hoc Networks. In: Stajano, F., Kim, H.-J., Chae, J.-S., Kim, S.-D. (eds.) ICUCT 2006. LNCS, vol. 4412, pp. 140–149. Springer, Heidelberg (2007)
156. Sen, J.: An efficient and user privacy-preserving routing protocol for wireless mesh networks. Journal Scalable Computing: Practice and Experience, Special Issue on Network and Distributed Systems 11(4), 345–358 (2010)
157. Cao, T., Lin, D., Xue, R.: Improved ring authenticated encryption scheme. In: Proceedings of the 10th Joint International Computer Conference (JICC), Kunming, China, pp. 341–346. International Academic Publishers World Publishing Corporation (November 2004)
158. Kidston, D., Li, L., Tang, H., Mason, P.: Mitigating Security in Tactical Networks. Communications Research Center. Defence R&D Canada (DRDC) Publication (September 2010)
159. Pirzada, A., McDonald, C.: Establishing trust in pure ad hoc networks. In: Proceedings of the 27th Australian Conference on Computer Science, Dunedin, New Zealand, pp. 47–54 (2004)
160. Yan, Z., Zhang, P., Virtanen, T.: Trust evaluation based security solution in ad hoc networks. In: Proceedings of the 8th Nordic Workshop on Secure IT Systems (NordSec 2003), Gjoevik, Norway (October 2003)
161. Sen, J., Chowdhury, P.R., Sengupta, I.: A distrusted trust establishment scheme for mobile ad hoc networks. In: Proceedings of the International Conference on Computation: Theory and Applications (ICCTA 2007), Kolkata, India, pp. 51–57 (March 2007)
162. Ben Salem, N., Hubaux, J.-P., Jakobsson, M.: Reputation-based Wi-Fi deployment. ACM SIGMOBILE Mobile Computing and Communication Review 9(3), 69–81 (2005)
163. Jarrett, M., Ward, P.: Trusted computing for protecting ad-hoc routing. In: Proceedings of the 4th IEEE Annual Communication Networks and Services Research Conference (CNSR 2006), Moncton, New Brunswick, Canada, pp. 61–68 (May 2006)
164. Sen, J.: A distributed trust management framework for detecting malicious packet dropping nodes in a mobile ad hoc network. International Journal of Network Security and its Applications (IJNSA) 2(4), 92–104 (2010)
165. Tang, C., Wu, D.: An efficient mobile authentication scheme for wireless networks. IEEE Transactions on Wireless Communications 7(4), 1408–1416 (2008)
166. Zhu, H., Lin, X., Ho, P.-H., Shen, X., Shi, M.: TTP based privacy preserving inter-WISP roaming architecture for wireless metropolitan area networks. In: Proceedings of the IEEE Wireless Communication and Networking Conference (WCNC 2007), Hong Kong, pp. 2957–2962 (March 2007)
167. Buchegger, S., Boudec, J.-Y.-L.: Performance analysis of the CONFIDANT (Cooperation Of Nodes-Fairness In Dynamic Ad-hoc NeTworks) protocol. In: Proceedings of the 3rd ACM International Symposium on Mobile Ad Hoc Networking and Computing (MobiHoc 2002), Lausanne, Switzerland, pp. 226–236 (June 2002)

168. Michiardi, P., Molva, R.: CORE: a COllaborative REputation mechanism to enforce node cooperation in mobile ad hoc networks. In: Proceedings of the 6th IFIP Communication and Multimedia Security Conference, Portoroz, Slovenia, vol. 228, pp. 107–212 (September 2002)
169. Ganeriwal, S., Srivastava, M.: Reputation-based framework for high integrity sensor networks. In: Proceedings of the 2nd ACM Workshop on Security of Ad Hoc and Sensor Networks (SASN 2004), New York, USA, pp. 66–77 (2004)
170. Srinivasan, A., Teitelbaum, J., Wu, J.: DRBTS: distributed reputation-based beacon trust system. In: Proceedings of the 2nd IEEE International Symposium on Dependable, Autonomic and Secure Computing (DASC 2006), Indianapolis, USA, pp. 277–283 (2006)
171. Bansal, S., Baker, M.: Observation-based Cooperation Enforcement in Ad hoc Networks. Research Report cs.NI/0307012, Stanford University, USA (2003)
172. Dong, J.: Secure and Robust Communication in Wireless Mesh Networks. Doctoral Thesis, Department of Computer Science, Purdue University, West Lafayette, Indiana, USA (December 2009)
173. Ahlswede, R., Cai, N., Li, S.-Y., Yeung, R.: Network information flow. IEEE Transactions on Information Theory 46(4), 1204–1216 (2000)
174. Yu, Z., Wei, Y., Ramkumar, B., Guan, Y.: An efficient signature-based scheme for securing network coding against pollution attacks. In: Proceedings of the 27th IEEE Conference on Computer Communications (INFOCOM 2008), Phoenix, AZ, USA, pp. 1409–1417. IEEE Press (April 2008)
175. Sen, J.: Cross-layer protocols for multimedia communications over wireless networks. In: Tarnay, K., et al. (eds.) Advanced Communication Protocol Technologies: Solutions, Methods and Applications, pp. 318–354. IGI-Global Publishers, USA (2010)
176. Gentry, C.: A Fully Homomorphic Encryption Scheme. Doctoral Thesis, Department of Computer Science, Stanford University, USA (2009)
177. Sen, J.: A survey on wireless sensor network security. International Journal of Communication Networks and Information Security (IJCNIS) 1(2), 59–82 (2009)

Trust Establishment Techniques in VANET

Jyoti Grover, Manoj Singh Gaur*, and Vijay Laxmi

Malaviya National Institute of Technology, Jaipur, India
{gaurms,vlaxmi,jgrover}@mnit.ac.in, gaurms@gmail.com

Abstract. Establishment of trust is amongst the most critical aspects of any system's security. For any network, trust refers to a set of relationships amongst the entities participating in the network operations. Trust establishment plays a key role in prevention of attacks in VANET. The nodes involved in defense of the network against such attacks must establish mutual trust for the network to operate smoothly. It is a major challenge as a receiving node needs to ensure authenticity and trust-ability of the received messages before reacting to them. It is assumed that each node in a VANET is equipped with a trust system to take such decisions. There are two options for trust establishment (1) Based on static infrastructure, (2) Dynamic establishment of trust in a self organized manner. Trust based on static infrastructure is more efficient and robust than dynamic infrastructure. The only concern using static infrastructure is the unavailability of fixed infrastructure in some locations. The main objective of this paper is to describe various trust establishment approaches for VANET. If all the nodes establish trust with other nodes in VANET, probability of occurrence of attacks can be drastically reduced.

Keywords: VANET, Trust, PKI, OBU, RSU.

1 Introduction

Vehicular Ad hoc network (VANET) is a specific type of Mobile Ad-Hoc Network (MANET) that provides communication between (1) nearby vehicles and (2) vehicles and nearby roadside equipments. These networks have no fixed infrastructure and rely on the vehicles themselves for implementing any network functionality. A VANET is a decentralized network as every node performs the functions of both host and router. The main benefit of VANET communication is the enhanced passenger safety by virtue of exchanging warning messages between vehicles. VANET differs from MANET as it provides higher mobility of nodes, larger scale networks, geographically constrained topology and frequent network fragmentation.

VANET Security is crucial because a poorly designed VANET is vulnerable to network attacks and this in turn compromises the safety of drivers. Security systems should ensure that transmission comes from a trusted source and is not tampered en-route by other sources. Trust and security are two interdependent concepts that

* Corresponding author.

S. Khan and A.-S.K. Pathan (Eds.): *Wireless Networks and Security*, SCT, pp. 273–301.
DOI: 10.1007/978-3-642-36169-2_8

cannot be de-segregated. For example, cryptography is used for implementing security in network system but it is dependent on trusted key exchange. In the same way, trusted key exchange cannot take place without requiring security services. While defining a secure system, these two terms can be used interchangeably due to their interdependency.

As VANET has an open and shared wireless ad hoc environment, it is possible that some nodes can be compromised. It is very important for a security solution to identify these malicious nodes and exclude them from the network. If the trust relationships are clarified in real time, it's easy to take appropriate security measures and take correct decisions regarding any security issue. A trust based model is required which can manage nodes dynamically and evaluate node's activities efficiently in a distributed manner. Illegitimate nodes can be detected based on their trust evaluations so that they cannot be used in any communication within the network. Thus trust values determination plays a very important role for improvement of the network security and reliability.

In this tutorial, we focus on various trust establishment techniques in VANET. Trust establishment process can be partitioned into two classes: Infrastructure based trust and self organized trust. We review these approaches and discuss the optimal approach that best fits VANET environment.

The rest of chapter is organized as follows. Various types of attacks are discussed in section 2 followed by trust establishment and evaluation approach in Section 3. Infrastructure based trust establishment is described in detail in Section 4. Dynamic trust establishment techniques are presented in section 5. Section 6 reviews trust establishment models in VANET. Open research issues are briefly discussed in Section 7 followed by concluding remarks in Section 8.

2 Attacks on Vehicular Networks

Due to the large number of autonomous network members and the presence of human factor, misbehavior of nodes in future vehicular networks cannot be ruled out. Several types of attacks have been identified and classified [3] on the basis of layer used by the attacker. At the physical and link layers, attacker can disturb the network system by overloading the communication channel with junk messages. Attacker can inject false messages [18] or rebroadcast an old message also. Some attackers can tamper with an On-Board Unit (OBU) or destroy Road-Side Unit (RSU). At network layer, an attacker can insert false routing messages or overload the system with routing information. Privacy of drivers can be disclosed by revealing and tracking the position of drivers. Some of these attacks are briefly explained:

1. **Bogus Information:** Attackers in this case are insiders, rational, and active. They can send wrong information in the network so that it can affect the behavior of other drivers. For example, an adversary can inject wrong information about a non-existent traffic jam or an accident diverting vehicles to other routes and freeing a route for itself.

2. **Cheating with Sensor Information:** This attack is launched by an attacker who is insider, rational and active. He uses this attack to alter the perceived position, speed and direction of other nodes in order to escape liability in case of any mishap.
3. **ID Disclosure:** An attacker is insider, passive and malicious. It can monitor trajectories of a target vehicle and can use this information for determining the ID of a vehicle.
4. **Denial of Service:** Attacker is malicious, active, and local in this case. Attacker may want to bring down the network by sending unnecessary messages on the channel. Example of this attack includes channel jamming and injection of dummy messages.
5. **Replaying and Dropping Packets:** An attacker may drop legitimate packets. For example, an attacker can drop all the alert messages meant for warning vehicles proceeding towards the accident location. Similarly an attacker can replay the packets after that event has been occurred to create the illusion of accident.
6. **Hidden Vehicle:** This type of attack is possible in a scenario where vehicles smartly try to reduce the congestion on the wireless channel. For example, a vehicle has sent a warning message to its neighbors and it is awaiting a response. After receiving a response, a vehicle realizes that its neighbor is in a better position to forward the warning message and stops sending this message to other nodes. This is because it assumes that its neighbor will forward the message to other nodes. If this neighbor node is an attacker, it can be fatal for the system.
7. **Worm Hole Attack:** It is challenging to detect and prevent this attack. A malicious node can record packets at one location in the network and tunnel them to other location through a private network shared with malicious nodes. Severity of the attack increases if the malicious node sends only control messages through the tunnel and not data packets.
8. **Sybil Attack:** In this attack, a vehicle forges the identities of multiple vehicles. These identities can be used to play any type of attack in the system. These false identities also create an illusion that there are additional vehicles on the road. Consequence of this attack is that every type of attack can be played after spoofing the positions or identities of other nodes in the network.

3 Trust Establishment

Trust [20] refers to the confidence of an entity of VANET on another entity. It is based on the expectation that the other entity will perform a particular action believed/expected/accepted by the originator. Trust is based on the fact that the trusted entity will not act maliciously in a particular situation. As no one can be absolutely sure of this fact, trust is completely dependent on the belief of the trustor. An entity is a physical device that participates in the communication process e.g. OBUs (On Board unit) and RSUs (Road side Unit) used in VANET. Trust represents the degree to which a node should be trustworthy, secure or reliable during any interaction with other nodes. A node can participate in the communication process of VANET only if this node is trustable for other nodes and satisfies the trust requirement. A node can have different trust values when evaluated by different nodes because requirement of trust evaluation may be different for individual nodes. Trust is time dependent as it can grow and decay over a period of time. Trust levels are determined by particular actions that the trusted party can perform for the trustee.

Suppose T(*i*,*j*) is a trust relationship between node *i* and *j*. If one node trusts another node to perform the expected operation, the trust relationship between these nodes can be established reliably from initiator's point of view. If node *i* wants some action performed by node *j* and if *j* is successfully performing this action, *j* is a trusted for node *i*. Node *i* will increase the trust value of *j* for its good behavior. So trust value keeps on increasing for each activity performed by a node that was expected by the initiator.

Number of node's activity is referred as node's trust evidence and it increases/decreases once it cooperates successfully with the communication initiator or not. Trust relations are based on the evidence created by the previous interactions of the entities within the application.

The specification, generation, distribution, discovery and evaluation of trust evidence are collectively called trust establishment. Trust establishment and management are essential components [7, 20, 25, 26] of a security framework of VANET. In wired networks, trust is achieved using indirect trust mechanisms. Establishing indirect trust requires some initial authentication mechanism like certification authorities authorizes all the communicating nodes. Establishing trust in ad hoc networks is a challenging task. It is based on establishing trust relationships with neighboring nodes. These trust relationships originate, develop and expire very frequently. Process of trust establishment is virtually impossible [17] due to absence of fixed trust infrastructure, short duration of links, shared wireless medium and physical vulnerability. For overcoming these problems, trust is established in ad hoc networks [7] using some assumptions like pre-configuration of nodes with secret keys and presence of centralized authority.

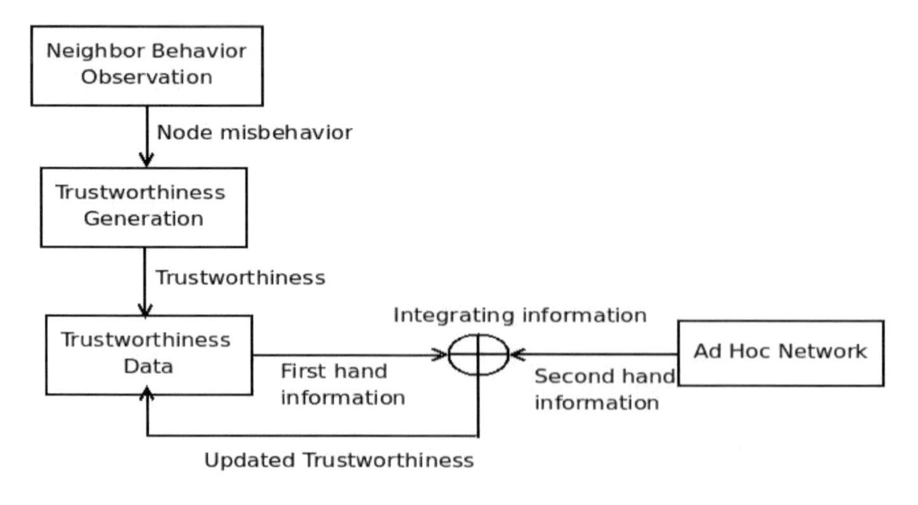

Fig. 1. Trust Management Approach for Ad hoc Networks

Trust management scheme for ad hoc networks is shown in figure 1. Initially, trustworthiness of all the nodes in network is set to some default value. This value is modified whenever a node obtains some information regarding its trustworthiness in terms of both direct and indirect observations. Whenever a node observes any type of misbehavior from neighboring node, it reduces its trust value according to the punishment factor. This punishment factor is different for different misbehaviors. Misbehavior includes dropping, modification and misrouting of packets at network layer and sending false RTS/CTS in the MAC layer etc. Most of the security schemes depend on predefined threshold or training data to build up the malicious behavior of attackers. However, it is very difficult to set thresholds and collect training data sets of attacks in ad hoc networks.

Since malicious nodes can participate in the trust management scheme by indirect observations, a proper method is required to combine multiple indirect observations from both trusted and untrusted neighbors. Two approaches [13] can be used to integrate multiple indirect observations in to direct information that each node directly observes. These are Bayesian approach and Dempster-Shafer Theory (DST) of evidence. There are two differences between these approaches- 1) DST approach does not regard negative evidence of an incident if there is no knowledge about it while Bayesian approach regards evidences either as positive or negative. 2) If there are two incidents inconsistent with each other, they will be considered either positive or negative according to Bayesian approach but DST approach can hold uncertain opinion toward any incident. DST approach is more suitable when there is uncertainty or no prior knowledge for the event.

There are two approaches for trust establishment (1) on the basis of static infrastructure, (2) dynamic establishment of trust in a self organized manner. Static infrastructure based trust relies on common, global, trusted parameters e.g. CA (Centralized Authority). This can be used for authentication of message. Trust knowledge is shared among all the nodes of network by using CA. In dynamic trust establishment, there is no global knowledge and point of control for controlling the whole system.

Safety related applications are very important in VANET. For example, vehicle on highway receives one safety message like change route in case of an accident. This vehicle forwards this safety message to other cars on same highway. After receiving this safety message, the vehicles need to check the trustworthiness of this safety message. Hence Trust management is required in safety applications. It is also required for detection of malicious nodes in case of any type of attack in VANET.

3.1 Requirement Analysis

There are two types of entities used in VANET- Vehicles and Road Side units (static). Vehicles are the communicating entities in VANET. These vehicles broadcast their current positions via beacon messages. There is a fixed interval during which vehicles broadcast beacon message. Being an Ad hoc network, there is no fixed infrastructure available at all the places. The trust management system should satisfy following requirements [22, 23, 35]:

1. **Distributed:** Trust management approach should be distributed to be applicable to the highly dynamic and distributed environment of VANET. All the vehicles should be able to evaluate their neighbors independently. Trust model discussed in [40, 41] uses direct interactions of peers to update ones belief in the trustworthiness of another.
2. **Dynamic:** The system must react immediately as enough evidences against are found. The system should not only keep the grading of nodes but should also be flexible to react quickly to any type of misbehavior.
3. **Fair:** The result of trust evaluation system should be meaningful. As long as there is no evidence for trustworthiness or untrustworthiness of vehicles, system remains neutral. Only the misbehaving vehicles should be detected *i*.e. there should not be any false positives and false negatives. If an honest vehicle is falsely detected as malicious, it represents a false positive. Similarly, a false negative is falsely detection of malicious node as a legitimate node.
4. **Properly Manageable:** Behavior analysis of all the nodes involved in trust management must be integrated. Although these nodes should be configured differently regarding their reliability, importance and output frequency. Trust evaluation unit must handle the loss of messages properly. A loss of beacon of an honest vehicle should not lead negative rating because communication may be unreliable.
5. **Independent Quality of Evaluation:** The quality of evaluation must be independent of different traffic scenarios. Capabilities of evaluation system may be different due to different traffic conditions. The evaluation system must exchange local positive ratings only in order to improve the view of the neighborhood in terms of trustworthiness.
6. **No Trust Distribution Loop:** The exchange of trust values should be limited to local ratings. Only one level reputation system is required. Aggregated local and cooperative trust ratings should not be sent to other nodes in the network as trust values can be falsely increased in loops.
7. **Unawareness of the Distrust to Attacker:** While exchanging the trust ratings with neighbors, attacker should not be aware when he is distrusted. For example, if an attacker is forging the position of some other node, he should not find enough time when he becomes distrusted and hence could change his identity to start from neutral rating.
8. **Scalable:** It is an important perspective in trust management in VANET environment. For example, in high density VANET scenario, number of vehicles reporting information can be very large. Whereas, in low density scenario, a node has to take decisions very quickly for critical situations. So, it has to consult or accept information from only a number of peers. This number is inconstant based on the dynamic VANET scenario. However, a efficient trust management system ensures that number is set to a small value to account for scalability.

3.2 Categories of Trust Establishment

There are various methods of establishment of trust in ad hoc networks. For example a node can establish trust periodically or on demand basis. Direct trust is established with one- hop neighbors and indirect trust is established with other nodes as in ad hoc networks. In some approaches, a centralized trusted entity is required. One approach focuses on trusting messages rather than their attributes. Figure 2. shows the taxonomy of trust establishment techniques.

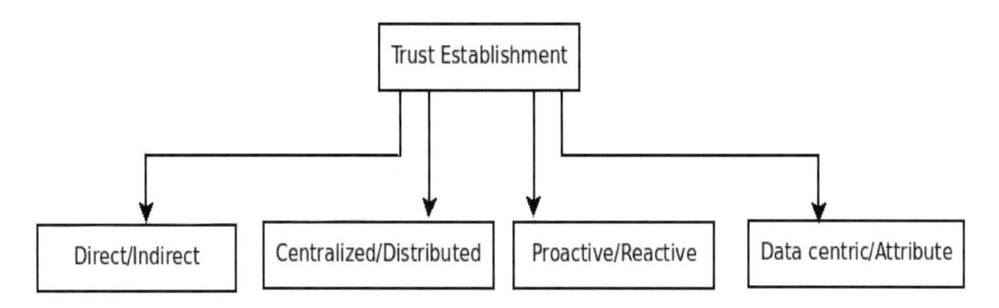

Fig. 2. Taxonomy of Trust Establishment Techniques

1. **Direct vs. Indirect (Recommendation) Trust:** It is defined as trustworthiness computed by one node for the other node. Direct trust can be derived based on previous similar experiences with the same party. It depends on the evidence captured during one to one interactions between other nodes. Evidence is attributed to positive/negative depending upon the neighboring node's behavior malicious/benign respectively. Magnitude of positive value is proportional to the type of the event like ROUTE-REQ, ROUTE-REP, and ROUTE-ERROR. Magnitude of negative value is a function of two factors - type of event and malicious behavior like flooding, packet dropping, packet modification, addition/deletion of routes, packet fabrication etc. Once the evidence is evaluated, node i revises its existing direct trust for node j. For future computations, revised trust becomes the existing direct trust. Indirect trust is based on recommendations [2] from other trusted parties. If there is a healthy relation between a pair of nodes, they can recommend each other to other members of the network. The node that provides recommendation is referred to as Recommender and recommended node is Recommendee. Recommendations are communicated among nodes by exchanging data packets.
2. **Centralized vs. Decentralized Trust:** Both of these approaches are based on infrastructure trust establishment. Global trusted entity is used in the centralized approach. It calculates trust value for all the nodes in the network. All the nodes of the network ask the trusted entity to provide information about other members of the network. Centralized trust has two implications- 1) failure of trusted entity can be fatal for the whole trust establishment approach. 2) Different entities in the

network have different opinions about the same target. So this fact is also not considered in centralized trust approach. In decentralized approach, each entity is responsible for maintaining trust in the system. Nodes calculate their own trust values for any target they want. This bottom-up approach is most widely used in ad hoc networks. The basic concept followed in this approach is [1] Pretty Good Privacy (PGP) for public key certification. Distinction between centralized and decentralized trust is in semantic terms only. The actual algorithm used for computation of trust executes at a single node. Computation can be done in a distributed manner throughout the network or trust computation algorithm can localized also *i*.e. each node interacts with its direct neighbor only without getting explicit cooperation from nodes further away.

3. **Proactive vs. Reactive Trust:** It is more related to communication efficiency of nodes in the network. Proactive trust computation uses more bandwidth for maintaining trust relationships accurate. The nodes periodically interact with other nodes for maintaining trust. So trust decisions are computed without any delay. More bandwidth is wasted to keep trust values up to date since most of the computed information will be obsolete before it is used. In reactive method, trust values are calculated when these are explicitly needed. If the local trust values (one hop neighbors) change more often than a trust decision needs to be made, then reactive computation is favored.
4. **Data-Centric vs. Attribute Based Trust Establishment:** In data centric approach [14], trust in each individual piece of data is computed. After that related or contradictory data are combined. Some evidence evaluation technique is applied for validating this data. In VANET, it is more necessary to establish trust in data rather than the nodes sending this data as in attribute based trust. For example, if a vehicle is receiving some safety information from other vehicles, it is not necessary to know who is sending this information rather this safety information is important. Node identities are irrelevant in VANET; safety messages with time and location are more relevant. In traditional trust establishment approach, entities (nodes) are the only parameter for establishing trust. Data-Centric trust establishment takes into account location and time of sending node as well as number and type of statements on data for deriving trust relations. Data-centric relations are appropriate for VANET as trust relations are established and re-established depending upon on the network topology and environment changes. For example in real time scenario, emergency information should be believed by receivers. Multiple time interaction between nodes is not possible in this case. In this approach trust is not established by only single source of data like some certification authority (CA) in traditional approach rather trust is derived from multiple evidence (messages from multiple vehicles). All the evidences are weighted according to some rules and also consider some trust metrics. Data and their weights are the input to decision logic and output is the level of trust in these data.

Roadside unit Aided Trust Establishment (RATE) scheme [38] is presented that executes data centric trust establishment in VANET which uses RSU as intermediator. RATE integrates direct observed data with feedback information while evaluating the trustworthiness of data, eventually improves the accuracy of evaluation result. RSU decouples data-consuming and data-providing entities, thus enables vehicles to diminish attacks launched by malicious nodes through directly linked communications.

3.3 Situation Aware Trust (SAT)

The concept of SAT is inspired by various VANET application situations. This approach is used to address important trust issues in VANET. SAT is based on three aspects for trust establishment.

- **Trust Built on Attributes:** In the concern of SAT [9], attributes are entities and data. Evaluation of the trustworthiness of an entity is performed by using authentication. Evaluation of trustworthiness of data is performed by using data integrity checking. Attributes can be classified as static and dynamic attributes depending on whether the attributes change frequently or remain same during long period of time compared to varied number of connections of VANET.
- **Proactive Trust:** Proactive trust is very useful for active safety applications such as cooperative collision warning system. Proactive means setting up trust in advance by predicting future trustworthy situations and pro-actively establishing the trust in advance in VANET.
- **Social Trust:** Social trust plays an important role to build up trust among human beings. Since VANET is driven by humans, social trust can be used for setting up trust among vehicles. Social network is useful when VANET application is running among people and in scenarios when road side network infrastructure is not available.

3.4 Trust Evaluation

The method of computation of trust in a network depends on the particular application [25] where trust values are used. The application determines the exact semantics of trust and the entity determines how trust relations will be used in application. Trust evaluation mechanism can be introduced in every node in VANET. Due to the dynamic topology and high mobility of vehicles, trust management is very difficult in VANET. Each node should evaluate trust on other nodes based on serious study and inference from trust factors like experience statistics, data value, intrusion detection result and reference of other nodes as well as node's own preference and policy. Some static units like RSU's can make the record of all the vehicles passing by. It can participate in evaluation of trust between different entities of VANET. But the restriction is non-uniform distribution of RSUs in some places. There is a trust matrix which stores the knowledge used for trust evaluation in every node. There are some factors [8, 28] that may affect the trust computation as follows:

- **Experience Statistics:** If there is some prior experience of communication between any pair of nodes then it is counted in trust index of these nodes. The communication success through some node will increase the trust index of that node. The communication failure through that node will decrease the trust index attached to that node.

  $$V_{es}(i, j) = F_{es}$$

  Value of function F is proportional to successful communication between node i and j and level of satisfaction from i to j. If two nodes (one hop neighbors) in VANET are directly communicating then they have direct experience of each other. Direct experience is given more trust value as compared to indirect evidences received from other nodes. PGP approach [1] uses only directly assigned trust values but this approach can be extended by using the trust evidences of neighboring nodes.
- **Data Value:** This is the value of communication data. Higher the value of data, the higher trust needed from one node to other.

  $$V_d(i, j) = F_d$$

  Here function F is proportional to importance of data transferred between nodes i and j.
- **Intrusion Black List:** Intrusion detection system of VANET provide blacklist of malicious nodes. The value of intrusion black list can be described as

  $$V_{ibl}(i, j) = 1/0$$

  Value 1 means j is a good node treated by node i while 0 denotes j is malicious node.
- **Reference:** If a node send recommendation or reputation of node j to node i, then it also impact the final trust evaluation result. The value of reference is expressed as

  $$V_r(i, j) = F_r$$

  Here function F is proportional to recommendation on node j and reputation of j.
- **Security Policy:** Network security requirements and policy also impact on trust evaluation system like node i's security policy on node j.

There are many ways to represent level of trust. Trust rings can be one way of representation, where each node creates several rings around it. These rings represent level of trust and number of rings is equal to number of neighboring nodes of this node. Immediate neighbors are placed in the inner-most ring followed by its indirect neighbors in the next ring. As the trust level increases/decreases, nodes on these rings keep on changing in case of mobility of nodes.

4 Infrastructure Based Trust Establishment

Trusted systems are able to prevent any type of attack on VANET. The system use secure and trusted communication infrastructure that is able to satisfy a set of security requirements: authentication, integrity, availability, no repudiation and privacy. Basically hierarchical trust model is followed in this case as shown in figure 3. In this section we are discussing various approaches [7, 27] for establishing infrastructure based trust.

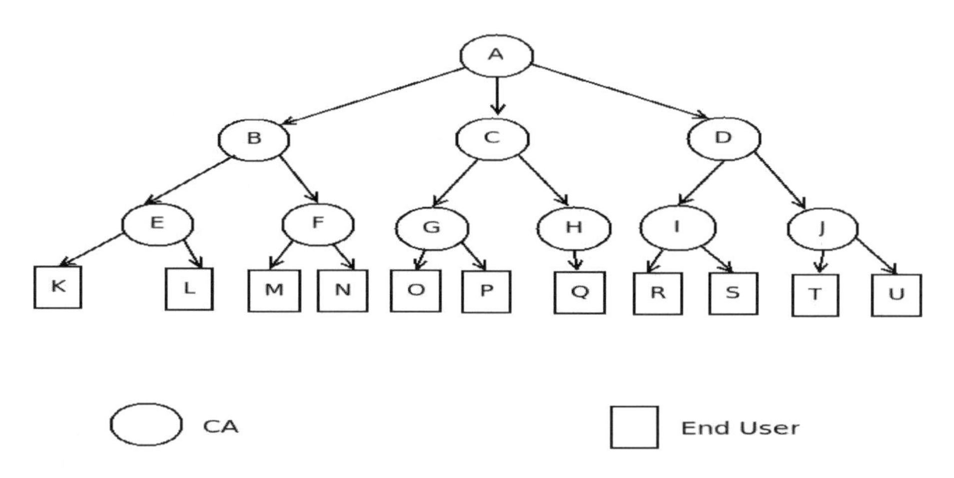

Fig. 3. Hierarchical Trust relationships

4.1 Signature and Certificate Based System

Network security solutions can be categorized depending on two methods a) Symmetric-key cryptography b) Asymmetric- key cryptography. Symmetric-key based security solutions require less computation as compared to asymmetric solutions which require certificates to establish trust between nodes. Symmetric-key based schemes have some drawbacks like high probability of compromise of shared key and scalability issues. As the cryptographic keys are shared among all the nodes in the network, the probability of shared key discovered also increase proportionally with the number of nodes. Symmetric-key based solutions are not suitable for large scale networks.

These are suitable only for small scale networks. Asymmetric-key based solutions are more secure as they eliminate the need for wide-scale key sharing.

Digital signature and certificate based system is asymmetric-key based solution. This is the most popular technique of infrastructure-based trust system [12]. Safety messages are not meant to be confidential so they do not need privacy. As a result, safety messages require authentication but do not require any encryption. A set of Public/Private key pairs is assigned to each vehicle to sign each message digitally and authenticate itself to receivers. Each message sent in the network contains a digital signature and corresponding certificate for the purpose of authentication and integrity. VPKI (Vehicular Public Key Infrastructure) is mostly used as self trust management technique. A centralized authority is required to issue digital certificates.

Every vehicle is registered with a national/regional authority and allocated a unique identifier called ELP (Electronic License Plate). This electronic identification is used for tracking of vehicles. In PKI solution, a safety message is signed with the private key of a vehicle and includes CA's certificate as follows. Here *V* is the sending vehicle, * stands for all the receivers, *M* is the message, T is the time-stamp to ensure the validity of the message, *PrKv* is private key, $Cert_v$ is the public key certificate of *V*.

$$V \rightarrow *: < M, Sign_{PrKv} [M|T], Cert_V >$$

For ensuring privacy, a vehicle has to store large key/certificate set and the keys need to be changed after an interval of time for cryptographic security [18]. All the secret information (Public/Private key pair) is stored in a TPD (Tamper Proof Device) to prevent duplication and modification by unauthorized vehicle. This device, also known as trusted platform module (TPM) [39, 44] offers physical protection of keys residing in it and ensures that they cannot be modified or read by a malicious outsider. It is also responsible for signing all outgoing messages. Access of this device is limited to authorized people. A new module is developed [31] that supports chain of trust to be built within components of network to achieve data flow integrity and handle all types of attacks.

In the case of VANET, there is no global trusted entity. It is up to individual nodes of VANET to establish trust. They themselves sign certificates for each other's keys and judge how much to trust these certificates and their issuer. If a node *i* has previous interactions with the issuing entities, then public keys and trustworthiness of issuer will be known to *i*. So *i* can decide whether to accept *j*'s key or not. Otherwise again trust relation is established in self organized manner. Node *i* first compute the trust values for one hop neighbors, then two hops and so on until destination is reached.

It is proposed [5] that for managing trust in ad hoc networks, certificates are issued by some centralized entity like by some CA in VANET and all other trust management tasks are performed by each node in the network e.g. storing, validating and revoking these certificates. Each node stores copy of all the valid certificates broadcasted by nodes. As each node has public key of trusted CA, it can perform the certificate validation process. Certificate revocation is the most difficult task to implement without any support of centralized entity. In this scheme, a node entering in the network broadcasts its certificate and requests for the profile tables of other nodes in the network. A profile table is like a packet of varied length depending upon the number of accusations launched against the nodes. This profile contains following attributes:

1. **Owner's ID:** It contains an integer indicating the serial number of owner's digital certificate.
2. **Peer ID:** This field contains the certificate serial number of a node that is accused of misbehavior.
3. **Certificate Status:** This is 1 bit flag. If it is set then it represents that certificate of peer *i* is revoked.
4. **Accusation Information:** It contains certificate serial number of a node that accused peer *i* of misbehavior and date of accusation.

If revocation of a node's certificate depends on a single accusation of misbehavior, then malicious nodes can be easily cut off the network access by trustworthy nodes. Trustworthy nodes can cause these certificates to be revoked through malicious accusations. All accusations should not be treated equally and if sum of weighted accusations is greater than a threshold, then certificate is revoked. In a scheme [34], a random password generator is used which generates and distributes the password to all child nodes. This scheme avoids maintaining long records of node details in central trusted authority. When a node is attempting to have communication with other node, it has to pass the password test by providing appropriate password. If a node passes the test, it is declared as legal and communication request is accepted by sharing key and message. A communicating node fails to pass the password test is declared as malicious and communication with that node is banned completely.

4.2 Pseudonyms

In a certificate based system, vehicle's identity can be revealed when it is interacting with other nodes through its public key. To protect privacy as required in VANET, using pseudonym, which can be changed over time, has been proposed [18, 27]. This would not establish anonymity but protect privacy. Public keys or ELP in case of VANET, need change at periodic intervals. This change is performed by some central authority (CA) which also grants pseudonym. Association between pseudonyms and real world entities is known to CA only. This solution is difficult to implement because of high mobility of nodes and dynamic nature of VANET. Binding of these pseudonyms with a particular vehicle at a particular time requires accurate synchronization. Revocation and reuse of these pseudonyms is another issue as numbers of vehicles on roads are increasing day by day.

4.3 Group Signatures

A group signature scheme [24] provides both security and privacy in VANET. This scheme allows a vehicle to sign the message on behalf of a group. A single group public key is used and it does not reveal the identity of the signer. It is not possible to verify if two signatures are issued by the same group member because each member of the group is assigned a unique private key. Private Key of group member is used for generating signatures with group public key. A vehicle outside this group can verify that the message is generated by this group but cannot detect which node has generated this message. A node designated as group manager is required to resolve the signatures of individual nodes. Group manager uses his secret key and given signature to determine the identity of group member who generated the message.

A vehicle which does not have group manager's secret key cannot determine identity of group member. This ensures that members of the group are anonymous within the group and also with other group members. In addition, no outsider can issue signatures. Only the group members can sign correctly. A group signature scheme consists of following procedures:

1. **Setup:** This protocol is used for interaction between a designated group manager and members of the group to decide group public key, private key of group manager and all the members of the group.
2. **Sign:** In this protocol, a group member signs a message *m* with its private key *p* and returns a signature *s*.
3. **Verify:** It is used for verification of a valid signature. It requires a message *m*, signature *s* and group's public key *Y*.
4. **Open:** This returns identity of the member who signed the message. This requires a signature s and group manager's private key.

This technique is quite promising in the case of VANET since it does not require permanent online connection to infrastructure. It works well in dynamic environment, privacy can be established and verification process is fast. Special attention should be given when vehicles leave one group and join other group. In this case cell dimension should be less than the diameter of transmission range of vehicles. There is a need for efficient group management, key certification and key revocation techniques because of dynamic nature of groups. As numbers of vehicles are increasing day by day, some kind of hierarchical structure is needed for efficient group management.

4.4 Blind Signature

This technique uses anonymous certificates in trust establishment system. Blind signature [27] allows a signer to digitally sign a statement without knowing the message. The requesting node uses a blinding function *F* with *b* as a random chosen blind factor. Message m is computed using this function *F*.

$$m' = F(m, b)$$

m is a simple message that is to be sent. *m'* is sent to CA. This centralized authority signs *m'* using ordinary signature algorithm *SA* and private key k_p for producing $Sig' = SA(m', k_p)$. CA sends Sig' back to the requesting node. The node then applies the reverse blinding function F^{-1} to compute $Sig = SA(m, k_p)$. Feature of this kind of the system is that a node requesting a certificate creates n blinded certificates with its attributes to be signed. Now the trusted authority can randomly ask the node in authenticated session to disclose *n-1* certificates and can check the attributes. If all the attributes are correct, the authority would sign last blind certificate. Thus authority does not know for which pseudonym it has signed hence providing anonymity to node. The authority can prevent nodes from attacking the trust system by remembering the frequency of node requesting for certificates and reject issuing the certificates in case of any abuse. Drawback of this approach is that a requesting node has to create multiple messages for this technique to work. This technique is not appropriate for VANET because of its restrictions.

4.5 Pair Wise Keys

If two nodes want to communicate with each other for a long time, they have to remain in the range of each other. For one-to-one communications of such nature, pair wise keys are used. Symmetric keys are more efficient in term of time and space overhead than asymmetric [12]. Main challenge is distribution of key pairs. It is very difficult to preload pair-wise shared keys to vehicles because of large scale and dynamic nature of VANET. One method for establishing pair wise key makes use of PKI and digital certificates. Vehicle *A* encrypts the message comprising of identity of *B*, time-stamp *T* and session key *K* with *B*'s public key *PuKB*. This message is also signed with *PrKA* private key of *A*. Subsequent message exchange can use this session key and integrity of message can be verified through HMAC (Hashed Message Authentication Code) with key *K*.

$$A \text{ -> } B\text{: } < E_{PuKB}\,[B|K|T],\ Sign_{PrKA}[B|K|T] >$$

$$A \text{ -> } B\text{: } < m,\ HMACK(m) >$$

This scheme does not scale well as number of digital signatures increase with growing number of vehicles leading to VANET. For only few numbers of vehicles, this scheme is not justifiable because of lack of congestion on the wireless channel. For critical safety applications, symmetric session keys are not used as non-repudiation property cannot be established.

4.6 Threshold Cryptography

The concept of threshold cryptography was first introduced by Adi Shamir [19, 27]. In this technique no centralized trust system is required. For example in *(n, t)* threshold cryptography, secret key is split in to *n* shares such that for a certain threshold $t < n$, any *t* components could combine and generate valid signature. This scheme is more secure as minimum *t* numbers of components are required to break the security of the system.

Threshold cryptography can be utilized to distribute trust in VANET. Key management challenges (issuing, revoking and storing certificates) in VANET can be resolved by distributing CA duties among the nodes of VANET. A CA signing key can be partitioned in to *n* parts and distributed to *n* nodes. Any *t* out of this *n* number of nodes can collaborate to sign and issue valid certificate. This scheme is quite elegant and offers a good measure of security. Due to the dynamic nature of VANET, this scheme is not applicable to VANET as threshold cryptography involves additional computations compared to other asymmetric-key protocols and it also requires unselfish cooperation of communicating nodes which is not realistic in VANET because every vehicle is in hurry to reach its destination.

4.7 Centralized Trust Management Problems

The operation of VANET is distributed and autonomous. Traditional authentication scheme in these networks would require centralized control structure and permanent on-line servers and support staff to manage them. A trust management system is needed that matches the nature of VANET and that is based on collaborative decision of individual nodes. Some of these limitations are:

1. Centralized authority has to be present at all the time.
2. It is very difficult for all the nodes to communicate with centralized authority as communication may happen only in local area. This technique forces all the users to contact the centralized authority for every enrollment and authentication operation.
3. Centralized authority uses single trust metric for the entire domain. In self organizing networks, trust evidence is not uniform. Evidence is in the forms of keys, node ID, hardware attributes and social relationships. Evidence evaluation cannot be uniform in ad hoc networks.
4. Centralized trust models require long lived evidence. Identity credentials are assigned at the time of enrollment of node to CA. If the credentials become compromised, it will be very costly to collect and maintain evidence again. Re-evaluating trust evidence is difficult. Self organized networks require frequent and on-line trust evidence re-evaluation.
5. Traditional centralized authority require the trust model to be established only at the central place but in self organized networks, trust relationships are built within the nodes so corresponding trust model needs to be built to match this requirement.

5 Infrastructure Based Trust Establishment

Trust establishment techniques [16] should adapt to dynamic environment of a VANET. All the techniques discussed in Section 4 failed to adjust with changes in VANET environment. Trust decisions must be made autonomously because fixed security infrastructure is not guaranteed at all times in VANET. Decisions must be based on the partial information collected for a short time from unknown nodes. Self organized trust establishment is required [21] as shown in figure 4 because of non-availability of infrastructure and shared global knowledge among the participating nodes.

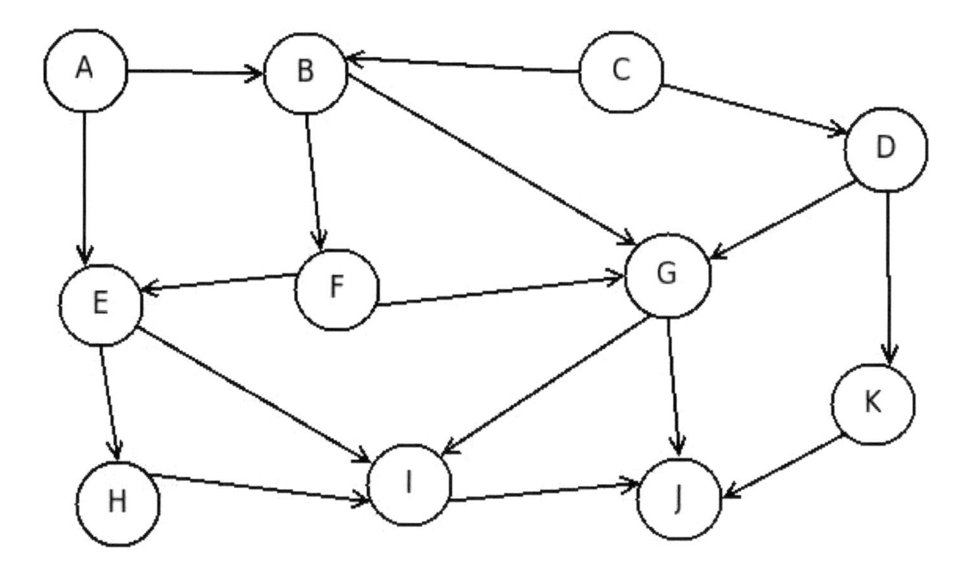

Fig. 4. Hierarchical Trust relationships

Self Organizing trust establishment mechanism can be direct, indirect and hybrid. In direct mechanism, trust is established by direct communication of nodes *i*.e. the nodes exchange trust values with all the nodes within its communication range. Trust relationships are transitive in indirect mechanism. For example node A is exchanging trust relations with its one hop neighbor B. Now while communication of nodes B and C, B is also exchanging the trust relations of node A apart from its own trust relations. So this trust establishment mechanism is indirect. Hybrid mechanism can also be used which is a combination of both direct and indirect trust establishment mechanisms.

5.1 Architecture of Ad Hoc Trust Framework

Being a self organized network, VANET trust framework is based on distributed and modular architecture [26]. Architecture of ad hoc trust framework is shown in figure 5. Each component of trusted framework resides in every node and performs set of actions to evaluate the reputation of other nodes. It incorporates self evidences, recommendations, and experience statistics for evaluating the trust level of other nodes. This framework is used to detect malicious, selfish and unreliable behavior of nodes communicating with each other. After detecting misbehavior in the network it provides feedback to other nodes in the network. Trust establishment framework for VANET is lightweight because it does not perform extensive risk and behavior analysis for trustworthiness assessment. It does not involve computationally heavy security tasks such as key generation, key revocation and cryptography. With minor integration efforts, it can be incorporated in the real network. This trust framework consists of following components:

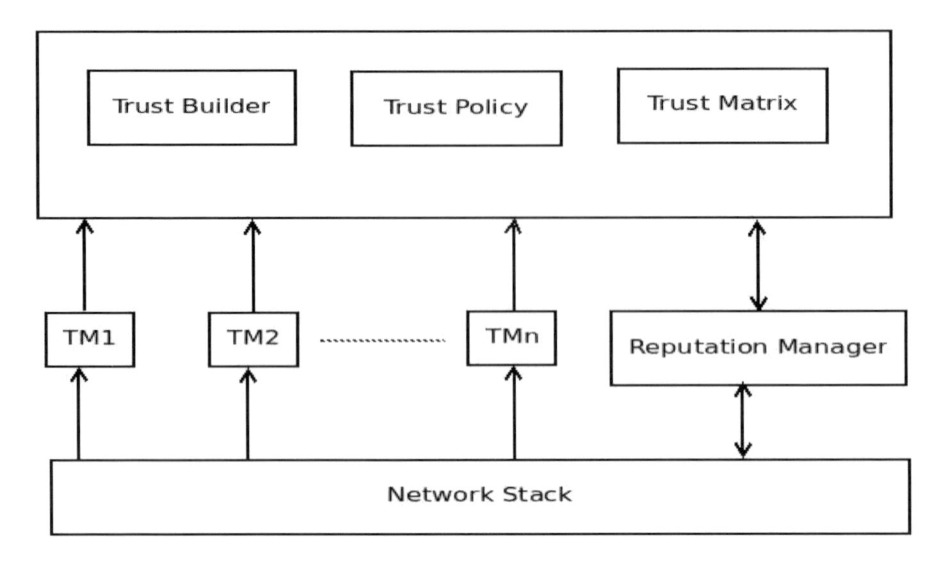

Fig. 5. Architecture of Ad hoc trust framework

- **Trust Monitor:** Monitoring process is performed independently by each node in the network. Trust monitor logs every activity performed by its neighboring nodes in 'evidence store'. Operation of the trust monitor is similar to common sensor *i*.e. translating the physical phenomenon or behavior of a node in to machine interpretable form. This phenomenon is trustworthiness in our case. Trust monitor compares its logged behavior to expected behavior and correspondingly increase or decrease the trustworthiness of neighboring nodes.
- **Trust Builder:** This component computes the trust value of other nodes with which there is established interaction. All these trust values are stored in trust matrix. The role of trust builder is to maintain and update trust matrix. Trust matrix is consulted before accessing application or service. Trust value computation depends upon several factors like interaction history and direct evidence as described in section 3.4. Direct evidence is given higher weightage as compared to recommendations. Weighting of these factors are defined after extensive experience in simulation.
- **Reputation Manager:** In order to compute the trust values, recommendations are exchanged between nodes and trust builder. The role of reputation manager is to manage the recommendation exchange procedure. Recommendations can be exchanged periodically or on demand basis. When a node has sufficient experience of interaction with other node in the network, then periodically they can exchange recommendations of each other. In on-demand approach, originator requests recommendation for a target node when it has insufficient experience with it. Reputation manager aggregates a recommendation collected from different nodes and returns a trust value to trust builder.

- **Trust Policy:** Trust policies are some predefined rules that can be applied to self organized networks in order to determine the trust values of nodes. Each node maintains a trust policy, set of parameter values which can define the functionality of reputation manager and trust builder.

5.2 CONFIDANT

Buchegger and Le Boudec [4] proposed the CONFIDANT (Cooperation Of Nodes-Fairness In Dynamic Ad hoc NeTworks) cooperative protocol for detecting and excluding the misbehaving nodes in ad hoc networks. In this technique, history of behavior of nodes is stored in every vehicle. Following components are involved in CONFIDANT protocol.

1. **Monitor:** This component runs in background and monitors all the neighbors by observing their routing protocol behavior using promiscuous mode.
2. **Trust Manager:** If this behavior is consistent for a specified period of time, the node is considered trusted otherwise labeled a malicious node. Trust manager stores trust value of the nodes. If the trust value falls below a threshold value, the path manager is involved.
3. **Reputation System:** This unit manages and updates the trust values of the nodes. These values are derived by evaluating some parameters like experience of observer, observations of neighborhood. These parameters are weighted according to their trustworthiness e.g. experience of direct communication between two nodes is given greater importance than observation of other nodes.
4. **Path Manager:** If trust value falls under a certain threshold value, then path manager is informed. It isolates malicious nodes by ignoring sending/receiving packets to/from these nodes.

This is a high level modular design but in real world it is difficult to deploy this solution as it is difficult to distinguish between misbehavior of nodes and errors due to fast changes in topology. Storage of history requires more space and computing power.

5.3 Location Limited Side Channels (LLSC)

This approach of establishing trust relations [27] between nodes in VANET uses location limited side channel. It is a special channel which is separated from the main communication channel. LLSC is designed in such a way that an attacker cannot gain physical access to the channel (for performing any operations like inserting or updating the messages). Critical information can be exchanged via secure channel. Applications of LLSC are authentication and pairing of previously unknown nodes in VANET. These nodes can exchange keys or hashes of keys over LLSC for pre-authentication. Authentication of less critical nodes can be done over normal wireless link. In VANET, infrared and radar communication can be used for establishing LLSC.

5.4 Self-certified Pseudonym Based Trust Establishment

This approach is used to provide privacy and Sybil-freeness [15] without requiring any continuous availability of centralized authority. Sybil-freeness means each entity is associated with only one identity rather than with multiple identities. In some attack scenarios, entities spoof the identities of one or more nodes and participate in the network using those identities. In this technique, users can compute pseudonyms from their cryptographic identities themselves. If initial identity domain is Sybil free, this Sybil freeness can be propagated to other identity domains even without continuous involvement of TTP (Trusted Third Party). CA will be needed only for initial setup of Sybil free domain. Initially user acquires membership certificate by enrolling with CA. User can create self certified pseudonym per identity domain by using this membership certificate. These pseudonyms are valid for the domains for which they were issued. Pseudonyms for different domains are unlink-able. Sybil attacker has to identify relationship between two pseudonyms generated for different identity domains. Attacker can eavesdrop on the wireless channel and find if some pseudonyms belong to same user. This technique can help preventing Sybil nodes in VANET.

6 Trust Models

Basic Components of Trust model are gathering information, information scoring, ranking and action execution. It collects behavior information from all the nodes in the network and scores this information according to various parameters. According to scores, trust values are weighted and if these trust values are below threshold then appropriate actions are taken.

When a node i has a long trust history with node j then they are more trustworthy. If node i does not have trust relationship with node j, then node j is removed from the network. Hence time is an important parameter in trust metric. Trust is a function that depends on both time that a node successfully participates in communication activity of the network and the past trust that a node has achieved in recent time.

6.1 Geo-location Based Trust Model for VANET

Trust can be propagated from one geographical area to another (e.g. when vehicles move from one city to other) using PKI infrastructure. Geo-location based trust approach [11] is dependent upon vehicles to perform authentication in untrusted domains by dynamically enabling cooperation among different CAs without explicit agreements. In VANET, vehicles belonging to different domains must have certificates issued by different CAs. In this section we are presenting the VANET trust management architecture that is based upon PKI infrastructure. Other trust management architectures are also possible where PKI infrastructure cannot be employed. Geo-location based trust model is shown in figure 6.

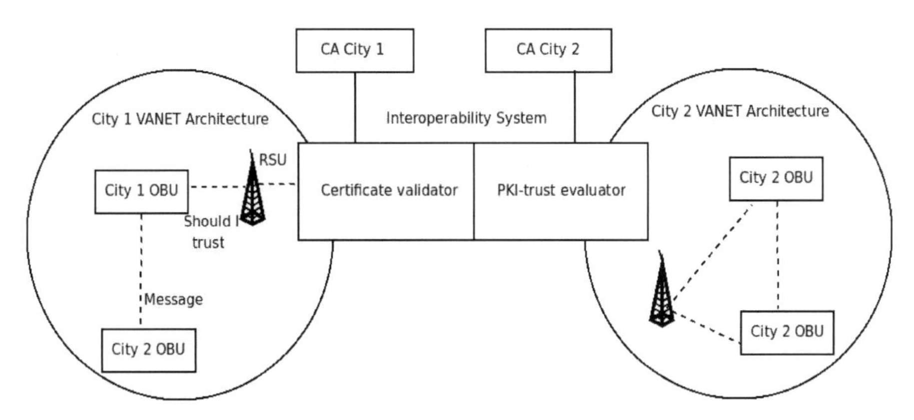

Fig. 6. Geo-location based trust model

1. **Interoperability System (IS)** is the main component of this architecture that cooperates between CAs of different geographical regions and takes appropriate decisions of validity of certificates issued by different CAs. It is intermediary between certificate verifier and the CA issuing certificates to different entities of VANET. It contains two subcomponents 1) certificate validator and 2) PKI-Trust evaluator. Real time status information of all the certificates issued by different CAs are validated by certificate validator.
2. **Trust Evaluator** computes CA security level by using PKI evaluation system. For example a vehicle from city 1 enters in city 2 and wants to communicate with the vehicles of city 2. If vehicle of city 2 receives some message from city 1 vehicle, it does not know whether to trust this message or not. So it will consult some RSU in its range. RSU further consults IS to validate and evaluate the sender's certificate and take some decision if it is possible to trust on certificate issuer or not. After getting the feedback from base station, city 2 vehicle will decide whether to trust the message or not.

A trust evaluation model [36] is presented that is based on location information and verification in Non Line Of Site (NLOS) condition. In this model, trust levels are evaluated by monitoring the behavior of neighboring nodes and their generated messages based on events. However, obstacles can create NLOS case which can interrupt direct communication among vehicles and prevent them from properly monitoring their neighboring nodes. We briefly describe the components of secure trust model:

1. **Gathering Localization Information:** Each vehicle is able to determine its own location information using localization service such as GPS device. It also collects its neighbor information from received beacons and group updates.

2. **Maintain Awareness of Neighborhood:** Each vehicle is able to maintain database of all its neighbors' location and mobility information. This database is periodically monitored to check for any data inconsistency.
3. **Location Verification:** If there is any inconsistency in neighborhood awareness service, it triggers location verification process. This module validates the claimed location of neighbors and updates the awareness database.
4. **Location Trust Evaluation:** Trust level of neighboring nodes are calculated based on location reachability and score is assigned to each node.

6.2 Self Organized Certificate-Based Trust Model

VANET is self organized network because of its multiple features. PGP- like mechanism can be used for initializing the system. Trust between nodes can be established through secure side channel communication (physical contact, infrared communication). It is assumed that social relationships among vehicles in VANET are same as those in PGP system. PGP model [1] is realistic model in which social relationships in real society is used for trust establishment. Trust relationships formed in VANET nodes exhibits the same feature as PGP system. A public key certificate based approach is used in this scheme. Every vehicle issue certificates to other nodes which are trusted from its domain. Nodes can authenticate each other using trust relationship chain established between the nodes in the network. A to-be member node can contact with many member nodes for getting their certificates to join the network and can be authenticated by other nodes in the network.

In threshold cryptography scheme, trust between a to-be member node and t member nodes in its neighborhood is established by human perception or biometrics. This method is practically not feasible for VANET. Whenever a node enters in the system, it has to acquire t nodes to trust it in its communication range. There should be offline trust relationship between this node and t member nodes. Due to ad hoc nature of VANET, this scheme cannot be applied in VANET. This scheme also suffers from high communication and computation overhead.

In self organized scheme [10, 21], trust is established from offline trust relationships which are generated from social relationships. Initially, nodes themselves issue certificates and form network of trust relationships. This scheme is very impractical and slow when used in VANET.

6.3 Distributed Trust Establishment Approach for VANET

In VANET, trust establishment is very difficult and challenging task because of lack of fixed infrastructure, very high mobility of vehicles, limited range of communication and lack reliability of wireless links. In this approach, operation of trust management process is distributed [6, 29] among all the nodes of network. Only at the time of bootstrapping, a centralized trust initiator is required to initiate the system.

Trust initiator is introduced in the bootstrapping phase of system to initiate the process of trust establishment. Trust relationships are established with this trust initiator so that any pair of nodes in the network can authenticate each other with higher probability via trust chain. This scheme is reliable as there exist at least two independent trust chains between any pair of nodes. Average length of the trust chain is made short for making the process efficient. Features of this scheme are:

1. Average length of shortest indirect trust paths between any pair of nodes is very small. It makes the scheme secure as there are fewer intermediate nodes in the trust paths.
2. Average number of shortest indirect trust paths between any pair of nodes is guaranteed to be at least 2, each with high probability ensuring robustness and reliability of the system.
3. This scheme is highly adapted to dynamic nature of VANET with high mobility and fast changing topology.

This approach [19] establishes trust relationships among members of VANET with minimum storage requirements for maintenance of trust information. It is highly scalable and adaptable to highly dynamic network topology. Two assumptions are made to develop trust establishment.

Every vehicle in VANET has its own private/public key pair and binding of node ID with public key is known to some centralized entity *i*.e. trust initiator. There are sufficient trust evidences between the member nodes and the trust initiator, so that all the nodes trust the trust initiator unconditionally at the time of system bootstrapping.

Suppose a node ‘P’ wants to enter in the system, it is required to obtain at least 2 certificates from current member nodes. Due to mobility of nodes and sparse social relationships existed among nodes, it is reasonable to acquire at least two independent certificates by node ‘P’. Node ‘P’ can contact any current member nodes within its communication range for requesting to join the network. Any of the available member nodes can reply the request. The corresponding member nodes communicate with other member nodes to authenticate the trust evidence provided by node ‘P’. If the trust evidence is authenticated by current member node, node ‘P’ obtains the certificate signed by this node. This process is repeated until node ‘P’ gets at least two independent certificates.

When a node ‘Q’ leaves the network, it broadcast its leaving information and signs on it so that other nodes revoke all the certificates issued after verifying the message. Any node which has detected the leaving process of node ‘Q’, broadcast leaving message to other member nodes of network. This broadcasted information is authenticated by each node independently. The certificates issued to node ‘Q’ are revoked if broadcasted information is authenticated. No certificate can reach the node after it leaves the network.

Distributed trust establishment model [35] can be grouped into two categories: (1) Entity-based trust model and (2) Data-based trust model. Entity-based trust models put emphasis on the trustworthiness of nodes. Data-based trust models focus on evaluating the trustworthiness of data. Distributed trust establishment models do not form long term relationships between peers. Hybrid trust models make use of peer trust to evaluate the trustworthiness of data. But at the same time, it also maintains peer trust over time.

6.3.1 Entity-Based Trust Model

Two entity-based trust models have been proposed: (1) Sociological and (2) Multi-faceted. The sociological trust model [42] is based on the concept of trust and confidence tagging. Various forms of trust have been identified: (1) Situational trust- which depends upon the situation only. (2) Dispositional trust- which is measured based on the peer's own beliefs. (3) System trust- Depends on the system. (4) Belief formation trust- which is evaluation of data based on previous factors. However, this model does not provide the formalization of architecture to combine different types of trust together.

Multi-faceted trust management model [40] combines the role-based and experience-based trust in order to formulate the evaluation metric for the integrated trustworthiness of vehicles. This model allows VANET entities to actively inquire about an event by sending requests to other entities, but restricts the number of reports that are received. For restriction of received reports, priority-based trust concept is used. It provides an ordering of the value of an information source within a role category, using the influence of experience-based trust. The trust of information sources and contextual information about an event (time and location) are integrated into a method for determining whether majority inputs have been reached. These inputs ultimately identify a action, a vehicular entity should follow.

6.3.2 Data-Based Trust Model

It is proposed [14] that data-based trust may be more appropriate for VANET. In this trust model, trustworthiness of data reported by peers is evaluated rather than the trust of peers. Prior trust relationships between entities are considered as one of the default parameter. Evidences regarding a particular event are evaluated considering different trust metrics using Bayesian inference and Dempster-Shafer theory. Final output indicates the level of trust in evaluated evidences and determines whether the event related with data has taken place or not.

Limitation of data-based trust model is that trust relationships between entities are never established, only ephemeral trust in data is formed. As data-based trust is based on per event basis, it needs to be established for every event again and again. In case of sparse VANET, this model would not perform well as enough evidences in support of an event may not be generated. A technique [43] to detect and correct malicious data in VANET is presented which uses the concept of data-based model. Each node maintains a model of VANET. This model contains all the knowledge that a particular node has about the VANET. Whenever a node receives any information, it is evaluated against the peer's model of VANET. If all the received data agrees with the model then peer accepts the validity of data, otherwise it is considered as malicious data. The main limitation of this approach lies in its assumption that each vehicle has global knowledge of the network, which may not be feasible in realistic environment.

Trust based message propagation and evaluation framework [30, 37] is presented which overcomes the problems in both above defined models. In this framework, peers share information regarding events (such as road condition or any safety information) and provide opinions about the trust level of information. This trust-based message propagation model collects and propagates peers' feedback in an efficient, secure and scalable way *i*.e. it dynamically controls information dissemination.

Several limitations [32] of trust management schemes for VANET are presented, particularly the problem of information cascading and oversampling, which commonly arise in social networks. Oversampling is a situation where a node observing two or more nodes, takes into consideration both their opinions equally without judging that they might influenced each other in decision making. All the approaches which use simple voting for decision making, leads to oversampling and gives incorrect results. A novel voting scheme is presented [33] to overcome the problem of information cascading and oversampling. In this scheme, each vehicle has different voting weight according to the distance from the event. The vehicle which is closer to the event has higher weight.

7 Open Issues

Trust establishment techniques play great role in security of any network system. In VANET, safety messages are broadcasted to nodes in specific area where the nodes can be directly affected from critical situations. Safety messages are more important rather than knowing who is sending this message. So time of sending the message, location from where the message is sent, number and type of statements on data are more important. Managing trust according to different situations of VANET is not widely explored area.

Trust management techniques of ad hoc networks cannot be directly applied in VANET. Due to the different characteristics of VANET from other ad hoc networks like high mobility of vehicles, large scale of networks, geographically constrained topology and frequent network partition, trust establishment approach should be combination of all the discussed approaches. For example, geographical based trust establishment approach can be combined with distributed and centralized approaches. Some new trust establishment approach is required in VANET that can combine existing approaches according to different situations in VANET and also focuses on safety messages rather than the nodes.

8 Conclusions

The broadcast nature of the wireless channels, the absence of a fixed infrastructure, the dynamic network topology, and the self-organizing characteristic of the network increase the vulnerabilities of a vehicular ad hoc network. Trust establishment is a critical task in VANET. In centralized trust establishment approach, the major problem is impersonation attacks, where one node can steal and use the identity of other nodes. In completely distributed trust establishment approach, the main problem is that a node might acquire multiple identities from different issuers. It is assumed in all the trust management approaches that distinct entities have unique identities and their opinions are independent. Without this assumption, nodes cannot be responsible for their actions or recommendations they provide to other nodes. Distributed trust calculation is sensitive to identity attacks e.g. masquerade, Sybil and whitewash attacks. In centralized trust management systems, it is easy to deploy prevention mechanisms for these attacks as compared to distributed systems.

Theoretically, centralized trust establishment technique (*i*.e. popularly used in fixed wired/wireless networks) can be used for maintaining trust in VANET but practically dynamic trust establishment is a better solution. Self organized trust establishment mechanism can build trust based on the mutual communication between the vehicles. Some efficient algorithms for dynamic trust establishment should be designed to deal with VANET characteristics.

In this paper, we presented various trust establishment mechanisms according to the characteristics of VANET. Most befitting technique for VANET security implementation is to use dynamic trust establishment. But this is very difficult to implement because of dynamic characteristics of VANET. Self Organized trust establishment mechanism can build trust based on the mutual communication between the vehicles. Overhead of information exchange, storage and analysis is high. Efficient algorithms for dynamic trust establishment should be designed to deal with VANET characteristics.

In threshold cryptography scheme, trust between a to-be member node and t member nodes in its neighborhood is established by human perception or biometrics. This method is practically not feasible for VANET. Whenever a node enters in the system, it has to acquire t nodes to trust it in its communication range. There should be offline trust relationship between this node and t member nodes. Due to ad hoc nature of VANET, this scheme cannot be applied in VANET. This scheme also suffers from high communication and computation overhead.

In self organized scheme, trust is established from offline trust relationships which are generated from social relationships. Initially, nodes themselves issue certificates and form network of trust relationships. This scheme is very impractical and slow when used in VANET.

In this tutorial, we have discussed situation aware trust establishment in VANET that is a combination of different trust management approaches. We also focused on data-centric trust as compared to entity-level trust. A robust trust establishment approach is required that can consider all the aspects of VANET and should be easy to integrate with the existing system.

Key Terminology

VANET	Vehicular Ad hoc Network
MANET	Mobile Ad hoc Network
ITS	Intelligent Transportation system
PGP	Pretty Good Privacy

GPS	Global Positioning System
RSU	Road Side Unit
OBU	On board Unit
TPD	Temper Proof Device
RTS	Request To Send
CTS	Clear To Send
CA	Central Authority
MAC	Medium Access Control
HMAC	Hashed Message Authentication Code
VPKI	Vehicular Public Key Infrastructure

References

1. Abdul-Rahman, A.: The PGP trust model. In: EDI-Forum (April 1997)
2. Aijazand, A., Bochow, B., Doetzer, F., Festag, A., Gerlach, M., Kroh, R., Leinmller, T.: Attacks on intervehicle communication systems - an analysis. WIT (2006)
3. Balakrishnan, V., Varadharajan, V., Tupakula, U., Lues, P.: Team: Trust enhanced security architecture for mobile ad-hoc networks. In: 15th IEEE International Conference on Networks ICON 2007, pp. 182–187 (November 2007)
4. Buchegger, S., Boudec, L., Jean-Yves: Performance analysis of the confidant protocol. In: MobiHoc 2002: Proceedings of the 3rd ACM International Symposium on Mobile Ad Hoc Networking & Computing, pp. 226–236. ACM, New York (2002)
5. Davis, C.: A localized trust management scheme for ad hoc networks, pp. 671–675 (2004)
6. Eschenauer, L., Baras, J.S., Gligor, V.: Distributed trust establishment inmanets' swarm intelligence. In: Collaborative Technology Alliances (CTA) Communications Networks (CN) Alliance-2003 Annual Symposium, pp. 125–129 (May 2003)
7. Eschenauer, L., Gligor, V.D., Baras, J.S.: On Trust Establishment in Mobile Ad-Hoc Networks. In: Christianson, B., Crispo, B., Malcolm, J.A., Roe, M. (eds.) Security Protocols 2002. LNCS, vol. 2845, pp. 47–66. Springer, Heidelberg (2004)
8. Theodorakopoulos, G., Baras, J.S.: On trust models and trust evaluation metrics for ad hoc networks. IEEE Journal on Selected Areas in Communications 24(2), 318–328 (2006)
9. Hong, X., Huang, D., Gerla, M., Cao, Z.: SAT: situation-aware trust architecture for vehicular networks. In: MobiArch 2008: Proceedings of the 3rd International Workshop on Mobility in the Evolving Internet Architecture, pp. 31–36. ACM, New York (2008)
10. Sen, J., Chowdhury, P.R., Sengupta, I.: A distributed trust establishment scheme for mobile ad hoc networks. In: International Conference Computing: Theory and Applications, ICCTA 2007, pp. 51–58 (March 2007)
11. Serna, J., Luna, J., Medina, M.: Geolocation-based trust for vanet's privacy. In: Fourth International Conference on Information Assurance and Security, ISIAS 2008, pp. 287–290 (2008)
12. Khalili, A., Katz, J., Arbaugh, W.A.: Toward secure key distribution in truly ad-hoc networks. In: SAINT-W 2003: Proceedings of the 2003 Symposium on Applications and the Internet Workshops (SAINT 2003 Workshops), p. 342. IEEE Computer Society, Washington, DC (2003)
13. Li, W., Joshi, A.: Outlier detection in ad hoc networks using dempster-shafer theory. IEEE International Conference on Mobile Data Management, 112–121 (2009)

14. Gligor, V.D., Raya, M., Papadimitratos, P., Hubaux, J.-P.: On data-centric trust establishment in ephemeral ad hoc networks (2007)
15. Martucci, L.A., Kohlweiss, M., Anderssonand, C., Panchenko, A.: Selfcertified sybil-free pseudonyms. In: WiSec 2008: Proceedings of the First ACM Conference on Wireless Network Security, pp. 154–159. ACM, New York (2008)
16. Papageorgiou, C., Birkos, K., Dagiuklas, T., Kotsopoulos, S.: Dynamic trust establishment in emergency ad hoc networks. In: IWCMC 2009: Proceedings of the 2009 International Conference on Wireless Communications and Mobile Computing, pp. 26–30. ACM, New York (2009)
17. Pirzada, A.A., McDonald, C.: Establishing trust in pure ad-hoc networks. In: ACSC 2004: Proceedings of the 27th Australasian Conference on Computer Science, pp. 47–54. Australian Computer Society, Inc., Darlinghurst (2004)
18. Raya, M., Hubaux, J.-P.: Securing vehicular ad hoc networks. J. Comput. Secur. 15(1), 39–68 (2007)
19. Ren, K., Li, T., Wan, Z., Bao, F., Deng, R.H., Kim, K.: Highly reliable trust establishment scheme in ad hoc networks. Comput. Netw. 45(6), 687–699 (2004)
20. Ren, Y., Boukerche, A.: Modeling and managing the trust for wireless and mobile ad hoc networks. In: IEEE International Conference on Communications, ICC 2008, pp. 2129–2133 (May 2008)
21. Repantis, Thomas, Kalogeraki, Vana: Decentralized trust management for ad-hoc peer-to-peer networks. In: MPAC 2006: Proceedings of the 4th International Workshop on Middleware for Pervasive and Ad-Hoc Computing (MPAC 2006), p. 6. ACM, New York (2006)
22. Samian, N., Maarof, M.A., Razak, S.A.: Towards identifying features of trust in mobile ad hoc network. In: AMS 2008: Proceedings of the 2008 Second Asia International Conference on Modelling & Simulation (AMS), pp. 271–276. IEEE Computer Society, Washington, DC (2008)
23. Schmidt, R.K., Leinmller, T., Schoch, E., Held, A., Schfer, G.: Vehicle behavior analysis to enhance security in vanets. In: Fourth Workshop on Vehicle to Vehicle Communications, V2VCOM 2008 (June 2008)
24. Sun, X., Lin, X., Ho, P.-H.: Secure vehicular communications based on group signature and id-based signature scheme. In: IEEE International Conference on Communications ICC 2007, pp. 1539–1545 (June 2007)
25. Theodorakopoulos, G., Baras, J.S.: Trust evaluation in ad-hoc networks. In: WiSe 2004: Proceedings of the 3rd ACM Workshop on Wireless Security, pp. 1–10. ACM, New York (2004)
26. Tsetsos, V., Marias, G.F., Paskalis, S.: Trust Management Issues for Ad Hoc and Self-organized Networks. In: Stavrakakis, I., Smirnov, M. (eds.) WAC 2005. LNCS, vol. 3854, pp. 153–164. Springer, Heidelberg (2006)
27. Wex, P., Breuerand, J., Held, A., Leinmullerand, T., Delgrossi, L.: Trust issues for vehicular ad hoc networks. In: Vehicular Technology Conference, VTC Spring 2008, pp. 2800–2804. IEEE (May 2008)
28. Virtanen, T., Yan, Z., Zhang, P.: Trust evaluation based security solution in ad hoc networks. In: Proceedings of the Seventh Nordic Workshop on Security IT Systems (2003)
29. Zouridaki, C., Mark, B.L., Hejmo, M., Thomas, R.K.: Robust cooperative trust establishment for manets. In: SASN 2006: Proceedings of the Fourth ACM Workshop on Security of Ad Hoc and Sensor Networks, pp. 23–34. ACM, New York (2006)

30. Chen, C., Zhang, J., Cohen, R., Ho, P.-H.: A Trust Modeling Framework for Message Propagation and Evaluation in VANETs. In: Proceedings of 2nd International Conference on Information Technology Convergence and Services (ITCS), pp. 1–8 (August 2010)
31. Sumra, I.A., Hasbullah, H., Ahmad, I., Bin Ab Manan, J.L.: Forming Vehicular Web of Trust in VANET. In: Proceedings of Saudi International Electronices Communications and Photonics Conference (SIECPC), pp. 1–6 (April 2011)
32. Huang, Z., Ruj, S., Cavenghi, M., Nayak, A.: Limitations of Trust Management Schemes in VANET and Countermeasures. In: Proceedings of 22nd International Symposium on Personal Indoor and Mobile Radio Communications (PIMRC), pp. 1228–1232 (September 2011)
33. Huang, Z., Ruj, S., Cavenghi, M., Stojmenovic, M., Nayak, A.: A social network approach to trust management in VANETs. Peer-to-Peer Networking and Applications Journal, 1–14 (2012)
34. Gowtham, G., Samlinson, E.: A secured trust creation in VANET environment using random password generator. In: Proceedings of International Conference on Computing, Electronics and Electrical Technologies (ICCEET), pp. 781–784 (March 2012)
35. Zhang, J.: A Survey on Trust Management for VANETs. In: Proceedings of International Conference on Advanced Information Networking and Applications (AINA), pp. 105–112 (March 2011)
36. Abumansoor, O., Boukerche, A.: Towards a Secure Trust Model for Vehicular Ad Hoc Networks Services. In: Proceedings of Global Telecommunications Conference (GLOBECOM 2011), pp. 1–5 (December 2011)
37. Zhang, J., Chen, C., Cohen, R.: A Scalable and Effective Trust based Framework for Vehicular Ad-Hoc Networks. Journal of Wireless Mobile Networks, Ubiquitous and Dependable Applications (JoWUA) 1(4), 3–15 (2010)
38. Wu, A., Ma, J., Zhang, S.: RATE: A RSU-Aided Scheme for Data-Centric Trust Establishment in VANETs. In: Proceedings of 7th International Conference on Wireless Communications, Networking and Mobile Computing (WiCOM), pp. 1–6 (September 2011)
39. Wagan, A.A., Mughal, B.M., Hasbullah, H.: VANET Security Framework for Trusted Grouping using TPM Hardware. In: Proceedings of Second International Conference on Communication Software and Networks, pp. 309–312 (February 2010)
40. Minhas, U.F., Zhang, J., Tran, T., Cohen, R.: Towards expanded trust management for agents in vehicular ad-hoc networks. International Journal of Computational Intelligence Theory and Practice (IJCITP) 5(1) (2010)
41. Minhas, U.F., Zhang, J., Tran, T., Cohen, R.: Intelligent Agents in mobile vehicular ad-hoc networks: Leveraging trust modeling based on direct experience with incentives for honesty. In: Proceedings of the IEEE/WIC/ACM International Conference on Intelligent Agent Technology, IAT (2010)
42. Gerlach, M.: Trust for Vehicular Applications. In: Proceedings of the International Symposium on Autonomous Decentralized Systems (2007)
43. Golle, P., Greenen, D., Staddon, J.: Detecting and Correcting Malicious Data in VANETs. In: Proceedings of VANET (2004)
44. Guette, G., Heen, O.: A TPM-based architecture for improved security and anonymity in vehicular ad hoc networks. In: Vehicular Networking Conference (VNC), pp. 1–7 (2009)

Improving the Security of Wireless Sensor Networks by Protecting the Sensor Nodes against Side Channel Attacks

Zoya Dyka and Peter Langendörfer

IHP GmbH, Im Technologiepark 25, 15236 Frankfurt (Oder), Germany
{dyka,langendoerfer}@ihp-microelectronics.com

Abstract. The intent of this chapter is to introduce side channel attacks as a significant threat for wireless sensor networks, since in such systems the individual sensor node can be accessed physically and analysed afterwards. Even though such attacks are known for some years, they have never been specifically considered before in the area of WSNs (Wireless Sensor Networks).

Keywords: Anti-Tampering, Attack tools, Blinding, Cryptography, Invasive Attacks, Masking, Non-invasive Attacks, Security, Semi-invasive attacks, Side Channel Attacks, Wireless Sensor Networks, Tampering.

1 Introduction

Wireless sensor networks (WSNs) are becoming an essential building block in application fields such as critical infrastructure protection, industrial automation and telemedicine to name a few areas in which security plays a central role. Potential attackers of those applications will most probably attack the most vulnerable part of the overall systems, i.e. the WSNs. The wireless sensor nodes can be attacked by "standard" network based approaches but also by physical means if they are left unattended in remote sites which is, after all, the preferred application for WSN. We are convinced that protecting the wireless sensor nodes is essential since compromised nodes put the whole system at risk. The challenge with sensor nodes is that they are low cost and running with extremely limited resources but are expected to be operational for long time intervals up to several years. The long life time provides potential attackers with a lot of time to execute an attack and even worse to benefit from a successful attack. Figure 1 illustrates the features of WSNs as well as potential ways to attack it such as network attacks and tampering attacks. During recent years much research effort has been spent on improving the network security of WSN, including research on secure protocols, efficient implementations of crypto operations etc. But, even though physical attacks are easy to execute only little research has been done in the area of protection against side channel attacks. The latter might even be simplified by highly optimizes implementations that do not take into account the observability of physical parameters such as execution time or power consumption, but focus on efficiency only. We are aware of only one project named TAMPRES [1] that aims at developing suitable means to protect WSNs against tampering attacks.

S. Khan and A.-S.K. Pathan (Eds.): *Wireless Networks and Security*, SCT, pp. 303–328.
DOI: 10.1007/978-3-642-36169-2_9

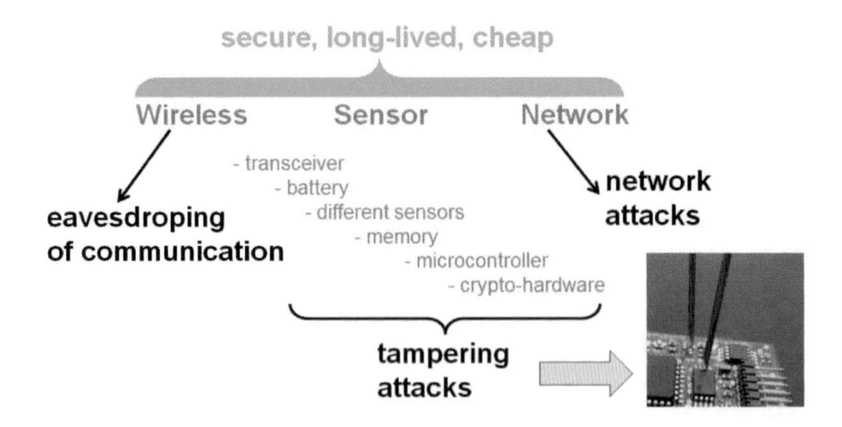

Fig. 1. Attributes of wireless sensor networks and indication of potential ways attacking them

In this chapter we will discuss side channel attacks against implementations of cryptographic functions as well as state of the art countermeasures. We aim not at providing full details on potential attacks but aim for a solid overview of attacks especially considering also attacks normally not taken into account since they are thought to be too expensive. The issue here is that expensive equipment that is needed for some attacks can be rented for a few hundred EUROs per hour, i.e. cost do no longer provide protection. Therefore we are convinced that also this type of attack and appropriate countermeasures need to be considered when designing "secure" sensor nodes.

The rest of this chapter is structured as follows. In the first section we will introduce cryptographic algorithms and explain what type of vulnerabilities in the implementations can be exploited. The following section will shortly introduce the physical parameters that can be used to gather information about intermediate states of the cryptographic functions that help to deduce the key. Then in the third section we will describe non-invasive, semi-invasive and invasive attacks against cryptographic devices using selected examples to illustrate the principles. The fourth section discusses countermeasures against the side channel attacks introduced previously. We will close that section and the chapter by setting up the countermeasures in contrast to the attacks.

2 Weaknesses of Cryptographic Algorithms

Cryptographic algorithms are the key for securing communication, if applied correctly they can ensure confidentiality, authentication, authorization and data integrity. The crypto systems mostly used today are RSA (Rivest, Shamir, Adleman) [2], ECC (Elliptic Curve Cryptosystem) [3], [4] and AES (Advanced Encryption Standard) [5]. The two former are so called asymmetric cipher systems. These systems use two keys per participant. One key needs to be published, whereas the other one needs to be kept secret. Using these public-private key pairs features such as digital signatures and data integrity can be ensured. In addition asymmetric cipher systems can be used to distribute keys for symmetric cipher systems. When it comes to securing wireless systems especially wireless sensor nodes that are battery powered ECC is the favorite system since it requires less computational effort and by far smaller keys than RSA. AES is a symmetric cipher system and is normally used to encrypt and decrypt bulk data, but cannot provide data integrity or digital signatures.

In order to guarantee the above mentioned security features cryptographic algorithms are evaluated by independent experts. The main point of the evaluation is the cryptographic strengths of a newly proposed approach. This feature is normally assessed by mathematical means. Thus, the concrete implementation of a certain algorithm is not taken into account, which is in principle fully okay since the broad variety of realizations clearly hinders thorough evaluation of the implementations themselves. Please note that as part of the selection of the advanced encryption standard (AES) implementation issues have been considered, but the focus was on throughput and energy efficiency, while resistance against side channel attacks was not taken into account [5]. The threat that results from the fact that the implementation is not part of the assessment of the security of cryptographic algorithms is, that potential attackers can exploit specific feature of a certain implementation.

In this chapter we are focusing on RSA and ECC due to the fact that they are used to distribute keys for symmetric approaches such as AES. In more clear words if an RSA or ECC key is lost, the damage is by far more severe than if an AES key is lost. Please note that AES suffers from similar problems when it comes to side channel attack as RSA and ECC and that such attacks and countermeasures are well reported in literature [6].

2.1 RSA

The RSA cryptosystem – as public key cryptosystem – is exploiting the fact that factorization of large integers is an extremely time consuming task. That means a potential attacker cannot gain any knowledge about the private key of a certain person/system even if he knows the public key of that person/system. The RSA key is composed of three integers:

- public exponent $\boldsymbol{e}$;
- private exponent $\boldsymbol{d}$;
- modulo $\boldsymbol{n}$

The integer pair $(\boldsymbol{e}, \boldsymbol{n})$ is used as the public and needs to be published. The private key is built of the pair $(\boldsymbol{d}, \boldsymbol{n})$ and needs to be kept secret. To be more correct only $\boldsymbol{d}$ needs to be protected since $\boldsymbol{n}$ is published anyway. For details concerning constraints for $\boldsymbol{e}$, $\boldsymbol{d}$, and $\boldsymbol{n}$ as well as correct key generation please refer to [7].

In order to exchange data confidentially two persons e.g. Alice and Bob need to know their public keys i.e. Bob needs to know $(\boldsymbol{e_{Alice}}, \boldsymbol{n_{Alice}})$ and Alice needs to know $(\boldsymbol{e_{Bob}}, \boldsymbol{n_{Bob}})$. The following lines describe the operations Bob needs to perform to send an encrypted message to Alice:

1. Represent the message as binary number x
2. Compute the modular exponentiation $y = x^{e_{Alice}} \bmod n_{Alice}$ and then
3. Send the result to Alice

Alice retrieves the encrypted message y and performs the following steps:

1. $y = x^{e_{Alice}} \bmod n_{Alice}$
2. Transform the resulting number into text and display the message

Even though Alice and Bob execute a private and a public key operation respectively, the basic mathematical operation is the same namely a modular exponentiation. This operation can be implemented using the "square-and-multiply" algorithm. When applying this algorithm squaring is done for each bit of the key independent of its value whereas the multiplication is executed only if the respective bit value is '1'. The negative aspect is that by that it reveals the number of '1' in the private key if the operation can be observed by the attacker. This is due to the fact that the difference in the number of calculations can be registered as difference in the computing time and/or consumed energy. This is especially true if the attacker knows the input.

Algorithm 1. RSA modular exponentiation implemented as "square-and-multiply"

Input: x - binary representation of message
(key, n) – exponent and modulo of the RSA-keys

Output : $y=x^{key} \bmod n$

1. $y=1$; $z=x$
2. **for** $i=0$ **to** $(key\_length-1)$
3. **if** $key_i = 1$ **then** $y = y \cdot z \bmod n$ *// this is the problematic part*
4. $z = z^2 \bmod n$
5. **Output** y

In the algorithm displayed above step number 3 is the one that reveals information about the RSA key. Other physical parameters than time and power that can be used to determine the key bits are discussed later in this chapter together with proper illustration how the key can be extracted. In addition we will introduce countermeasures in that section of this chapter.

2.2 ECC

There are two types of elliptic curves (EC) used for standardized cryptographic systems. These are elliptic curves over prime fields $GF(p)$ and curves over binary fields $GF(2^n)$ [8], [7]. The latter are best suited for hardware implementations. Elliptic curve cryptography uses mathematical operations that are defined in finite fields, i.e. Galois fields (*GF*). The cryptographic protocols ECDH [9], ECAES [10], [11] und ECDSA [12] that define key generation, de- and encryption and the generation/verification of digital signature are all based on the EC point multiplication denoted as ***kP*** operation. Coefficient ***k*** is a large integer and ***P=(x,y)*** is a point on the elliptic curve. All three integers ***k***, ***x***, ***y*** are according to the NIST standardization large numbers of more than 200 bit length to provide security until 2030 [8]. As an asymmetric crypto system ECC uses two keys a private and a public one. The private key is an integer here denotes as ***k*** and the public key is composed of the parameters of the selected curve and a specific point ***P*** that is calculated by multiplying the private key with the base point ***G*** of the elliptic curve i.e. ***P=k·G***. All parameters of the curve i.e. its equation, base point ***G***, number of points etc. are not only part of the public key but also part of the private key ***k***.

As for RSA two entities that want to exchange messages using ECC need to know the public key of each other. I.e. Bob needs to know all parameters of EC $\boldsymbol{E_{Alice}}$ as well as the EC point $P_{Alice} = (x_{Alice}, y_{Alice})$ and Alice knows all parameters of EC $\boldsymbol{E_{Bob}}$ and the EC point $P_{Bob} = (x_{Bob}, y_{Bob})$. In order to send an encrypted message to Alice Bob needs to perform the following steps:

1 – Transform the message to be sent in a binary number
2 – Interpret this number as the ***x***-coordinate of a point on EC $\boldsymbol{E_{Alice}}$ and search for the corresponding ***y***-coordinate. To do this Bob needs the equation of EC $\boldsymbol{E_{Alice}}$ e.g. $\boldsymbol{y^2=x^3+x+1}$ that is part of the public key of Alice. The result of this step is a point $M=(x,y)$.
3 –generate a random number ***d*** and calculate the following two EC-points: $R=d{\cdot}G_{Alice}$ und $S=d{\cdot}P_{Alice}+M$. These two points – ***R*** and ***S*** – constitute the message that is sent to Alice.

When Alice receives the two points she needs to perform the following calculations in order to retrieve the plain text:

1 – calculate point ***M***: $M=S\text{-}\boldsymbol{k_{Alice}}{\cdot}R=(x,y)$ using her private key $\boldsymbol{k_{Alice}}$
2 – transform the ***x*** coordinate of the resulting point ***M*** into text and display it

In such a message exchange the sender (Bob) performs ECC public key operations and the receiver computes an ECC private key operation. Both operations are using the same basic mechanism i.e. the EC point multiplication ***kP***. The cryptographic operations using the public key i.e. encryption and signature verification require two *kP* operations, whereas signature generation and decryption require only a single *kP* operation.

The *kP* operation is a complex computation that can be realized using the „double-and-add"-algorithm [13] in which the result is computed as a sequence point doubling ***2P*** operations and point additions ***P+Q***. Each bit in ***k*** triggers a point doubling, whereas the point addition is execute if and only if the current bit is '1'.

Algorithm 2. EC point multiplication implemented as "double-and-add"

Input: $P=(x,y)$ - binary representation of the message
k – private key or generated random number depending on the type of operation

Output: $Q=k \cdot P$

1. $Q=O; R=P$
2. **for** $i=0$ **to** $(k\_length-1)$
3. **if** $k_i = 1$ **then** $Q = Q+R$ *// here is the operation depending on individual bits*

   //of the number k
4. $R = 2P$
5. **Output** Q

The weakness of „double-and-add" is pretty similar to the one of „square-and-multiply". The number of calculations to be performed per key bit depends on the value of the individual key bits. In both algorithms both operations are executed if and only if the bit value is '1'. That means the time and power consumed to compute "double-and-add" for a zero in the key is by far less than for a one in the key. These parameters can be observed and analyzed by an attacker revealing at least the number of ones and zeros in the key and by that extremely speeding up the determination of the key.

Another issue is the quality of the random numbers used. For ECC it has an even more important role than for other crypto graphic approaches since each message exchange requires fresh random numbers. Secure random number generators are out of the scope of this chapter but are discussed in [14], [15], [16]. Additional issues of ECC implementations are analyzed in [17].

3 Physical Parameters and Their Influence on Key Extraction

The cryptographic strengths of a cipher algorithm may depend according to the definition of Kerckhoff [13] only on the used key that is kept secret. This means a potential attacker may know the algorithm itself, the plain text, the encrypted text and even the length of the key. In such a situation the attacker can test different numbers in order to reveal the key. Such an attempt is called brute force attack and the attacker needs to test 2^n number in the worst case to get a key of length n. The average number of attempts is 2^{n-1}.

The situation changes dramatically, if the attacker gets physical access to device running the cipher algorithm. In such a case the attacker can record not only the input and output values of the completed cryptographic operation but also intermediate values of the cryptographic operation that provide additional information and by that simplify the determination of the key. Each physically measureable parameter represents such information: execution time of individual steps of the operation, average energy consumption during operation, distribution of the energy consumption during the operation, temperature, electromagnetic emission etc. Figure 2 illustrates the knowledge of the attacker as well as the physical parameters that can be exploited to reveal the key.

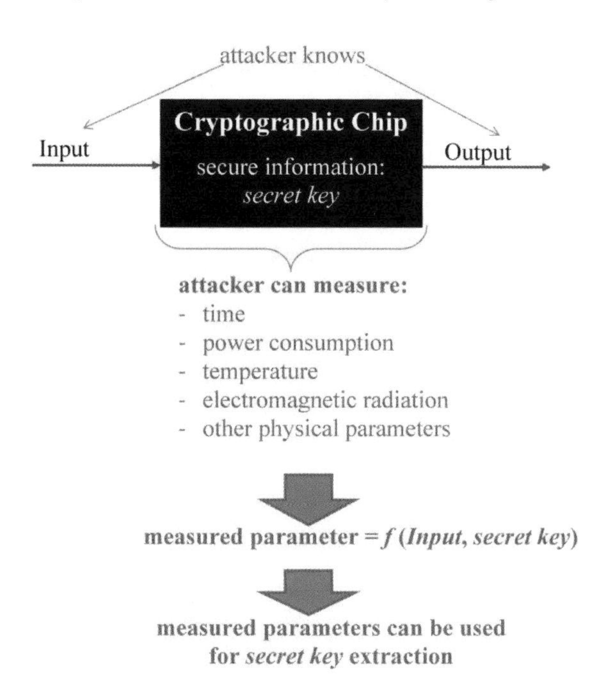

Fig. 2. Implementation of a crypto-system as „black box“ and data known by the attacker: input data, output data, measurable physical parameters and their relation to the key

Any implementation of a crypto-system can be interpreted as a „black box“ that manipulates input data to get corresponding output data according to the implemented algorithm *g* and a secret key: *Output = g(Input, secrete key).* The attacker knows input, output and *g* his aim is to reveal the secrete key. Achieving this by just analyzing pairs of input and output data is a very time intensive task and the time required grows exponentially with the length of the secrete key. The time to reveal a key is often used to determine the security of a cipher algorithm for a given key length. [18], [19] state that a computational effort of 10^{24} MIPS years is sufficient to protect a key till 2030, but please note that this assessment is made under the assumption that an attacker knows only input and output data, i.e. nothing about intermediate values, the consumed energy, execution time or any other additional data. This means that the attacker has no physical access to the implementation.

Please note that the whole situation changes dramatically if the attacker gets physical access to the device under attack. In such a situation the assumption that the attacker knows only input and output data as well as the cipher algorithm does no longer hold true. The physical access allows to gather additional information about the computation e.g. intermediate results i.e. values for which the en-/decryption was not yet fully completed or execution times for example. This information is then exploited to reveal the secret key. Due to the fact that such data is available the attacker does no longer need to run a brute force attack.

The fact that intermediate steps of a cipher algorithm can be observed is due to the properties of the CMOS (Complementary Metal Oxid Semiconductor Technology) logic that is used to build cryptographic devices. Each manufacturer provides a set of gates that realizes Boolean functions such as XOR, AND or OR. In addition to its logical inputs and outputs each gate has a connection to the Vdd (power supply). As long as the state of the input values does not change the value of the output also does not change the gate is inactive and the power consumption is negligible. But if the input value(s) change the gate becomes active and its transistors (or at least a part of them) switches. This causes current to flow through the gate. This current leads to a set of observable physical side effects: large power consumption, changes in the electromagnetic field near by the gate, optical emission etc.

The vast majority of today's cryptographic devices uses a synchronous design. This means all gates are retrieving new input values and are switching at the same time triggered by the clock signal. The actual number of switching gates depends on the input values and varies from clock signal to clock signal. In more clear words the number of switching gates is directly connected to the secrete key and to the input data that can be controlled by the attacker. Thus, by knowing the algorithm and the input data the attacker can use the measured physical side effects to reveal the secret key. In order to reveal the secret key the attacker can additionally try to influence the measurable physical parameters. This is due to the fact that the CMOS logic reacts on changes in its environment, i.e. it works reliable and stable only if the predefined/specified conditions are kept. Deviations of temperature, clock frequency, electromagnetic field, light etc. may alter the states of individual transistors of the device under attack. This leads to the fact that the calculated output value is no longer correct. Analyzing pairs of such faulty value may provide additional information to the attacker that simplify to reveal the key.

Attacks can be classified into active and passive attacks depending on whether or not the device under attack is manipulated. The following paragraphs provide a short introduction of the different types of attacks.

Passive Attacks

- Timing attack

This attack exploits the variations in computational time for secret key operations. The time elapsed between providing a certain input and getting back the corresponding output is measured. This time depends on the input forming a time distribution. If it depends also on the key, it may become feasible to reveal the secret key.

Successful attacks are report for:

- RSA, DSA, Diffie-Hellman [20],
- ECC [21]
- AES [22]

- Power Analysis attacks [23], [24]

While the chip is operating the current power consumption is measured. This can be done for the whole chip but also for selected part of it.

They are two kinds of Power Analysis (PA):

- SPA (Simple Power Analysis): only very few measurements are done and primarily visual inspection is used to identify relevant power fluctuations.
- DPA (Differential Power Analysis): a lot of measurements is executed and afterwards statistical analysis and error correction techniques are applied to extract information correlated to secret keys.

Successful attacks are report for:

- RSA [24]
- ECC [25]
- AES [26]

- Electromagnetic Analysis (EMA)

In this attack the electromagnetic emission of a chip is measured and recorded. This can be done for the complete chip but also for selected parts of it. The latter may provide better results for later analysis.

Similar to PA two kinds of EMA can be differentiated:

- SEMA (Simple Electro Magnetic Analysis) only few measurements are needed to reveal the key e.g. by optical inspection of the measured traces.
- DEMA (Differential Electro Magnetic Analysis) many traces are recorded and analyzed in a similar ways as for DPA.

Successful attacks are report for:

- RSA [27], [28]
- ECC [29], [30]
- AES [31]

- Optical emission analysis

Photon emission is related to CMOS gate switching activities. This effect can be recorded and visualized with photosensitive charge-coupled device camera (CCD). The spectrum that is most affected is from 500 nm to above 1200 nm.

Successful attacks are report for:

- Memory cells [32]
- AES [33]

- Optical or infrared imaging

Optical imaging of a chip can reveal significant information about the structure of the chip. E.g. it immediately shows where memory blocks are located since they have a regular structure in contrast to logic blocks. An additional source of information are pictures taken with microscope that provide data with respect to the on chip wiring on different metal layers. This type of attack can be executed from the front or backside of the chip. The specific issue with this attack is that the backside of the chip is used to extract information. This simplifies especially optical attacks against the structure since the metal layer do not block the view on the structure as it is the case from the front side [34]. This type of attack can also be used to extract data from memory cells.

- Data remanence analysis

Residue charge of transistors can be measured. This attack exploits the fact that the charge of transistors building the memory does not vanish immediately when the voltage is switched off in case of volatile memory and that it does not fully vanish in case of non-volatile memory. The really challenging issue here is that even erasing does not fully hide the original content of the memory cell.

Successful attacks are report for:

- SRAM [35], [36]
- erased EPROM, EEPROM and Flash memory [37]

Active Attacks

- Laser scanning

In order to improve the success rate of attacks based on laser scanning techniques the effect that photons can ionize active areas inside the chip (photovoltaic effect) can be exploited. The photon injection increases the current noticeable for closed transistor channels, but for open channels this effect is negligible.

They are two laser scanning techniques described in the literature:

- optical beam induced current (OBIC) [38]: photocurrents are used directly to produce the image.
- light-induced voltage alteration (LIVA) [39]: images are produced by monitoring the voltage changes of the constant current power supply as the optical beam is scanned across the IC surface.

Successful attacks are report for extraction of information from SRAM [40], [34]

- Glitch and Fault injections

The main idea of glitch and fault attacks is to induce a fault in the chip in order to bring it into an undefined state that reveals additional information an attacker can exploit to reveal the key or to get access to the stored data.

- Clock glitch

By increasing the clock rate above the specified working frequency the chip might no longer work correctly. As a result it might be that output registers are not fully updated by the last operation since it did not complete and by that the registers contain intermediate results that are normally not accessible. Thus the attacker can gain access to these data.

- Power glitch [41]

Alteration of the voltage changes the switching times of the transistors, this means the signal delay of the gates changes which can lead to similar effects as those described for clock glitches.

- Electromagnetic impulse

Strong changes in the electromagnetic field in near vicinity of a chip can lead to changes in the state of gates: it can induce electrical current or change the strength of current. This can influence the states of transistors (open or closed) leading to a faulty behavior of the chip, with the same consequences as discussed above.

- Ultra-Violet (UV) flashes

The photon effect changes the intensity of the current in a certain number of transistors across the whole chip what will lead to similar effects as those discussed already.

- Laser fault

Laser fault attacks are taking advantage of the precise positioning of the laser beam in comparison to UV flashes e.g. if each memory cell shall be altered individually. The photon effect induced a local current in the effected transistors that leads to state changes of the flip flops [42].

- Thermal fault

Also changing the temperature of the chip may cause a faulty state and by that help to extract key material [43].

Figure 3 displays a low cost device that can be built for less than 2500 Euros (600 Euros equipment/1500 Euros manufacturing cost) and used to run different types of attacks. The board displayed was developed in the TAMPRES project for supporting glitch and fault attacks [44].

The board consists of:

- A socket for a daughter board on which the device under attack (DUA) needs to soldered. This architecture allows to attack several devices independent of their form factor using the same and may be optimized main board.
- A set of different power supplies that can be controlled digitally or manually
- A set of different shunts for measuring the power consumption
- Different clocks.
- Power supply for laser.
- An FPGA to control the whole board including the communication with the device under attack.
- In addition the design was intentionally done pretty spacious in order to provide easy manual access to all parts.

This equipment enables the attacker to:

- Control and record the communication with the DuA.
- Generate power and clock glitches.
- Measure power consumption i.e. record traces for SPA and DPA.
- Shoot laser impulses on the DuA, that needs to be depackaged before hand.

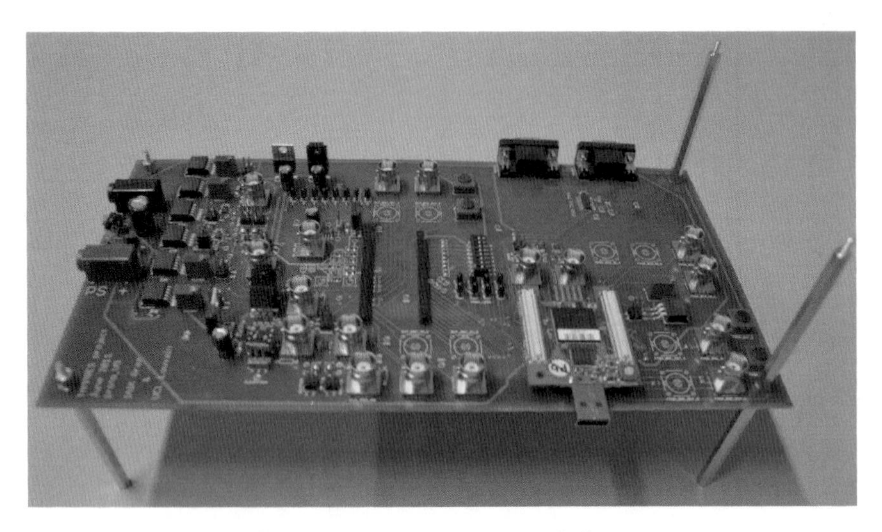

Fig. 3. Board for running glitch and fault attacks

For analyzing the structure of the ASIC using optical imaging or laser based fault injection the chips needs to be decapsulated at least some parts. Depackaging allows the attacker to run more and more sophisticated attacks and can be done mechanically or by using acids.

The equipment for the decapsulation is cheap. The process using nitric and/or sulphuric acid can be done in any chemical laboratory and requires not more than a few days and a few devices. At IHP some chips have been depackaged using these acids, the results i.e. the different levels of destruction of the ASICs are shown in Figure 4 - Figure 6.

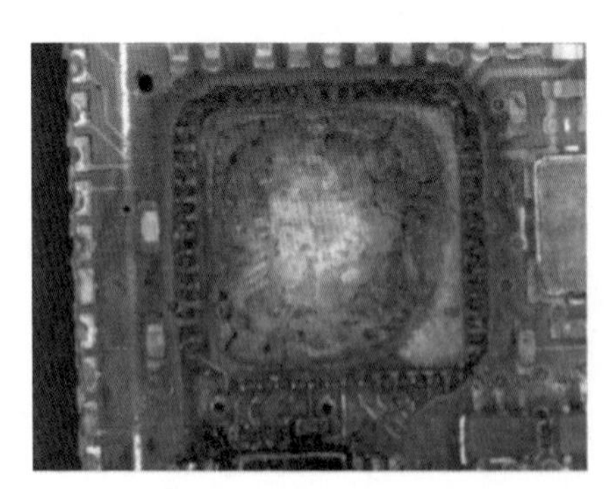

Fig. 4. Two different chips fully destroyed after initial attempt to depackage them, the right chip called TSN was designed and manufactured at IHP and used for further experiments shown in the following figures

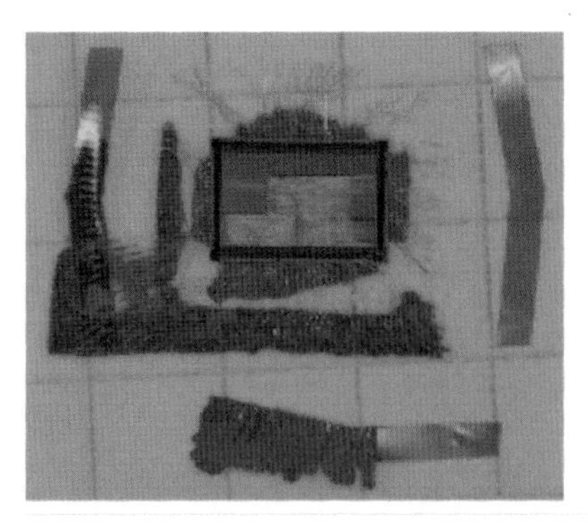
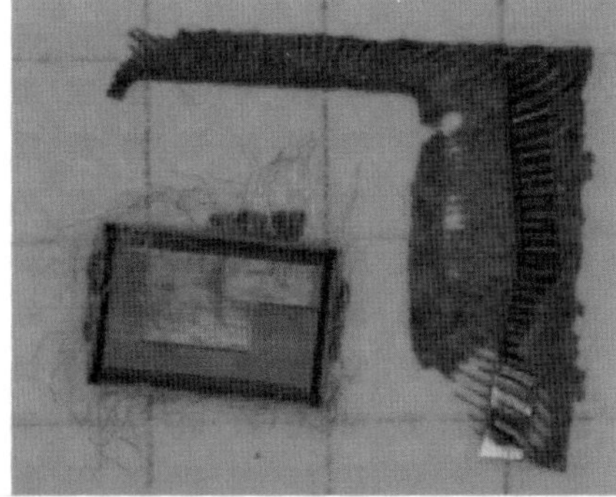

Fig. 5. Somewhat successful attempt to depackage the TSN, at least the ASIC is undamaged but bond wires are destroyed

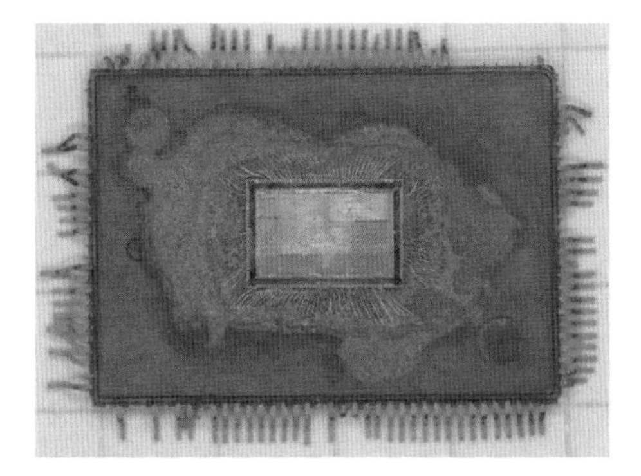

Fig. 6. Successfully depackaged TSN, ASIC and bond wires are undamaged

Some attacks e.g. EMA or clock glitches do not really require the ASIC to be decapsulated. There are some attacks that can be run on a still packaged device but also after depackaging. Power analysis is such an attack. In case the chip is not decapsulated the power consumption of the whole ASIC is measured at its pins. If the ASIC is still working properly after decapsulation, the power measurements can be done directly at the upper metal layers using a micro probing station. To run this type of attack it is necessary to remove also the highest layer of the chip, i.e. the passivation layer, at least in parts. The degree of alterations on the device under attack (DuA) is the basis for one of the most used classifications of attacks. The following three classes can be distinguished:

- *Non-invasive attacks*

Attacks that can be executed without any alteration of the chip are denoted as non-invasive attacks. When executing the attack all necessary values are measured and recorded using an intact chip. The special danger of these attacks is that they can be executed without leading to any type of traces in more clear words they can go undetected.

- *Invasive* attacks

This type of attacks requires the most complex alterations of the DuA. In these cases the chips needs to be decapsulated and at least the passivation is removed.

- *Semi-invasive* attacks

In this type of attacks the chip is decapsulated e.g. in order to improve the quality of some measurements, but they can be done without any additional changes on the DuA, i.e. even the passivation does not need to be removed.

The following section introduces some examples of successful attacks.

4 Examples of Attacks

In this section we will introduce some successful attacks that show how the effect of the cryptographic operations on measurable values can be visualized and used to extract the secret key material.

One of the most reported attacks is the power analysis. It is based on the analysis of the current power consumption, while the chip is operating. Usually these measurements are performed on the pins of the chip. Decapsulation of the chips is not necessary in this case. Figure 7 shows the typical electrical circuit for this type of measurements. Figure 8 shows the measurement set-up in IHP.

Figure 9 and Figure 10 show the measurements result of an RSA-chip and of an ECC-chip, respectively. In both graphs the current power consumption of the cryptographic chip under attack is represented as a function of time while an operation with the private key – decryption or signature generation – is done. Only part of the full time diagram is shown on each picture.

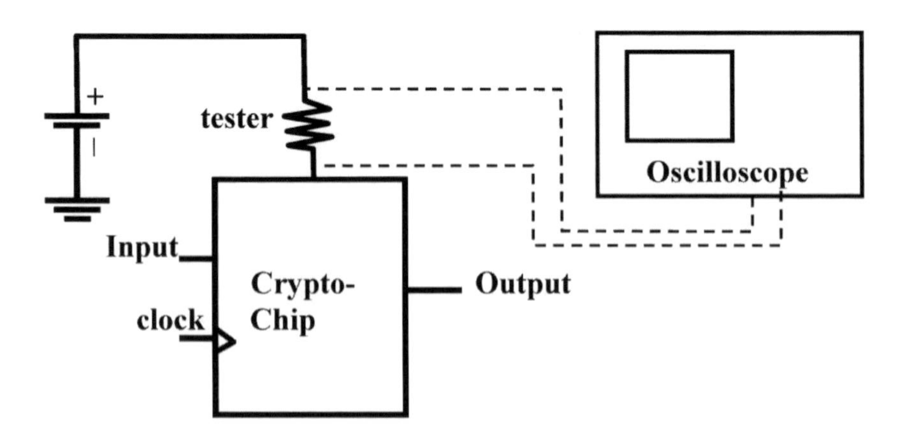

Fig. 7. Typical measurements circuit for running a power analysis attack

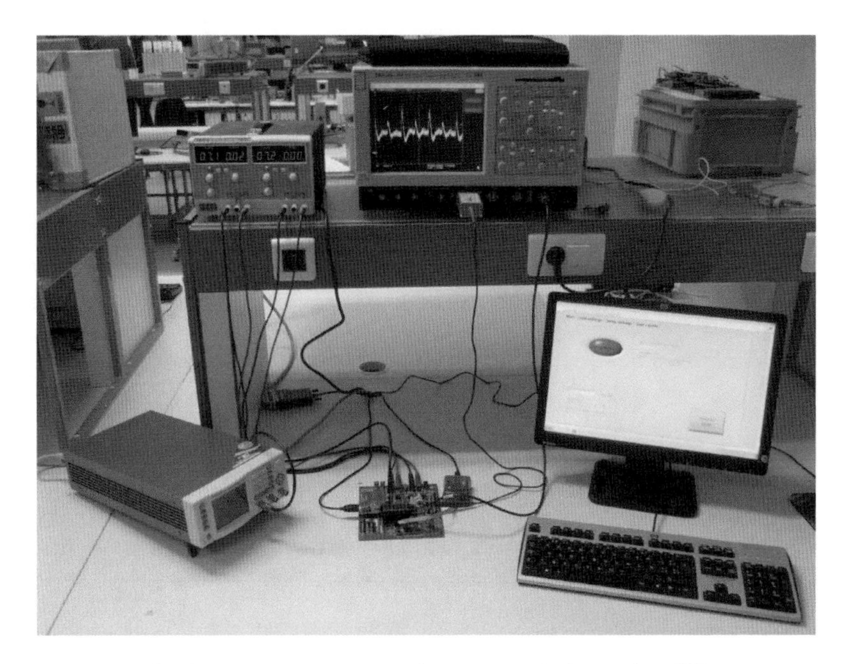

Fig. 8. Measurement set-up of a power analysis attack at IHP

Under the assumption that the RSA private key operation is implemented using the "square-and-multiply" algorithm for the modular exponentiation (see section 1.1) the attacker can easily obtain the part of the private exponent as it is shown on picture 6 with following considerations:

- Only one of the operations – either multiplication (see step 3 of the algorithm) or squaring (see step 4 of the algorithm) can be performed in a single clock.
- The multiplication will be performed less often than squaring.

Thus, the attacker can assume, that each large impulse in figure 6 represents the multiplication, i.e. the step 3 of the "square-and-multiply" algorithm. This impulse together with the following impulse caused by the squaring in step 4 of the algorithm corresponds to a "one" bit of the private key. The case with squaring impulse only corresponds to a "zero" bit of the private key. The attacker can always verify the correctness of the extracted key because he knows the input and output.

The similar considerations can be applied for the key extraction of the ECC-chip.

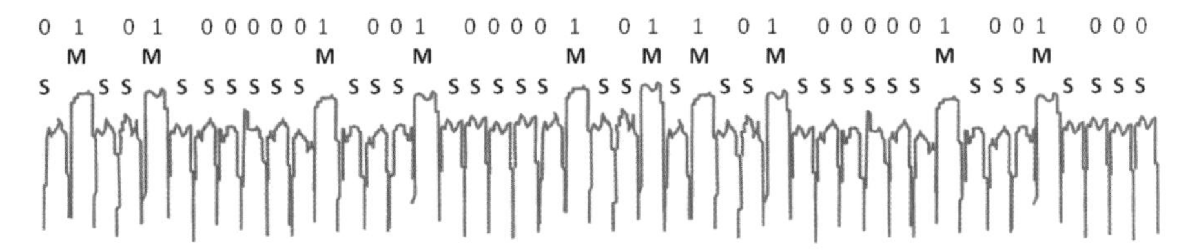

Fig. 9. The principle of SPA-attacks against RSA implementations with an interpretation of the power trace according to algorithm 1. An SPA-example trace similar to the graph shown here is published in [24].

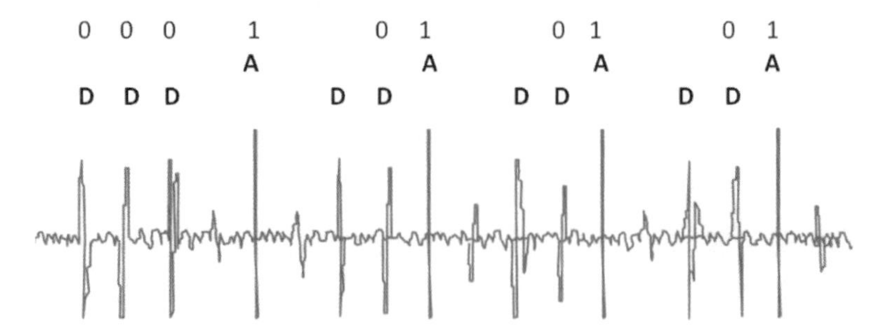

Fig. 10. Example of a power trace of an SPA-attack against ECC (An SPA-example trace similar to the graph shown here is published in [25])

After decapsulation of the chip the current power consumption can be also measured directly on the chip wires using special equipment. This invasive technique is called "microprobing".

Figure 11 shows the microprobing station in IHP. It contains a set of different active and passive probes and manipulators, a high resolution microscope, device test socket and precision *x-y* stepper table.

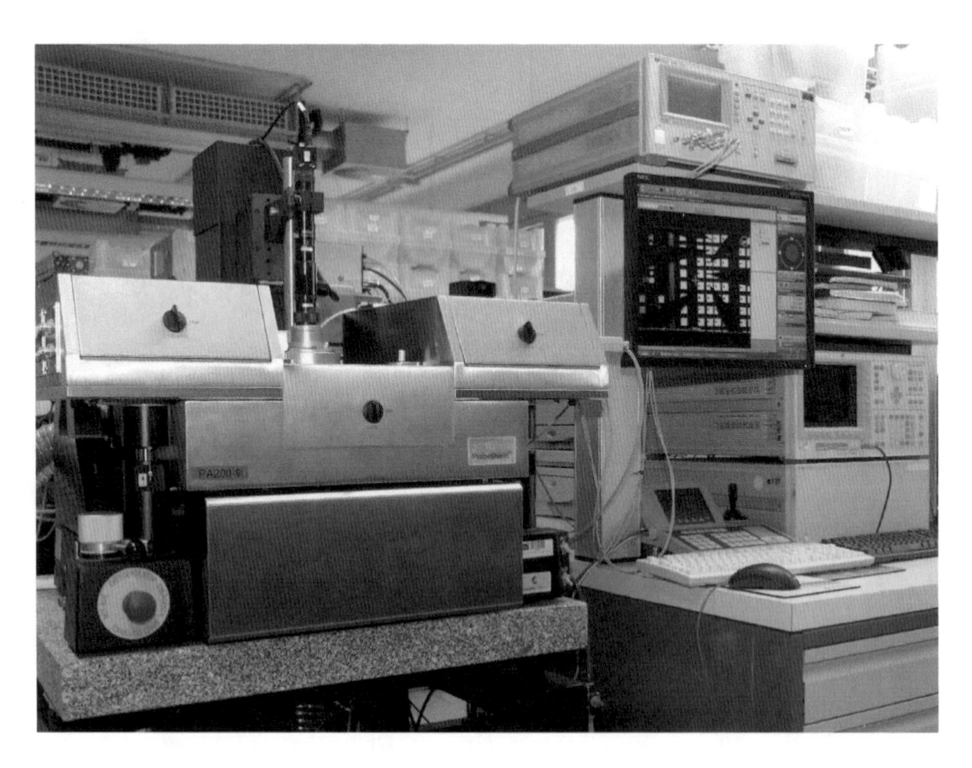

Fig. 11. Microprobing Station at IHP

In order to get an electrical contact to the measurement points the surface of the chip need to be prepared properly i.e. the passivation need to be removed. This can be done in two different ways:

- Chemical etching, in this case the passivation will be removed in a pretty large area.
- Using a laser-cutter allows to remove the passivation at selected points only.

The latter is by far more suited for microprobing since it provides a better contact to the measurement point as shown in Figure 12. Figure 13 shows the measurement process using a microprobing station at IHP.

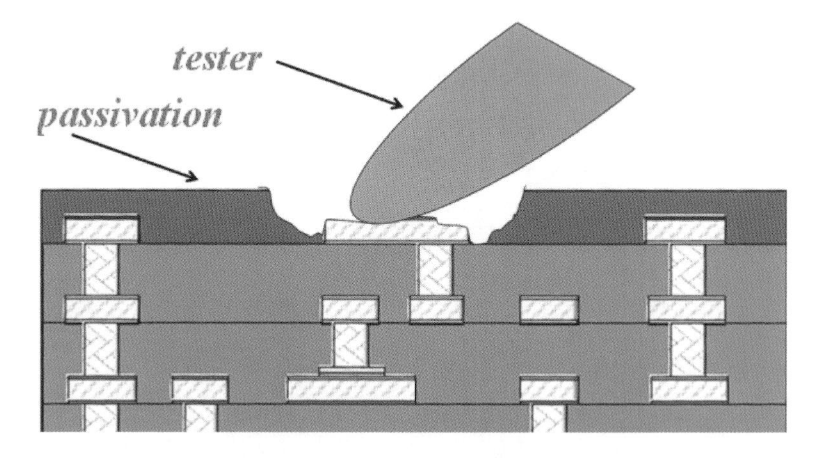

Fig. 12. Microprobing: in order to do measurements directly at the metal layer the passivation need to be removed. Using a laser cutter the area in which the passivation is removed can be kept small, this helps to create a stable electrical contact since the needle cannot move around.

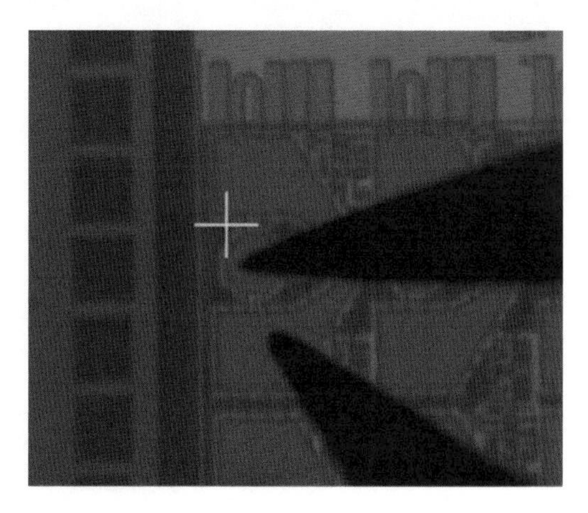

Fig. 13. Microprobing: two needle probers with contact to the chip for measurements

The measurements can be done also on other metal layers if additional special equipment is ready to use. If other layers than the two top metal layers shall be attacked a Focused Ion Beam station (FIB) is needed. It consists of a vacuum chamber with an ion source that normally accelerates Gallium ions and bundles them to a beam. This ion beam can be used to modify on chip wiring i.e. those wires can be cut but new wires can be implanted as well. The current power consumption of the cryptographic chip can be measured after the changing its structure with a FIB. These measurement results can provide the attacker additional helpful information for the key extraction.

Modern FIB stations come with an additional electron source in order to provide visualization according to the scanning electron microscope (SEM) principle. The lateral resolution of the ion beam is about 4nm and the one of the SEM about 2 nm. Thus, the granularity of resolution is by far smaller than the currently used CMOS technologies.

Figure 14 depicts the IHP FIB station, and Figure 15 shows changes in the chip structure realized using a FIB.

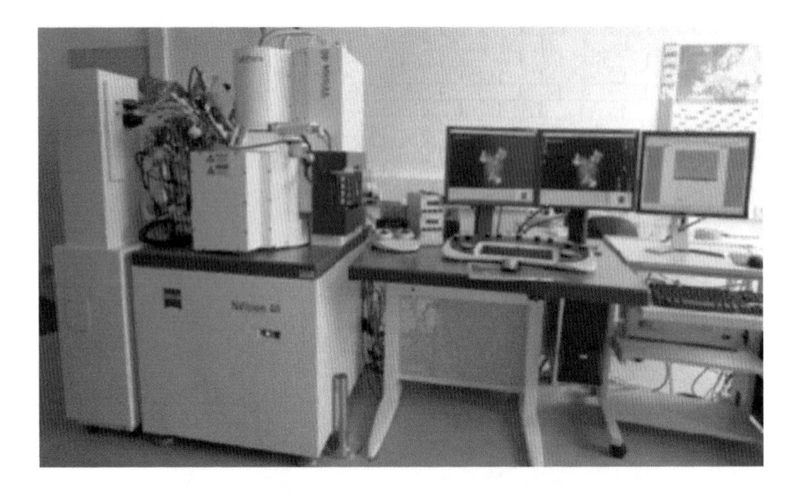

Fig. 14. Focused Ion Beam Station at IHP

The typical examples of semi-invasive attacks are different imaging techniques for passively obtaining information (microscope and infrared imaging, optical emission analysis) or influence on the chip with light or temperature. These attacks are possible only after decapsulation of the chip since they require access to its surface. The electrical contact to internal lines of the chip is not required and the structure of the chip is not destroyed or modified, what differentiates semi-invasive from invasive attacks.

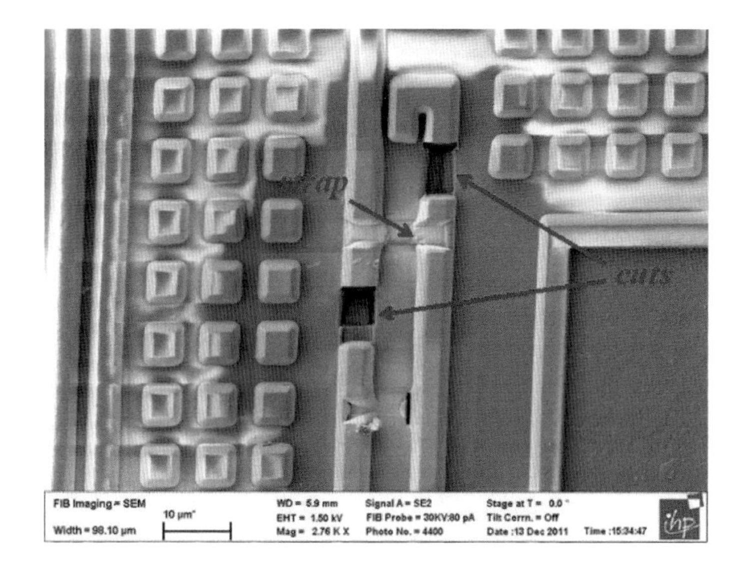

Fig. 15. Picture taken with the IHP FIB-station showing modification of the structure of chip manufactured at IHP

Semi-invasive attacks can be applied for modification of SRAM and EEPROM content, or for obtaining information about any individual transistor as well for changing states of transistors [34].

A practical low-cost attack is described in [42]. The attack exploits the sensitivity of transistors to light: illuminating a transistor causes it to conduct. It can cause a security fault in chip that can result in access to the memory cells that store the private key. For this attack a cheap photoflash lamp was mounted on top of a manual probing station that contains a microscope with the maximum magnification of 1500x. Authors [42] experimented with the 8-bit microcontroller PIC16F84 of Microchip Technology Inc. [45]. According to [42] it was possible to change any individual bit of the SRAM array. These results were published 10 years ago. Semi-invasive attacks are currently becoming a very serious threat to hardware security. For example, [46] presented a successful semi-invasive two-fault laser attack on a protected RSA implementation running on a 32-bit ARM Cortex M3 core [47].

In next section we give a short overview of principles of countermeasures.

5 Principles of Countermeasures

In this section we focus on basic principles of countermeasures since an exhausting discussion of specific countermeasures will never be complete and new types of attacks will require new individually designed countermeasures. But we will provide an example of a successful countermeasure pattern i.e. the use dummy operations.

The following enumeration provides a solid overview on principles of countermeasures as well as individual means:

- **Prevent Access to Measurement and Manipulation Relevant Parts of the Chip**
  - Shielding and covering of the chip or even the PCB (printed circuit board) to defeat non-invasive attacks and to increase the complexity of the chip preparation for invasive and semi-invasive attacks
    - Different types of housing
      - Weather and/or intrusion resistant casing
      - Integrated Faradey-cage to prevent EMA
      - etc.
    - resin-, foam covering including PINs
    - passive shielding of the chip structure e.g. use of additional metal layers to prevent optical imaging and EMA
  - technological approaches to defeat invasive and semi-invasive attacks
    - etching resistant passivation layer
    - manufacturing in smaller technologies
    - increase number of layers per chip
    - hide memory in lower layers of the chip
    - protect or destroy test structures and/or scan chains

- **Detection of - and Reaction on Manipulations**
  - Anti-Tampering means
    - Sensors e.g. light, power, voltage, frequency, temperature to detect attempts to open the chip
    - Active shielding of the chip structure: additional metal layers or wires that cover the ASIC and that are connected to a certain voltage or signal, alterations in the voltage or the signal indicate a manipulation
    - Error detection and correction: redundant implementation of blocks with identical functionality. Manipulations are detected if those blocks provide different results
  - Typical reactions after detecting manipulations
    - Overwriting the memory (best with random numbers)
    - Switching off the device
    - Chip self destruction
- **Reduce the Information Provided by the Measurement Values**
  - Security by obscurity
    - No information about the used crypto algorithm
    - No standardized solution (own elliptic curve)
    - Unmarking, remarking and repackaging of the chip

- Flawless Implementation
  - algorithmic:
    - flawless re-design of crypto implementations (e.g. use „dummy operations“)
    - PUFs (Physically Unclonable Functions) to store the private key
    - etc.
  - hardware:
    - integration of „dummy“ gates
    - gates with power consumption independent of the input values
    - asynchronous design
    - dual-rail logic
- Information concealment
  - Adapt the signal to noise ratio
  - Execute program code in random order
  - Use modules with random power consumption
  - Randomization, e.g. blinding, doubling, masking

In the following paragraph we will use the "double-and-add" and the "square-and-multiply" algorithms to illustrate an algorithmic approach to increase the effort for key extraction.

In section 1 of this chapter we already explained the weaknesses of "double-and-add" algorithm used for implementing the EC point multiplication and the one of the "square-and-multiply" algorithm used for RSA exponentiation. The major issue of both algorithms is the different number of operations executed depending on the bit values of the key. I.e. if the current key bit is a '0' only one of the two operations is performed, point doubling in case of EC point multiplication and squaring in case of RSA. The time difference or the difference in power consumption in comparison to the case when the current key bit value is a ‚1‘ i.e. when two operations are executed is measurable. Figure 16 shows a successful example of an SEMA against an ASIC, providing an ECC implementation. The private key can be extracted by simple optical inspection of the power consumption.

In order to avoid key extraction by analyzing the execution time or power consumption the algorithms can be modified. The modified algorithms are denoted as "double-and-add-always" and "square-and-multiply-always" respectively. The main idea is to introduce additional operations that are not needed to compute the correct result but that are used only to ensure that the calculation time and the energy consumption are independent of the key bits. The drawback of such means is the increased resource consumption, chip area and/or energy consumption.

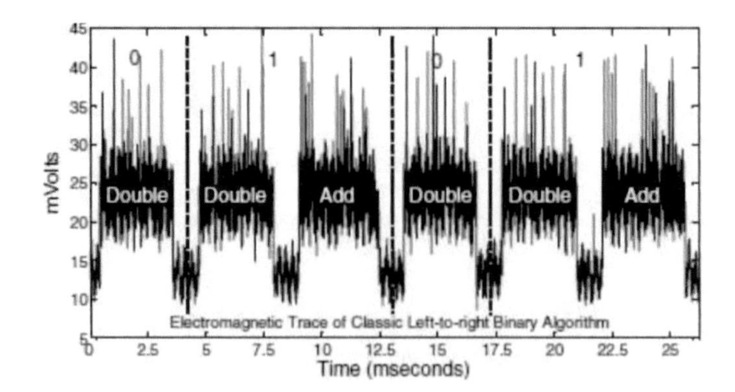

Fig. 16. Successful SEMA-Attack against an ECC implementation: current power consumption as a function of time during the execution of the "double-and-add" algorithm, (source [29])

Figure 17 shows the measurement results for the "double-and-add-always" algorithm. In this implementation the point doubling and the point addition are executed independent of the value of the individual key bits. So, the key extraction can be avoided, see Figure 17 now there is a point addition executed after each point doubling operation.

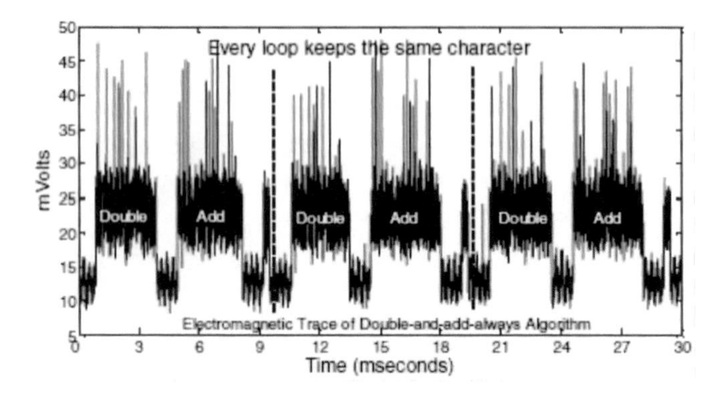

Fig. 17. Current power consumption as a function of time during the execution of the "double-and-add" algorithm, (source [29]) SEMA-resistant Implementation, i.e. "double-and-add-always"

The following table shows attacks, exploited effects countermeasures, and provides an assessment whether or not a certain countermeasure helps against a certain attack. No value in the table means no effect of the countermeasure on the attack, 'y' means helpful means, +- means the attack becomes more difficult if such a means is deployed but is still feasible. 'y[1]' means that the counter measure is successful if the attack is detected. We are aware of the fact that this is extremely difficult for non-invasive attacks.

Table 1. Attacks and countermeasures

Attacks		non-invasive					decapsulated chip (semi invasive, invasive)									
Name and short description of countermeasures		Timing attack	Power Analysis (PA)	Electro Magnetic Analysis	Analysis of test structures (Scan chains)	Glitch	Electro Magnetic Analysis	Imaging	Optical emission Analysis (backside)	Data Remanence Analysis	Fault Injection (front side)	Fault Injection (back side)	PA with Microprobing	Alteration of the chip structure	Reverse-Engineering	Analysis of test structures
Smaller technology	Smaller voltage → side-channel leakage less informative		±	±			±	±	±	±	±	±	±	±	±	
Additional metal layers	Access to transistors from the front side is more difficult			±			±	±		±	±		±	±	±	
Destroying test structures (HW)	Avoiding Access to measurement point used for test purposes												±			y
Dummy Gate	side- channel leakage per clock less informative since some not really use gates are switching		y	y			±	±	±		±	±	±	±	±	±
Internal Clock	Aims at preventing manipulation (acceleration and deacceleration) of the clock		±	±		y	±		±				±			
Internal voltage control	Aims at preventing manipulation of the voltage		±	±		y	±		±							
Asynchronous design	Analysis of measurement results is more difficult, the alignment due to the clock is missing		±	±		±	±						±			
Signal-Noise Ratio	side-channel leakage less informative, peaks in consumption not that easy detectable		y	y			±						±			
Dummy Operations	Create constant side- channel leakage per processed bit of the key i.e. side-channel leakage≠f(key) →chip will consist of more gates	y	y	y			±		±				±			
Blinding, Duplication, Masking	side-channel leakage is no longer a direct function of the inputs; it does not necessarily mean that the side-channel leakage is also independent of the key	y	±	±			±		±				±			
Coating with integrated destruction of the correct function chip	Aims at preventing measurements at the correct working chip		y	y	y	y	y		y		y	y	y	±	±	±
Reaction on detected tampering e.g. invalidating sensitive parts of the memory	No longer correct data e.g. secret keys stored in the memory	y[1]	y[1]	y[1]	y[1]	y	y		y	y	y	y	y			y
Reaction on detected tampering: switching off the device	No functionality of the chip if an attack was detected, i.e. access to secret data is prohibited	y	y	y	y	y	y		y		y	y	y			y
	Exploited physical effect for attack	Execution time of the algorithm as a function of time (time =f(inputs))	Differences in the power consumption between active and inactive gates	EM-radiation that is emitted when transistors are switching	Intermediate measurement points implemented for test purposes and/or scan chains	Dependency of transistors on environmental conditions	EM-radiation that is emitted when transistors are switching	Structure of a non-working chip cannot be analyzed	Luminescence that comes with transistors switching t	Remaining charge of the transistors	Dependency of transistors on environmental conditions (at a working device)		Measurement at the on chip wiring are source of different intermediate values	The structure of the chip is altered using a FIB e.g. implanting new contacts	The visible structure of the chip: transistors, on chip wiring etc. will be recorded e.g. by photos	Intermediate measurement points implemented for test purposes and/or scan chains

6 Conclusion

In this chapter we introduced side channel attacks as a significant threat for wireless sensor networks, since in those systems the individual sensor node can be accessed physically and analysed afterwards. Even though such attacks are known for years, they have never been considered before in the area of WSNs. This might be partly due to the fact that other security problems have gained more attention since they have been considered more likely than the more complex and more expensive side channel attacks. However the cost of side channel attacks can no longer be considered as a protection means since high end equipment can be rented for a few hundred EURO per hour. Thus we tried to create more awareness for side channel attacks and potential countermeasures in this chapter.

Acknowledgement. The work presented here was partially funded by the EU as part of the TAMPRES project (258754).

References

[1] Project: tamper resistant sensor nodes, http://www.tampres.eu/
[2] Rivest, R.L., Shamir, A., Adelman, L.M.: A method for obtaining digital signatures and public key cryptosystems. Technical Report MIT/LCS/TM-82, Laboratory for Computer Science, MIT, Cambridge (1977)
[3] Koblitz, N.: Elliptic curve cryptosystems. Mathematics of Computation 48(177), 203–209 (1987)
[4] Miller, V.S.: Use of Elliptic Curves in Cryptography. In: Williams, H.C. (ed.) CRYPTO 1985. LNCS, vol. 218, pp. 417–426. Springer, Heidelberg (1986)
[5] NIST Computer Security Division: Advanced Encryption Standard, FIPS 197 (2001), http://csrc.nist.gov/publications/fips/fips197/fips-197.pdf
[6] Zhou, Y., Feng, D.: Side-Channel Attacks: Ten Years After Its Publication and the Impacts on Cryptographic Module Security Testing, Cryptology ePrint Archive. Report 2005/388, http://eprint.iacr.org/2005/388.pdf
[7] NIST Computer Security Division: Digital Signature Standard (DSS), FIPS 186-3 (2001), http://csrc.nist.gov/publications/fips/fips186-3/fips_186-3.pdf
[8] Second Standards for Efficient Cryptography Group (SECG), SEC 2: Recommended Elliptic Curve Domain Parameters (2010), http://www.secg.org/download/aid-784/sec2-v2.pdf
[9] Barker, E., Johnson, D., Smid, M.: Nist special publication 800-56a, recommendation for pair-wise key establishment schemes using discrete logarithm cryptography (2007), http://csrc.nist.gov/publications/nistpubs/800-56A/SP800-56A_Revision1_Mar08-2007.pdf (revised)
[10] Kaliski, B.: Elliptic Curve Cryptography, RSA Labor (1999), http://www.scribd.com/doc/59254287/48/ECAES-Encryption
[11] ANSI X9.63: Public Key Cryptography for the Financial Services Industry: Elliptic Curve Key Agreement and Key Transport Schemes (1998), ftp://ftp.iks-jena.de/mitarb/lutz/standards/ansi/X9/x963-7-5-98.pdf

[12] Certicom Research, Standards for efficient cryptography group (secg) Sec 1: Elliptic curve cryptography (2009), http://www.secg.org/download/aid-780/sec1-v2.pdf
[13] Hankerson, D., Menezes, A., Vanstone, S.: Guide to Elliptic Curve Cryptography. Springer-Verlag New York, Inc. (2004)
[14] Drutarovskı, M., Fischer, V.: True Random Number Generator Embedded in Altera ACEX Devices. In: Proceedings of DCIS 2002, pp. 587–592 (2002)
[15] Fischer, V., Drutarovskı, M.: True Random Number Generator Embedded in Reconfigurable Hardware. In: Kaliski Jr., B.S., Koç, Ç.K., Paar, C. (eds.) CHES 2002. LNCS, vol. 2523, pp. 415–430. Springer, Heidelberg (2003)
[16] Schellekens, D., Preneel, B., Verbauwhede, I.: FPGA Vendor Agnostic True Random Number Generator. In: Field Programmable Logic and Applications (FPL 2006), pp. 1–6 (2006)
[17] Fan, J., Guo, X., Mulder, E.D., Schaumont, P., Preneel, B., Verbauwhede, I.: State-of-the-art of Secure ECC Implementations: A Survey on Known Side-channel Attacks and Countermeasures. In: Proceedings of the 2010 IEEE International Symposium on Hardware-Oriented Security and Trust (HOST 2010), Anaheim Convention Center, California, USA, June 13-14, pp. 76–87. IEEE Computer Society (2010)
[18] Eberle, H., Shantz, S.C., Gupta, V., Gura, N.: Accelerating Next-Generation Public-key Cryptography on General-purpose CPU. In: Hot Chips 16, IEEE Symposium on High Performance Chips. Stanford University (2004)
[19] Giry, D., Quisquater, J.-J.: Cryptographic key length recommendation, BlueKrypt - v 26.6 (2010), http://keylength.com
[20] Kocher, P.C.: Timing Attacks on Implementations of Diffie-Hellman, RSA, DSS, and Other Systems. In: Koblitz, N. (ed.) CRYPTO 1996. LNCS, vol. 1109, pp. 104–113. Springer, Heidelberg (1996)
[21] Brumley, B., Tuveri, N.: Remote Timing Attacks are Still Practical, Cryptology ePrint Archive, http://eprint.iacr.org/2011/232
[22] Koeune, F., Quisquater, J.-J.: A Timing Attack against Rijndael, Katholische Universitaet Louvain, Crypto Group. Technical report CG-1999/1 (1999), http://citeseerx.ist.psu.edu/viewdoc/summary?doi=10.1.1.42.679
[23] Kocher, P., Jaffe, J., Jun, B.: Differential power analysis. Technical report (1998), http://www.cryptography.com/public/pdf/DPA.pdf
[24] Kocher, P., Jaffe, J.: Introduction to differential power analysis. Journal of Cryptographic Engineering 1(1), 5–27 (2011)
[25] Kadir, S.A., Sasongko, A.: Simple power analysis attack against elliptic curve cryptography processor on FPGA implementation. In: International Conference on Electrical Engineering and Informatics, July 17-19, pp. 1–4 (2011)
[26] Mangard, S.: A Simple Power-Analysis (SPA) Attackon Implementations of the AES Key Expansion. In: Lee, P.J., Lim, C.H. (eds.) ICISC 2002. LNCS, vol. 2587, pp. 343–358. Springer, Heidelberg (2003)
[27] Perin, G., Torres, L., Benoit, P., Maurine, P.: Amplitude Demodulation-based EM Analysis of Different RSA Implementations. In: Proceeding of DATE 2012, March 12-16, pp. 1167–1172 (2012)
[28] Heyszl, J., Mangard, S., Heinz, B., Stumpf, F., Sigl, G.: Localized Electromagnetic Analysis of Cryptographic Implementations. In: Dunkelman, O. (ed.) CT-RSA 2012. LNCS, vol. 7178, pp. 231–244. Springer, Heidelberg (2012)

[29] Wu, K., Li, H.: Electromagnetic analysis on elliptic curve cryptosystems: Measures and counter-measures for smart cards. In: Third International Symposium on Intelligent Information Technology Application, pp. 40–43. IEEE (2009)
[30] De Mulder, E.: Electromagnetic Techniques and Probes for Side-Channel Analysis on Cryptographic Devices. Dissertation, Katholieke Universiteit Leuven (2010), http://www.cosic.esat.kuleuven.be/publications/thesis-182.pdf
[31] Carlier, V., Chabanne, H., Dottax, E., Pelletier, H.: Electromagnetic side channels of an FPGA implementation of AES. Technical report, IACR Cryptology ePrint Archive (2004), http://eprint.iacr.org/2004/145.pdf
[32] Skorobogatov, S.P.: Using optical emission analysis for estimating contribution to power analysis. In: Fault Diagnosis and Tolerance in Cryptography (FDTC), pp. 111–119. IEEE Computer Society (2009)
[33] Ferrigno, J., Hlaváč, M.: When AES blinks: introducing optical side channel. IET Information Security 2(3), 94–98 (2008)
[34] Skorobogatov, S.P.: Semi-invasive attacks - a new approach to hardware security analysis, Computer Laboratory, University of Cambridge. Technical report ucam-cl-tr-630 (2005)
[35] Tuan, T., Strader, T., Trimberger, S.: Analysis of Data Remanence in a 90nm FPGA. In: IEEE 2007 Custom Integrated Circuits Conference (CICC), pp. 93–96 (2007)
[36] Skorobogatov, S.: Low Temperature Data Remanence in Static RAM. Technical Report UCAM-CL-TR-536, University of Cambridge, Computer Laboratory (2002)
[37] Skorobogatov, S.Y.: Data Remanence in Flash Memory Devices. In: Rao, J.R., Sunar, B. (eds.) CHES 2005. LNCS, vol. 3659, pp. 339–353. Springer, Heidelberg (2005)
[38] Wills, K.S., Lewis, T., Billus, G., Hoang, H.: Optical Beam Induced Current Applications For Failure Analysis of VLSI Devices. In: Proceedings International Symposium for Testing and Failure Analysis, pp. 21–26 (1990)
[39] Ajluni, C.: Two New Imaging Techniques Promise To Improve IC Defect Identification. Electronic Design 43(14), 37–38 (1995)
[40] Samyde, D., Skorobogatov, S.: On a new way to read data from memory. In: SISW 2002 First International IEEE Security in Storage Workshop, USA (2002)
[41] Kaliski, B., Robshaw, M.: Comments on some new attacks on cryptographic devices, RSA Laboratories. Technical report Bulletin Number 5 (1997)
[42] Skorobogatov, S.P., Anderson, R.J.: Optical Fault Induction Attacks. In: Kaliski Jr., B.S., Koç, Ç.K., Paar, C. (eds.) CHES 2002. LNCS, vol. 2523, pp. 2–12. Springer, Heidelberg (2003)
[43] Skorobogatov, S.: Local Heating Attacks on Flash Memory Devices. In: Proceedings of the 2009 IEEE International Symposium on Hardware-Oriented Security and Trust (HOST 2009). Moscone Center, San Francisco (2009)
[44] Schmidt, J.-M., Kirschbaum, M.: Analysis of attacks on sensor nodes software and hardware. TAMPREs - Tamper Resistant Sensor Node - Project, Deliverable D1.2 Report (2011), http://www.tampres.eu/
[45] Microchip Technology Incorporation, http://www.microchip.com/
[46] Trichina, E., Korkikyan, R.: Multi Fault Laser Attacks on Protected CRT-RSA. In: Fault Diagnosis and Tolerance in Cryptography (FDTC) 2010, Workshop, August 21-21, pp. 75–86 (2010)
[47] 32-bit ARM Cortex M3 core documentation, http://www.arm.com/products/processors/cortex-m/cortex-m3.php

Intrusion Detection in Wireless Sensor Networks: Issues, Challenges and Approaches

Amrita Ghosal and Subir Halder

Department of Computer Science and Engineering
Dr. B. C. Roy Engineering College, Durgapur, India
ghosal_amrita@yahoo.com,
sub.halder@gmail.com

Abstract. Wireless sensor networks (WSNs) have generated immense interest among researches for the last few years motivated by several theoretical and practical challenges. The increase in interest is mainly attributed to new applications designed with large scale networks consisting of devices capable of performing computations on the sensed data and finally processing the data for transmitting to remote locations. Providing security to WSNs plays a major role as these networks are generally deployed in inaccessible terrain and also for their communication being in the wireless domain. These reasons impose security mechanisms to be employed on the highly vulnerable sensor networks that are robust enough to handle attacks from adversaries. WSNs consist of nodes having limited resources and therefore classical security measures applicable in traditional networks cannot be applied here. So the need of the hour is using systems that lie within the boundary of the sensor nodes resource potential as well competent enough to handle attacks. Intrusion detection is one such defense used in sensor networks having the ability to detect unknown attacks and finding means to thwart them. Researches have found intrusion detection system (IDS) to be very much compatible in sensor networks. Therefore intrusion detection holds a very prominent research area for researchers. So familiarity with this promising research field will surely benefit the researchers. Keeping this in mind we survey the major topics of intrusion detection in WSNs. The survey work presents topics such as the architectural models used in the different approaches for intrusion detection, different intrusion detection techniques and highlights intrusion detection methods applicable for the different layers in sensor networks. The earlier achievements in intrusion detection in WSNs are also summarized along with more recent works and existing problems are discussed. We also give an insight into the possible directions for future work in intrusion detection involving different aspects in sensor networks.

1 Introduction

In recent years wireless sensor networks have emerged as a promising research platform with various interesting application areas such as battlefield surveillance, traffic monitoring, healthcare, environment monitoring, etc [1]. Sensor networks are

S. Khan and A.-S.K. Pathan (Eds.): *Wireless Networks and Security*, SCT, pp. 329–367.
DOI: 10.1007/978-3-642-36169-2_10

generally deployed in areas where human accessibility is restricted and they use the unguarded wireless medium for communication. These factors are largely responsible for making security one of the prime factors of importance in WSNs. The wireless sensor network (WSN) consists of nodes that are resource-constrained in terms of computational and communication abilities. Therefore during the design of security systems for sensor networks the design guidelines should conform to the resources of the nodes and their limited capabilities. In contrast to traditional networks where very strong security mechanisms are implementable researchers working in the sensor network platform have always tried to use lightweight security schemes that are robust enough to handle the attacks faced by these networks.

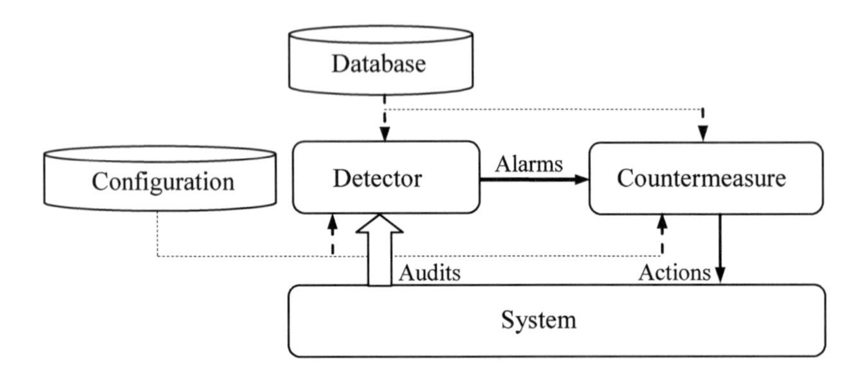

Fig. 1. Block diagram of basic intrusion detection system

Wireless sensor networks are exposed to inside attacks and outside attacks. It is very difficult for a single security technique to thwart the above mentioned two types of attacks in sensor networks. So intrusion detection which is an important aspect of network security came to be used in sensor networks. Intrusion detection system (IDS) is defined as a system that tries to detect and alert attempted intrusions into a system or a network [2]. Intrusion detection is a set of actions that discover, analyze, and report unauthorized and damaging activities. Intrusion detection uses the IDS with the aim of detecting any breach in confidentiality and integrity, and reduced availability of resources. IDS monitor the network and improves the user's activity to detect intrusion. Intrusion detection systems must be able to distinguish between normal and abnormal activities in order to discover malicious attempts in time. The fundamental components of a traditional IDS are: 1) sensors or agents for monitoring and analyzing activity, 2) a management server for centralizing information collected by the sensors or agents and managing them, 3) a database server for storing all the data produced by the IDS and 4) a console for providing interface to users and administrators for checking the status of the system monitored, receive alerts, investigate events and configure the system. One of the earliest works on intrusion detection is generally considered to be the one reported by Anderson [3], that introduced the idea of doing anomaly detection by creating profiles of normal use and detecting deviations from those profiles. This idea was later formally presented by

Denning [4] and this work is considered to be the stepping stone for modern intrusion detection. A block diagram representing basic IDS is shown in Fig. 1 [5], [6]. From Fig. 1 it can be seen that the IDS receives audit information from the system it is protecting. There are several inputs that include- a database containing presently known attacks, the current configuration of the system and audit information that describes the events as they are happening in the system. When the detector has access to all the required data, it decides which information is important and deduces the possibility of normal actions that can be considered as indications of intrusions.

The contributions of this survey can be summarized as:

— Describing the fundamentals of intrusion detection in WSNs including motivation, challenges and implementations (Section 2).
— Discussing previous works based on different approaches (Section 3).
— Providing a classification of existing intrusion detection architectures and models developed for WSNs (Section 4).
— Addressing existing intrusion detection techniques along with detailed descriptions (Section 5).
— Discussing handling of attacks in the different OSI layers using intrusion detection (Section 6).
— Potential research areas for future research (Section 7).

2 Fundamentals of Intrusion Detection in WSNs

Literally the term intrusion means both intrusion by outsider and insider abuse. Authors in [7] have categorized intrusions into two types as listed below-

- *Misuse or Signature-based detection*: The attacker detects the weakness in the system and based on that finds a way to get into the system [8]. The attack patterns used in this type of detection are known as signatures. If any malicious node uses known attacks for intruding, it will be captured if its pattern of attack matches some signature. It is necessary to update the signatures periodically to make this effective. But the major drawback of misuse detection systems is their failure to detect unpublished attacks.
- *Anomaly detection*: In this type of intrusion detection, the normal behaviour is defined and the intrusion detection system tries to detect anything that is suspicious. Anomaly detection assumes that intrusion is a kind of anomalous activity. So if it detects anomalous behaviour, it can detect an intrusion.

2.1 Motivation of Intrusion Detection in WSNs

The deployment of sensor networks in inaccessible areas coupled with the nature of their communicating medium and resource constraints pose difficulties for existing classical security techniques such as steganography to prevent all kinds of intrusions. Several works in sensor network security have focused on specific types of attacks and ways of preventing them using various techniques. One of the security technique used is cryptography that is used for ensuring authentication and integrity by verifying the data source and its contents. The cryptographic operations are based on primitives such as hash functions, symmetric encryption and public key cryptography [9] which can protect WSN against external attacks. But these cryptographic techniques are unable to detect internal attacks when the attacker knows the keys and uses them to perform encryption/decryption. This technique is defined as the first line of defense. As mentioned earlier due to several inherent factors it is quite impossible to guarantee full prevention in sensor networks. Also attackers always try to launch new attacks unknown to the protection system of the network. Therefore for certain environments it is necessary to establish a second line of defense: An IDS that can detect an attack (known or unknown) and notify the sensors about it. This mechanism allows detecting abnormal or suspicious activities on the analyzed target and triggers an alarm when intrusion occurs. As far as existing security techniques are concerned, none of them can provide security in WSNs from both internal and external attacks, thus IDS is strongly preferred for sensor networks that can handle both internal and external attacks.

2.2 Challenges of Intrusion Detection in WSNs

Designing intrusion detection systems in WSN involves many challenges, mainly due to the lack of resources. Intrusion detection works on the basic principle that the behaviour of a network under attack is different from that of a normal working network. WSNs have many configurations which makes it difficult to generate a clear cut demarcation between normal and abnormal working of a network. Since common nodes are designed to be cheap and small, they do not have enough hardware resources. As compared to entities of traditional wired networks, the nodes in sensor networks are highly susceptible to failures, thereby making implementation of intrusion detection systems further difficult. Moreover the vast differences in network characteristics of sensor networks are also responsible for accomplishing the task of intrusion detection in such type of networks [10]. As cited in [11] there are several factors responsible for selecting the appropriate IDS technique that is applicable for a particular scenario. These factor are described below-

Network Topology- The deployment of sensor nodes is an important aspect of sensor networks. Placing of nodes in appropriate locations requires in depth study so as to ensure proper connectivity and enough redundancy. If the topology of the sensor network supports measurements for redundancy, it is expected that it also provides each node opportunity for validating measurements reported by other sensor nodes. It has been found that a mesh topology is more robust to node failures or compromise. Node failure or compromise is more prevalent in tree topology type networks leading to frequent disconnections in the network.

Mobile vs. Stationary- Two types of nodes exists in sensor networks- one is the normal nodes with limited computational abilities and the other is the base station node having powerful resources. Both the base station and the ordinary sensor nodes can be mobile or static. Networks having mobile nodes have more capabilities than network with static nodes. Also more capabilities mean resource rich thereby making the mobile nodes costlier with respect to static ones.

Open vs. Closed Sensor Networks- A closed network allows limited number of nodes to work under the same administrator and does not allow any node to join the network without any prior authorization. On the other hand, closed networks permit any node to join the network in an ad hoc manner. A closed network involves more administrative control over each node, while an open network supports standard protocols and interoperability so as to allow nodes without any prior security authorization to join the network. Nodes in the closed network require authenticating themselves using some cryptographic support. The adopted security measure in the closed network can come under the attack of an adversary leading to disruptions in the network.

Physically Accessible or Inaccessible- Attacking a sensor network depends on the factor whether the malicious node is close enough to the network which it is targeting. Physical proximity ensures that the adversary can place their own sensor, depending on the application, in the same location. If the sensors have high complexity with respect to replication, the adversary targets the base station or the sensor communication. Therefore, every sensor application needs to be analyzed properly for identifying the risks involved and selecting proper countermeasures.

Critical or Non-critical Application- The collection of real-time data for critical applications has an impact on the types of countermeasures that can be used. For example a dangerous situation occurring in an industrial application requires real-time detection and corrective measures to defend the same whereas a long-term environmental monitoring application may not require an immediate response. The environmental monitoring application may employ sensors that are unattended for long periods of time, while the industrial application may have the sensors confined within the physical perimeter of a factory and thus face a different hazardous environment. Sensor networks consist of nodes having different degrees of complexity, network connectivity and cost that depend on the application. Also, the criticality of the application is responsible for the factor that sensor networks are deployed along with other technologies and human processes for ensuring the robustness of the monitoring and control processes.

Hazardous or Non-hazardous Environment- Placing sensor networks in non-hazardous environment exposes the network to threats such as node tampering, trying to obtain the keying material from the nodes and also placing new or cloned node for generating false incorrect data. Sensor networks deployed in hazardous environments are open to attacks such as denial of service or eavesdropping. These environments also make the base station, from where the final data is collected, a more attractive target. If sensors are deployed in non-hazardous environments and no physical countermeasures are available to prevent physical tampering with the devices, then the sensors themselves and the sensor network application must be able to detect tampering. Tamper-resistant and tamper-evident technologies can be employed to help detect and diagnose physical attacks against the sensor network.

Routing Algorithms- Routing algorithms use network traffic analysis for detecting malicious activity. Mostly used sensor routing technologies are ZigBee, TinyOS, and IEEE 802.15.4. TinyOS is an event based operating environment designed for supporting dissemination and collection protocols of sensor networks. The Zigbee standards define the network, security and application software layers for wireless sensor networks. The IEEE 802.15.4 standard defines MAC and PHY layers for wireless sensor networks. The IEEE 80.15.4 standard has 16 channels in the 2.4GHz ISM band, 10 channels in the 915MHz and one channel in the 868MHz band. The IEEE 802.15.5 standard has CSMA-CA channel access, supports data rates of 250kbps, 40kbps, and 20kbps, provides automatic network establishment by the coordinator, and incorporates power management for ensuring low power consumption.

Cryptographic Support- Public key cryptography is generally not suitable for sensor networks because of their complex computations leading to huge amount of energy consumption as sensor nodes have limited battery power. Security mechanisms are developed using efficient symmetric key cryptography that is more energy efficient. So whether sensor networks are using public key cryptography or symmetric key cryptography is a critical factor for deciding the IDS applicable for the network.

Designing IDS for WSNs keeping these challenges in mind leads to two constraints- 1) the IDS should be highly accurate in detecting any security breaches that includes unknown attacks and 2) it should be lightweight so as to ensure minimum overhead on the infrastructure and management processes.

2.3 Implementation of IDS in WSNs

The intrusion detection system has become a significant component of wireless sensor networks with relation to the security issues. But, deployment of intrusion detection systems in such networks is a prime research area as their implementation introduces a number of potential drawbacks that can have negative impact on security. In this section we discuss about the shortfalls faced during IDS implementation in sensor networks.

It is quite impossible to use IDS in every sensor node as a fully powered agent due to the resource constrained nature of such nodes unless high end nodes are used. Therefore, each node acts independently and sends/receives data to/from the base station. IDS operating for sensor networks send alerts to the base station as warnings. IDS should be developed in such a way that it is highly specialized for defending against particular threats to the network. Also if IDS is used in sensor networks where huge traffic is handled, there is a possibility that traffic causing intrusion can be missed. This is because nodes in sensor networks generally have several constraints with regard to handling of large volume of data in the network.

Intrusion detection systems have not achieved full automation till date. It becomes necessary for monitoring the IDS logs at regular intervals. Periodic monitoring is required for analyzing the different types of malicious activities detected by the IDS. It is not possible for IDS to provide analysis of the detected intrusions taking place over a period of time. This procedure is still done manually. This is quite difficult to achieve in sensor networks as human intervention is mostly limited in such networks and therefore problem arises for implementation of IDS in WSNs. So implementation schemes for IDS designs in sensor networks should be automated in such a way that overhead incurred is minimum. Also as IDS mechanisms work on attack signatures, there is need for updation of the signature database periodically. This is also a bottleneck for sensor networks to assign the task of updating the database as well as maintaining the database because sensor nodes are equipped with very limited memory.

Implementation of IDS in sensor networks is presently in a premature stage because of the resource limitations of the nodes present in such networks. But with evolvement of new technologies such as mobile sensor nodes, neural network implementation in sensor nodes, use of IDS technology in sensor networks will become more efficient.

3 Literature Review Based on WSNs IDS Approaches

IDS approaches proposed for safeguarding sensor networks can be classified into four distinct categories [11]:

— IDS using routing protocols
— IDS based on neighbour monitoring
— IDS based on innovative techniques
— IDS based on fault tolerance

Apart from the above mentioned categories, we have also included another subsection on recent works in intrusion detection covering topics ranging from hybrid intrusion detection techniques to different types of distribution affecting IDS mechanisms.

3.1 IDS Using Routing Protocols

In this category the goal of an attacker, either being insider or outsider, is to manipulate user data directly or try to affect the underlying routing protocol. Several types of routing protocols are available for sensor network applications, some focus on energy saving, others on resource awareness or in-built security mechanisms. However, there is no perfect routing protocol yet which has proven to be robust against all attacks e.g. flooding, selective forwarding, packet dropping. Several intrusion detection systems have been proposed to detect routing attacks in WSNs.

Loo et al. [12] have proposed distributed IDS for detecting routing attack in WSNs. In the proposed system, IDS agent is installed in every node for detecting anomaly in routing. The authors have identified various fields e.g. number of packets received, sent or broadcast, route request sent or forwarded or received, etc for the detection mechanism. These fields are used by each node for determining standard deviation of normal messaging by each neighbouring node. Based on the standard deviation of each node a fixed width cluster is formed. Finally, clusters are analyzed for identifying compromised nodes. Several challenges are there for implementation of the system. One of them is identification of various features and using those features for identifying the particular attack.

Krontiris and Dimitriou [13], have proposed IDS based on set of rules for detecting routing attacks in WSNs. In the proposed work every node maintains a failure counter for its every neighbouring node for counting the number of times each neighbouring node fails to broadcast data packet. Initially every node senses the environment and sends the sensed data packet to a neighbouring node. The sending node keeps the data packets in its buffers for a while. It waits for that neighbouring node to forward the data packet towards base station. If the neighbouring node fails in forwarding the data packet, it increments a failure counter corresponding to the neighbouring node. If forwarding takes place, then data packet is removed from the buffer. This operation continues, and after some time if it is found that for a particular node failure, counter crosses a certain limit, neighbouring nodes generate alerts against that node and voting takes place according to the given rules. In another work Krontiris et al. [14] discuss about the possibility of the sinkhole attack in routing protocol. In [14] authors extended their previous work [13] and added another set of rules for the sinkhole attack too.

H. Hai and E. N. Huh [15] proposed a lightweight intrusion detection scheme for defending routing attack in WSNs. In the deployment phase of the proposed scheme each node collects the two hop neighbourhood information by exchanging hello packets and creates a malicious counter for each of them. After collecting the information about their neighbour, each node maintains a neighbourhood information list. During the process of exchanging the data packets each node checks both source and destination of the received packets. If both source and destination nodes are in its neighbourhood information list then data packets are forwarded towards base station. If either source or destination nodes are not in the neighbourhood information list of the node, then based on certain rules the malicious counter value is increased. If the malicious counter crosses the certain threshold value, then the node is set as malicious and revoked from the neighbouring list.

S. Misra et al. [16] proposed a self-learning, distributed, energy-aware routing protocol for detecting intrusion in WSNs. The distributed nature of the proposed routing protocol enables functioning of each node independently without any knowledge about the adjacent nodes. Also it avoids all other nodes being sacrificed when a single node is compromised. To make the protocol energy aware, the authors incorporated stochastic learning automata on packet sampling mechanism. They show that the proposed protocol provides promising results for intrusion detection.

3.2 IDS Based on Neighbour Monitoring

In this type of IDS, nodes work collaboratively for malicious node or attacker detection. Generally in WSNs nodes located spatially close to each other tend to have similar behaviour. A node is considered as malicious if its behaviour significantly differs from its neighbours. Several works has been proposed for intrusion detection in WSNs based on neighbour monitoring. The state of the art work in G. Li et al. [17], I. Krontiris et al. [18], [19], A. Stetsko et al. [20], and Hassanzadeh and Stoleru [21] have proposed IDS based on this principle.

G. Li et al. [17] have proposed IDS for WSNs where nodes are partitioned into groups for monitoring the neighbourhoods. Initially nodes in a network are partitioned into number of groups in such a way that the nodes in a group are physically close to each other and collect similar data. After data collection by each group, the proposed intrusion detection algorithm is run on each group. Through careful monitoring, if some nodes find that there is noticeable difference between their sensed data and those of some other nodes in the same group, they conclude that some group members have been compromised. After detecting the compromised node, that node is segregated from the network.

I. Krontiris et al. [18] have proposed a generalize architecture for intrusion detection that can operate under any circumstances based on neighbour monitoring. According to the proposed approach, each node hosts an IDS agent. The IDS agent performs neighbour monitoring, decision making and response. During the neighbour monitoring task, every node monitors each immediate neighbour and collects audit data. During the decision-making task every node based on local audit data determines the existence of possible intrusions and forwards their conclusions to each immediate neighbour in order to make the final collective decision. The local detection engine applies the defined specifications about what is normal behaviour and monitors audit data according to these constraints. The cooperation between neighbouring nodes is performed by applying the majority vote rule in order to determine the existence of an attack. Furthermore, when an attack is detected the local response module is activated. Depending on the severity of the attack the response might be direct or indirect. The direct response excludes the suspected node from the routing paths and forces regeneration of cryptographic keys for the rest of the neighbours. The indirect response notifies the base station about the suspected behaviour of the possible intruder and reduces the reputation of the link to that node so as to gradually characterize it as unreliable. The proposed approach is evaluated against the sinkhole attack. In voting based IDS there is possibility of drastic flooding over the network caused by broadcasting local detection results. In the proposed work to prevent message flooding, alarm messages are restricted to a region formed only by the alerted nodes.

In another work I. Krontiris et al. [19] have proposed an IDS for WSNs for detecting single attacker based on neighbour monitoring. The backbone of the proposed IDS is the voting module which is responsible for executing the protocol in the voting phase. The attacker in the network is identified by the nodes through exchange of their suspect list by collaborative voting. The final intrusion detection result comprises either the number of times each node appears in the suspect list or the number of votes it collects. The node with the majority of votes is declared as the attacker and isolated from the network. However, voting based IDS is costly because it requires every neighbour to multicast an authenticated ballot.

A. Stetsko et al. [20] have designed an IDS for WSNs based on neighbour monitoring principle. In the proposed strategy every node runs an agent, which monitors the information flowing in its neighbourhood. The agent consists of data acquisition, statistics, detection, alert database and collaboration components. The data acquisition component gathers data from packet headers and stores the processed information in the statistics component. The detection component analyzes the information stored by the statistics component and stores information about suspicious or malicious nodes in the alert database component. Therefore, by checking the alert database one can retrieve the information about a node whether it is malicious or not. If a node wants to share an event with its neighbours or the base station, it activates the collaboration component. Finally authors have shown that the jamming, hello flood, selective forwarding, sinkhole, sybil and packet alteration attacks can be detected using the proposed neighbour-based IDS.

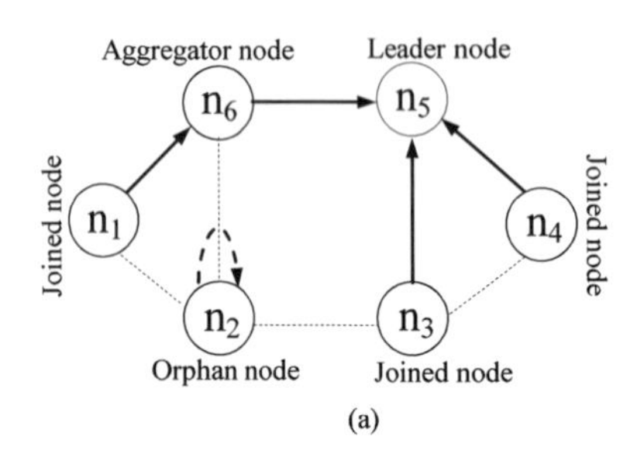

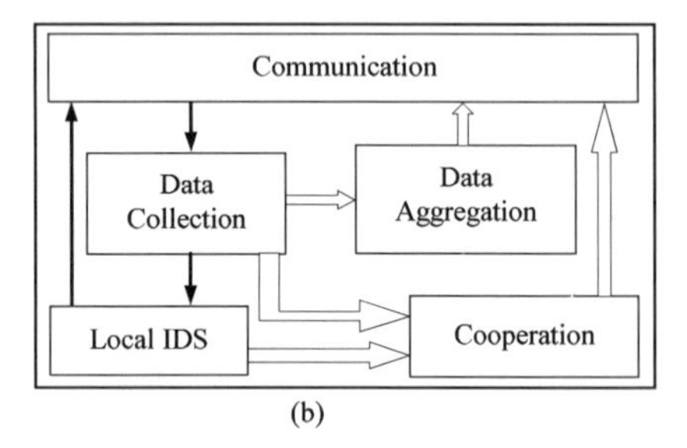

Fig. 2. (a) Network model of cooperative IDS with various nodes responsibilities. (b) General node architecture in cooperative IDS.

Hassanzadeh and Stoleru [21] have proposed cooperative intrusion detection functions amongst nodes. The objective of the proposed solution is minimizing energy consumption and event reporting delay in the nodes, while maximizing network coverage and data accuracy in the network. Authors have considered the cooperative IDS architecture where nodes are organized in cluster trees with a single base station. Figure 2(a) depicts an example of a network with cooperative IDS. Figure 2(b) shows generic architecture for a node. Based on the roles of the nodes, authors have identified four types of nodes e.g., joined, aggregator, leader and orphan. Joined nodes are the leaves in a cluster tree. They monitor local activity e.g., communication, processes running, data produced and run a local IDS. Results are reported to the parent, which can be an aggregator or a leader. Aggregator nodes also monitor local activity, receive reports from children, either joined or other aggregator nodes, and aggregate received data with their information, using the data aggregation module. The aggregated data is reported to a parent, either another aggregator or leader; leader is the root of a cluster tree. A leader receives reports from its children, either joined or aggregator nodes, and executes intrusion detection functions as part of a cooperation module. The results are reported to the base station. Leaders of all cluster trees form a connected graph, which contains the base station. Orphan nodes are not part of a cluster tree. They run local IDS and do not forward their observations to their neighbours.

In order to minimize energy consumption and event reporting delay in the nodes, while maximizing network coverage and data accuracy in the network, the authors have formulated it as a multi-objective optimization problem. A genetic algorithm based on penalized function has been developed to solve the multi-objective optimization problem. Finally the authors have validated the superior network performance and intrusion detection rates obtained by the proposed collaborative IDS.

3.3 IDS Based on Innovative Techniques

In this type of IDS, different innovative techniques have been proposed by the authors for detecting intrusion in WSNs. We have elaborated few state of the art works in this section which are based on certain innovation techniques.

Agah et al. [22], [23] proposed a game theory based protocol for IDS in WSNs. The proposed approach is formulated as a repeated game between an intrusion detector and the nodes of a WSN. The proposed protocol recognizes the nodes that agree to forward data packets, but fail to do so. The approach categorizes different nodes based upon their dynamically changing behaviour and enforces cooperation among nodes. Any non-cooperative behaviour is punished. The intrusion detector situated at the base station monitors the collaboration of other nodes and builds up a history that represents their reputation. Sensor nodes that contribute to common network operation increase their reputation. The reputation is used as a metric of trustworthiness and is used to statistically predict the future behaviour of sensor nodes. The advantage of the proposed approach is that using the history created by the base station for each sensor node, and the negative reputation the base station assigns to any malicious behaviour, it is possible to create routing paths consisting of less malicious nodes for more secure transmissions. Thus the malicious nodes are isolated. The main disadvantage of the proposed approach is that when the number of malicious nodes in the sensor network increases, the success rate of the IDS decreases. This can be explained if we consider the fact that the IDS attempts to lower false positive and false negative rates and as a result the detection rate is decreased since it misses more malicious nodes. This technique may not be suitable for certain environmental monitoring applications, but may be considered in more critical applications in which an intrusion is likely and cannot be tolerated.

Premkumar and Kumar [24] have proposed a quickest intrusion detection scheme in WSNs by keeping minimum number of nodes active. In order to quicken intrusion detection authors have modeled the intrusion detection problem as a Markov Decision Process (MDP). In the proposed work there is a fusion center based on MDP that decides how many sensors need to be turned on after each time slot. The optimization problem studied by the authors is to minimize a linear combination of the detection delay and energy consumption, subject to a constraint on the probability of false alarm. A Lagrangian version of the problem is posed within the framework of dynamic programming for the classical quickest change detection. It is shown that the posteriori probability p_k of the change happening before time k, serves as a sufficient statistic for both the stopping rule as well as the control law that determines how many sensors need to be turned on at each time k. Some structural properties of the optimal stopping rule and control law are also proposed. In particular, as is to be expected, it is optimal to stop and declare the change when p_k exceeds a certain threshold. It is also conjectured, based on numerical results, that the number of sensors to be turned on increases with p_k initially, peaks at a certain value, and then decreases with p_k; the intuition behind this behaviour being that it is not necessary to waste sensing resources if the change is either very unlikely or very likely, given the observations. The authors show that the proposed algorithms are capable of detecting intrusion using minimal number of observations and keeping minimum number of nodes active.

Servin and Kudenko [25] have proposed a distributed reinforcement learning-based IDS architecture for detecting flooding-based Distributed Denial of Service (DDoS) attacks in WSNs. In the proposed work, groups of agents are distributed in the network to detect normal or abnormal state of the network. Identification of abnormal network state leads to the detection of flooding-based DDoS attacks. Through reinforcement learning, each agent learns to act optimally via observation and taking feedback from the environment. In the proposed IDS architecture, there are two types of agent viz. sensor agents (SA) and decision agents (DA). SA collects and analyzes state information from environment, and sends action-signals to the DA. After receiving action-signal form SA, DA sends an action-signal to a central DA. After analyzing the receive signal central DA triggers the appropriate action, if action is correct both SA and DA receive a positive reward. If the action is not correct, both SA and DA receive a negative reward. Here reward is use to coordinate the signals sent by the SA to the DA. Therefore, these reward collection process continues and after a certain number of iterations every agent learns about the action that needs to be executed in a specific state for obtaining positive rewards.

Kaltiokallio and Bocca [26] have proposed IDS for real-time intrusion detection and tracking by distributed processing of the received signal strength indicator signals in WSNs. In the proposed IDS, a distributed algorithm enables each node to transmit only those alert messages, which are related to significant events. Therefore, it reduces the amount of alert messages the nodes have to transmit to the base station resulting in enhancement of the overall lifetime of the system. The alert messages received at the base station are than combined and processed in real-time to produce accurate estimates of the current position of the intruder. Further to improve lifetime of the system they used TDMA communication among the nodes which allows the nodes to keep the radio off for most of the time. They show that using the proposed IDS, intrusion inside monitored area is not only successfully detected but with moderate location error.

3.4 IDS Based on Fault Tolerance

Several fault tolerant techniques have been proposed by the authors related to intrusion detection in WSNs. Here, we have elaborated some of the intrusion detection works based on fault tolerance.

The first work towards an intrusion fault tolerant protocol for WSNs is INSENS [27].The INSENS protocol is more capable of tolerating the intrusions than detecting the intrusions. In this protocol, base station maintains a complete view of the communication topology. To achieve this, each node sends the list of its neighbours to the base station, with proofs of neighbourhood. These neighbourhood proofs allow the base station to eliminate unnecessary communication links that may be injected by intruders. After reception of these proofs, the base station builds a map of the existing topology. Thereby using this centralized approach, INSENS constructs the routing table for each node. Moreover, the base station has full control on the routes' quality and can easily build any kind of multi-path topology, including node disjoint paths. Nevertheless, INSENS is not scalable to large networks since it requires a large amount of communication between sensors and the base station.

Lee et al. [28] proposed SeRINS, an intrusion fault tolerant secure multi-path protocol for WSNs. The proposed protocol requires less message communication than INSENS. To achieve this base station maintains a hop count metric using a set of one way hash chain. Three types of chains are present in the proposed scheme-

- One chain is used for base station authentication and round identification. At each round, the base station reveals a new value of the chain in the reverse order of its generation.
- Another, on the fly chain is used during the relay of the route request (RREQ) message to prevent an intruder from decreasing its hop count. This chain is implemented by adding a field in the RREQ message.
- The last chain is similar to the previous one and uses a field named that aims to detect a malicious node trying to relay a field of its parent without applying the one way function.

Each node in the network has the capability for partial verification of any neighbouring node. After partial verification, if any suspicious node is found, it is reported to the base station. Base station takes the decision of choosing alternative path for exchanging message with the nodes. Hence, instead of fully centralized approach, the proposed protocol employs a tradeoff between fully centralized and distributed approaches for intrusion detection. However, during the process of creating hop count metric base station dos not takes care about the node disjoint path, so there is a possibility of duplicate entries for the same node.

Y. Challala et al. [29] proposed an intrusion-fault tolerant protocol for WSNs based on total in-network verification. In contrast to other solutions, the proposed protocol provides an efficient and secure method to build node disjoint paths in a totally distributed manner without referring to the base station. The path construction in the proposed protocol is based on the idea of branch-aware route discovery [30]. It consists of tagging the exchanged RREQ messages with the identifier of the first relaying node after the base station. These nodes are known as root nodes, and their sub-trees are known as branches. Using these tags, a node can easily decide whether two RREQ messages came from disjoint routes by comparing their branch ids. Figure 3 shows an example of a branch-aware discovery process. Nodes n_1, n_2, and n_3 represent the root nodes of the branches. Two neighbouring nodes belonging to two distinct branches discover a new disjoint path through each other. For instance, as shown in Fig. 3 node n_9 possesses a main route through its main branch rooted at node

n_1, and also has an alternative route through its neighbour n_5 in the branch rooted at node n_2 (and both routes are node disjoint).

To secure this, the proposed protocol replaces the branch ids with one way hash chains in order to avoid fabrication of bogus branches. In fact, using plain text identifiers for root nodes allows an intruder to attract more routes by injecting non-existent branches. The one way hash chains prevent such malicious misbehaviour by allowing legitimate sensors to authenticate the root nodes.

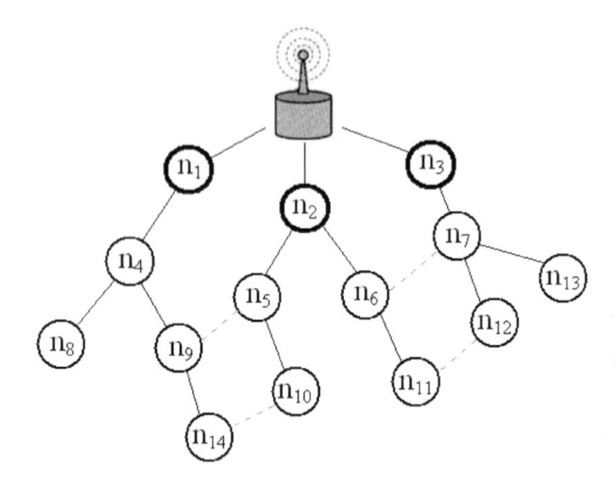

Fig. 3. Node disjoint path discovery using branch-aware discovery

By analyzing the above existing solutions, it is concluded that intrusion-fault tolerant approaches do not provide an acceptable trade-off between the level of fault tolerance and the induced communication overhead. So there is scope for further improvement where a scheme will be more fault tolerant with minimum communication overhead.

3.5 Recent Works on IDS

In this sub-section we have elaborated number of recent potential works in intrusion detection. The proposed works cover topics ranging from hybrid intrusion detection techniques to varied distributions affecting IDS mechanisms.

T. H. Hai et al. [31] have initially studied the problem of intrusion detection in WSNs and proposed a hybrid intrusion detection scheme for clustered WSNs. The objective of using clustered WSNs is to reduce the energy consumption so that network lifetime can be prolonged. In the proposed scheme, an IDS agent is located in every sensor node. Each sensor node has two intrusion modules-local IDS agent and global IDS agent. On the basis of requirement each agent is activated. The local IDS agent module is responsible for monitoring the information sent and received by the sensor. The global IDS agent module is responsible for monitoring the communication of its neighbour nodes.

Authors have used the watchdog monitoring mechanism, pre-defined routing attack based rules along with the two-hop neighbours' list for monitoring packets in their neighbourhood. If the monitor nodes discover a potential attack e.g., selective forwarding, wormhole, sinkhole and hello flood attack in their radio range, they create and send an alert to the CHs. If the number of alerts about a suspicious node crosses a threshold, the CHs create a rule and propagate it to every node in the cluster. As the proposed intrusion detection scheme requires every node to be active and send alert packets to the CHs for intrusion detection, therefore large number of alert packets are transmitted through out the network. Hence the proposed intrusion detection scheme is not energy efficient.

Further to make the intrusion detection scheme more energy efficient authors have proposed two algorithms for reducing the energy consumption related to processing the alert packets. First algorithm is trust based where each node calculates the average trust of its neighbour nodes. If average trust for a neighbour node is below a threshold, the CHs drop the alert packets received from them without further processing. Second algorithm is over-hearing based on the fact that if a monitor node is aware about any malicious activity within its transmission range, an alert packet is prepared to be sent to the CHs. If a monitor node does not obtain the medium to send an alert packet, it knows there is transmission taking place within its range. The monitor node buffers the alert packets and over-hears the packets sent within range. If the monitor node detects a neighbour sending the same alert packet, it drops the alert packet in its buffer. Thus using the two algorithms authors have ensured the reduction of transmission of alert packets by monitor nodes. Finally authors have shown that the proposed intrusion detection scheme can detect more than 90% of malicious nodes.

I. R. Chen et al. [32] have developed a probability model for analyzing how often code attestation should be performed to maximize the expected lifetime of a sensor. A sensor node fails when either the sensor node's energy is depleted, or it is compromised before energy depletion and returns incorrect sensor readings during a reading event. The code attestation is invoked for verifying the memory content of a sensor node by computing the checksum of program code and data. If code attestation happens too often, the energy consumption may drain the battery quickly such that the reliability of the sensor node decreases and this offsets the benefits of code attestation. On the other hand, if it is not done frequently enough, an intrusion may not be detected in time such that a compromised sensor may return incorrect sensor readings resulting in system failure. Therefore, authors have provided the probability of invoking code attestation probabilistically considering, triggering a periodic sensor reading event. Authors have concluded from their simulation results that code attestation can be executed more often whenever the node compromising rate is high, the false result probability (negative or positive) is low, the energy consumption for running code attestation is low, the energy consumption for code recovery is low compared with the energy consumption for sensor data reading, or the sensing interval is long. To measure the reliability of a node, authors have used the parameter mean time to failure and it is defined as the number of periodic sensor reading events that the sensor node is able to return the sensor readings correctly before failure.

Shen et al. [33] have proposed an intrusion detection game for optimal intrusion detection based on the signaling game. Authors have considered distributed-centralized network in which each node is equipped with an IDS agent, but only the IDS agent in CH will launch. The signaling game refers to a class of two-player game in which one player (called the Sender) is informed and the other (called the Receiver) is not. Generally, in a signaling game, the Sender has private information about its type set while the Receiver has the common information about its type only. Authors have modeled the interactions between a member node and a CH-IDS agent with signaling game. In addition, authors seek pure-strategy Bayesian–Nash equilibrium (BNE) and mixed-strategy BNE for the stage game, and the mixed-strategy Perfect Bayesian equilibrium (PBE) for the multi-stage dynamic intrusion detection game. Authors have set up and proved the theories of equilibriums for their stage intrusion detection game, which provide the optimal strategy for the CH-IDS agent for deciding whether to defend or remain idle.

Also authors have developed the stage intrusion detection game into a multi-stage dynamic intrusion detection game in which, based on Bayesian rules, the beliefs on the malicious sensor node can be updated. Upon the current belief and the Perfect Bayesian equilibrium (PBE), the best response strategy for the CH-IDS agent can be gained. Also, authors have proposed an intrusion detection mechanism and the corresponding algorithm. Finally through simulation authors have shown the effectiveness of the proposed game, thus, the CH-IDS agents are capable of selecting their optimal strategies for defense against malicious sensor nodes' attacks.

Wanga and Lun [34] have proposed a novel k-Gaussian sensor deployment scheme and investigated the intrusion detection probability in WSN under the multi-level probabilistic sensing model. The main idea of the k-Gaussian node deployment strategy is employing multiple deployment points in the network area and a subset of the total nodes are deployed around each deployment point following a Gaussian distribution to form a k-Gaussian distributed WSN. Initially authors have developed an analytical model for intrusion detection where nodes are deployed in network area using k-Gaussian distribution. Authors have calculated the intrusion detection probability under various application scenarios. Here the different scenarios are considered based on the distance between the intruders current and target locations.

The analytical results shows that the intrusion detection probability depends on the following set of network parameters i.e., number of sensors, sensing range, deployment points, distribution deviation, the intruder's behavior i.e., the starting distance, and the application requirements such as maximal allowable intrusion distance. Finally authors have used Monte Carlo simulation to measure the performance of k-Gaussian distributed WSN and shown that the k-Gaussian distributed WSN statistically outperforms the uniform distributed WSN as well as Gaussian distributed WSNs.

S. S. Wang et al. [35] have proposed a mechanism of IDS in a cluster-based wireless sensor network. The proposed IDS is integrated and is able to resist intrusions, as well as does real time processing by analyzing the attacks. According to the different capabilities and probabilities of attacks on the sink, cluster head and sensor node, authors have proposed three separate IDS for each of them.

The proposed IDS for sink is named as Intelligent Hybrid Intrusion Detection System (IHIDS), which has learning ability. The IHIDS combines anomaly and misuse detection, and goals for high detection rate and low false positive rate. The anomaly detection model can filter a large number of normal packets first, and then the abnormal packets are forwarded to the misuse detection model to identify the type of attacks. However, if the misuse detection model cannot identify the type of attack, it is then forwarded to the learning mechanism of IHIDS for learning the new classes of attacks. For CHs, authors have proposed a Hybrid Intrusion Detection System (HIDS), which has the same detection model as IHIDS, but there is no learning ability in HIDS. The goals of HIDS are to detect attacks efficiently and avoid resource wasting. However, HIDS would retrain the behaviour of new attacks, which have been detected and classified from IHIDS. Finally for SNs, authors have proposed a misuse IDS, which uses the attack model for matching the packets fast and then to find attacks. Because the resources of sensor node are less than other devices, e.g., the sink, CHs, etc., authors have adopted a simple and fast detection method in SN to avoid overwork, and to save resources for the purpose of safety. Finally authors have measured the performance of the proposed IDS using Back Propagation Network (BPN) and Adaptive Resonance Theory (ART) analytical tools of intrusion detection. The authors have concluded that ART outperforms BPN.

Y. Wang et al. [36] have first analyzed the problem of intrusion detection in WSNs when nodes are deployed using Gaussian distribution. It is known that Gaussian distribution allows the deployment of nodes in an unbounded network area while most real life WSN applications take place in a bounded network area of interest. Therefore, in order to make the node deployment suitable for real life applications, authors have developed an analytical model for intrusion detection where nodes are deployed in the network area using truncated Gaussian distribution. Authors have derived the intrusion detection probability analytically with respect to various network parameters considering both single-sensing detection and multiple-sensing detection models. In a single-sensing detection model, at least one node must be located in the intrusion detection region for detecting the intruder(s). On the contrary, in multiple sensing detection model, at least a threshold number of nodes must reside in the intrusion detection region for recognizing the intruder.

Authors have identified three sets of parameters that determine the intrusion detection probability in a given truncated Gaussian distributed WSN. The first set is network based parameter i.e., number of nodes deployed in the network area, sensing range of a node and standard deviation of Gaussian distribution. The second set is application based parameter i.e., threshold number of nodes in multiple sensing detection model and maximal allowable intrusion distance. The third set is intruder based parameter i.e., intruder's starting distance. Authors have used the Monte-Carlo simulations for validating the analytical results obtained for intrusion detection probability by varying the three sets of parameters and provided the guidelines in selecting and determining critical network parameters.

4 Classification of IDS Architectures and Models for WSN

This section classifies the different sensor network IDS architectures proposed till date. Also the various existent IDS models that have been taken into consideration for designing of different network models are discussed here.

4.1 IDS Architectures for WSN

IDS architectures are classified into two basic categories depending on the data collection mechanism: host-based and network-based [37]. Host-based IDS check several types of log files such as kernel, system, application, etc. and compare the logs against an internal database of common signatures for known attacks. The operation of network-based IDS differs from host-based IDS. The tasks of network-based IDS consist of scanning network packets, auditing packet information, and logging any suspicious packets. Also, IDS architectures can further be classified based on the detection techniques. Commonly used detection techniques are- signature-based IDS, anomaly-based IDS and specification-based IDS. Out of these three techniques, the first two are widely used while the last one is used in few applications. Signature-based IDS tries to find the existence of predefined signatures or behaviours that matches a previously known malicious action or indicates an intrusion. Anomaly-based IDS checks for any behaviour that is not within the predefined or accepted model of behaviour. Authors in [38] have defined another type of IDS known as specification-based IDS that identifies a set of constrains which reflect the correct behaviour of a program or protocol. The authors have also divided ad-hoc network IDS architectures into three categories and these categories can be modified according to the needs of wireless sensor network IDS. A brief description of the categories is given below:

Stand-alone- Every node functions as independent IDS and is responsible for detecting attacks only for itself. In this category, IDS does not share any information or co-operate with other systems. In this architecture all the nodes of the network are capable of running IDS.

Distributed and Cooperative- This architecture is similar to the previous one with respect to every node managing their independent IDS but the IDS of all nodes cooperate among themselves for establishing a global intrusion detection mechanism.

Hierarchical- In this type of architecture the network is divided into clusters and each of these clusters has cluster head (CH). The CH serve as the backbone of the routing infrastructure. The nodes in every cluster are responsible for routing within the cluster and accept allegation messages from the other cluster members representing something malicious. The cluster heads (CHs) are also responsible for detecting attacks against the other CHs of the network.

4.2 IDS Models for WSN

This section describes some of the existing IDS models for WSNs [39]. The different models use several methods and architectures to build the IDS. Given below is the description of basic features for each of the IDS in order to have a clear conception of their logic.

Self-organized Criticality & Stochastic Learning Based IDS- Authors in [40] proposed anomaly detection based on the structure of events occurring naturally. This approach uses the importance of self organized characteristics of a particular location. Environmental parameter such as temperature is used for evaluation purpose for detecting future anomalies by comparing recent data with existing data. It also makes use of a Hidden Markov Model, which was earlier, implemented in network-based IDS for wired systems [41], [42]. This model is memoryless as the probability of being in a certain state for the Hidden Markov Model depends only on the previous state.

IDS for Clustering-Based Sensor Networks- The security of cluster-based sensor networks using intrusion detection systems has been enhanced in [43]. Two techniques are proposed in this work. The first one uses a model that depends on authentication and can defend outside attackers only. It basically appends a message authentication code (mac) to every message. Whenever a node sends a message, a time stamp is attached to it and a mac is generated using the pairwise key or individual key depending whether the sender is CH, member node (MN), or base station. The receiver verifies the sender using the LEAP [44] security mechanism. The second technique is termed energy-saving and like the previous one also withstands outside attackers. It focuses on detecting misbehaviour in both MN and CH. The monitoring of member nodes (MNs) is done by their respective CHs. On detection of any misbehaviour, the CH broadcasts an encrypted alarm message to control the specific misbehaving node. Monitoring of the CH is also done by some of the MNs under it. Monitoring is done with the following algorithm. The CH decides which MNs are energy capable of monitoring the CH. This is implemented by sending messages that query the energy status of every MN. The nodes having low energy are ignored by the CH. Rest of the MNs is divided into groups. Each group is responsible for monitoring the CH turn wise. At any time only one group (the active group) monitors the CH. If misbehaviour is detected by some threshold numbers of monitor MNs, then the CH is discarded.

A Non-cooperative Game Approach- In [22] the authors devised a game theoretic approach for thwarting intrusion in sensor networks. Three schemes have been used for achieving this. Every mechanism divides the sensor network into clusters which have CH as one of the nodes. The commonality between these three mechanisms is that each of them with the help of IDS identifies the most suitable CH capable of protecting the network. The first scheme defines one non-operative game between the attacker and the nodes. The desired CH is found out using game theory and Nash equilibrium methods. The second scheme takes the help of Markov Decision Process (MDP) for determining the CH that the IDS will protect. From the knowledge of past events, using the MDP the most vulnerable CH can be identified. In the third scheme intuitive metric is devised that uses traffic load or activity load as the measuring parameter. At a particular time slot, the CH having the highest traffic load is protected by the IDS.

Decentralized IDS- Authors in [45] proposed IDS based on certain rules such that the criteria of sensor networks are fulfilled. The rules defined by them are as follows: 1) pre-selection is done from the available set of rules; 2) the existing information from the network is compared with the information required from the pre-select rules for obtaining the final rules and 3) the setting of parameters of the final rules are done with the values of the design definitions. An algorithm that should be followed by the IDS has been devised. This scheme involves three phases. Phase one performs the task of collecting data from the incoming messages. The accumulated data is then analyzed with the help of the rules in phase two. If the analysis results are unsuccessful, a failure is generated. The phase three checks whether the number of the failures is greater than the expected number of failures in the network. An affirmation in the checking procedure raises an alarm for intrusion detection alarm.

5 State of the Art Intrusion Detection Techniques

This section outlines the existing detection techniques as illustrated in Fig. 4 along with the new advancement in intrusion detection technique introduced recently. The next part of this section provides a vivid description of the underlying techniques. There are mainly two types of detection techniques that exist for sensor networks: signature detection and anomaly detection [46].

Signature detection basically creates a report of known attack signatures based on which IDS performs signature detection by comparing present activity with each of the stored attack reports. An alarm is generated by the system on finding a match. The major disadvantage of signature-based detection technique is that it fails to detect new types of attack. On the other hand, normal reports of system behaviour are prepared in case of anomaly detection. Here comparison is done between the system's normal reports and the current activity. The drawback of anomaly-based technique is that it generates of lots of false alarms [47].

A new approach known as specification-based technique [13], has been established recently and combines the best of both signature-based and anomaly-based intrusion detection systems. It develops specifications manually for describing the normal system behaviour, thereby, reducing the probability of false alarms. It is easier to apply this detection technique in sensor networks as normal behaviour is not possible to be simply defined by machine learning techniques and training. This technique requires norms to be defined that are used to describe normal operation.

In another work, [48], intrusion detection techniques have been divided into single-sensing detection and multi-sensing detection. In single-sensing detection, the attacker is successfully detected by one sensor node while in multi-sensing detection, several sensor nodes act together for detecting the intrusion.

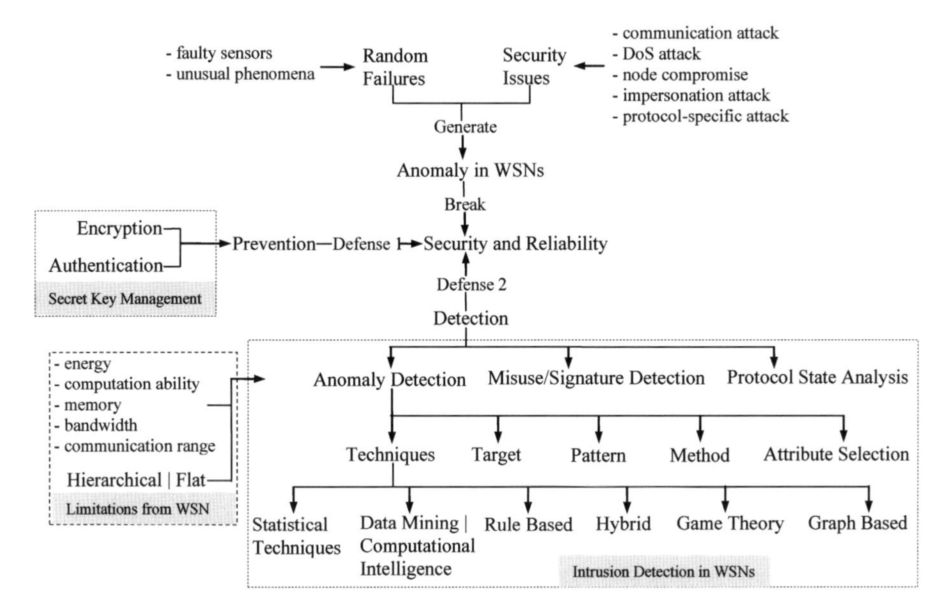

Fig. 4. Flowchart of Intrusion Detection Techniques in WSN

5.1 Misuse/Signature Detection

Misuse/signature detection is defined as a process of comparing signatures with known attack patterns (signatures) [49], [50]. Each signature is a pattern corresponding to a known threat. The misuse detection system tries to recognize any 'bad' behaviour according to these patterns. Misuse/signature detection involves complicated computations that require sizeable memory. Also this IDS requires large amount of memory for storing attack signatures. All the requirements stated above for misuse/signature-based intrusion detection does not make it a suitable choice as detection systems for WSNs that have very limited storage, communication and computational capabilities. Therefore, anomaly detection has been used for enhancing the security and reliability of WSNs.

5.2 Anomaly Detection

Here we discuss anomaly detection based on heterogeneous and homogeneous sensor networks. There are several aspects of key design principles that are followed such as target, typical security threats, detection pattern, detection method and attribute selection [10]. We mainly concentrate on the detection methods that are followed.

5.2.1 Anomaly Detection for Heterogeneous WSNs

Three types of detection methods exist for heterogeneous WSNs. In the first type the CH alone is responsible for the data processing task while in the second type CH along with the common sensor nodes accomplishes this and in the third type the base station solely carries out this procedure. For the first type the common sensor nodes are responsible for collecting the input datasets and/or partially participate for analysis and decision procedures [43], [51]. But this procedure leads to excess energy exhaustion for the CHs alone. This made way for introduction of the second [52], [53], [54], [55], [56] and third [57], [58] types of data processing procedures. In hierarchical WSNs several methods such as statistical techniques, data mining and computational intelligence, game theory, and hybrid detection have been employed for implementing detection schemes.

Every common sensor node collects the inputs, followed by a preprocessing procedure. The original/preprocessed inputs or local normal profiles are then sent to the CH or base station, where the global normal profile is produced either with a training algorithm, some prior-knowledge or a combing algorithm during the data processing procedure. The procedure of analysis and decision is carried out at each common sensor node or the CH respectively, or both. Finally, anomaly detection output is produced as a specific form having the analysis and decision procedure. These detection schemes follow a commonality in using hierarchical architecture for implementing detection within a distributed manner, which spreads the energy overhead around the entire network and reduce the communication burden on a single node. In a distributed detection scheme, as the common sensor nodes also contribute in the data processing procedure, therefore the burden on the CH is lightened.

Given below are some specific detection methods that also include the three major types of intrusion detection techniques in heterogeneous networks as mentioned above along with other detection methods.

5.2.1.1 Statistical Techniques

Distributed detection using kernel density estimator- Distributed detection is done using a kernel density estimator. Anomaly is identified by estimating the underlying distribution of sensed data. The local detection is accomplished by each common sensor node. The CH performs the task of collecting all local normal profiles for carrying out global detection within its group. Here timing parameters are used to make sure smooth delivery of streaming data.

Online detection using kernel density estimator- This detection technique uses online approximation of sensed data in a sliding window with 'chain-sample' algorithm. With respect to the previous technique several improvements are done in terms of reducing the size of the resulting set from two sensor nodes by warehousing of samples and also computation of the data by facilitating the combination of bandwidths.

Detection using statistical measures- To combat insider attacks such as exceptional message and abnormal behaviour detection mechanism is designed using statistical measures. Here two types of detection procedures are introduced. One of them designates the CH to watch over the other nodes in the cluster while in the other each common sensor node watches its one-hop neighbours. A random secret key pre-distribution mechanism is used for this detection scheme.

Detection using rules based on probability- One existing work uses a probability model as detection technique [59] in the rule-based scheme [13], for defending black-hole and selective forwarding attacks. The probability model ensures accurate measurement of traffic behaviours, leading to the decline in the false alarm rate of the rule-based detection scheme. Some of the common sensor nodes are selected as watchdogs, for monitoring the neighbouring nodes within their communication range while the CH is responsible for the analysis and decision procedure.

Advantages- The main advantages of statistical methods are- these techniques employ various measurements such as mean, variance, standard deviation that can be used in a wide variety of statistical distributions in networks. These methods are very flexible for use.

5.2.1.2 Data Mining and Computational Intelligence-Based Techniques

Distributed detection using K-means clustering- A K-means clustering algorithm is designed for a distributed detection scheme based on data mining and computational intelligence-based techniques [55]. Every common sensor node locally collects the input dataset for obtaining a normal profile. The CH is responsible for collecting all local normal profiles for data processing the output of which is a global normal profile. The global normal profile is sent to every common sensor node for initiating the analysis and decision procedure for detection.

Distributed detection using clustering ellipsoids- The work [59] proposed a distributed detection scheme based on clustering ellipsoids. The base station handles the computation of the global hyper-ellipsoid for accommodating non-homogenous data underlying the distributions. Here the common sensor nodes are involved in performing detection with the help of global hyper-ellipsoid.

Distributed detection using support vector machines- Another technique uses one-class quarter-sphere Support Vector Machines (SVM), as a representative algorithm of SVM, for distributed anomaly detection [56]. Each common sensor node evaluates the local quarter-sphere while the CH collects these locally computed radii for producing a global radius. After this, detection commences at each common sensor node using the global normal profile.

Detection using multi-agent and refined clustering- Multi-agents based detection scheme is introduced in [51] where advantage of self-organizing map neural network algorithm and K-means clustering algorithm are used. All nodes in the network are attached with detection agents including sentry, analysis, response, and management. The CH monitors the common sensor nodes under it, while a few of them are activated in terms of their remaining energy for playing the role of monitoring the CH. This scheme is advantageous from the point of speeding up detection accuracy as well as bringing down the false alarm rate.

Advantages- Data mining and computational intelligence algorithms-based detection schemes characterize by strong detection generality, meaning effective to defense against a wider range of security threats even if unknown. The tempting detection generality, of course, comes along with high complexity, such that these schemes' best effort are tried to operate in distributed manner.

5.2.1.3 Game Theory-Based Techniques

Non-cooperative game theory- Another approach based on game theory is introduced in the work [23] for locating the vulnerable areas in a WSN. It is based on many risk factors such as reliability of a sensor node, different types of attack, and past behaviours of the attacker. The identified areas are provided with the protection of detection instead of the entire network for preserving energy. Here intrusion detection is modeled as a game played between detection system and adversary based on some pre-defined strategies.

Advantages- Similar to the genetic algorithm (GA)-based scheme [57], non-cooperative game theory-based schemes are not concerned with detection immediately; however, it could assist detection schemes in advancing their performance as well as efficiency. The design of the payoff function is crucial to the forecasting accuracy, which is worth more studying. Moreover, if the GA-based scheme which is capable of optimizing the placement of the monitoring nodes could cooperate with the game theory-based scheme which enables identifying the vulnerable areas, it is expected that the detection schemes can achieve better performance.

5.2.1.4 Hybrid Detection Techniques

Detection with prevention technique- A hybrid detection framework is a combination of the energy-saving detection technique and the authentication based prevention technique. This detection scheme imparts the task of monitoring the common sensor nodes to the CH. A portion of the common senor nodes are selected on the basis of their residual energy for keeping a watch over their CH. Secret keys are established during initialization process. A secret key exists between the base station and common sensor nodes. Also the common sensor node shares a set of pairwise secret keys with its neighbouring nodes and a cluster secret key within a cluster. Another secret key known as the group secret key is shared among all sensor nodes in the network. The packets transmitted in the network are categorized as control messages and sensed data. When the base station, cluster head, or any intermediate node forwards a control message, a message authentication code (mac) is appended with proper secret key. The intermediate nodes forwarding this control message verify the appended mac and replace it with a new mac. The verifying and replacing of mac continues until this control message arrives at its destination.

Advantages- Few schemes mentioned in [54] to cooperate with a prevention-based technique in hierarchical WSNs. Moreover, the security foundation established with a prevention technique is only served as enhancing the security of the network, instead of taking advantage of the functions brought by the availability of secret keys. WSNs should have been protected by a security foundation [60]. Apparently, the detection scheme will be more efficient if capable of utilizing the functions provided by this security foundation, rather than making use of prevention and detection separately.

5.2.2 Anomaly Detection for Flat/Homogeneous WSNs

Similar to detection techniques in heterogeneous networks, the detection pattern in homogeneous/flat networks can also be classified into three types. The first category assigns some nodes designated as active nodes to watch over other nodes in their neighbourhood. The neighbourhood of these nodes can be one hop [61], [62], radio range [63], [64] or based on some other specifications [65], [66], [67]. The nodes monitoring their neighbours are also responsible for accomplishing data processing. The analysis and decision procedure is either done the active nodes or through cooperative manner. In the second type the base station performs anomaly detection in the network [68], [69], [70]. The third type partitions the network into groups. A portion of the sensor nodes in each group is activated to take charge of the monitoring and data processing procedure [17], [71].

Several specific detection methods are also used for anomaly detection in homogeneous sensor networks that also include the three major types as mentioned above have been discussed below-

5.2.2.1 Rule-Based Detection Techniques

Decentralized detection using rules- A decentralized rule-based scheme has been proposed [45], where a rule union is used to fulfill the specific demands of application scenarios. In this case the WSN comprises of common nodes, monitor nodes, intruder nodes, and base station. Each monitor node monitors the neighbours within its radio range. This scheme uses two phases. In the first phase, each monitor node collects messages and filters off the important information for subsequent analysis. The applicable rules are selected according to requirements during the second phase. This scheme provides a good framework for rule-based detection. But, the details of determining monitor nodes with regard to how many and which sensor nodes should be on duty for ensuring protection of the entire network has not been elaborated.

Detection using multi-hop acknowledgement - The mechanism of multi-hop acknowledgement (ACK) is used in a detection scheme for defending selective forwarding attack [67]. Detection remains active while packets are transmitted from the source node to the base station. The base station, intermediate nodes and source node are part of this communication. During node initialization every sensor node is loaded with a secret unique key that is shared by the base station and the node. One-to-many authentication is achieved using a one way hash function. The detection is carried out in two directions- upstream (from the source node to the base station) and downstream (from the base station to the source node). In the upstream detection, each intermediate node performs report packet, ACK packet, and alarm packet. As the report packet is forwarded in the upward direction, ACK packets are sent in the downward direction. If fewer amounts of ACK packets are received over a certain time period, an alarm packet is generated. This alarm packet reports that the next downward stream is doubtful. In the downstream detection, the intermediate node identifies the adversary node from discontinuous packet-ids of a particular source node. This scheme is simple and fast, but depends on a security foundation based on secret key management.

Detection using rules- For detecting black-hole and selective forwarding attacks, a detection scheme was proposed in [13]. Here some nodes called watchdog nodes are activated for monitoring and together they take the ultimate decision. A set of detection rules are followed. One rule states that if the packet drop rate of a node is more than a certain threshold limit within a time period, alarm is generated. Also if more than half of the watchdog nodes generate alarm against a specific node, that node is designated as a malicious one. This mechanism is energy efficient, as simple detection rules are used that require less communication overhead. But this scheme is unsuitable for application scenarios that require a high detection accuracy and low false alarm rate.

Detection using group deployment knowledge- Defense against node replication attacks was developed in a scheme [72] taking advantage of group deployment knowledge. The detection procedure starts at a sensor node when this node receives a request from its neighbour for forwarding a message. The use of a group deployment strategy is the key assumption in this scheme. In this strategy, sensor nodes are deployed in close proximity with the identical location termed as group deployment point using a probability density function. Similar to the multi-hop ACK-based scheme, here also a specific protocol is designed. But the performance of this scheme is not steady as it largely depends on the accurate deployment of the nodes towards the group deployment point.

Advantages- In general, these schemes rely on prior-knowledge, restricting their detection generality. But, the detection speed certainly benefits from no explicit training procedure.

5.2.2.2 Statistical Techniques

Detection using radio model- In one work a scheme has been proposed that deals with HELLO flood and wormhole attacks [64]. Every node monitors its neighbouring nodes within its communication range. If a node hears a message transmission from one of its neighbours, it starts detection. A message transmission is considered suspicious if its signal power does not match with its sender's geographical position. The final decision is then made by a vote mechanism.

Detection using packet arrival process- A detection scheme is proposed in [61] based on the statistical measure of the packet arrival process. Each sensor node keeps the normal traffic profile of its one-hop neighbour nodes, with the help of which anomaly is detected. Each sensor node separately maintains two buffers- one received buffer and another intrusion buffer for its one-hop neighbours. Evaluation of mean and standard deviation is performed. If the evaluation results in surpassing a threshold value an alarm is generated.

Detection using packet power levels and arrival rates- In another work [62], a scheme is developed taking into account packet power levels and arrival rates for identifying any anomalies. Here also each sensor node takes care of its one-hop neighbour nodes. A main packet buffer is maintained for recording the arrival times and received power of the latest packets. If the received power of an incoming packet is below the minimum value or above the maximum value of the received powers currently recorded in the main packet buffer, that particular incoming packet is regarded as anomalous and likewise an alarm is generated. Similarly arrival rate of packets are also checked and if found below the necessary threshold value, alarm is generated.

Detection using statistical distribution- Liu et al. [63] proposed an insider attacker detection scheme exploiting statistical distribution of the spatial correlations that exist among the networking behaviours of the sensor nodes lying close to each other. Each sensor node collects information from the neighbours within its communication range. A suspicious node is detected based on a voting decision. This detection scheme consists of four phases- information collection, false information filtering, outlier detection, and majority vote. The attributes chosen for this scheme are flexible, thus making it possible for the detection mechanism to be extended using other attributes. Further a false information filtering procedure is adapted so that the interference introduced by unattended adversaries can be prevented.

Detection using auto-regression model- An auto-regression (AR) model was proposed as the detection method in [68]. The detector is installed in the base station for examining whether the real value of the data measured by a sensor node is equal or approximately equal to the value predicted by the AR model. This model is chosen as it is efficient, accurate and flexible.

Detection using hop count- This method uses hop-count monitoring as the detection scheme for defending against sinkhole attack [65]. The nodes keep count of the hop-counts of the packets transmitting through them. The ad hoc on-demand distance vector routing protocol is used here. The nodes collect the hop counts of their neighbouring nodes when the base station initiates the network and periodically maintain routes by broadcasts.

Detection using quantitative measure- In this detection mechanism a data transmission quality (DTQ) function is proposed for identifying compromised nodes [17]. Also a voting procedure is used for taking the final decision. The entire network is divided into several groups. Every node maintains a DTQ table of its neighbours belonging to its group. Two communication scenarios are considered: intra-group and inter-group. The DTQ function varies steadily for legitimate nodes, whereas decreases continuously for suspicious nodes. This function identifies compromised nodes quantitatively taking attributes such as energy cost, data transmission quality, slack variable, etc into consideration. The use of weight-based voting mechanism increases the reliability of the proposed scheme.

Detection using grouping and statistical distribution- In [69] a group-based detection scheme has been proposed using a statistical distribution-based technique. The network is divided into a set of groups and the nodes within the same group are physically close to each other. Detection is carried out in two steps, using attributes such as sensed data, packet sending rate, packet dropping rate, packet mismatch rate, packet receiving rate, and packet sending power. At first, if a node detects a deviation, it alerts the other nodes in its group. After that if alert messages are received from the same node frequently, that node comes under the suspicion of other nodes in the group. This scheme is quite efficient as it employs interval estimation technique derived from statistics as well as a weight-based voting mechanism.

Advantages: Detection schemes based on statistical techniques are most popular. As these schemes go through a relatively complex training procedure, they achieve stronger detection generality than rule-based schemes.

5.2.2.3 Graph-Based Techniques

Detection using routing pattern- For defense against sinkhole attacks a detection scheme based on routing patterns was introduced by [70]. At first a lightweight algorithm is used for identifying the area under attack by the base station using the network flow's information. After that the base station localizes the intruders by modeling the attacked area with a graph, according to the routing pattern. This scheme uses a secret key-based security foundation thereby increasing path redundancy. Also an algorithm using hop counts is used for detecting many malicious nodes.

Advantage: Graph-based technique is presently not very much accepted for anomaly detection in WSNs. If it is supported by specifically designed routing protocol and security foundation, then graph-based detection schemes can be mad more applicable in sensor networks.

5.2.2.4 Data Mining and Computational Intelligence-Based Techniques

Detection using rule learner- A detection scheme based on association rule learning was proposed in [66]. Sensor nodes are equipped with an intrusion detection agent, which consists of a local intrusion detection component monitoring its host node and a packet-based intrusion detection component identifying malicious nodes using the communication activities of its neighbours. This method introduces an efficient strategy of rule evaluation as well as a model tuning algorithm that generates lower energy overhead and makes it robust enough.

Advantage: Data mining and computational intelligence-based IDS techniques are applicable in flat WSNs as they provide the detection service with good generality.

6 Intrusion Detection Based on OSI Layer

This section provides vivid discussion on the role played by IDS for defending different attacks affecting the working of the different protocol layers in WSNs. In [9] the authors have detailed various methodologies for detecting and protecting the layers from attacks using IDS.

6.1 Physical Layer

The physical layer which is used as the radio interfacing layer comes under attacks such as jamming and tampering. Mechanisms such as spread spectrum techniques and frequency hoping have been devised for prevention of such attacks. The authors [9] have used the value of Received Signal Strength Indicator (RSSI). After deployment the nodes initiate neighbour discovery during which the RSSI value of the neighbouring nodes are recorded by each node. The RSSI values provide indication for any kind of intrusion in the network. Any node after receiving a packet, checks the RSSI value of it. If the RSSI value of the received packet is within the correct range, the packet is accepted otherwise it is rejected.

Another work [54] provides intrusion detection technique for detecting jamming attack. Here the authors use two variables namely timer and counter along with the RSSI value for detecting this attack. A threshold is selected for RSSI indicated by TR. The value of counter and timer is indicated by n. If the RSSI value is greater than the threshold value TR, the counter value increases and thereby the value of n changes. If n goes on increasing and exceeds a particular threshold (T_n), the IDS detects jamming attack and alarm is generated by the system.

6.2 MAC Layer

Data link layer or MAC layer uses scheduling-based protocols for accessing the medium. The scheduling algorithms allots time schedule for each node. Scheduling algorithm such as TDMA designates specific slots for every node whereas SMAC algorithm provides sleep and wake up schedules for every node. The mechanism that has been used for detection is that if a particular node receives a packet from another node when the transmitter node is supposed to sleep, an alarm is generated i.e. the adversary node tries to possess as another normal node for launching attacks in the network. The authors in [9] have shown how intrusion detection is done in TDMA and SMAC algorithms which have been described below.

TDMA- Nodes use specific time slots allocated to them during which they transmit and receive data. As shown in Fig. 5 the available bandwidth is divided into frames and each frame is divided into time slots. The number of time slots of the TDMA frame depends on the length of the frame. Data is transmitted in the form of bursts in the time slots. The number of data bits i.e. the length of the data bursts transmitted through a time slot depends on the length of the time slot and the transmission bit rate of the system.

SMAC- This algorithm provides sleep and wake schedule to nodes as depicted in Fig. 6. During the wake schedule the nodes perform their tasks while they sleep during their sleep schedule. If a node performs activities during the time it should be sleeping, an intrusion is detected. The advantage of defending intrusion detection in MAC layer with the help of these scheduling algorithms is that very low overhead is required which makes it ideal for resource constrained networks.

Authors in [73] have proposed IDS for detecting attacks such as collision attack, exhaustion attack and unfairness attack that place in the data link layer. For detecting collision attack, the number of collisions taking place in the packets are observed

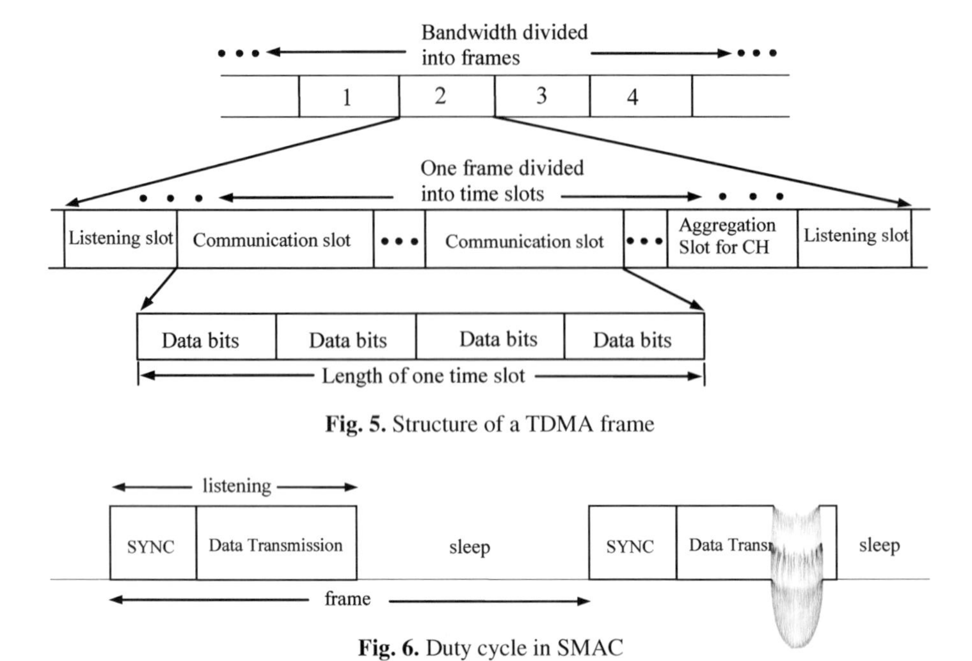

Fig. 5. Structure of a TDMA frame

Fig. 6. Duty cycle in SMAC

using variables timer and counter. If the number of collisions per second denoted by C exceeds a threshold (T_c), collision attack is detected by the IDS and the system is notified.

Exhaustion attack takes place if collision occurs during the end of the frame resulting in exhaustion of battery in the nodes. Similar to collision attack, here also the authors [54] have used the variables timer and counter. The battery is checked for any exhaustion that is indicated by the battery level. If the battery level goes low, the number of packets that are useless are checked. Thresholds for battery level and number of useless packets are kept. If the battery level becomes less than its threshold value while the number of useless packets exceeds the threshold, the counter is incremented and the value of timer also increases. Continuing in this manner if the value of the timer exceeds the threshold T_n, IDS detects exhaustion attack and informs the system.

Using IDS for detecting another weaker version of DoS attack i.e. unfairness attack has also been done. Unfairness attack occurs when collision attacks or exhaustion attacks or improper treatment of a MAC-layer priority scheme occurs frequently. Then number of collisions occurring in the packets per second (C) are checked along with the battery. It is checked whether C exceeds the threshold T_c, after which the battery level TL is checked followed by the number of useless packets TP. If C exceeds the threshold T_c, and if the battery gets lower than the threshold TL, as well as number of useless packets is also exceeds the limit TP, then counter increments and n also increases which is equal to timer value. After a point of time if n exceeds the threshold, unfairness attack is detected resulting in alarm generation in the system.

6.3 Routing Layer

Intrusion detection in the network layer has been done using a protocol named as information authentication for sensor network (IASN) by the authors in [9]. The IASN protocol works with routing protocols such as Dynamic Source Routing, Destination Sequenced Distance Vector and Directed Diffusion. Authors have shown how intrusion detection is achieved using each of these routing protocols. Authentication is done on the received data instead of user authentication. Every node keeps track of its neighbours and is responsible for knowing what type of information it is supposed to receive from its neighbours. When any information is received from a neighbouring node, that information is matched against that neighbour. If that information is not supposed to come from that particular node, it is assumed that adversaries are trying to inject malicious packets in the network.

Authors in [73] have devised IDS for selective forwarding attack where adversary nodes drop certain packets, refuse to forward or propagate them any further [74]. Timer and counter are kept as variables. The number of packets dropped by nodes is detected first. After that it is checked whether a node is dropping all the packets by observing the sequence numbers or some of the incoming packets. Based on this the counter is incremented as well as the timer value n. If n exceeds the threshold value after a period of time, selective forwarding attack is detected. Otherwise if all the packets are dropped black hole attack is detected.

Homing Attack- Some powerful nodes might be served as cryptographic key managers, query or monitoring access points, or network uplinks. Once found, these nodes can be attacked by collaborators or mobile adversaries using other active means. For detecting homing attacks, the IDS first checks whether the sensor network has monitoring nodes or not. If monitoring nodes are present, two parameters- packet sending time and packet sending rate are taken into account. It is checked whether packet monitoring node is monitoring the rate of sending the packets to the neighbouring nodes or it is monitoring the time between each packet that is sent out by the neighbouring nodes. If rate and time are monitored the counter increases by one along with the timer value. After a point of time if the timer value crosses the threshold value T_n, IDS detects homing attack taking place and alerts the system.

Misdirection Attack- This type of attacks the malicious node forwards messages along wrong paths causing flooding to occur in the network. Hello flood attack falls under this category where the attacker creates a situation where the legitimate nodes believe that the attacker node is their neighbouring node. For detecting this attack the authors [54] have used a checking mechanism where the number of the hello packets received by a node is noted. If the number of hello packets received by a node is more than a specific threshold, the IDS detects hello flood attack in the network.

Another type of attack that also belongs to this category is known as sinkhole attack. Here the adversary compromises a node and makes it highly attractive as a routing node for neighbours. The neighbouring nodes are duped by the malicious node which advertises itself as a high quality route to the base station. The neighbouring nodes then forward their packets through the attacker node resulting in sinkhole attack. This attack is detected by the IDS using a checking method whether any node has high quality route, strong transmission, or bandwidth. It is verified with the next hop for the route. If the next hop confirms the route to the base station as good, the packet passes through the node. Otherwise no packets are forwarded through that route as the IDS detects sinkhole attack in the network.

Wormhole Attack- Authors [74] have detected another type of attack known as the wormhole attack where the attacker receives packets from one part of the network, and tunnels them through a private channel or wormhole to a different part of the network where the packets are, then replayed. This attack has been detected by initially using leashes. The generation of leashes must maintain that all nodes tightly synchronize their clocks and the maximum difference between the clocks of any two nodes must be Δ. The value of Δ is known to all nodes in the network. The sending node while sending the packet includes in it the time (t_s) at which it is sent. The receiver compares the value of t_s with the time at which it receives the packet denoted by t_r. Based on calculations using transmission time and speed of light the receiver is able to detect whether the packet has traveled too far. If the result of calculation by the receiver is positive no further packets are accepted by the receiver and IDS detects the occurrence of wormhole attack.

6.4 Transport Layer

A number of attacks exist in transport layer in WSNs e.g., flooding attack, de-synchronization attack etc. Several mechanisms using IDS have been proposed to combat these attacks which we have briefly discussed below.

Flooding Attack- Flooding attack takes place in the transport layer where the adversary sends several connection establishment requests to the victim node, making the node exhaust its energy [75]. Two variables counter and timer are used by the IDS for detecting this attack. The detection is done on the basis of whether the number of requests for connection per second is exceeding a particular threshold. If the threshold is exceeded, flooding attack is detected and the IDS generates an alarm for the network.

De-synchronization Attack- De-synchronization is another transport layer attack where message packets are captured by the adversary and it fakes these using wrong sequence numbers before releasing them in the network. This leads to increase in network traffic ultimately causing congestion. Here also the two variables counter and timer are used both of which are initialized to zero. The difference between the missing sequence numbers are calculated for every receiving node. If the difference in sequence numbers is more than a specific threshold, de-synchronization attack is detected by the IDS.

Time Synchronization Attack- Time synchronization attack taking place in the transport layer is detected using IDS by comparing the time between the suspected captured node and its neighbouring nodes. If time difference occurs between these nodes it can be concluded that time synchronization attack has occurred and the network is notified by the IDS.

6.5 Application Layer

At the application layer the authors in [9], have proposed mutual guarding techniques and use of round trip times for enhancing security. Unfortunately, round trip times have very high false positive rates because of background noise, weather, etc. In the mutual guarding technique, the authors described how nodes guard each other and give an example of four nodes guarding each other. If an intruder tries to attack from the mutual area of these four nodes, then other three nodes will detect an anomaly. Their paper does not describe how the sensor nodes will be organized in the network.

Node Capture Attack- Authors [54] have used IDS for detecting node capture attack where an attacker captures one or more of these nodes and tries to send malicious messages throughout the network. Here whether a node has been captured is checked using mobile agent nodes. Node capture attack takes place either in JTAG, bootstrap loader or external flash of the sensor nodes. If the code in the node has been modified JTAG type of node capture attack is detected by the IDS. If the node has USB access, it is checked whether the source code and compiler has changed, if so bootstrap attack is detected. If neither JTAG nor bootstrap attack is detected, the IDS checks the node's I/O pin connector for an overwritten microprocessor. The result of the checking if positive the IDS alarms the network of external flash type node capture attack.

7 Potential Research Areas

This section provides an insight into the research areas of intrusion detection of sensor networks where very little has been done till date. Areas where further improvements can be done are also described briefly.

Cooperative Intrusion Detection- Cooperative intrusion detection envisages the intrusion detection mechanism among few nodes also known as monitoring nodes that collectively work for defending intrusions in the network. Few works [19], [21] have been proposed for cooperative intrusion detection in sensor networks. Potential research area includes defensive mechanism for multiple attacker scenarios. Also balancing the tradeoff among network topology and different objectives such as power consumption, security effectiveness etc. for cooperative intrusion detection is coming up as a prospective area of future research.

Intrusion Detection for Advanced Metering Infrastructures- Another area where not much has been done regarding intrusion detection is advanced metering infrastructures (AMI) [76]. These infrastructures include several communication networks such as wide area networks, neighbourhood area networks etc. Providing flawless security is of utmost important for this architecture which is one of the promising areas that needs the attention of researchers. The detection techniques should be such that the false alarm rate is nearly negligible as these infrastructures are used in real time. Also how automated response and recovery actions can be employed needs to be developed.

Hierarchical Intrusion Detection for Wireless Industrial Sensor Networks- Intrusion detection in hierarchical wireless industrial sensor networks [77] is an evolving area that needs much attention. How the different attacks prevalent in sensor networks can be defended in wireless industrial sensor networks is an upcoming research area. Moreover how the usage of collaborative cluster environments can influence the working of wireless industrial sensor networks needs to be highlighted.

Energy Efficient Learning Solution for Intrusion Detection- We are very much aware about the fact that sensor nodes are highly resource constrained and energy conservation is the most important criteria that needs to be fulfilled by the sensor network. Therefore mechanisms should be developed such that the learning solutions used by the intrusion detection systems [78] are as much energy efficient as possible. Therefore mechanisms should be devised for reducing the overhead incurred due to expense of energy in the network.

Intrusion Detection Systems in Mobile WSNs- Very few works have been done on intrusion detection in mobile wireless sensor networks [79] considering the characteristics of sensor nodes which gets more complicated in the mobile environment. For IDS to be implemented in mobile WSNs needs to be decentralized, minimize the traffic overhead and address the mobility problem. Addressing the mobility issues for intrusion detection in WSNs is an open research problem.

8 Conclusion

In this survey work we address the different issues of intrusion detection system in wireless sensor networks related to the different architectures, models along with the various detection mechanisms prevalent in both homogenous and heterogeneous sensor networks. Also the underlying challenges for implementing intrusion detection systems in WSNs are also dealt with. The different attacks taking place in the various layers of the sensor networks and also how proficient IDS is to be able to thwart them are also discussed. The literature review section deals with detailed description of previous works based on different approaches.

Finally the paper concludes highlighting the important research areas that hold potential for future works. Areas where very less work has been done and how improvisation can be brought about has also been discussed in this work.

References

1. Halder, S., Ghosal, A., DasBit, S.: A Pre-determined Node Deployment Strategy to Prolong Network Lifetime in WSN. Computer Communication 34(11), 1294–1306 (2011)
2. Heady, R., Lugar, G., Servilla, M., Maccabe, A.: The Architecture of a Network Level Intrusion Detection System. Technical Report, Computer Science Department, University of New Mexico (1990)
3. Anderson, J.P.: Computer Security Threat Monitoring and Surveillance. Technical Report, James. P. Anderson Co., Fort Washington, Pennsylvania (1980)

4. Denning, D.: An Intrusion Detection Model. IEEE Transactions on Software Engineering 13(2), 222–232 (1987)
5. Beyah, R.A., Holloway, M.C., Copeland, J.A.: Invisible Trojan: An Architecture, Implementation and Detection Method. In: Proceedings of 45th Midwest Symposium on Circuits and Systems, vol. 3, pp. 500–504 (2002)
6. Debar, H., Dacier, M., Wespi, A.: Towards a Taxonomy of Intrusion-Detection Systems. Computer Networks 31(8), 805–822 (1999)
7. Kumar, S.: Classification and Detection of Computer Intrusions. In PhD Thesis, Purdue University (1995)
8. Kaplantzis, S.: Security Models for Wireless Sensor Networks. In PhD Conversion Report, Centre of Telecommunications and Information Engineering, Monash University, Australia (2006)
9. Ghosal, A., Halder, S., DasBit, S.: A Dynamic TDMA Based Scheme for Securing Query Processing in WSN. Wireless Networks 18(2), 165–184 (2012)
10. Xie, M., Han, S., Tian, B., Parvin, S.: Anomaly Detection in Wireless Sensor Networks: A Survey. Journal of Network and Computer Applications 34(4), 1302–1325 (2011)
11. Mitrokotsa, A., Karygiannis, A.: Intrusion Detection Techniques in Sensor Networks. In: Lopez, J., Zhou, J. (eds.) Wireless Sensor Network Security, pp. 251–272. IOS Press, Amsterdam (2008)
12. Loo, C.E., Ng, M.Y., Leckie, C., Palaniswami, M.: Intrusion Detection for Routing Attacks in Sensor Networks. Int'l Journal of Distributed Sensor Networks 2(4), 313–332 (2006)
13. Krontiris, I., Dimitriou, T., Freiling, F.C.: Towards Intrusion Detection in Wireless Sensor Networks. In: Proceedings of 13th European Wireless Conference, pp. 1–7 (2007)
14. Krontiris, I., Dimitriou, T., Giannetsos, T., Mpasoukos, M.: Intrusion Detection of Sinkhole Attacks in Wireless Sensor Networks. In: Kutyłowski, M., Cichoń, J., Kubiak, P. (eds.) ALGOSENSORS 2007. LNCS, vol. 4837, pp. 150–161. Springer, Heidelberg (2008)
15. Hai, T.H., Huh, E.N.: Detecting Selective Forwarding Attacks in Wireless Sensor Networks Using Two-hops Neighbor Knowledge. In: Proceedings of 7th IEEE Int'l Symposium on Network Computing and Applications, pp. 325–331 (2008)
16. Misra, S., Krishna, P.V., Abraham, K.I.: A Simple Learning Automata-Based Solution for Intrusion Detection in Wireless Sensor Networks. Wireless Communications and Mobile Computing, Special Issue: Architectures and Protocols for Wireless Mesh, Ad Hoc, and Sensor Networks 11(3), 426–441 (2011)
17. Li, G., He, J., Fu, Y.: Group-Based Intrusion Detection System in Wireless Sensor Networks. Computer Communications 31(18), 4324–4332 (2008)
18. Krontiris, I., Giannetsos, T., Dimitriou, T.: LIDeA: A Distributed Lightweight Intrusion Detection Architecture for Sensor Networks. In: Proceedings of 4th Int'l Conference on Security on Privacy for Communication Networks, article 20 (2008)
19. Krontiris, I., Benenson, Z., Giannetsos, T., Freiling, F.C., Dimitriou, T.: Cooperative Intrusion Detection in Wireless Sensor Networks. In: Roedig, U., Sreenan, C.J. (eds.) EWSN 2009. LNCS, vol. 5432, pp. 263–278. Springer, Heidelberg (2009)
20. Stetsko, A., Folkman, L., Matyáš, V.: Neighbor-Based Intrusion Detection for Wireless Sensor Networks. Technical Reports: FIMU-RS-2010-04, 33 pages (2010)
21. Hassanzadeh, A., Stoleru, R.: Towards Optimal Monitoring in Cooperative IDS for Resource Constrained Wireless Networks. In: Proceedings of 20th Int'l Conference ICCCN, pp. 1-8 (2011)

22. Agah, A., Das, S.K., Basu, K., Asadi, M.: Intrusion Detection in Sensor Networks: A Non-cooperative Game Approach. In: Proceedings of 3rd IEEE Int'l Symposium on Network Computing and Applications, pp. 343–346 (2004)
23. Agah, A., Das, S.K.: Preventing DoS Attacks in Wireless Sensor Networks: A Repeated Game Theory Approach. Int'l Journal of Network Security 5(2), 145–153 (2007)
24. Premkumar, K., Kumar, A.: Optimal Sleep–Wake Scheduling for Quickest Intrusion Detection using Sensor Networks. In: Proceedings of Int'l Conference IEEE INFOCOM, pp. 2074–2082 (2008)
25. Servin, A., Kudenko, D.: Multi-Agent Reinforcement Learning for Intrusion Detection: A Case Study and Evaluation. In: Bergmann, R., Lindemann, G., Kirn, S., Pěchouček, M. (eds.) MATES 2008. LNCS (LNAI), vol. 5244, pp. 159–170. Springer, Heidelberg (2008)
26. Kaltiokallio, O., Bocca, M.: Real-Time Intrusion Detection and Tracking in Indoor Environment Through Distributed RSSI Processing. In: Proceedings of 17th Int'l Conference on Embedded and Real-Time Computing Systems and Applications, vol. 1, pp. 61–70 (2011)
27. Deng, J., Han, R., Mishra, S.: INSENS: Intrusion-Tolerant Routing for Wireless Sensor Networks. Computer Communications 29(2), 216–230 (2006)
28. Lee, S.B., Choi, Y.H.: A Secure Alternate Path Routing in Sensor Networks. Computer Communications 30(1), 153–165 (2006)
29. Challala, Y., Ouadjaoutb, A., Laslab, N., Bagaab, M., Hadjidj, A.: Secure and Efficient Disjoint Multipath Construction for Fault Tolerant Routing in Wireless Sensor Networks. Journal of Network and Computer Applications 34(4), 1380–1397 (2011)
30. Lou, W., Kwon, Y.: H-Spread: A Hybrid Multipath Scheme for Secure and Reliable Data Collection in Wireless Sensor Networks. IEEE Transactions on Vehicular Technology 55(4), 1320–1333 (2006)
31. Hai, T.H., Huh, E.N., Jo, M.: A Lightweight Intrusion Detection Framework for Wireless Sensor Networks. Wireless Communications and Mobile Computing 10(4), 559–572 (2010)
32. Chen, I.R., Wang, Y., Wang, D.C.: Reliability of Wireless Sensors with Code Attestation for Intrusion Detection. Information Processing Letters 110(17), 1–9 (2010)
33. Shen, S., Li, Y., Xua, H., Cao, Q.: Signaling Game based Strategy of Intrusion Detection in Wireless Sensor Networks. Computers and Mathematics with Applications 62(6), 2404–2416 (2011)
34. Wanga, Y., Lun, Z.: Intrusion Detection in a K-Gaussian Distributed Wireless Sensor Network. J. Parallel Distributed Computing 71(12), 1598–1607 (2011)
35. Wang, S.S., Yan, K.Q., Wang, S.C., Liu, C.W.: An Integrated Intrusion Detection System for Cluster-based Wireless Sensor Networks. Expert Systems with Applications 38(12), 15234–15243 (2011)
36. Wang, Y., Fu, W., Agrawal, D.P.: Gaussian Versus Uniform Distribution for Intrusion Detection in Wireless Sensor Networks. IEEE Trans. on Parallel and Distributed Systems (2012), doi:10.1109/TPDS.2012.105
37. Strikos, A.A.: A Full Approach for Intrusion Detection in Wireless Sensor Networks. In: Wireless and Mobile Network Architectures, School of Information and Communication Technology KTH - Royal Institute of Technology (2008) (online course material)
38. Brutch, P., Ko, C.: Challenges in Intrusion Detection for Wireless Ad-Hoc Networks. In: Proceedings of Symposium on Applications and the Internet Workshops, pp. 368–373 (2003)

39. Abuhelaleh, M.A., Elleithy, K.M.: Security in Wireless Sensor Networks: Key Intrusion Detection Module in SOOAWSN. In: Proceedings of the 14th Communications and Networking Symposium, vol. 3, pp. 56–61 (2011)
40. Doumit, S., Agrawal, D.P.: Self-organized Criticality & Stochastic Learning Based Intrusion Detection System for Wireless Sensor Network. In: Proceedings of Int'l Conference IEEE MILCOM, vol. 1, pp. 609–614 (2003)
41. Ourstou, D., Matzner, S., Stump, W., Hopkins, B., Richards, K.: Identifying Coordinated Internet Attacks. In: Proceedings of 2nd SSGRR Conference (2001)
42. Park, H.-J., Cho, S.-B.: Privilege Flows Modeling for Effective Intrusion Detection Based on HMM. Department of Computer Science. Yonsei University, Seoul
43. Su, C.-C., Chang, K.-M., Kuo, Y.-H., Horng, M.F.: The New Intrusion Prevention and Detection Approaches for Clustering-based Sensor Networks. In: Proceedings of Int'l Conference IEEE Wireless Communications and Networking Conference, vol. 4, pp. 1927–1932 (2005)
44. Zhu, S., Setia, S., Jajodia, S.: LEAP: Efficient Security Mechanisms for Large-Scale Distributed Sensor Networks. In: Proceedings of 10th ACM Int'l Conference on Computer and Communications Security, pp. 62–72 (2003)
45. Silva, A.D., Martins, M., Rocha, B., Loureiro, A., Ruiz, L., Wong, H.: Decentralized Intrusion Detection in Wireless Sensor Networks. In: Proceedings of 1st ACM Int'l Workshop on Quality of Service & Security in Wireless and Mobile Networks, pp. 16–23 (2005)
46. Zhang, Y., Lee, W., Huang, Y.: Intrusion Detection Techniques for Mobile Wireless Networks. Wireless Networks 9(5), 545–556 (2003)
47. Cuppens, F., Miege, A.: Alert Correlation in a Cooperative Intrusion Detection Framework. In: Proceedings of IEEE Symposium on Security and Privacy, pp. 202–215 (2002)
48. Wang, Y., Wang, X., Xie, B., Wang, D., Agrawal, D.P.: Intrusion Detection in Homogeneous and Heterogeneous Wireless Sensor Networks. IEEE Transaction on Mobile Computing 7(6), 698–711 (2008)
49. Mishra, A., Nadkarni, K., Patcha, A.: Intrusion Detection in Wireless Ad Hoc Networks. IEEE Wireless Communications 11(1), 48–60 (2004)
50. Farooqi, A.H., Khan, F.A.: A Survey of Intrusion Detection Systems for Wireless Sensor Networks. Int'l Journal of Ad Hoc and Ubiquitous Computing 9(2), 69–83 (2012)
51. Wang, H.B., Yuan, Z., Wang, C.D.: Intrusion Detection for Wireless Sensor Networks Based on Multi-agent and Refined Clustering. In: Proceedings of Int'l Conference on Communications and Mobile Computing, pp. 450–455 (2009)
52. Palpanas, T., Papadopoulos, D., Kalogeraki, V., Gunopulos, D.: Distributed Deviation Detection in Sensor Networks. ACM SIGMOD 32(4), 77–82 (2003)
53. Subramaniam, S., Palpanas, T., Papadopoulos, D., Kalogeraki, V., Gunopulos, D.: Online Outlier Detection in Sensor Data using Non-parametric Models. In: Proceedings of 32nd Int'l Conference on Very Large Data Bases, pp. 187–198 (2006)
54. Zhang, Y.-Y., Yang, W.-C., Kim, K.-B., Park, M.-S.: Inside Attacker Detection in Hierarchical Wireless Sensor Network. In: Proceedings of 3rd Int'l Conference on Innovative Computing Information and Control, p. 594 (2008)
55. Rajasegarar, S., Leckie, C., Palaniswami, M., Bezdek, J.C.: Distributed Anomaly Detection in Wireless Sensor Networks. In: Proceedings of 10th IEEE Int'l Conference on Communication Systems, pp. 1–5 (2006)

56. Rajasegarar, S., Leckie, C., Palaniswami, M., Bezdek, J.C.: Quarter Sphere Based Distributed Anomaly Detection in Wireless Sensor Networks. In: Proceedings of IEEE Int'l Conference on Communications, pp. 3864–3869 (2007)
57. Rahul, K., Liu, H., Chen, H.-H.: Reduced Complexity Intrusion Detection in Sensor Networks using Genetic Algorithm. In: Proceedings of IEEE Int'l Conference on Communications, pp. 1–5 (2009)
58. Mashtaghis, M., Rajasegarar, S., Leckie, C., Karunasekera, S.: Anomaly Detection by Clustering Ellipsoids in Wireless Sensor Networks. In: Proceedings of 5th Int'l Conference on Intelligent Sensors, Sensor Networks and Information Processing, pp. 331–336 (2009)
59. Tiwari, M., Arya, K.V., Choudhari, R., Choudhary, K.S.: Designing Intrusion Detection to Detect Black Hole and Selective Forwarding Attack in WSN Based on Local Information. In: Proceedings of 4th Int'l Conference on Computer Sciences and Convergence Information Technology, pp. 824–828 (2009)
60. Perrig, A., Szewczyk, R., Wen, V., Culler, D.E., Tygar, J.D.: SPINS: Security Protocols for Sensor Networks. In: Proceedings of 7th ACM Int'l Conference on Mobile Computing and Networks, pp. 189–199 (2001)
61. Onat, I., Miri, A.: A Real-time Node-based Traffic Anomaly Detection Algorithm for Wireless Sensor Networks. In: Proceedings of Int'l Conference on Systems Communications, pp. 422–427 (2005)
62. Onat, I., Miri, A.: An Intrusion Detection System for Wireless Sensor Networks. In: Proceedings of IEEE Int'l Conference on Wireless and Mobile Computing, Networking and Communications, vol. 3, pp. 253–259 (2005)
63. Liu, F., Cheng, X., Chen, D.: Insider Attacker Detection in Wireless Sensor Networks. In: Proceedings of Int'l Conference 26th IEEE INFOCOM, pp. 1937–1945 (2007)
64. Júnior, W.R.P., Figueiredo, T.H.P., Wong, H.C., LoureiroPires, A.A.F.: Malicious Node Detection in Wireless Sensor Networks. In: Proceedings of 18th Int'l Conference on Parallel and Distributed Processing Symposium, p. 24b (2004)
65. Dallas, D., Leckie, C., Ramamohanarao, K.: Hop-Count Monitoring: Detecting Sinkhole Attacks in Wireless Sensor Networks. In: Proceedings of 15th IEEE Int'l Conference on Networks, pp. 176–181 (2007)
66. Yu, Z., Tsai, J.J.P.: A Framework of Machine Learning Based Intrusion Detection for Wireless Sensor Networks. In: Proceedings of IEEE Int'l Conference on Sensor Networks, Ubiquitous and Trustworthy Computing, pp. 272–279 (2008)
67. Yu, B., Xiao, B.: Detecting Selective Forwarding Attacks in Wireless Sensor Networks. In: Proceedings of 20th Int'l Symposium on Parallel and Distributed Processing (2006)
68. Curiac, D.-I., Banias, O., Dragan, F., Volosencu, C., Dranga, O.: Malicious Node Detection in Wireless Sensor Networks using an Autoregression Technique. In: Proceedings of 3rd Int'l Conference on Networking and Services, p. 83 (2007)
69. Li, T., Song, M., Alam, M.: Compromised Sensor Nodes Detection: A Quantitative Approach. In: Proceedings of 28th Int'l Conference on Distributed Computing Systems Workshops, pp. 352–357 (2008)
70. Ngai, E.C.H., Liu, J., Lyu, M.R.: On the Intruder Detection for Sinkhole Attack in Wireless Sensor Networks. In: Proceedings of IEEE Int'l Conference on Communications, vol. 8, pp. 3383–3389 (2006)
71. Ngai, E.C.H., Liu, J., Lyu, M.R.: An Efficient Intruder Detection Algorithm Against Sinkhole Attacks in Wireless Sensor Networks. Computer Communications 30(11-12), 2353–2364 (2007)

72. Ho, J.-W., Liu, D., Wright, M., Das, S.K.: Distributed Detection of Replica Node Attacks with Group Deployment Knowledge in Wireless Sensor Networks. Ad Hoc Networks 7(8), 1476–1488 (2009)
73. Qureshi, R., Chang, M., Weerasinghe, H., Fu, H.: Intrusion Detection System for Wireless Sensor Networks. In: Proceedings of Int'l Conference on Security & Management, pp. 582–585 (2008)
74. Karlof, C., Wagner, D.: Secure Routing in Wireless Sensor Networks: Attacks and Countermeasures. Ad Hoc Networks, Special Issue on Sensor Network Applications and Protocols 1(2-3), 293–315 (2003)
75. Roosta, T., Shieh, S., Shastry, S.: Taxonomy of Security Attacks on Sensor Networks. In: Proceedings of IEEE Int'l Conference on System Integration and Reliability Improvements (2006)
76. Berthier, R., Sanders, W.H.: Specification-Based Intrusion Detection for Advanced Metering Infrastructures. In: Proceedings of 17th IEEE Pacific Rim Int'l Symposium on Dependable Computing, pp. 184–193 (2011)
77. Shin, S., Kwon, T., Jo, G.-Y., Park, Y., Rhy, H.: An Experimental Study of Hierarchical Intrusion Detection for Wireless Industrial Sensor Networks. IEEE Transactions on Industrial Informatics 6(4), 744–757 (2010)
78. Misra, S., Krishna, P.V., Abraham, K.I.: Energy Efficient Learning Solution for Intrusion Detection in Wireless Sensor Networks. In: Proceedings of 2nd Int'l Conference on Communication Systems and Networks, pp. 1–6 (2010)
79. Mostarda, L., Navarra, A.: Distributed Intrusion Detection Systems for Enhancing Security in Mobile Wireless Sensor Networks. Int'l Journal of Distributed Sensor Networks 4(2), 83–109 (2008)

Network Coding for Security in Wireless Reconfigurable Networks

Rafaela Villalpando-Hernández[1], Cesar Vargas-Rosales[2], David Muñoz-Rodríguez[2], and Fernando Ruiz-Trejo[2]

[1] Tecnológico de Monterrey Campus Laguna
[2] Tecnológico de Monterrey Campus Monterrey
{rafaela.villalpando,cvargas,dmunoz}@itesm.mx, ferjorz@gmail.com

Abstract. Wireless Reconfigurable Networks (WRN) adapt rapidly and flexibly to network variations, providing advantages to establish efficient communication for emergency operations, disaster relief efforts, and military networks. Security is a necessity where data integrity and confidentiality are exposed to attacks. Security schemes are based on cryptography, providing an expensive and partial defense, since high processing needs are inconvenient for WRN. Hence, the design of a distributed, low cost detection and defense mechanism is important.

In this chapter, we present the fundamentals of network coding in WRN, its advantages and how particular problems in wireless networks limit those. We provide an algebraic representation of a distributed, low cost Detection and Defense Mechanism (DDM) that responds to the WRN demands. We evaluate quality of routes involved in the security mechanism, as well as make a selection of the best route for the DDM. The DDM uses network coding to distribute information, and to detect and defend from sink holes and selective forwarding attacks. For performance, we include the number of successful packets, overhead and accuracy in terms of detected attacks and false detections.

1 Introduction

A Wireless Reconfigurable Network (WRN) is a collection of two or more wireless communications devices or nodes, with networking capabilities. Such devices can communicate directly with other nodes that are located within their coverage area defined by their radio range. For those nodes outside their radio range, multi-hop style communication can be established. For the latter scenario, an intermediate node is used to relay or forward the information toward the destination. Since the communication takes place between a pair of devices and every destination can be reached by a multi-hop trajectory, WRNs are wireless networks that do not need infrastructure such as the base stations for the cellular networks. This lack of infrastructure defines the network since its nodes will move independently from each other, and this makes reaching the destinations a self-organizing and self-administer task.

A WRN is self-organizing and adaptive wireless network where the term reconfigurable implies that the network can take different forms, can be mobile,

S. Khan and A.-S.K. Pathan (Eds.): *Wireless Networks and Security*, SCT, pp. 369–402.
DOI: 10.1007/978-3-642-36169-2_11

standalone, or networked. Reconfigurable nodes or devices must be aware of their environment since they should be able to detect the presence of other devices and perform the necessary handshaking to allow communications and the sharing of information and services. Applications for WRNs can include the military, disaster areas, commercial sector due to the miniaturization of electronic devices, their proliferation and the growing desire of people to be connected all the time, [1]. WRNs present several challenges for any security mechanism due to multiple impairments that these networks face. WRNs have nodes that communicate through wireless links in a point to point fashion that modify the topology upon the requirements of the current network conditions, and at that accommodate large number of nodes adapting rapidly and flexibly to the constant variations of the network topology. The nodes are generally heterogeneous mobile devices with the responsibility of discovering the current network topology and performing basic networking functions like packet forwarding and route discovery without the intervention of a central unit. Significant applications include establishing survivable, efficient, dynamic communication for emergency/rescue operations, disaster relief efforts, and military networks. Such network scenarios cannot rely on centralized and organized connectivity because they are subject to periodical topology changes, these can be conceived as applications of Mobile Ad Hoc Networks (MANET). A mobile *ad hoc* network (MANET) consists of a set of mobile nodes that perform basic networking functions like packet forwarding, routing, and service discovery without the intervention of a fixed infrastructure (e.g. base station), [1], [2]. All network activity including discovering the topology and delivering messages must be executed by the nodes themselves. A sensor network is a heterogeneous system combining tiny sensors and actuators with general purpose computing elements. Sensor networks may consist of hundreds or thousands of low-power, low-cost nodes, possibly mobile but more likely at fixed locations, deployed en masse to monitor and affect the environment, [3].

Ad hoc/sensor networks require efficient distributed protocols to determine network organization, link scheduling, routing, information capacity, security, position location, etc. Several impairments such as variable wireless link quality, propagation path loss, channel fading, multiuser interference, power and topological changes, become relevant issues at the time of designing required network protocols. Security in WRNs is a significant requirement to implement several applications where data integrity and confidentiality can be exposed to a wide variety of attacks. Applications requiring data confidentiality and integrity have increased significantly, especially for wireless networks. Some examples of such applications are military tactical operations, law enforcement and financial operations. To secure a WRN, several attributes must considered such as availability, confidentiality, integrity, authentication, and non-repudiation [4]. Also, several characteristics of wireless reconfigurable networks must be taken into account for the design of security algorithms, [2], and these define several challenges:

- First, communication through wireless links makes wireless networks vulnerable to link attacks, active impersonation, and message distortion. Secret information, violating confidentiality can be achieved by eavesdropper since information is exposed through the wireless link. Messages can be deleted modified or injected by active attacks, thus violating availability, and integrity, [5].

- Second, in sensor networks we find nodes with relatively poor physical protection, these nodes present high probability of being compromised by a more robust attacking node. Therefore, we should take into account the attacks commanded by compromised nodes in the network.
- Third WRNs are subject to periodical and constant changes in topology (i.e., nodes frequently join and leave the network). Therefore trust relationship among nodes must also be modified properly.

There are other types of security attacks, such as attacks intended to modify the routing information of the nodes in the network, in such a way that the information may never arrive to its destination. Routing protocols may be susceptible to several routing attacks like: spoofing (altering the routing information), selective forwarding, sinkhole attacks, wormholes, HELLO flood attacks, and others, [6]. These security attacks can make impossible to attain communication between nodes in the network by corrupting the routing information, without the recognition of the nodes affected. In general, all the discussed security problems negatively interfere in the performance and in the quality of service that the network can provide to mobile users, because of the contamination of the information to be transmitted. If one or several security problems are present in the network, there are no guarantees that ensure that information will be delivered to desired destinations with the appropriate security level. This causes several problems such as network congestion, corruption of routing tables, and so on. It is possible to help to attenuate such problems by introducing a central entity into a security solution. However, this implies that if the central unit is compromised by an adversary the entire security solution will be compromised. Therefore, any security mechanism that depends on a central unit with static configuration is vulnerable in WRN environments. For WRNs, it is desirable to design security mechanisms capable to adapt on-the-fly to security attacks and network conditions. Several security algorithms have been proposed in the past, however they provide a partial solution to some of the security problems discussed, see [5] and [6]. Current security schemes for wireless reconfigurable networks base their solution on cryptography primitives providing not only a partial defense to security problems, but an expensive solution. The elevated cost is due to the high demand of processing node capabilities that cryptography mechanisms usually require which results inconvenient for networks with nodes limited in processing and power capacities. Even more, cryptography primitives usually rely on a central trusted unit and on an initial key distribution phase, which results in a difficult requirement to fulfill for a reconfigurable network where nodes are in constant motion. Therefore, a distributed, low expensive detection and defense mechanism for security attacks in WRNs is an important subject to study.

1.1 Components and Characteristics of WRNs

WRNs have two main components which are shown in Figure 1, [3], the mobile host and the wireless link. A mobile host or node constitutes the physical interface between the user and other network devices. A node contains a control unit, a transceiver, and an antenna system, and it is able to establish direct communication with any other network device that is within its power-transmission range or coverage area, usually considered a circle surrounding it. The other components are the wireless links that must provide high speed connections that enable communication between nodes.

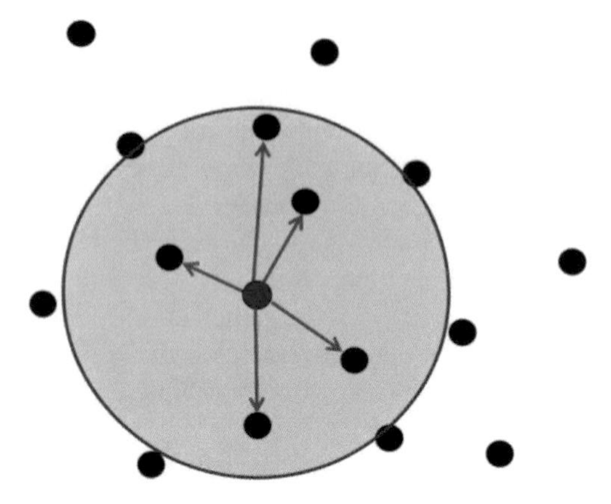

Fig. 1. Wireless Reconfigurable Network

The three main characteristics of WRNs are the node-to-node communication that allows the devices to organize themselves in a cooperative form dealing with topology changes due to mobility. In cellular networks, communications between two cellular devices is made throughout an infrastructure where the base station connects to. In contrast, wireless reconfigurable networks achieve Point-to-Point communication between nodes, this means that two nodes within their sensitivity area can communicate directly without the help of a base station. The WRNs are self-organizing and adaptive to the communications requirements of the current network conditions. Self-organizing is a fundamental characteristic of reconfigurable networks. A given node in the network has to be capable of detecting nodes that are within its transmission range, and it also must discover which other nodes can be reached through immediate connected nodes that are reached by a direct wireless link. Reconfigurability is also about detecting services available in the networks. Because of this characteristic most of the protocols for traditional networks can not directly be implemented in WRNs. Nodes in a WRN are free to move, hence there is no static infrastructure. The network is reconfigurable upon mobility and communication needs since links are being established and dropped as network evolves in time and deployment space, also, nodes can be turning on and off, causing topology changes just as mobility can cause. Thus topology is dynamic and nodes must reconfigure in order to cope with such changes as fast as they can.

1.2 Security in WRNs

Security is an important topic within the area of WRNs,, especially for applications that need security for information integrity. Important issues to be considered in a WRN security scenario, contains the following aspects: availability, confidentiality, integrity, authentication and non-repudiation, [5]. *Availability* ensures the survivability of network services despite denial of service attacks. A denial of service attack could be launched at any layer of a WRN. On the physical and medium access control layers, an intruder node could employ jamming to interfere intentionally with communication on physical channels. On the network layer, an adversary could disrupt the routing protocol and disconnect the network. On the higher layers, an adversary could bring down high-level services. *Confidentiality* ensures that certain information is never disclosed to unauthorized entities. Network transmission of sensitive information, such as strategic or tactical military information, requires confidentiality. Leakage of such information could have devastating consequences. Routing information must also remain confidential in certain cases, because the information can be used to identify targets for attacks. *Integrity* guarantees that a message being transferred is never corrupted. A message could be corrupted because of transmission failures, such as signal propagation impairments, or because of attacks on the network. Nodes must be identifiable somehow and must be authenticated in order to achieve integrity of the information being transmitted. *Authentication* enables a node to ensure the identity of the peer node it is communicating with. Without authentication, an adversary could mask a node, thus gaining unauthorized access to resource and sensitive information and interfering with the operation of other nodes. *Non-repudiation* ensures that the node origin of a message cannot deny having sent the message. Non-repudiation is useful for detection and isolation of compromised nodes. When a node receives an erroneous message from another node, non-repudiation allows the origin node to accuse the destination using the message and to convince others that such destination is compromised, [7].

The methodology presented here, proposes a Detection and Defense Mechanism (DDM) for routing security attacks formulated with random network coding. The concept of network coding was proposed by [8]. Network coding can be interpreted as network distributed processing due to the processing that each node is able to command over the incoming information. These processing capabilities range from insertion and deletion of bits, to the linear coding of data, shifting of bits and storage. The main concept in which network coding is based, is the mixing of incoming information at intermediate nodes in the network. A receiver is capable to deduce from coded information, the data packets that were originally sent to the destination node. In contrast to the traditional operation of the nodes in which data collisions in intermediate nodes are not desirable, in networks commanding network coding data collision is seen as an opportunity to improve several performance parameters of the network. One of the most important opportunities of network coding is the random mixing of data stream. Network coding represents an innovative and useful tool that has the potential to be applied in order to improve several performance parameters and also to implement distributed mechanisms in areas like security. It has been proved that random network coding helps to improve several network performance parameters, such like throughput, bandwidth savings, network monitoring, etc., [9], [10]. Interference and channel fading are two important network factors that must be taken into account when we evaluate a security method, because they may have a direct negative impact on the performance of such method in wireless reconfigurable networks. The proposed method is a novel and inexpensive method that implements network coding not only to collaborate in the information distribution over the network, but as a tool for monitoring and detecting these security problems demonstrating robustness under several network conditions, such as interference, channel fading and mobility. The proposed method combines network coding and a basic knowledge of the network topology to formulate a distributed solution that results on the increase of successfully received packets on the destination without a significant sacrifice of the bandwidth usage, [2].

2 Network Coding Fundamentals

The multicast communication process in a conventional wireline network consists in a source sending data packets throughout intermediate nodes to a set of destination nodes. Therefore, intermediate nodes only send information to the next node in the route. Network coding is a technique that can improve the efficiency of the communication process and take advantage of it by giving to intermediate nodes another function besides that of store and forward incoming data packets. Intermediate nodes are also instructed to encode data packets in order to improve the network performance in several ways as explained in the following subsections. Network coding as a concept was first introduced in [11] for satellite communications systems, where the authors obtain the inner and outer bounds of the admissible coding rate region. However, the term network coding was first introduced in [8] by Ahlswede, where a full analysis in a point to point communication network is developed. Due to its applicability, network coding can be used to improve several performance parameters in both wired and wireless communication networks, such as bandwidth consumption, throughput maximization, quality of service and security.

Nodes in a network such as routers carry out the store and forward tasks. Thus, we can see that a traditional network element such as a router produces at its output ports, after a processing delay, copies of incoming data packets through its input ports. In network coding, the routers will be able to perform operations on the incoming data packets, so that, the output data packets produced by the router, now will consist of the result of a function of one or several input data packets, [12]. In Figure 2, we show an intermediate node i in a network, where incoming data packets from different sources are combined and processed to produce three different outputs to be delivered to different destinations. In general, *network coding*, consists in the encoding or combination through functions of different data packets at different intermediate nodes, so that information is delivered to destinations, and such destinations can decode the messages intended for them. The functions performed by the intermediate nodes are generally binary operations on the bits of the incoming data packets, defined on Galois Fields, see [13]. In order to generalize network coding and with the purpose of showing its advantages, it is better to consider a network model as described in the following subsection.

2.1 Network Model

In order to develop the models for the remaining of this chapter, we introduce some basic notation considering a network that has a set of nodes V, performing point to point communication throughout a set of links or edges E. The number of links in the network is $\|E\| = L$, and every link $i \in E$ has associated a capacity C_i where C_i is a nonnegative real number and C is the set of link capacities in the network, i.e., $C = \{C_i \in \mathbb{R} : i \in E\}$. Such a network is represented by a directed, acyclic graph $G = (V, E, C)$ as in [8]. Data packets $X_{1i}, \ldots, X_{ni}$ are sent from mutually independent source nodes $s_1, \ldots, s_n$ to a set of destination nodes $d_1, \ldots, d_n$, through intermediate node i, see Figure 2. X_{1i} is the information flow produced by source 1 and sent to intermediate node i. Y_{i1} is the information flow obtained by the encoding function of the intermediate node i that is sent to destination node 1. A fundamental network topology where network coding shows its advantages is *the butterfly topology* as shown in Figure 3.

In the butterfly topology of Figure 3, data packets travel throughout intermediate nodes $\{1, 2, 3, 4\}$. In general, intermediate node i encodes incoming information generating outgoing flows $Y_{i1}, \ldots, Y_{in}$, where Y_{ij}, $j = 1, \ldots, n$, is the result of the

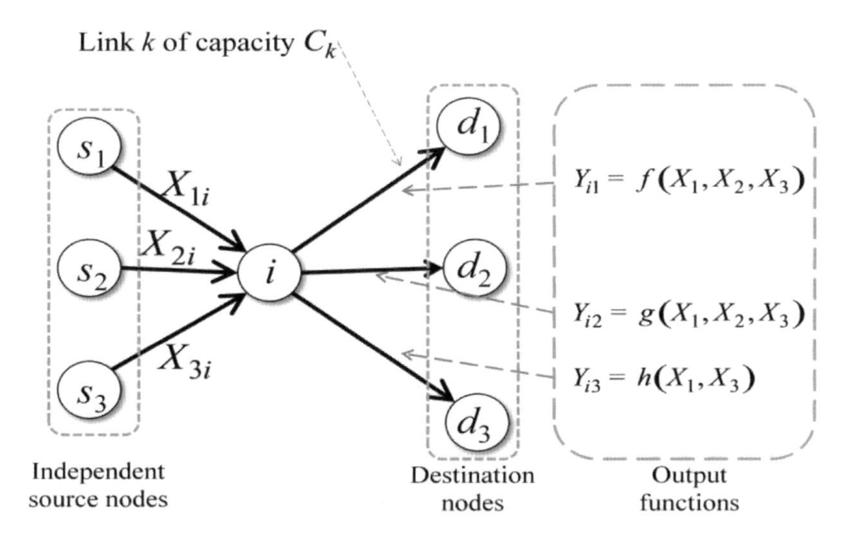

Fig. 2. Concept of network coding

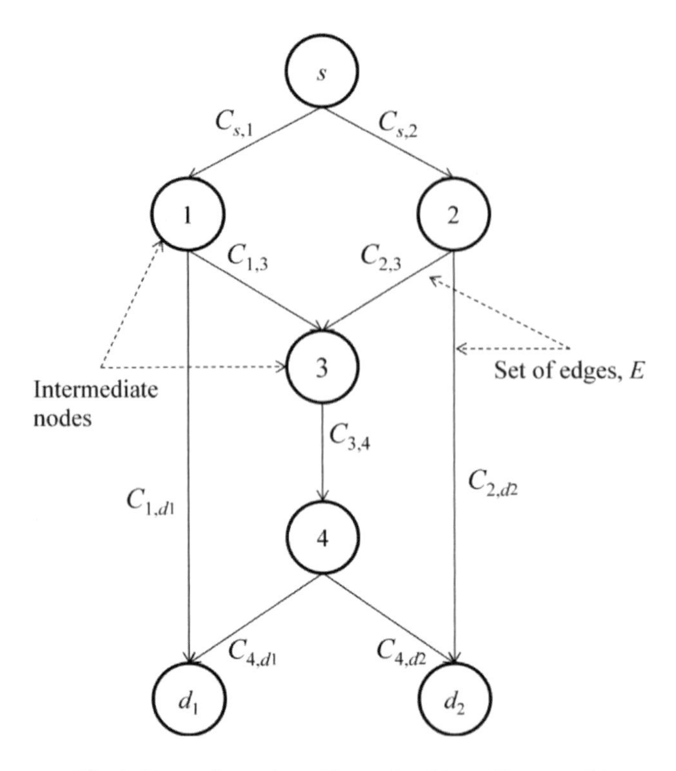

Fig. 3. Network topology illustration (Butterfly network)

output function of the intermediate node *i* that encodes the incoming data packets and sends this result to node *j*. This encoding function can be considered as a linear combination of incoming information flows, [14], [15], in other words

$$Y_{ij} = \sum_{s} \beta_{si} X_{si} + \sum_{k \in I(i)} \beta_{ki} Y_{ki}, \tag{1}$$

where data packets X_{si} are encoded by a random linear coding coefficient given by β_{ki} used by node *k* for outgoing information flows for link (*k*, *i*). The set of linear coding coefficients is $\boldsymbol{\beta}$. Equation (1) will be generalized in Section 4 of this Chapter, where the security mechanism is introduced.

2.2 Advantages of Network Coding

Network coding is known for its advantages that can be grouped in two directions, the first is the reduction of bandwidth used for the delivery of information, and the second is a consequence of the first which is the throughput improvement.

Let us use the butterfly topology illustrated in Figure 4 to explain some of the advantages provided by network coding, the scenario is that all links in the network have a capacity of one bit per transmission, and that nodes are only allowed to transmit one bit in each outgoing link in the whole example, except for intermediate node 3. First, consider the reduction of bandwidth consumption. For the system in Figure 4(a), where network coding is not being used at intermediate nodes, the source s generates two bits of information, b_1 and b_2, and needs to send them to two different destinations d_1 and d_2. Since network coding is not allowed in this example, we rely on the basic routing being used in the network. Bit one is sent to node 1, and since it needs to reach both destinations, node 1 decides to send bit one through both outgoing links $(1,d_1)$ and $(1,3)$. Similarly, bit two is sent to node two and since bit two needs to be delivered at both destinations, node two sends bit two through both outgoing links $(2,d_2)$ and $(2,3)$. At this moment, destination d_1 has information bit b_1, destination d_2 has information bit b_2, and intermediate node 3 has both information bits b_1 and b_2. Node 3 needs to send bit b_1 to destination d_2 and bit b_2 to destination d_1. Node 3 needs to make a decision to transmit only one of the bits and to hold the other bit for a later transmission, then node 3 will use two transmission times to send both bits to node 4, and node 4 will transmit the corresponding bit needed to each destination. The total number of transmissions taken in the network in order to deliver both bits to both destinations is ten.

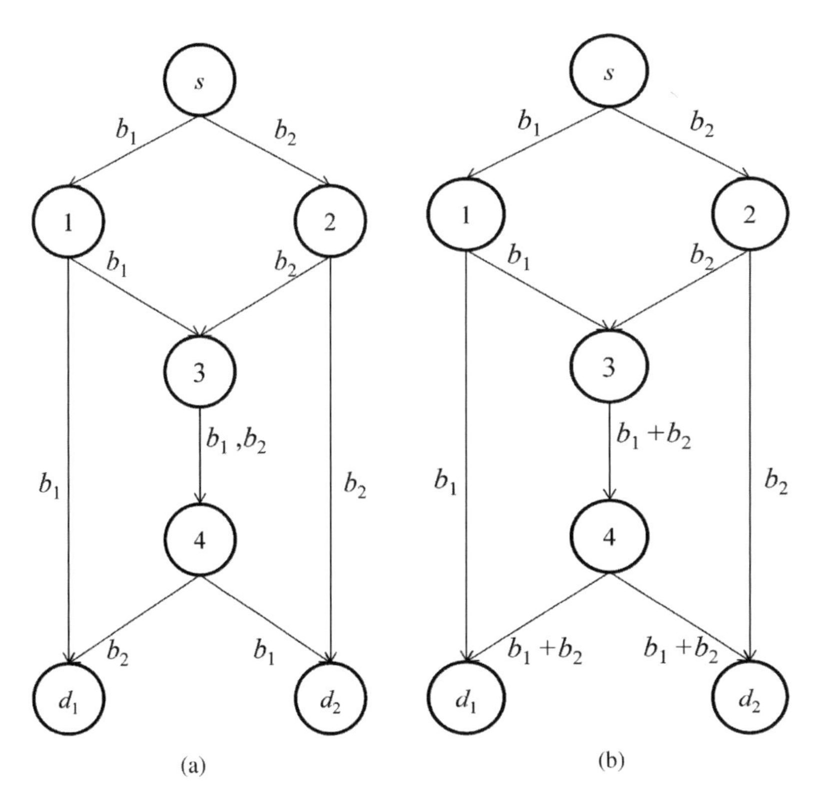

Fig. 4. Bandwidth improvement using network coding

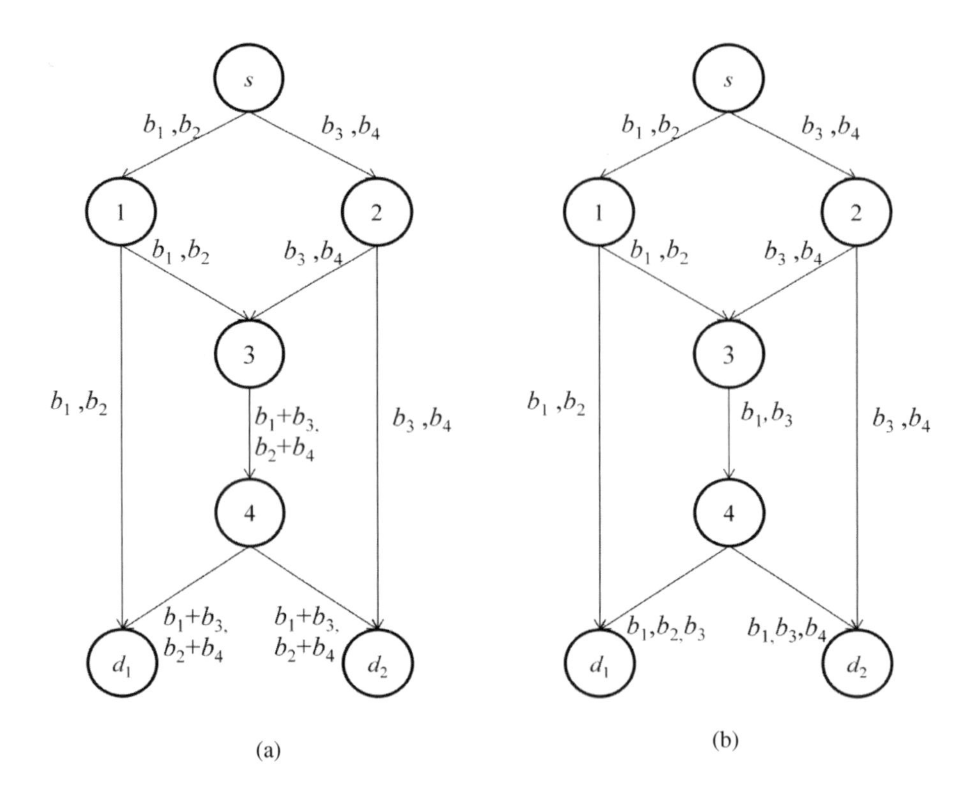

Fig. 5. Throughput improvement using network coding

In Figure 4(b), network coding is allowed at intermediate nodes. Following the same procedure as that for the network in Figure 4(a), we observe that node 3 encodes two data packets b_1 and b_2, and the outgoing data packet is b_1+b_2, (recall that these operations are binary or modulo-2 operations). Therefore a total of 9 bits are transmitted in the network to have d_1 and d_2 recover b_1 and b_2 at both destinations by a simple decoding operation, e.g., destination d_1 has bit b_1 and receives (b_1+b_2), then it performs the decoding operation with the information bit that has at that time, i.e., $(b_1+b_2)+b_1=b_2$, see [16]. Since in Figure 4(a) at least one more bit must be sent in the system to achieve the same goal, thus with a very simple network code we can save 10% of bandwidth consumption.

With the same butterfly topology, we can investigate the throughput improvement that can be achieved allowing network coding. An improvement is achieved by delivering more information bits during the same number of transmission attempts in the network. In Figure 5(a), we can see that if two bits are allowed to be sent in every link, four bits (b_1, b_2, b_3, b_4) are recovered in every destination node when network coding is used in the intermediate node 3. However if network coding is not used throughout the intermediate nodes in the network (see the network in Figure 5(b)), only three bits can be recovered in every destination node, this statement is mathematically proved in [8]. Therefore, we can see that an improvement of 25% in throughput (amount of successful information bits per transmission delivered to the destination) is achieved when network coding is allowed in the system as illustrated in Figure 5(a).

3 Network Coding in Reconfigurable Networks and Its Applications

Network coding has many advantages from the point of view of wireline networks as introduced in Section 2 of this chapter, but for wireless networks, some different considerations are needed since the existence of a connection or a route in a network not only depend on the topology, but also on the wireless channel conditions and the wireless transceiver characteristics of the devices. We could end up with a network where the nodes are reachable from the point of view of coverage or signal range, but due to the existence of interference or outage conditions, such links might not be working and reachability is not possible.

Regardless of the network scenario, it has been shown that linear network coding satisfies the conditions to attain communication capacity in networks, see [14] and [17]. One of the advantages of linear network coding is its simplicity since the output flow from a node is a linear combination of the input flows to that node.

As stated in the list of advantages presented in [12], network coding can help reduce the energy required to transmit the same amount of information, since the use of link bandwidth is reduced, we can also see that the amount of signals being transmitted in order to deliver the information to all the destinations requesting it, is reduced. This reduction of the use of link bandwidth is also translated to the reduction of interference in the wireless environment which helps the signals being transmitted to be received at higher levels of power, which translates to a possible reduction of the transmission power of the devices and hence on the energy savings of the device.

Also, since network coding reduces the amount of packets needed to transmit in order to achieve the same delivery of information, the delay experienced by the end-users is reduced from the point of view of the transmission and delivery processes. The drawback is that now the end-user devices need to perform more processing tasks related to the coding taking place, which also increases the energy use of each device.

In wireline and wireless networks, an intruder can be considered to be represented as a wiretap channel being used by the source and destinations, see [18], where the main purpose is that the wiretapper does not obtain relevant information from the communication taking place. Also in [18], security is seen and discussed from three different point of views, the physical, the computational and the information theoretic.

Security in wireless reconfigurable networks (WRN) faces several challenges due to the nature of these systems. WRN topology is always changing due to the movement of nodes, communication is carried out point to point, there is constant emigration and immigration of nodes to the network, the transmission medium is the air and nodes have different processing capabilities. These characteristics imply several restrictions and security problems that wired networks do not present.

In WRN security becomes a requirement where data integrity and confidentiality must be protected from a variety of attacks. Conventional security techniques used in wired networks such as public and private key cryptography result too expensive in terms of processing capacity and power consumption to be used in WRN.

Security attacks in WRN may be directed to impair information or to impair network performance. Some examples of security attacks intended to impair information are confidentiality attacks, data integrity attacks and data authenticity attacks. One example of data confidentiality attacks is eavesdropping, where unauthorized nodes have access to the information transmitted between a pair of nodes. The data integrity attacks occur when an intruder alters data before transmits it to the following node in the route. Finally, when an adversary declares itself with fake network identification a data authenticity attack is commanded. In the other hand, we have attacks directed to impair network performance parameters such as bandwidth and routing. For example, the sinkhole attack occur when an intruder attracts network traffic by advertising itself as having a better path from a source to a destination. Also, selective forwarding attack occurs after an adversary creates a sinkhole, and it refuses to forward a selection of data packets to the destination node. The black hole attack is a type of selective forwarding attack, occurs when the intruder does not forward any incoming data packet to the destination node.

With the rapid development of wireless communication technology, wireless reconfigurable networks have been widely used in military, emergency rescue, and personal communications. WRN is characterized by highly dynamic, unpredictable channel quality and limited node energy. Packet losses in WRN are often due to these inherent characteristics, rather than congestions. For non-congestion caused packet loss, the correct reaction is to increase the transmission rate to overcome the lossy links. However, the traditional TCP protocol cannot tell the reason of packet losses are seen as signs of congestion.

Network coding was initially proposed as a way to reduce the number of multicast transmissions over wired networks. After that network coding has gained much interest. J.K. Sundararajan, [7], first put forward the idea of interfacing network coding with TCP to improve the TCP performance in wireless networks. By means of a coding layer between TCP and IP, original TCP packets are encoded at the sender and decoded at the receiver with a certain redundancy. The redundancy of information covers the non-congestion caused losses and TCP performance is improved without any disturbance to other layers. But the synchronization of data transfer and decoding operation which has a major impact on TCP/NC performance is not guaranteed [19].

The TCP/NC protocol introduces a network coding layer between TCP and IP in the protocol stack, where an encoder module lies on the sender side and a decoder module lies on the receiver side, see [7], as shown in Figure 6.

The main idea behind TCP is to use ACK's of newly received packets as they arrive in correct sequence order in order to guarantee reliable transport and also as a feedback signal for the congestion control loop. This mechanism requires some modifications for systems using network coding. The key difference to be dealt with is that under network coding the receiver does not obtain original packets of the message, but linear combinations of the packets that are then decoded to obtain the original message once enough such combination have arrived, see [7].

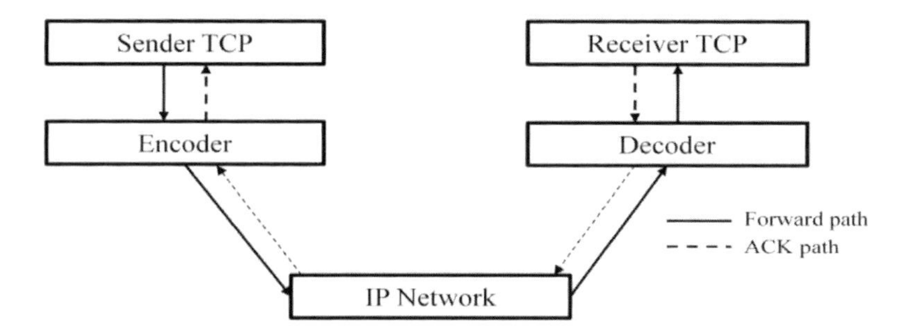

Fig. 6. TCP/Network Coding architecture

The encoder buffers packets generated by TCP and for every arrival from TCP, it transmits random linear combinations of the buffered packets on average, the amount of combinations generated depends on the redundancy necessary for protection. The contents of a coded packet represent a linear combination of the original uncoded packets. This form of network coding requires a different header in the packets which is added to the TCP network packets. The original uncoded packets are retained in the encoding buffer until an appropriate TCP ACK arrives from the receiver side. The purpose of adding redundancy is to separate the loss recovery aspect from the congestion control aspect. Losses can now be recovered without forcing TCP retransmissions and the associated congestion window size reductions.

This would ensure that the number of equations reaching the receiver will match the number of packets entering the encoder. On the decoder side, upon receiving a new linear combination the decoder places it in a decoding buffer, appends the corresponding coefficient vector to the decoding matrix, and performs Gaussian elimination. This process helps identify the "newly seen" packet. We can say that a packet is seen if after Gaussian elimination of the coefficient matrix, the packet corresponds to one of the pivot columns. The decoder then sends a TCP ACK to the sender requesting the first unseen packet in order. Thus, the ACK is a cumulative ACK like in conventional TCP. The Gaussian elimination may result in a new packet being decoded. In this case, the decoder delivers this packet to the receiver TCP. Any ACKs generated by the receiver TCP are suppressed and not sent to the sender. These ACKs may be used for managing the decoding buffer. Packets are treated as vectors over a finite field $\mathbb{F}_q$ of size q. The kth packet that the source generates is said to have an index k and is denoted as p_k, see [7].

4 Detection and Defense Mechanism Based on Network Coding

In this section, we present the network coding based mechanism for security in WRNs. In [20], a Detection and Defense Mechanism (DDM) is proposed for detecting sinkhole and selective forwarding attacks that uses network coding for the content distribution, and also as a tool for detection purposes. The basic scenario for the analysis in [20] is as that illustrated in Figure 7, where a trustable route from source s to destination d and a new route of lower cost are shown. It is assumed that routes may change according to node availability. There exists the possibility that nodes may be compromised by an intruder that attempts to create a sinkhole or a selective forwarding attack from the new route.

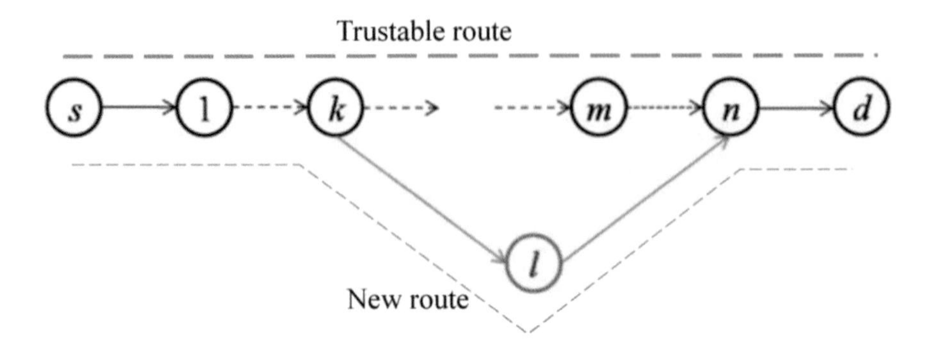

Fig. 7. Illustration of a Network Topology Change

For the analysis of the DDM technique we follow notation similar to that introduced in Section 2.1 of this chapter, we also follow the notation in [17]. The incoming flows of packets to node v are seen as a set of discrete independent random processes with information packets $\boldsymbol{X}(v)=\left\{X_{v1},X_{v2},\ldots,X_{v\mu(v)}\right\}$ seen as the collection of $\mu(v)$ discrete random processes that are observable at node v. We define a connection (path) as $c=(s,d,X_{sd})$, we call s a source and d a sink of connection c, note that a flow is generated by the source and its destination is node d. We always assume that $s\neq d$. The set of links incident to node v and generating input flows to node v is denoted as $I(v)$, similarly, the set of links outgoing from node v is denoted as $O(v)$. The outgoing flows of packets from node v are denoted as $\boldsymbol{Z}(v)=\left\{Z_{v1},Z_{v2},\ldots,Z_{vw(v)}\right\}$ and are seen as the collection of $w(v)$ discrete random processes that are observable at the output of node v, with this, we can see that $\mu(v)=\|I(v)\|$, and that $w(v)=\|O(v)\|$. The network has links denoted as j=1,2,..., also let R_{sd} denote the set of links that conform the path or connection between origin s and destination d. For the DDM we consider the unicast problem, therefore we redefine the set of discrete packets $\boldsymbol{X}(v)$ over a finite field F_q of size q, observable at the source. For link $e=(i,j)$, we define Y_e as the random process carried by link $e=(i,j)$, which is given by

$$Y_e = \sum_{l=1,}^{\mu(i)} \alpha_{il} X_{il} + \sum_{\substack{k \in I(i), \\ e'=(k,i)}} \beta_{e',e} Y_{e'} - \sum_{\substack{k \in O(j), \\ e'=(j,k)}} \beta_{e,e'} Y_{e'}, \tag{2}$$

where the coefficients α_{il}, and $\beta_{e,e'}$ and $\beta_{e',e}$ are elements of the finite field F_{2^m}. The negative sign in Equation (2) is for the returning links. The output Z_{ik} of any node *i* is formed from incoming random processes Y_e, $e \in I(i)$ to node *i* as linear combinations as follows

$$Z_{ik} = \sum_{e \in I(i)} \varepsilon_{e,k} Y_e, \quad k = 1, 2, \ldots, w(i), \tag{3}$$

where the coefficients $\varepsilon_{e,k}$ are elements of F_{2^m}. For an illustration of equations (2) and (3), see Figure 8.

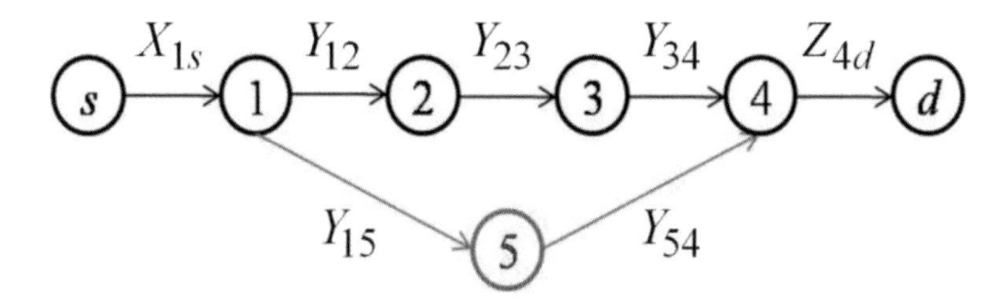

Fig. 8. Network information flow for the former attack case

At any given time, let the active links be identified by a number, and given equations (2) and (3), we can define the adjacency matrix with returning links as *H* which is given by the number of links *L* active in the network, i.e., the dimension of the matrix *H* is *L*x*L*. We define the adjacency matrix *H* of the network with elements $H_{i,j}$ as follows

$$H_{ij} = \begin{cases} \beta_{i,j}, & i \in I(v) \text{ and } j \in O(v), \text{ for some common node } v, \\ -\beta_{i,j}, & i \in O(v) \text{ and } j \in i(v), \text{ for some common node } v, \\ 0, & \text{otherwise.} \end{cases} \tag{4}$$

The matrix $(I - H)$ has a polynomial inverse with coefficients $\beta_{i,j}$. Let A be defined as the input coding coefficient matrix, which contains the input coefficients α_{il}.

Each node in the network can have different number of input flows, each with its own input coefficient, hence matrix A will have as many rows as input flows there are in the network, i.e., a total number of rows given by $\sum_{v \in V} \mu(v)$. The *(i,j)*-th element of the matrix will be given by

$$A_{i,j} = \begin{cases} \alpha_{ji}, & X_{vi}, \text{ for some node } v, i = 1,2\ldots,\mu(v), j \in I(v), \\ 0, & \text{otherwise.} \end{cases} \tag{5}$$

Let the matrix B be defined as the output coefficient matrix, with a number of rows given by $\sum_{v \in V} w(v)$. The *(i,j)*-th element of the matrix will be given by

$$B_{i,j} = \begin{cases} \varepsilon_{ji}, & Z_{vi}, \text{ for some node } v, i = 1,2\ldots,w(v), j \in O(v), \\ 0, & \text{otherwise.} \end{cases} \tag{6}$$

Let a network be given by matrices *A, B* and *H,* based on these, we can formulate the transfer matrix of the network with returning links in a similar way as that in [17] for networks without returning links.

$$Q = A(I - H)^{-1} B^T \tag{7}$$

The DDM assumes an initialization phase, where every node in the network has been authenticated with an Access Point (AP) or Base Station (BS). In this phase, a mechanism for trusted node classification like that in [21] is considered. Then we have a route discovery phase based on trusted nodes that produce what we call trustable routes. After this phase, we suppose that nodes can move, therefore some routes may change and scenarios such as that presented in Figure 7 may happen over the network operation time. We also assume that nodes in the network update their routing tables every *t* seconds, in order to find new available trustable routes. At this point, it is better to consider Figure 8, which is a simplified version of the network in Figure 7. Node 1 is on charge of starting the DDM, and detects that it has to send the coded information over the two available routes, and sends an alarm message to the next node, which propagates it to the rest of the nodes in the trustable route. Node 4 in Figure 8 is in charge of receiving the two coded information flows from nodes 3 and 5. Then node 4 requests to the suspicious node 5 its coding coefficient and then node 5 is supposed to send it correctly (if it is not then it will be considered an intruder). When this information is received by node 4, then it processes a detection packet W_4. Detection information calculated by node 4 is sent to node 3, where node 3 is on charge of processing the detection packet W_3. This information is sent to node 2, then if node 2 obtains W_2=0 node 2 sends to node 1 a packet verifying the new route as a trustable route, i.e., no attack has been detected. However, if $W_2 \neq 0$ then node 2 sends to node 1 a packet indicating that a possible selective forwarding attack has been detected on the new route. This process is explained in detail in [20].

4.1 DDM Algebraic Feasibility

As stated before, the DDM depends on the existence of at least one trusted route. In Figure 9, we present a network topology with a set of trusted routes from source *s* to destination *d*. For the deployment of the DDM, we need a topology where return links are allowed as we can see in the figure. Recall that nodes 1 and 4 in Figure 8, are the nodes in charge of the verification process. In Figure 9, we show the selection of those nodes, recall that these nodes can be different depending on the new route to be checked. In the network to be analyzed, identify all the origin-destination pairs *s-d*, then each of its routes in order to form the sets of routes R_{sd} where each route is defined by the identification of the links that form it. We define links depending on the nodes that join and route that belong to, i.e., $e_{ij}^{k} = (i, j) \in r_k, \quad r_k \in R_{sd}$ is the link that joins nodes *i*, *j* and is traversed by route r_k from the set of routes for that particular origin-destination pair. We can distinguish the following routes, r_1 composed by the set of links $e_{13}^{1}, e_{35}^{1}, e_{57}^{1}$; r_2 composed by links $e_{24}^{2}, e_{46}^{2}, e_{68}^{2}, e_{89}^{2}$; r_3 composed by $e_{13}^{3}, e_{36}^{3}, e_{68}^{3}, e_{89}^{3}$. For r_1 the checking nodes will be node 1 and node 7, for route r_2, nodes 2 and 9 and finally for r_3, nodes 1 and 9, respectively.

Recall that the adjacency matrix with returning links is defined by *H* and that for coding coefficients $\beta_{i,j}$ of returning links, we associate a minus sign in order to know that it is a returning link from edge *i* to edge *j*. For the network in Figure 9 with route order r_1, r_2 and r_3, ($e_{13}^{1}, e_{35}^{1}, e_{57}^{1}$, $e_{24}^{2}, e_{46}^{2}, e_{68}^{2}, e_{89}^{2}$, $e_{13}^{3}, e_{36}^{3}, e_{68}^{3}, e_{89}^{3}$) we have that *H* is given by the block matrix

$$H = \begin{bmatrix} H_{r_1} & 0 & 0 \\ 0 & H_{r_2} & 0 \\ 0 & 0 & H_{r_3} \end{bmatrix}. \tag{8}$$

In Equation (8), the block matrices with dimensions indicated, is obtained by using Equation (4), so for example we have

$$H_{r_1} = \begin{bmatrix} 0 & \beta_{e_{13}^{1} e_{35}^{1}} & 0 \\ 0 & 0 & \beta_{e_{35}^{1} e_{57}^{1}} \\ 0 & -\beta_{e_{13}^{1} e_{35}^{1}} & 0 \end{bmatrix} \tag{9}$$

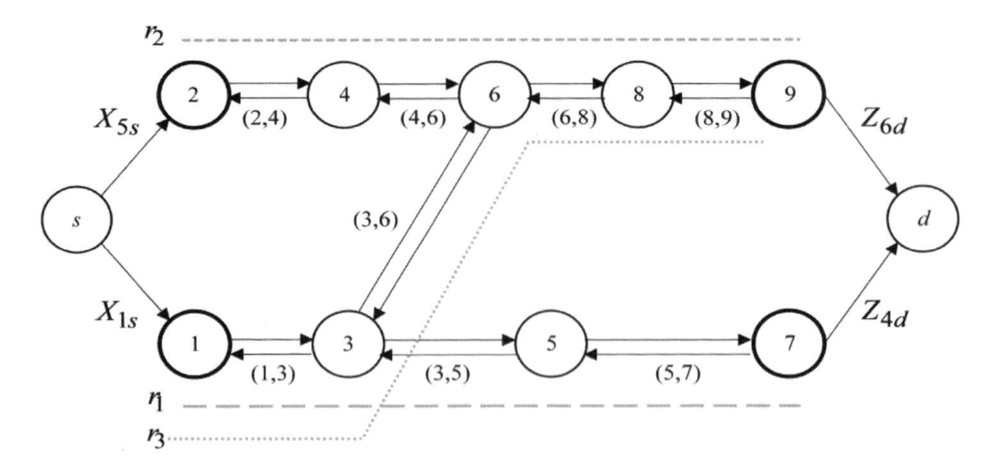

Fig. 9. Network topology for set of not mutually exclusive trusted routes with different returning links

The coding coefficient input matrix A and B for the network in Figure 9 is the following

$$A=\left[\alpha_{e_{13}s}^{1} \quad 0 \quad 0 \quad \alpha_{e_{24}s}^{2} \quad 0 \quad 0 \quad 0 \quad \alpha_{e_{13}s}^{3} \quad 0 \quad 0 \quad 0\right]=\left[A_{r_1} \quad A_{r_2} \quad A_{r_3}\right]. \tag{10}$$

The output of the checking process performed by the DDM is located at a node after node 3 for r_1 and r_3. Therefore the output matrix B for the network in Figure 9 is

$$B=\left[0 \quad 0 \quad \varepsilon_{e_{57}d}^{1} \quad 0 \quad 0 \quad 0 \quad \varepsilon_{e_{89}d}^{2} \quad 0 \quad 0 \quad 0 \quad \varepsilon_{e_{89}d}^{3}\right]=\left[B_{r_1} \quad B_{r_2} \quad B_{r_3}\right]. \tag{11}$$

Note in equations (8) to (11) that matrices H, A and B are block matrices subdivided in blocks corresponding to every feasible route. Therefore, the transfer matrix Q of the network given in Equation (7) can also be subdivided into blocks by routes $Q=\left[Q_{r_i}, Q_{r_i+1}, \ldots, Q_{r_k}\right]$ where Q_{ri} is given by the following equation

$$Q_{r_i}=A_{r_i}\left(I-H_{r_i}\right)^{-1}B_{r_i}^{T}. \tag{12}$$

By means of matrix Q, we are now able to evaluate the feasibility of every set of routes [17] as done with M in [17]. For matrix Q, we can calculate the determinant det(Q) and find that a sufficient and necessary condition for the feasibility of a given connection set is that det(Q) results positive, i.e., $|Q|>0$. For the network in Figure 9, a feasible connection can be found if we let $\beta_{e_{12}^{1}e_{35}^{1}}, \beta_{e_{35}^{1}e_{57}^{1}}, -\beta_{e_{57}^{1}e_{35}^{1}}, \alpha_{e_{12}^{1}s}, \varepsilon_{e_{35}^{1}d}=1$. This assignment is going to return det(Q_{r1})=0.5.

An important result obtained in this work is that the value of det(Q_{ri}) for a given set of routes does not depend on the length of the connection set, but on the number of returning links on the route. In other words, for routes with arbitrary length, but with the same number of returning links we obtain the same value of det(Q_{ri}). It is possible to verify this result for routes in the network illustrated in Figure 9. In Table 1, we show the det(Q_{ri}) obtained for different number of returning links for routes with arbitrary length. In this table, we can see that as the number of returning links increases det(Q_{ri}) converges to 0.618.

Table 1. Number of returning links against det(Q_{Ri})

Returning Links	2	3	4	5	6	7	8	9
det(Q_{Ri})	0.5	0.667	0.6	0.625	0.617	0.619	0.618	0.618

4.2 Network Impairments

In this work, we take into consideration two main impairment factors that impact the performance of the DDM. These impairments have been already evaluated throughout simulations in [22]. The outage probability is calculated over an entire alternative route based on the assumption of multicast channel Rayleigh fading [23]. Define $P_{jk,out}$ as the outage probability that the message is transmitted unsuccessfully from node j to the node k and is given by

$$P_{jk,out} = 1 - \left[\frac{\varphi/2}{\varphi/2 + \varepsilon/(SNR)_{T_x}} \right]^{n-1} \tag{13}$$

where $(SNR)_{Tx}$ is defined as P_T/N , where N is the noise power, P_T is the node power transmission and ϕ is the transmission sector. The outage probability is defined when the received SNR is less than the defined threshold ε in a given condition of distance D. Finally n is the number of nodes in the route.

The interference probability P_{if} is also calculated over the entire route, according to the distance of the nearest neighboring node to any node in the route. This probability is given according to a Frechet pdf and is given by

$$f_{If}(\eta) = \frac{2\lambda\pi}{\gamma P_T}\left(\frac{P_T}{\eta}\right)^{2/\gamma+1} e^{-\lambda\pi(P_T/\eta)^{2/\gamma}}, \tag{14}$$

where γ is the path loss, λ is the node density, P_T is the node power transmission and η is the power threshold. In [22], authors present in detail the formulation of these probabilities.

We take a different approach from the network impairments exposed in [22]. We obtain the algebraic representation of the network including outage and interference probability. For the network in Figure 9, we have the following probability matrix where $P_{R_i} = P^i_{jk,out} P^i_{if}$

$$P = \begin{bmatrix} P_{R_1} & 0_{R_2} \text{x} 0_{R_1} & 0_{R_3} \text{x} 0_{R_1} \\ 0_{R_1} \text{x} 0_{R_2} & P_{R_2} & 0_{R_3} \text{x} 0_{R_2} \\ 0_{R_1} \text{x} 0_{R_3} & 0_{R_2} \text{x} 0_{R_3} & P_{R_3} \end{bmatrix}. \quad (15)$$

Note that network matrices H, A and B remain the same as those presented in equations (8) to (12). However the transfer matrix with returning links is now reformulated in order to introduce the probability matrix in (15). Therefore the probability transfer matrix of the network will be

$$Q^P = A\left(I - (1 - P)IH\right)^{-1} B^T = \left[Q^P_{R_i} \quad Q^P_{R_{i+1}} \quad \cdots \quad Q^P_{R_k} \right]. \quad (16)$$

In Table 2, we can see the determinant of Q^P for different number of returning links with several probabilities P_{R_i}. We can see that for all the probabilities det(Q^P) converges to a constant value as the number of returning links increases. As the probability P_{R_i} grows, the convergence is faster than for lower probabilities. In Table 2, we can see that det(Q^P) provides knowledge about the connection feasibility and even more, we can see that for lower probability we obtain a higher value of det(Q^P), therefore this table also provides a partial knowledge of the connection quality. However, in order to obtain a complete knowledge of the quality of the route we need to consider that for a larger number of links in the route (including returning links) the quality of the route will be more compromised than for shorter routes, this is not obtained directly from det(Q^P).

Table 2. Number of returning links against det(Q^P)

Returning Links	2	3	4	5	6	7	8	9
det(Q^P) for P=0.2	0.487	0.575	0.547	0.556	0.553	0.554	0.554	0.554
det(Q^P) for P=0.4	0.441	0.474	0.467	0.468	0.468	0.468	0.468	0.468
det(Q^P) for P=0.6	0.344	0.351	0.350	0.350	0.350	0.350	0.350	0.350
det(Q^P) for P=0.8	0.192	0.192	0.192	0.192	0.192	0.192	0.192	0.192

4.3 Algebraic Route Selection

Obtaining the algebraic representation for a network with returns taking into account physical impairments such as interference and outage due to channel fading, we are able to find the best trusted route to command de DDM based on the transfer matrix Q^P in Equation (16). As seen in Table 2, det(Q^P) provides knowledge of the connection feasibility and a partial knowledge about the quality of the route. Therefore, we propose to find the best feasible route based on the determinant of Q^P, or which we need to introduce some other available information as the number of forward links l_i and the number of returning links J_i in the trusted route r_i (excluding source and destination). By inspection we find that the best feasible route is given by

$$I_A = \max_{R_i} \left(\frac{\det\left(Q_{R_i}^P\right)}{l_i J_i} \right). \tag{17}$$

By Equation (17), we try to select the shorter route with the minimum amount of returning links based on the existence of the connection set, given by $\det\left(Q_{R_i}^P\right)$. We introduce such results in Section 5 of this Chapter.

5 Some Results on Network Coding in WRN

Most of the simulation results where Network Coding is applied are not in WRN, the main advantages in the implementation of Network Coding in WRN are mainly in throughput when the loss rate gets higher and the packet loss along the links improves.

In this section, we present simulation results and experimental results aimed at establishing the fairness properties and the throughput benefits. The simulations are based on TCP-Vegas and the experimental results use the TCP-Reno.

5.1 Simulation Results

In this section, we compare the performance of the DDM under Dijkstra routing analyzed in [22], against the performance of the DDM under the route selection based on the formulation proposed in Equation (17). By means of this route selection that we are going to call "*Best Route Selection*," we are able to take into consideration several terms like feasibility of the connection, route quality and the total length of the route. We follow the same criteria as that in evaluations carried out in [22]. Our simulations have the base environment of a network scenario with n nodes which are uniformly distributed in a square area of A=10,000m^2. We assume a homogeneous transmission range $r_{max}=r$. We set C=2, as in [20]. Also the path loss exponent of the wireless channel environment is considered to be γ=2, and the node density $\lambda=n/A$. The necessary power received is set to PR=-100dBm for the interference calculations. For the outage probability, we set the sector $=\pi/2$, and the relation $\varepsilon_{(SNR)Tx}$ =26dB, as in [22]. We select randomly source and destination nodes and find the best route (trusted route) between them based on two criteria to be compared, the first is the shortest path length (Dijkstra) and the second "*Best Route Selection*". After the trusted route is found, the adversary is introduced and generates a new route connecting origin and destination. The methodology proposed to find the best route to command the DDM based on Equation (16) is the following

- Find a set of trusted routes from source *s*, to destination *d*. Some routes may not have the same length.
- Update the routing tables of nodes involved.
- Calculate P_{R_i} of every route in the set.
- Select the best route to connect *s* with *d*, based on Equation (17).
- Carry out a communication session.
- At the end of every communication session, check if any link in the actual route is lost based on the outage probability of the route $P^i_{jk,out}$. If this happens then, movement is considered and a new set of trusted routes need to be found.
- If a new route is announced to be better than the routes in the set of trusted routes, then the connection between *s* and *d*, is carried out through this new route, and the DDM is executed over the best route in the set of trusted routes.
- If an attack is detected over the new route, then the communication is closed, and the new route is isolated and the communication between *s* and *d* is resumed over the best of the routes in the set of trusted routes.
- If no attack is detected over the new route, a monitoring process is activated by the DDM as in [20].

We evaluate the DDM for this scenario under different attack probabilities $p_{a_{\max}}$ and different maximal transmission ranges *r*, and node densities. The metrics used to evaluate the performance of the DDM are the following:

- **Successfully Received Packets (SRP)** defined as ratio of the number of packets sent by the source *s* to the number of packets received by the destination *d*, without retransmission.
- **Packet Overhead (PO)** which gives a measure of the total communication overhead in the network due to the DDM. It is defined as the ratio of the number of packets sent by the system to carry out the DDM to the number of packets sent by a system without the DDM.
- **Detection Rate (DR)** defined as the ratio of the number of authentic attacks detected by the DDM to the number of authentic attacks executed by an adversary.
- **False Positive Rate (FPR)** provides a measure of the ratio of the number of false attack declarations (this is, when data errors were caused by interference) to the total number of detected attacks.
- **False Negative Rate (FNR)** provides a measure of the ratio of the false negative detected attacks (the attacks that the DDM declares as no attacks when they are authentic attacks), to the total detected attacks. The last metric is introduced in order to complete outage evaluations.
- **No Applicability Rate (NAR)** due to outage probability (route reliability). This is the ratio of the number of times that the DDM is no longer applicable because a trusted or a new reliable route is not found due to route outage or mobility to the number of times that a set reliable route is found.

In the following figures, we make the comparison between the DDM using Dijkstra routing against the DDM using "*Best Route Selection*". We also present a node density and transmission range analysis for the DDM under two physical impairments which are NNN interference and outage relaying routing. In figures 10 to 14, we set the maximal transmission range to r=20m and several node densities. In these figures, we evaluate the improvement provided by the Best Route Selection algorithm over the Dijkstra routing algorithm. In addition from figures 10 to 14, we carry out a density analysis of the DDM using the Best Route Selection in order to prove that robustness in the performance of the DDM can be maintained under the increase of the number of nodes in the network, as already shown in [22] using Dijkstra. Note that same network conditions were taken into account for the system using Dijkstra, (r=20m and N=100).

In Figure 10, we present the PO imposed by the DDM under two routing criterions. We can see that with the Best Route Selection method the PO caused by the DDM is substantially lower than that obtained the Dijkstra routing, for $p_{a_{\max}}$ =1 we have that the decrease of PO is about 8%. Moreover, we can observe that the PO with Best Route Selection slightly decreases as the attack probability increase, which can be considered a desirable behavior. At the same time we present the PO obtained by the DDM using Best Route Selection under different node densities. Here, we can observe that the PO is not increased if the node density is increased. Even more we can see that for some values of the attack probability it is possible to slightly decrease the PO, which is a good result.

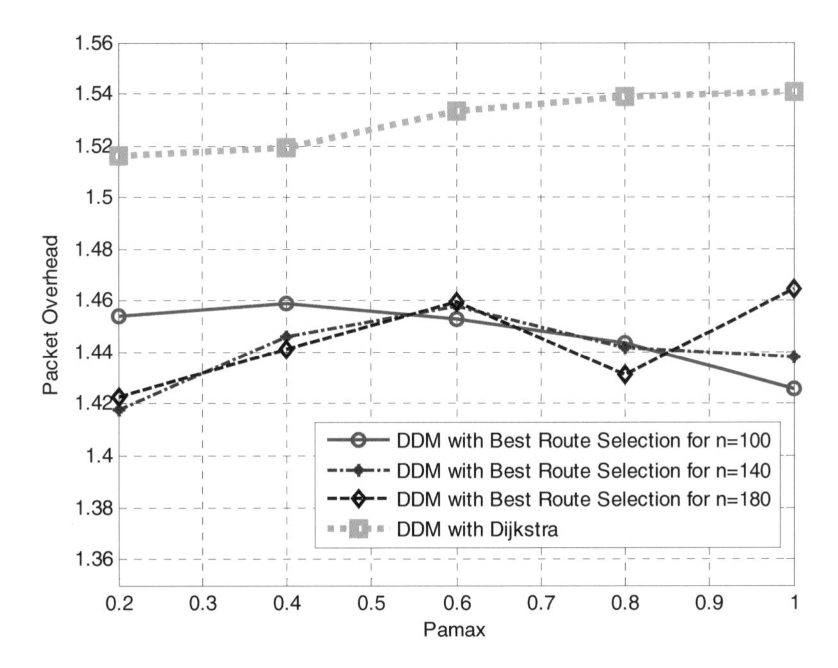

Fig. 10. Comparison of the PO under Dijkstra Routing and Best Route Selection. Node density analysis of the PO given by the DDM using Best Route Selection.

In Figure 11, we present the %SRP obtained by the DDM using Dijkstra Routing and that obtained using Best Route Selection. We can see that a significant improvement is achieved using Best Route Selection and that with this routing criterion the %SRP is more stable against attack probability that that obtained with Dijkstra routing. Also we can see that the best improvement is achieved for $p_{a_{max}}$ =1, and it is about 7%. In Figure 11, we present the density analysis of the %SRP using Best Route Selection. Here, we can see that the DDM still presents a robust response under node increasing. We also can observe some improvement in the performance for some attack probabilities; this may be explained by the increment of route options.

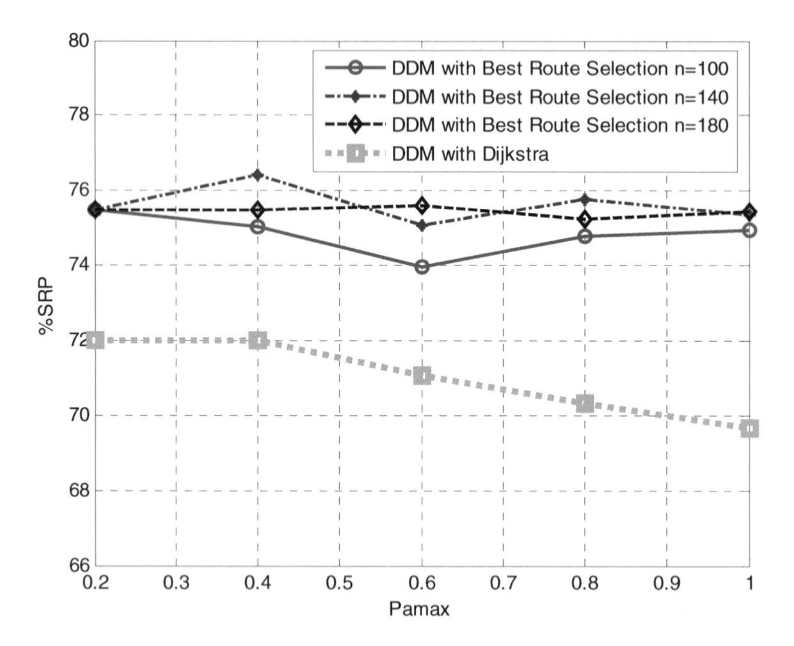

Fig. 11. Comparison of the %SRP under Dijkstra Routing and Best Route Selection. Node density analysis of the %SRP given by the DDM using Best Route Selection.

In Figure 12, we present the comparison of the %DR of the DDM using Dijkstra with that using Best Route Selection. It is possible to observe that using Best Route Selection the detection capabilities of the method are improved significantly for all the ranges of $p_{a_{max}}$ as an example, $p_{a_{max}}$ =1 the improvement is about 5%. In the same figure we can observe the %DR of the DDM using Best Route Selection for different node densities and we can deduce that detection capabilities of the method are not negatively affected by the increment of nodes in the network, the robustness of the method is maintained. We also can observe an increment on the %DR for some attack probabilities.

The %FPR for both routing schemes can be observed in Figure 13, where we can see that it is reduced by using the Best Route Selection for all $p_{a_{max}}$. For $p_{a_{max}}$ =1 we have an improvement about 5% using the proposed routing algorithm. The node

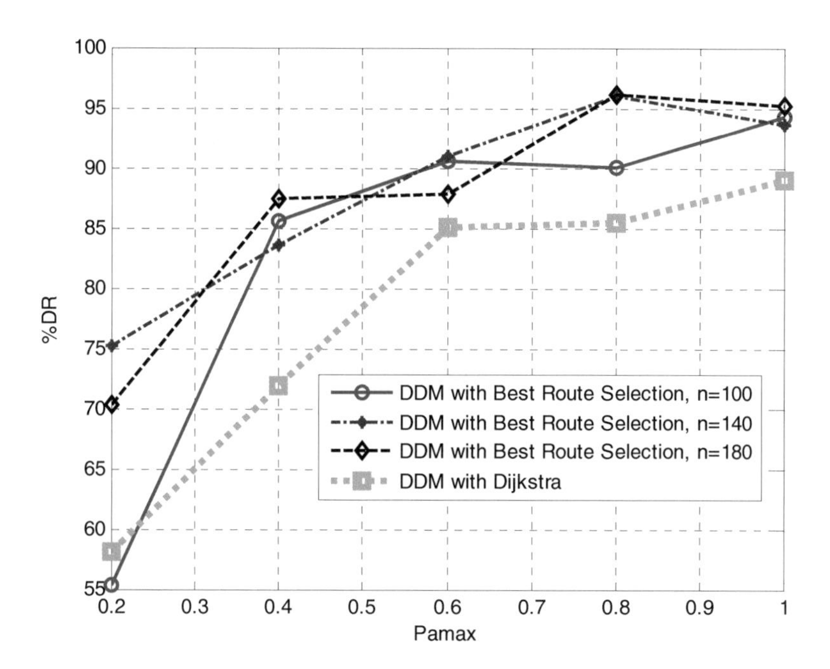

Fig. 12. Comparison of the %DR under Dijkstra Routing and Best Route Selection. Node density analysis of the %DR given by the DDM using Best Route Selection.

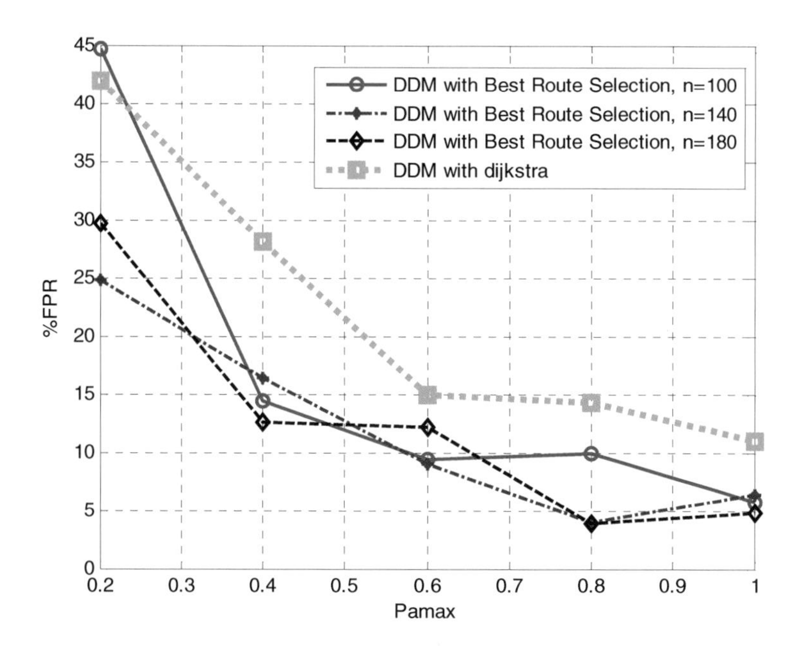

Fig. 13. Comparison of the %FPR under Dijkstra Routing and Best Route Selection. Node density analysis of the %FPR given by the DDM using Best Route Selection.

density analysis of the %FPR using Best Route Selection can also be observed in Figure 13, where we are able to see that the node increment does not increase the false positive declarations. We can even see that some improvements are achieved in several points of attack probability. Note that the %FNR plot is not included because this parameter is zero for all $p_{a_{max}}$, for the node densities analyzed.

Finally in Figure 14, we present the % NAR for both routing schemes and we can see that the Best Route Selection give us a better performance than the Dijkstra routing, because a lower %NAR is obtained with the Best Route Selection than with Dijkstra. Improvement attained by the Best Route Selection is about 60% for $p_{a_{max}}$ =1. We also present the %NAR obtained for different node densities using Best Route Selection, where we can see that node increment improves this parameter. As noted before, improvement may be explained by the increment of route options.

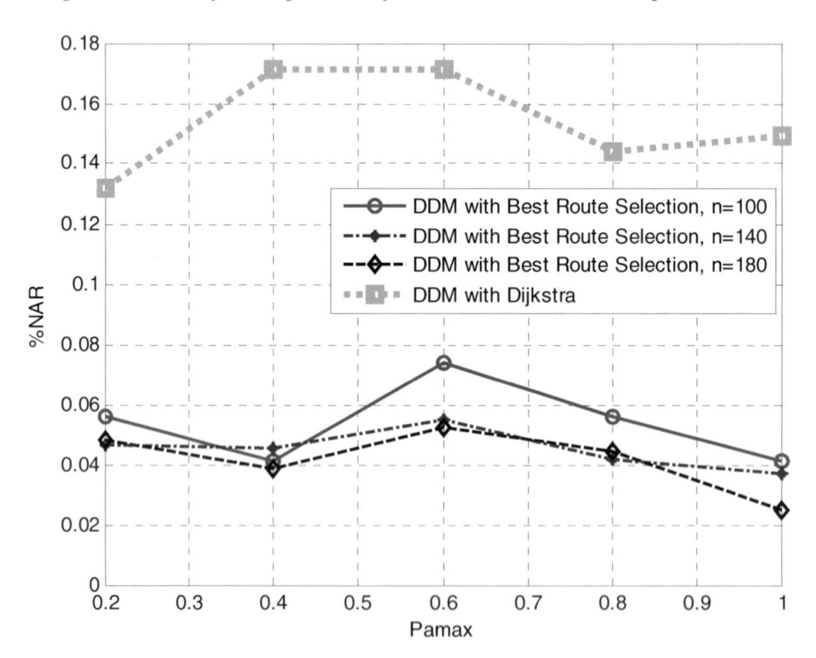

Fig. 14. Comparison of the %NAR under Dijkstra Routing and Best Route Selection. Node density analysis of the %NAR given by the DDM using Best Route Selection.

From Figures 10 to 14, we have demonstrated that the Best Route Selection based on the algebraic representation of the DDM proposed in this chapter, gives us a more robust route selection than the Dijkstra Routing. In the following figures we present a range analysis of the DDM using Best Route Selection with *n*=180 nodes. We use this node density because in figures above was observed a slight improvement from *n*=100 nodes to *n*=180 nodes.

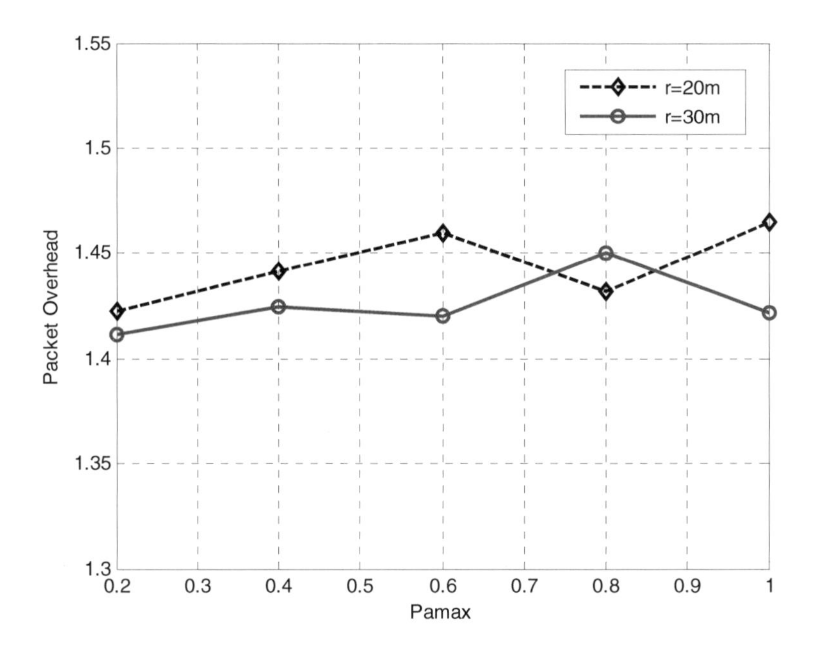

Fig. 15. PO with Best Route Selection, for r=20m and r=30m. n=180.

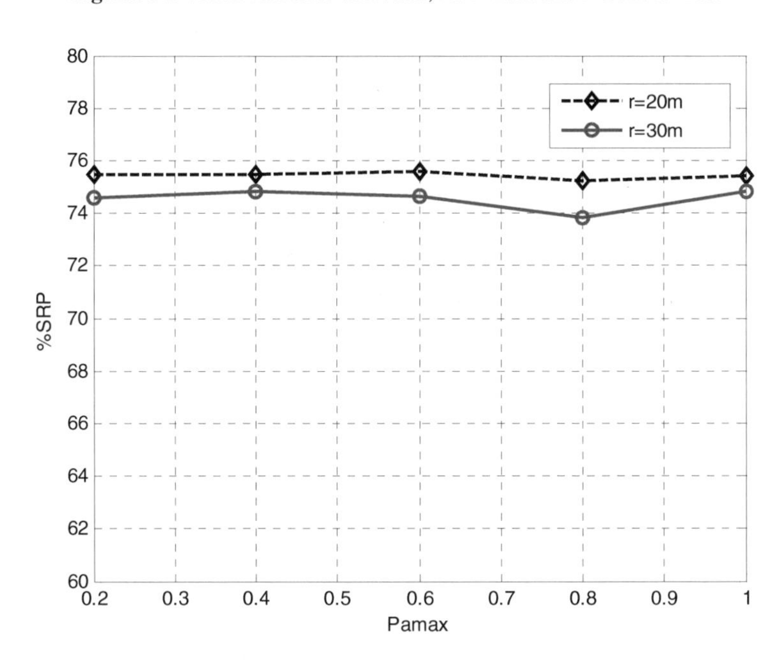

Fig. 16. %SRP with Best Route Selection, for r=20m and r=30m. n=180.

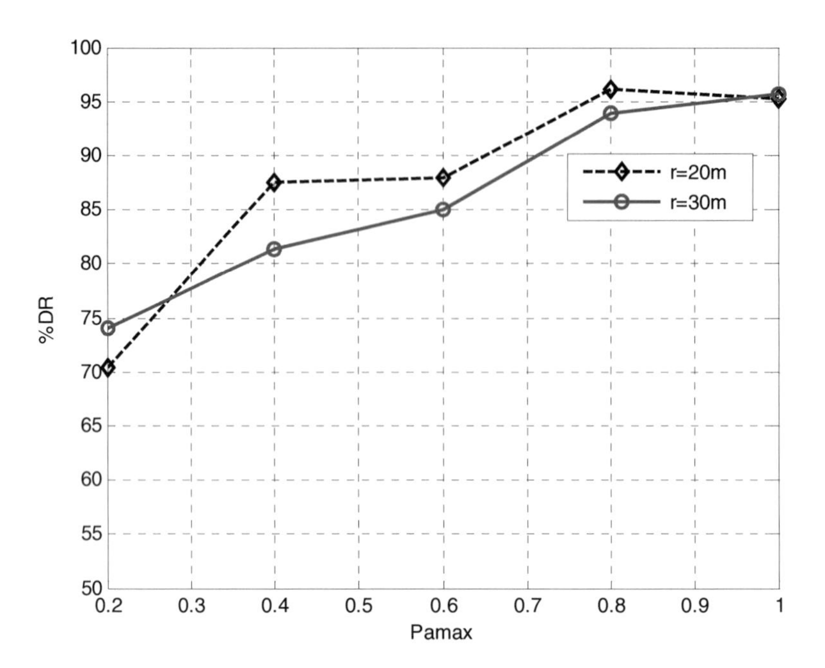

Fig. 17. %DR with Best Route Selection, for r=20m and r=30m. n=180.

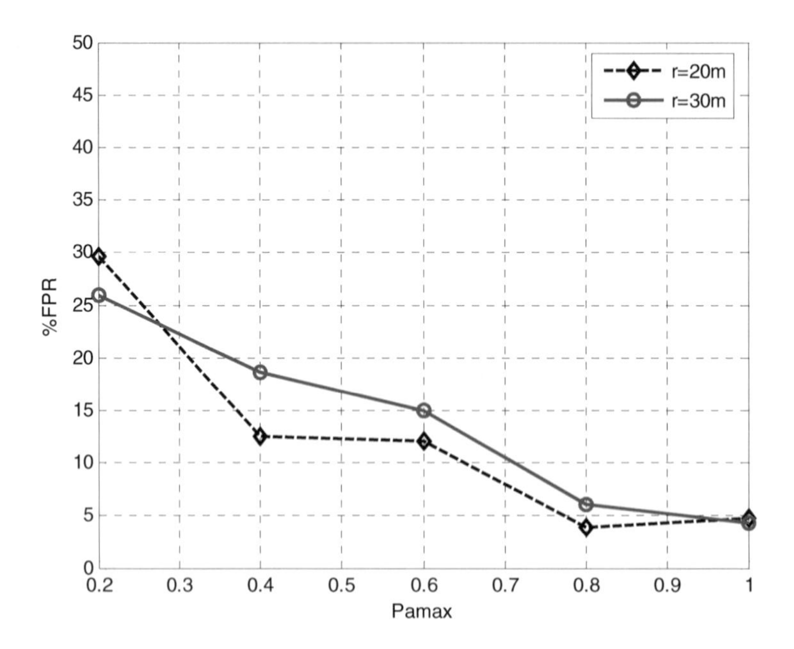

Fig. 18. %FPR with Best Route Selection, for r=20m and r=30m. n=180.

In Figure 15, we plot the PO against the maximal attack probability for two different transmission ranges r=20m and r=30m. In this figure we can observe that for almost every $p_{a_{\max}}$ the PO for r=30m is lower than for r=20m. This may be, because the best route selected tend to be shorter for r=30m than for r=20m.

In Figure 16, we present the %SRP given by the DDM using best route selection with n=180 nodes, for r=20m and r=30m. In this figure we can see that there is not improvement attained with the increment of the transmission range r. The same happens with %DR and %FPR in figures 17 and 18 respectively, the performance obtained with r=30m is close to that obtained with r=20m, and no improvement is observed. This may be explained because with such node density probably we let out of the route selection good routes for r=30m.

Finally, in Figure 19 we plot the %NAR against $p_{a_{\max}}$ for n=180 and for two transmission ranges, r=20m and r=30m. In this figure we can see that there is a better performance for r=20m than that obtained with r=30m. The explanation to this result may be the same that the given before for the figures above. The %FNR plot is not included because this parameter is zero for all $p_{a_{\max}}$, for the transmission ranges analyzed.

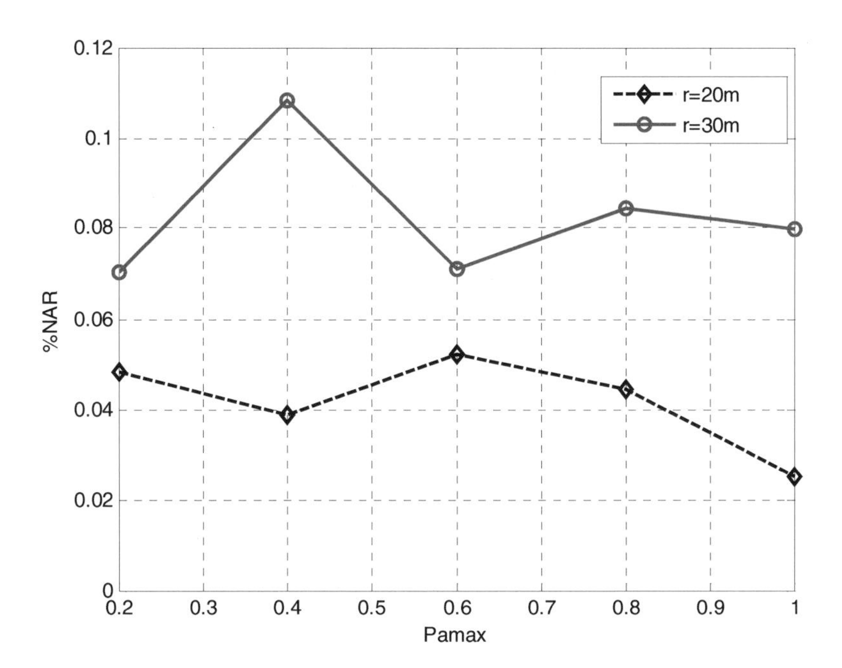

Fig. 19. %NAR with Best Route Selection, for r=20m and r=30m. n=180.

In this Section, we have presented an algebraic representation of a network commanding a detection and defense mechanism (DDM) for sink holes and selective forwarding attacks based on network coding. We extend the work done in [17] for networks allowing returning links by obtaining the transfer matrix H. We also give the sufficient and necessary conditions for a feasible connection with returning links based on the determinant of the transfer matrix Q. A probabilistic approach of the network is also presented based on the transfer matrix Q and on the interference and outage probabilities. We also propose to use this representation in order to select the best route for commanding the detection and defense mechanism following the proper methodology. Finally, it is demonstrated throughout simulations that the proposed routing selection method is a reliable routing tool for commanding the DDM given better results than those obtained with Dijkstra routing.

5.2 Some TCP Results in WRN

The results introduced in this section were obtained from a wireless reconfigurable network scenario, where the topology for the simulation is a network where a main route connecting an origin and destination is generated consisting of seven hops (eight nodes). The source and destination nodes are at opposite ends of the chain. The objective is to communicate a set of packets from the source to the destination. The source node emits packets of information continuously until the end of the simulation (500 packets) according to a Bernoulli process of rate λ packets per slot, the system has a capacity μ packets per slot and the load factor is defined by $\rho = \lambda / \mu$. In this simulation, intermediate nodes have a continuous cross traffic of packets, thus intermediate nodes in the network have queues. The packets are sent using a sliding window TCP protocol of size ten packets, when the destination node receives a successful packet an acknowledgment (ACK) is send to the source node. During the simulation packet losses can occur due the lack of capacity in the buffers of intermediate nodes or due to a probability of packet loss across the seven hops that considers outage, interference and fading for the wireless channel environment.

When the simulation finishes sending the 500 packets, we obtain a matrix of data with parameters of our interest as packet loss, round-trip time of each packet, buffer occupation in each node, and especially the time it took for the message to be delivered. Important results that were obtained are shown in Figure 20, the manner in which the graph was obtained is explained below. Figure 20 shows the percentage of packet losses throughout the simulation for different values of the capacity of the buffers in each node. We can see for example that for a traffic load of 0.7 overload will occur whenever the capacity of the buffers is less than 25, i.e., a buffer size of 30 will suffice to have practically zero packet losses.

In each of the simulations we obtained, we generated a buffer matrix which represents the number of packets on the buffer at each node (columns) along all the simulation time (rows). After that we obtained a histogram for each node (column) which provides the number of times that these nodes had a specific buffer size throughout the simulation. The matrix obtained is formed by the number of possible buffer occupancy states (columns) and the number of simulation times (rows). It is worth mentioning that the simulations were done for different values of the traffic load ρ , ranging from 0.5 up to 4, and different values of the capacity of the buffers

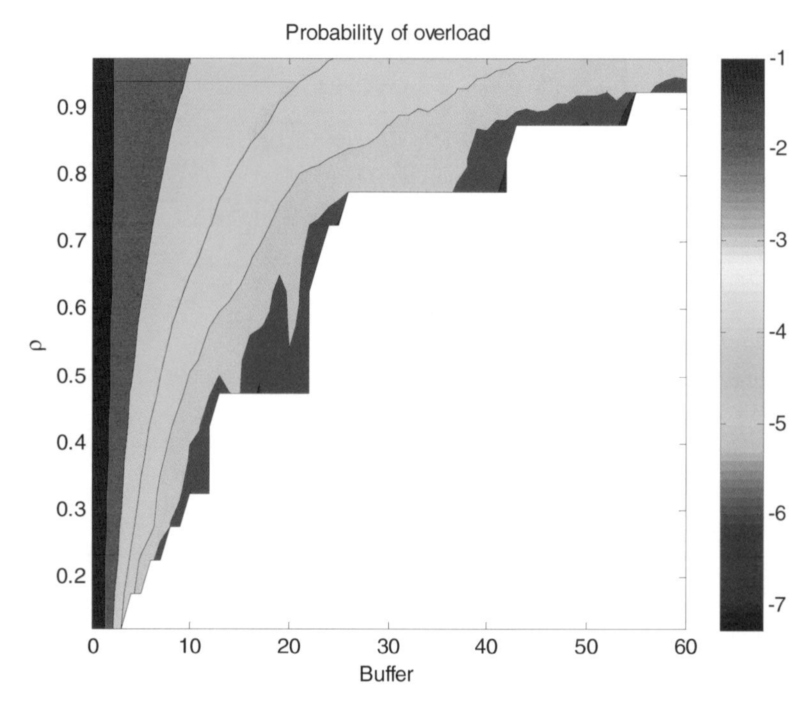

Fig. 20. Probability of overload

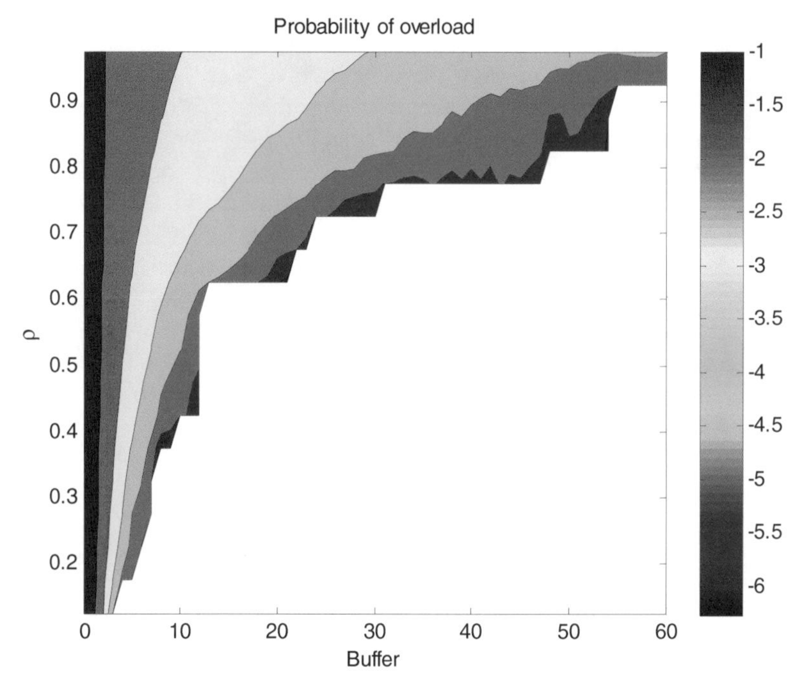

Fig. 21. Probability of overload, 4-hop route

(20, 30, 40, 50, 60 packets). After obtaining the histogram we got a mean value for each possible capacity of buffer (columns), this means that we obtained the mean value of how many times the nodes has occupancy of x packets in the buffer. These occupancy values are then referenced to their maximum values in order to obtain percentages. The final matrix represents the percentage of nodes that have a specific buffer size for a given value of ρ and vice versa.

The same results as those shown in Figure 20 were obtained for a scenario of a 4-hop route, but the values of buffer size and traffic load do not change much, hence the use of a buffer size as a function of traffic load in order to obtain determined packet loss levels remains practically the same for short routes as for long routes in WRNs, this is shown in Figure 21.

In conclusion, the tendency in WRNs would be to use large buffer sizes independently of the traffic load experienced.

In [7], TCP/NC was tested on a TCP-Reno flow running over a single-hop wireless link. The experiment is performed over 802.11a with a bit rate of 6Mb/s and a maximum of 5 link layer retransmission attempts. The quantity measured during the experiment is the goodput, we can define a goodput as the application level throughput, i.e. the number of useful information bits, delivered by the network to a certain destination, per unit of time. The amount of data considered excludes protocol overhead bits as well as retransmitted data packets. This is related to the amount of time from the first bit of the first packet is sent (or delivered) until the last bit of the last packet is delivered.

The principal observation in the results was that in the lossless case, TCP performs better than TCP/NC. This could be because of the computational overhead that is introduced by coding and decoding operations, and also the coding header overhead. However, as the loss rate increases, the benefits of coding begin to outweigh the overhead. The goodput of TCP/NC is therefore higher than TCP, and hence the fall in goodput is more gradual with coding than without, [7].

The main results obtained were the variation of throughput with loss rate for both TCP and TCP/NC. The loss rate of all links was kept at the same value, and this is varied from 0% to 20 %. Two scenarios were compared: two TCP flows competing with each other and two TCP/NC flows competing with each other. in this simulation was observed that TCP's throughput falls rapidly as loss increase, however, TCP/NC is very robust to losses and reaches a throughput that is very close to the link capacity, we can say that TCP/NC has a linear behavior. An important observation is that TCP/NC increases almost in a 90% the throughput than the TCP normal.

6 Conclusions and Future Work

In this chapter, we introduced the fundamentals of wireless reconfigurable networks and network coding. We provided examples of the advantages of using network coding in wireline networks and discussed the extension of its use to wireless networks. We discussed some of the basic security problems in WRNs and how those could be detected by the use of network coding. A detection mechanism is proposed, and results are presented for this mechanism in a WRN scenario with outage and interference. The effects of TCP in packet loss throughout a WRN are also introduced, and the importance of choosing the correct buffer capacity for the nodes as a function of the traffic load is also discussed. As we discussed previously, it has been shown that network coding helps improve capacity especially for multicasting. One future direction would be to provide load balancing using network coding, which could be implemented by using different coding coefficients for different network regions keeping packet flows from traversing some parts of the network. Another future direction of network coding is the integration to routing algorithms for ad-hoc and sensor networks.

References

[1] Toh, C.K.: Ad-Hoc and Mobile Wireless NEtworks: Protocols and Systems. Prentice Hall, New Jersey (2002)
[2] Villalpando Hernandez, R.: Detection and Defense Mechanism Against Security Attacks in Reconfigurable Networks: Network Coding Approach, Electrical and Computer Enginnering, Tecnológico de Monterrey, Campus Monterrey, Monterrey, NL, Mexico (2008)
[3] Perkins, C.E.: Ad Hoc Networking. Addison Wesley (2001)
[4] Molva, R., Michiardi, P.: Security in Ad Hoc Networks. In: Conti, M., Giordano, S., Gregori, E., Olariu, S. (eds.) PWC 2003. LNCS, vol. 2775, pp. 756–775. Springer, Heidelberg (2003)
[5] Zhou, L., Haas, Z.J.: Securing Ad-Hoc Networks. IEEE Network 13(6), 24–30 (1999)
[6] Papadimitratos, P., Haas, Z.J.: Secure Routing for Mobile Ad Hoc Networks. In: SCS CNDS, San Antonio, TX, pp. 193–204 (2002)
[7] Sundararajan, J.K., Shah, D., Me, et al.: Network Coding Meets TCP: Theory and Implementation. Proceedings of the IEEE 99(3), 490–512 (2011)
[8] Ahlswede, R., Cai, N., Li, S.-Y.R., et al.: Network Information Flow. IEEE Transactions on Information Theory 46(4), 1204–1216 (2000)
[9] Ho, T., Médard, M., Koetter, R., et al.: A Random Linear Network Coding Approach. IEEE Transactions on Information Theory 52(10), 4413–4430 (2006)
[10] Baochun, L., Di, N.: Random Network Coding in Peer-to-Peer Networks: From Theory to Practice. Proceedings of the IEEE 99(3), 513–523 (2011)
[11] Yeung, R.W., Zhang, Z.: Distributed Source Coding for Satellite Communications. IEEE Transactions on Information Theory 45, 1111–1120 (1999)
[12] Chou, P.A., Wu, Y.: Network Coding for the Internet and Wireless Networks. IEEE Signal Processing Magazine 24(5), 9 (2007)
[13] McEliece, R.J.: Finite Fields for Computer Scientists and Engineers. Kluwer Academic Publishers, Boston (1987)
[14] Li, S.-Y.R., Yeung, R.W., Cai, N.: Linear Network Coding. IEEE Transactions on Information Theory 49(2), 11 (2003)
[15] Li, S.Y.R., Sun, Q.T., Ziyu, S.: Linear Network Coding: Theory and Algorithms. Proceedings of the IEEE 99(3), 372–387 (2011)
[16] Yeung, R.W.: Information Theory and Network Coding. Springer, New York (2008)
[17] Koetter, R., Médard, M.: An Algebraic Approach to Network Coding. IEEE/ACM Transactions on Networking 11(5), 14 (2003)

[18] Ning, C., Chan, T.: Theory of Secure Network Coding. Proceedings of the IEEE 99(3), 421–437 (2011)
[19] Li, J., Ge, W.: Enhanced Network Coding for TCP in Wireless Networks. In: 7th International Conference on Wireless Communications, Networking and Mobile Computing, WiCOM (2011)
[20] Villalpando, R., Vargas, C., Munoz, D.: Network Coding for Detection and Defense of Sink Holes in Wireless Reconfigurable Networks. In: 3rd International Conference on Systems and Networks Communications, ICSNC 2008, pp. 286–291 (2008)
[21] Cox, L.P., Noble, B.S.: Honor Among Thieves in Peer-to-Peer Storage. In: 19th ACM Symposium on Operating Systems Principles, SOSP (2003)
[22] Villalpando, R., Vargas-Rosales, C., Munoz, D.: Interference and Outage Evaluation of a Network Coding Based Detection and Defense Mechanism for WRN. In: IEEE ICDS, Malta (2009)
[23] Soo-Kim, N., Beongku, A., Do-Hyeon, K., et al.: Wireless Ad-Hoc Networks Using Cooperative Diversity-Based Routing in Fading Channel. Communications, Computer and Signal Processing (2007)

A Secure Intragroup Time Synchronization Technique to Improve the Security and Performance of Group-Based Wireless Sensor Networks

Miguel Garcia[1], Diana Bri[1], Jaime Lloret[1], and Pascal Lorenz[2]

[1] Integrated Management Coastal Research Institute, Universitat Politecnica de Valencia, C/Paranimf, n° 1, 46730, Grao de Gandia. Spain
{migarpi,diabrmo}@upvnet.upv.es, jlloret@dcom.upv.es
[2] Network and Telecommunication Research Group, University of Haute Alsace, 34 rue du Grillenbreit, 68008, Colmar. France
lorenz@ieee.org

Abstract. Time synchronization is required in wireless sensor networks in order to improve its performance. This improvement could be noticed in terms of energy, storage, computation, shared resources or bandwidth. One of the main applications in WSNs has been to decrease the energy consumption. A wireless network can save energy with this feature, but if this synchronization is corrupted, it could cause a worse behavior. Firstly, we will analyze which are the most important time synchronization issues. Then, a secure time synchronization method will be presented for group-based wireless sensor networks to avoid malicious attacks. The synchronization technique is based on a system model for secure intra-group synchronization. This system will use simple messages, where nodes of each group will exchange several parameters like time stamps, groupID, etc. in order to make a secure system. The system proposed has high scalability, due to group-based feature, while saves energy thanks to the designed synchronization technique. In order to test our proposal we will simulate several situations to show the performance of our synchronization algorithm.

1 Introduction

Nowadays, wireless sensor networks (WSN) are used in several application fields [1]. For instance, in energy efficiency control, high-security environments, environmental monitoring, industrial control, automotive industry, medical field, automation, etc. The implementation of these networks is following an exponential growth [2] due to their own features. On the one hand, these networks are deployed easily and they are configured by themselves. So, each sensor node can become a transmitter, receiver or gateway between two other nodes without direct sight or register information of neighboring nodes at any time. On the other hand, they support energy efficient management which allows them to get a high rate of autonomy to work by themselves.

S. Khan and A.-S.K. Pathan (Eds.): *Wireless Networks and Security*, SCT, pp. 403–422.
DOI: 10.1007/978-3-642-36169-2_12

In order to deploy a WSN, it is necessary to analyze its lifespan, coverage, budget, the best way to install it, the suitable response time, accuracy and frequency of measurements and security. Moreover, in order to select the most suitable sensor nodes, we should take into account its energy, way of communication, processing capacity, way of synchronization, size and cost [3].

However, these networks have some limitations such as:

- It is necessary to optimize the energy consumption of nodes to achieve maximum autonomy.
- Bandwidth and network coverage are limited, so the way of communication, transmission and reception of data is very different in such networks.

Therefore, the main feature to consider in these networks is the energy saving. This is the reason why the nodes are awake in short periods of time, only when they are taking data or measuring some variable. It depends on the frequency between measurements. The three possible states are shown in Fig. 1.

Synchronization is important for several distributed systems [4]. In particular it is essential for an efficient use of energy, data fusion, localization and many other applications. Synchronization algorithms for traditional networks are inefficient for wireless sensor networks, for instance features assumed by the traditional time protocol NTP (Network Time Protocol) [5], they are not applicable for WSNs. It assumes a static network topology where nodes are installed before the synchronization, configured by hand and the delay between nodes can be estimated with high precision.

In contrast, WSNs have dynamic topologies and nodes can be attached to anything: animals, buoys, plants, etc., and anytime new nodes can be added to the network. Features of WSNs cannot be predicted in advance and energy is a resource very limited. So, although synchronization of nodes is essential, it is difficult to get it.

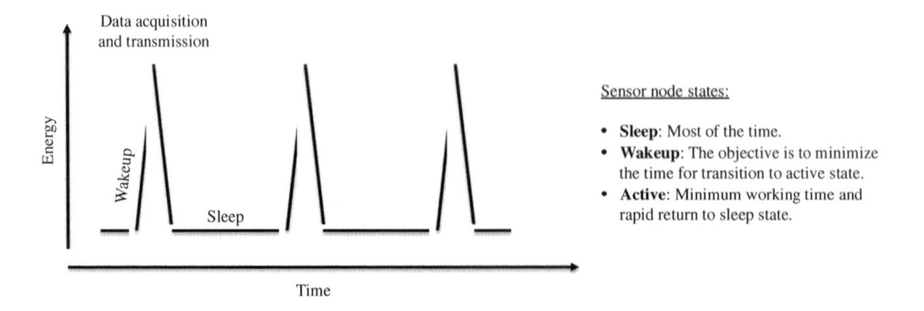

Fig. 1. Three possible states of energy consume in a sensor node

The synchronization protocol should use low energy, be scalable, robust and assume an ad-hoc dynamic topology. Moreover, sensor network is so vulnerable to attacks that both an internal and external adversarial can easily corrupt the synchronization mechanism in sensor networks.

For these reasons, in this paper we are going to show an algorithm to know the offset time between nodes inside a group. This algorithm is based on principles of receiver-receiver synchronization protocols, but it includes security. All messages exchanged in the synchronization process will be encrypted using group keys. Our objective is that our algorithm exchanges fewer timing messages between sensor nodes, which means saving energy. Receiver-receiver protocols require a super node, which broadcasts a time reference message, so it is a centralized protocol. In contrast, in our proposal all nodes send messages about timing information. So, there is not an only one super node that manages the synchronization.

The remainder of the paper is organized as follows. Section 2 examines some related works about time synchronization protocols used in wireless sensor networks. The main aspects related to time synchronization in WSN are explained in Section 3. Then, in Section 4 is developed our secure algorithm and the method used to manage the keys in a group. Our proposal of secure time synchronization is explained in Section 5, in this section we show a mathematical study to know the offset times between sensor nodes. Several simulations are presented in Section 6. Finally, in Section 7 we summarize our results and conclude our paper.

2 Related Work

Many protocols have been proposed for time synchronization in wireless and wired LANs. Network Time Protocol (NTP) [5] has been used on the Internet for many years. We can consider nodes on the Internet use a GPS system [6] in order to synchronize them, but this system does not work in indoor environments. Moreover, this protocol is not useful for WSNs because it consumes quite energy, CPU process and data storage and these features are limited in WSNs.

Several time protocols to synchronize nodes in WSNs have been shown lately [7]. In this paper, authors review the problem of wireless sensor network synchronization and associated key issues, including the fundamental signal processing tasks needed, such as estimation of the skew and offset between two clocks, round trip delay estimation, extension of synchronization to multiple nodes, as well as performance analysis and bounds.

One of the early works on synchronization in WSN was called post facto synchronization proposed by Elson and Estrin [8]. In this approach, each node is normally clock unsynchronized with the rest of the network, a reference node periodically broadcasts a reference broadcast message to all other nodes of the network. When an event is detected, each node stores the time received (time-stamp with its local clock). After receiving the message reference nodes will use this time as a reference and adjust their timestamps with respect to that reference. The same authors present RBS in [9]. It is a receiver-receiver protocol that requires beacon nodes that transmit reference packets (but without time stamps) to multiple nodes. Nodes record time-of-receipt and exchange these time-stamps to determine clock offsets and skews using standard linear regression techniques. However, it is not robust to clock skew, typically due to frequency offsets, and it assumes that propagation delays are negligible.

TimingSync Protocol for Sensor Networks (TPSN) [10] is based on the main idea of sender-receiver synchronization protocols. It has two phases: level discovery and synchronization. In the level discovery phase, TPSN is level-by-level time synchronization; it leads to a tree structure. In the synchronization phase, each sensor performs pairwise time synchronization with its parent in a top-down fashion. The nodes in level 2 are synchronized with each node at level 1 and so on until all nodes in the network are synchronized with respect to the root.

Flooding time synchronization protocol (FTSP [11], designed by Maroti et al., floods the whole network with messages, which contain values of the global time, i.e. the time of an elected synchronization leader. The synchronization leader periodically broadcasts synchronization messages containing its time-stamps. A network node records its local time upon reception of a synchronization message (using MAC-layer time-stamping), thus obtaining one reference point, which contains two time values: global time and local time. When a node collects an enough number of reference points, it computes the drift rate of its clock with respect to the leader clock using linear regression.

Tiny-sync and Mini-sync [12] are proposed to keep a global time in WSN by synchronizing any two nodes in the whole network. A pair of nodes uses bi-directional time-stamped packet transmissions to estimate the clock offset between them so that the two nodes are synchronized. However, since every pair of nodes must perform two-way message exchanges to get synchronized, a significant amount of communication traffic will be generated. In addition, a certain number of control packets have to be transmitted across the whole area of WSN to organize the overall network into several node-pair groups.

Also, there are time synchronization algorithms for cluster-based WSNs. In [13] H. Kim et al. propose the cluster-based hierarchical time synchronization protocol (CHTS) for wireless sensor networks. They provide network wide time synchronization through the introduction of abstractions for cluster and hierarchy. Authors introduce a novel way to calculate synchronization period to meet the required synchronization error level by using clock drift value, deterministic delays and nondeterministic offset. Finally, they simulate their algorithm in several scenarios in order to find the best environment where CHTS could work properly.

But as we see in [14], time synchronization methods needs security. There are different attacks that can lead to errors in synchronizing the clocks of sensor nodes. The primary goal of the attackers is to make other sensors set a wrong clock time. The secondary goal of the attackers is to affect more sensors by making themselves locate close to the root of the sync-tree, i.e., having low levels. Therefore synchronization protocols should be safe. In [14] the authors propose a scheme with three phases: a) abnormality detection performed by individual sensors, b) trust-based malicious node detection performed by the base station, and c) self-healing through changing topology of the synchronization. This algorithm is tested under two attacks and the conclusions are that their system can quickly detect and defeat the misleading attack and the wormhole attack, with reasonable implementation overhead tree.

Another example is [15]. S. Ganeriwal et al. propose a suite of protocols for secure pairwise and group synchronization of nodes that lie in each other's power ranges and of nodes that are separated by multiple hops. Their protocols offer different points of operation in the energy-accuracy subspace and the choice of the specific protocol should be made by the network designer depending on his application needs.

Finally, [16] is another paper where a secure clock synchronization process is proposed. The idea behind their scheme is using node's neighbors as verifiers to check if the synchronization is processed correctly so that it can detect the attacks launched by compromised node. This scheme is based on three phases: a) Level Discovery Phase, b) Synchronization Phase and c) Verification Phase. According to the simulations presented by the authors, this scheme guarantees that normal nodes can synchronize their clocks to global clock even if each normal node has up to t colluding malicious nodes during synchronization phase.

After analyzing several related works about time synchronization in WSNs, we are going to show the most important issues about this subject. In next section, we will describe the need of use time synchronization techniques in WSN, which the basic techniques are and the main sources of errors related to this topic.

3 Time Synchronization: Techniques and Weaknesses

WSNs have the need to coordinate the communication, computing, sensing, and performance of distributed nodes, so an accurate and consistent system to synchronize them is essential to keep the sensor network in working order. This section discusses about the main motivations for time synchronization, the challenges that it involves and some solutions to solve them.

3.1 Why a WSN Needs Time Synchronization?

WSNs need time synchronization among sensor nodes due to several reasons. Here is a list with the main ones:

1. Time-stamping in measurements. Even the simplest data collection often requires readings with date/time and location information. This is especially important when a significant delay between the transmitter and the receiver node or the base station is possible.
2. Signal processing in the network. Date/time information is necessary to determine which information from several sources can be added within the network. Many collaboration algorithms of processing signals, like tracking phenomena or unknown objectives, require a constant and exact synchronization.
3. Localization. There are several techniques in order to locate nodes, which use the propagation time required to transmit a message to a reference node, so these techniques need proper time synchronization.
4. Cooperative communication. Some techniques of multi-node cooperative communication involve multiple transmitters broadcasting in phase to a given receiver. Such techniques have the potential to provide energy savings and significant strength, but require precise timing.

5. Media Access. Schemes based on TDMA (Time Division Multiple Access) require node synchronization to assign them to different slots and guarantee collision-free communication.
6. Sleep scheduling. One of the most significant sources of energy savings is the inactivity of the radio device, which is called sleep state. However, the synchronization is necessary to coordinate the sleep schedules of neighboring devices, so they can communicate with each other efficiently.
7. Coordinated action. In advanced applications in which the sensor network includes distributed sensor nodes, besides detection it is necessary the synchronization to coordinate the distributed sensor nodes with distributed control algorithms.

3.2 Synchronization Types Used in WSN

There are different ways to classify the synchronization techniques in WSN, but it is often represented by several pairs of contrasting distinctive features [17].

Synchronization algorithms can be proactive or reactive. The proactive algorithms carry out synchronization tasks continuously to keep the offset delay bounded. For example a reference node sends periodically a message with its time-value to other nodes. This allows knowing the timing offset respect to the reference time. Reactive algorithms act when an event is detected; a node sends its time to the base station that performs the necessary corrections.

Moreover, these synchronization schemes are divided into adaptive and non-adaptive. Adaptive schemes can modify important parameters of the synchronization procedure, such as frequency, response to certain changes in the network or the environment. On the other hand, non-adaptive schemes never change their way of working.

Synchronization schemes introduce computational overhead. They increase network traffic due to exchange messages and this means more power consumption. It is important to maintain a balance between the disadvantages to incorporate a synchronization system in WSNs and the advantages of keeping bounded the errors introduced by offset time clocks in the sensor nodes.

Among the most common synchronization procedures we can mention:

- Sender-receiver synchronization: Pairwise sender-receiver synchronization is performed by a handshake protocol between a pair of nodes. This protocol is executed in three steps (see Fig. 2a). T1, T4 represent the time measured by the local clock of node A. Similarly T 2, T3 represent the time measured by B's local clock. At time T1, A sends a synchronization pulse packet to B. Node B receives this packet at T2, where T2 is equal to T1+δ+d. Here, δ and d represent the offset between the two nodes and end-to-end delay respectively. At time T3, B sends back an acknowledgement packet. This packet contains T2 and T3. Node A receives the packet at T4. Similarly, T4 is related to T3 as T4 = T3-δ+d. Node A can now calculate the clock offset and the end-to end delay.

- Receiver-receiver synchronization: In receiver-receiver model, it synchronizes receivers by recording the arrival time of a reference message in a broadcast medium, and then exchanging this arrival time to determine the discrepancy between the respective clocks of the receivers. This technique is shown in Fig. 2b. The approach avoids various sources of synchronization error because a broadcast message is received at exactly the same time by all receivers, at least in terms of the granularity of their clocks.
- WSN nodes can be synchronized either to some external time reference, or by means of some intra-network synchronization procedures. A GPS receiver can serve as an external source of time for a wireless network, where the time values can be disseminated by a special node attached to the GPS receiver. In this case, network nodes are able to transform their time values to the reference time values and vice versa, thus achieving the time synchronization. Alternatively the network can be synchronized internally, i.e. using the clock of some node as a reference clock.

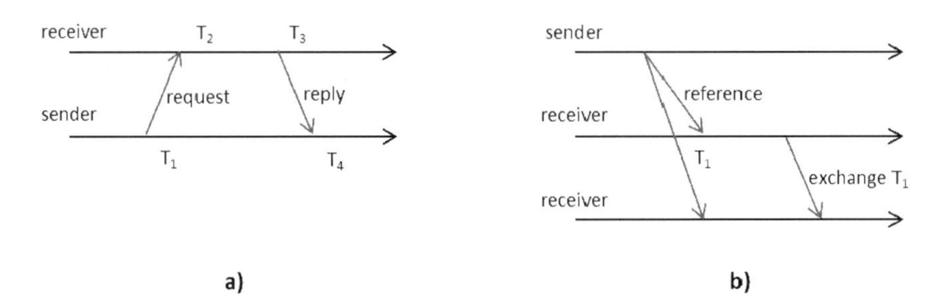

Fig. 2. a). Sender-receiver synchronization. b). Receiver-receiver synchronization.

3.3 Error Sources in Time Synchronization

Time synchronization algorithms in WSNs use the measured time in the exchange of messages between nodes to synchronize the clocks on different nodes, estimating their offsets.

An important aspect that we must take into account when we use algorithms based on the exchange of messages is that this exchange generates delays. Moreover of these delays, the exchange of messages can be corrupted, so they could inject errors in the synchronization algorithms. In order to know the origin of these errors we should identify some phenomena that take into the process for sending data. The most important are:

- Send Time: It is the time used to load the message and to make the request and to send it to the MAC layer in the transmitter side. Depending on the overhead of system and the current processor load, the send time is not deterministic and it can be as high as tens of milliseconds.
- Access Time: It is the delay caused by the access to transmission channel at the beginning of transmission. Access time is the least deterministic message when a

message is delivered in a WSN. It varies from tens to hundreds of milliseconds, depending on the current network traffic.

- Transmission Time: It is the time used by the transmitter to send a message. It is around ten milliseconds, but it depends on the length of the message and the transmission speed.
- Propagation Time: It is the time required by the message to propagate it from sender to receiver. This time is quite deterministic in WSN and it only depends on distance between nodes.
- Reception Time: It is the time taken by the receiver to receive the message. It equals than transmission time. The transmission and reception times can overlap in WSN.
- Receive Time: It is the time required to process the incoming message and notifying to this receptor. Its characteristics are similar to Send Time.
- Attackers can be either outsiders or insiders. The outsiders do not possess keys for encryption and authentication. Thus, they can only jam the transmission of time synchronization messages, but not modify or generate those messages. On the other hand, the compromised sensors, which are insiders, can manipulate the time synchronization protocols in many ways.

4 Secure Group-Based Wireless Sensor Network

A group-based WSN is network, which is divided into several logical groups. Every group does the same function (collecting data, creating alarms, sending messages, etc.). Everyone has a central node that limits the zone where the node from the same group will be placed, but its functionality is the same as the rest of the nodes and moreover it manages its group. Every node has a nodeID that is unique on the network. The first node of each group of the network acquires a group identifier (groupID).

All nodes in each group have the same functionality, but according its location can be divided into Border nodes because they are in the border between two groups, Central node which helps to create each group and Regular nodes that are the rest of nodes which make up the group along with the Central node and Border nodes (see Fig. 3). The process of creating groups is based on sending messages and according to neighboring nodes' replies and a group maximum radius (R_{group}), which is previously saved into all nodes of network, these groups limit their size. Each one knows its neighboring group using these messages, for more information about group's creation you can refer to [18].

Border nodes are, physically, the edge nodes of the group. When there is an event in a node, this event is sent to all the nodes in its group in order to take the appropriate actions. All nodes in a group know all the information about their group. Border nodes have connections with other border nodes from neighbor groups and they are used for sending information to other groups or receiving information from other groups.

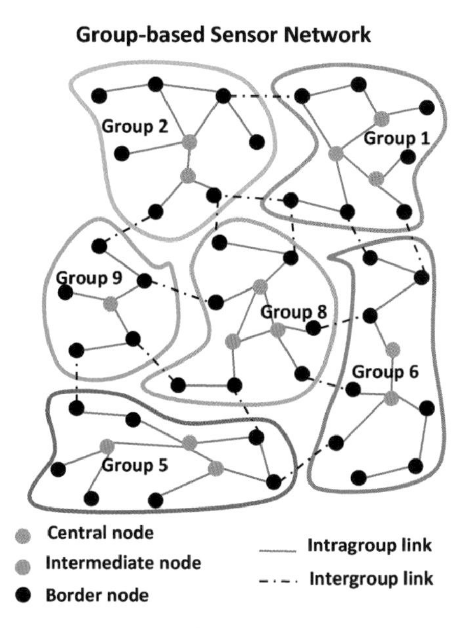

Fig. 3. An example of group-based sensor network

In order to add security in our group-based network, we have defined two security zones: the intragroup security and the intergroup security. The intergroup security is related with the secure communications between different groups of the same network. In contrast, the intragroup security keeps the security inside the group.

The intergroup security is based on the symmetric cryptography with a single key for all members of the network. This feature has been chosen due to the limited local memory and process operation of the nodes of these networks. In contrast, the intragroup security is different. In this case, a node can belong to a group, and, later, it can change to other group. For that reason the keys have to be different depending on the group. Communications within each group are closed and only known by the nodes that belong to that group. When the information is sent to other group, it this is encrypted through the symmetric key known by all the nodes of the network. The communication is more effective because the nodes use less cycles to process the security and therefore they consume less energy.

The intragroup communication security is more complex than the intergroup one. There are several parameters that must be taken into account. In addition, a member is not able to access the network information before it joins the group; neither the nodes can access the information after they leave the group. The central node of each group is responsible of managing the group key. It is also responsible of saving the key logical tree structure of the group members (Fig. 4).

The key tree is created when the group is being formed. Each border node knows the keys from them to the central node. The operation can be shown in [19]. In order to reduce the memory wasted by the central node (this node must store and maintain

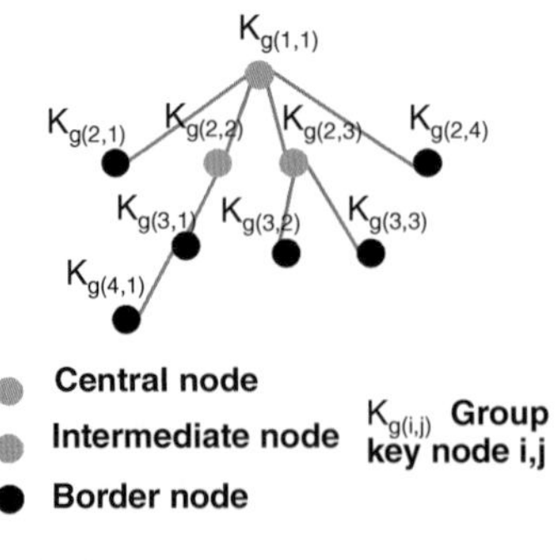

Fig. 4. Key distribution in a group

the key tree), we use pseudorandom functions to generate the keys. For that reason, a node only needs to know some rules, but not all keys. The group key of a node (i, j) is defined by equation 1.

$$K_{g(i,j)} = F_p(2^i + j) \oplus r \tag{1}$$

Where the group key of the node (i, j) is equal to a pseudorandom function (F_p) plus an update factor r. A pseudorandom function is a family of functions with the property that the input-output behavior of a random instance of the family is "computationally indistinguishable" from that of a random function.

The operation mode is as follows: Let us suppose that the key $K_{g(i,j)}$ must be updated. The central node must send P (see equation 2) to all nodes that had stored the key $K_{g(i,j)}$.

$$P = r \oplus r' \tag{2}$$

Each authorized node will calculate the new key through equation 3.

$$K'_{g(i,j)} = K_{g(i,j)} \oplus P = F_p(2^i + j) \oplus r \oplus r \oplus r' = F_p(2^i + j) \oplus r' \tag{3}$$

When all upgrades have been finished, all keys have the structure expressed in the equation 4. Using this method, the central node must only save in memory the used pseudorandom function (F_p) and the update factor (r').

$$K'_{g(i,j)} = F_p(2^i + j) \oplus r' \tag{4}$$

The entire secure process is explained in detail in the work [19].

5 Intragroup Secure Time Synchronization

Our proposal is based on RBS protocol commented on section 2. Our objective is that our algorithm exchanges fewer timing messages between sensor nodes, which means saving energy. RBS protocol requires a super node, which broadcasts a time reference message, so it is a centralized protocol. In contrast, in our proposal all nodes send messages about timing information. So, there is not a super node which manages the synchronization.

We propose a receiver-receiver synchronization scheme, where each message exchanged between nodes inside a group is encrypted. The encryption keys are explained in the previous section. Each sensor node encrypts its message with its key and any node of logic branch that joins with central node (see Fig. 4) can decode this message. Moreover, all nodes can decode messages encrypted by central node. For example, sensor node in Fig. with the key ($K_{g(4,1)}$) encrypts a message. This message can be decoded by all nodes which connect it to the central node. In contrast, when this message arrives to this one and it should be sent to other branch, the message should be encrypted with its own key.

In order to explain our algorithm we use an example scheme illustrated in Fig. 5. It shows our intragroup receiver-receiver synchronization scheme.

We suppose that the group identification is group_ID=1. Firstly, the head node (HN_1) starts the kth synchronization process sending a synchronization beacon ($Beacon_{syn}$) with the information $T^1_{syn,k}$.

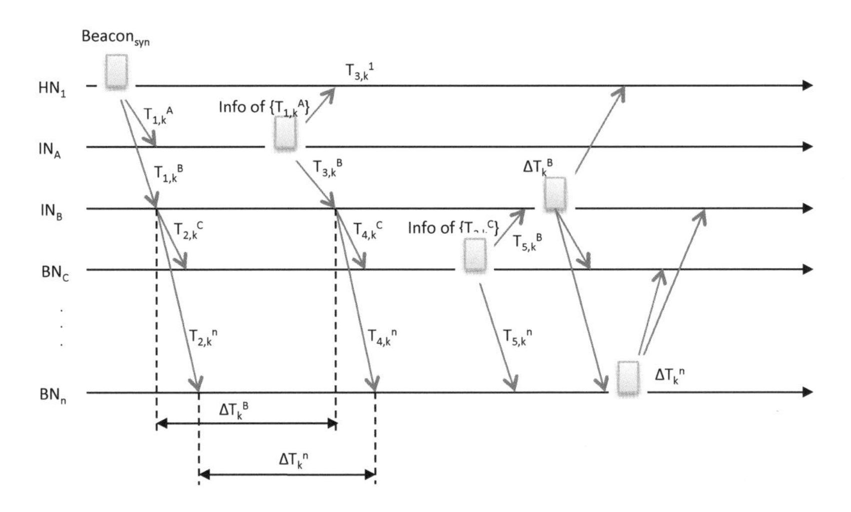

Fig. 5. Intragroup synchronization scheme

This message only arrives to sensor nodes under head node coverage. These nodes record the message arrival time, following our example ($T^A_{1,k}$, $T^B_{1,k}$). These times are explained on equations 5 and 6, where $T^1_{syn,k}$ is a reference time, φ^{1-A}_{offset} is clock offset between the node HN_1 and the node IN_A, δ^{1-A}_{prop} is the propagation delay between both nodes and τ^{1-A}_{rx} is the receive delay at these nodes.

Then, these nodes (IN_A, IN_B) resend this message to others nodes (BN_C, BN_n), which are farther from the head node. These last nodes record the new times ($T^C_{2,k}$, $T^n_{2,k}$), which are detailed on equation 7 and 8. In these cases, both equations depend on $T^A_{1,k}$, $T^B_{1,k}$.

$$T_{1,k}^{A} = T_{syn,k}^{1} + \varphi_{offset}^{1-A} + \delta_{prop}^{1-A} + \tau_{rx}^{1-A} \tag{5}$$

$$T_{1,k}^{B} = T_{syn,k}^{1} + \varphi_{offset}^{1-B} + \delta_{prop}^{1-B} + \tau_{rx}^{1-B} \tag{6}$$

$$T_{2,k}^{C} = T_{1,k}^{B} + \varphi_{offset}^{B-C} + \delta_{prop}^{B-C} + \tau_{rx}^{B-C} \tag{7}$$

$$T_{2,k}^{n} = T_{1,k}^{B} + \varphi_{offset}^{B-n} + \delta_{prop}^{B-n} + \tau_{rx}^{B-n} \tag{8}$$

Then, node IN_A sends a message similar to $Beacon_{syn}$ with the information $T_{1,k}^{A}$ to HN_1 and IN_B. They record the arrival time in each node ($T_{3,k}^{1}$, $T_{3,k}^{B}$). INB resend this information to BN_C and BN_n. These last nodes record also the times $T_{4,k}^{C}$, $T_{4,k}^{n}$. Similarly, BN_C sends a beacon with information of $T_{2,k}^{C}$. This beacon is received by IN_B and BN_n, so they record the times $T_{5,k}^{B}$ and $T_{5,k}^{n}$ too. Finally, nodes IN_B and BN_n, selected randomly in this example, calculate the differences ΔT_{k}^{B} and ΔT_{k}^{n}. With these data we can know the offset times between these nodes, as we show as follows.

We assume that one process can be finished in a relatively short time, so we can neglect the effect of clock skews in one process.

Equation 9 is deduced from equation 5 and 6. φ_{offset}^{A-B} is clock offset between node IN_A and IN_B, it is used to correct the clock of node B.

$$\varphi_{offset}^{A-B} = \varphi_{offset}^{1-B} - \varphi_{offset}^{1-A} = T_{1,k}^{B} - T_{1,k}^{A} - \left[\left(\delta_{prop}^{1-B} - \delta_{prop}^{1-A}\right) + \left(\tau_{rx}^{1-B} - \tau_{rx}^{1-A}\right)\right] \tag{9}$$

All sensor nodes are similar, so we can deduce that τ_{rx}^{1-B} is equal to τ_{rx}^{1-A}. So we obtain the equation 10, where φ_{offset}^{A-B} is the difference between both times plus an increment of propagation delay. This propagation delay can vary from few picoseconds until 333 picoseconds (see equation 11), it depends on radio coverage of each sensor node.

$$\varphi_{offset}^{A-B} = T_{1,k}^{B} - T_{1,k}^{A} \pm \Delta\delta_{prop} \tag{10}$$

$$\Delta\delta_{prop} = [0\,ps, 333\,ps[\tag{11}$$

These propagation delays do not contribute significantly to the overall error. According to reference [5], we can model these delay using a Gaussian distributed random function with mean zero and variance σ^2. For these reasons, it could be dismissed. Finally, φ_{offset}^{A-B} can be estimated from this synchronization process from equation 12.

$$\hat{\varphi}_{offset}^{A-B} = \frac{1}{M}\sum_{k=1}^{M}\left(T_{1,k}^{B} - T_{1,k}^{A}\right) \tag{12}$$

From equations 7 and 8 we can obtain φ_{offset}^{C-n} (see equation 13). With the same reasoning explained in equations 11 and 12, we have a simple unbiased estimator (see equation 14).

$$\varphi_{offset}^{C-n} = \varphi_{offset}^{B-n} - \varphi_{offset}^{B-C} = T_{2,k}^{n} - T_{2,k}^{C} - \left[\left(\delta_{prop}^{B-n} - \delta_{prop}^{B-C}\right) + \left(\tau_{rx}^{B-n} - \tau_{rx}^{B-C}\right)\right] \quad (13)$$

$$\hat{\varphi}_{offset}^{C-n} = \frac{1}{M}\sum_{k=1}^{M}\left(T_{2,k}^{n} - T_{2,k}^{C}\right) \quad (14)$$

From the message sent by IN_A, we get equation 15 and 16:

$$T_{3,k}^{1} = T_{syn,k}^{A} + \varphi_{offset}^{A-1} + \delta_{prop}^{A-1} + \tau_{rx}^{A-1} \quad (15)$$

$$T_{3,k}^{B} = T_{syn,k}^{A} + \varphi_{offset}^{A-B} + \delta_{prop}^{A-B} + \tau_{rx}^{A-B} \quad (16)$$

Where the used variables have the same meaning as in equations 5 and 6. Using equation 15 and 16, we obtain φ_{offset}^{A-1} (see equation 17).

$$\varphi_{offset}^{A-1} = T_{3,k}^{1} - T_{3,k}^{B} + \varphi_{offset}^{A-B} - \left[\left(\delta_{prop}^{A-1} - \delta_{prop}^{A-B}\right) + \left(\tau_{rx}^{A-1} - \tau_{rx}^{A-B}\right)\right] \quad (17)$$

Taking into account equation 9 to know the value of φ_{offset}^{A-B} and equation 18, we can obtain equation 19.

$$\Delta T_{k}^{B} = T_{3,k}^{B} - T_{1,k}^{B} \quad (18)$$

$$\varphi_{offset}^{A-1} = T_{3,k}^{1} - \left(T_{1,k}^{B} + \Delta T_{k}^{B}\right) - \left[\left(\delta_{prop}^{A-1} - \delta_{prop}^{A-B} + \delta_{prop}^{1-B} - \delta_{prop}^{1-A}\right) + \left(\tau_{rx}^{A-1} - \tau_{rx}^{A-B} + \tau_{rx}^{1-B} - \tau_{rx}^{1-A}\right)\right] \quad (19)$$

Similar to equations 12 and 14, we can get an estimate of φ_{offset}^{A-1} from equation 20.

$$\hat{\varphi}_{offset}^{A-1} = \frac{1}{M}\sum_{k=1}^{M}\left(T_{3,k}^{1} - \left(T_{1,k}^{B} + \Delta T_{k}^{B}\right)\right) \quad (20)$$

Following the same process explained before we obtain the next equations:

$$T_{4,k}^{C} = T_{3,k}^{B} + \varphi_{offset}^{B-C} + \delta_{prop}^{B-C} + \tau_{rx}^{B-C} \quad (21)$$

$$T_{4,k}^{n} = T_{3,k}^{B} + \varphi_{offset}^{B-n} + \delta_{prop}^{B-n} + \tau_{rx}^{B-n} \quad (22)$$

$$T_{5,k}^{B} = T_{syn,k}^{C} + \varphi_{offset}^{C-B} + \delta_{prop}^{C-B} + \tau_{rx}^{C-B} \quad (23)$$

$$T_{5,k}^{n} = T_{syn,k}^{C} + \varphi_{offset}^{C-n} + \delta_{prop}^{C-n} + \tau_{rx}^{C-n} \quad (24)$$

With equations 23, 24 and 13 we can get equation 25. Moreover, the node BN_n can calculate ΔT_{k}^{n} using equation 26. Using these data we obtain φ_{offset}^{C-B} depending on $T_{5,k}^{B}$, $T_{2,k}^{n}$ and ΔT_{k}^{n} (see equation 27).

$$\varphi_{offset}^{C-B} = T_{5,k}^{B} - T_{5,k}^{n} + \varphi_{offset}^{C-n} - \left[\left(\delta_{prop}^{C-n} - \delta_{prop}^{C-B}\right) + \left(\tau_{rx}^{C-n} - \tau_{rx}^{C-B}\right)\right] \quad (25)$$

$$\Delta T_k^n = T_{4,k}^n - T_{2,k}^n \tag{26}$$

$$\varphi_{offset}^{C-B} = T_{5,k}^B - \left(T_{2,k}^n + \Delta T_k^n\right) - \left[\left(\delta_{prop}^{C-n} - \delta_{prop}^{C-B} + \delta_{prop}^{B-n} - \delta_{prop}^{B-C}\right) + \left(\tau_{rx}^{C-n} - \tau_{rx}^{C-B} + \tau_{rx}^{B-n} - \tau_{rx}^{B-C}\right)\right] \tag{27}$$

Finally, we can estimate φ_{offset}^{C-B} using the equation 28.

$$\hat{\varphi}_{offset}^{C-B} = \frac{1}{M}\sum_{k=1}^{M}\left(T_{5,k}^B - \left(T_{2,k}^n + \Delta T_k^n\right)\right) \tag{28}$$

In this way, all sensor nodes are synchronized. There is a secure synchronization between several pair of nodes, which involves a secure synchronization between all nodes inside a group. The next section shows that this process is improved according to number of repetitions.

6 Test Performance

In order to check our proposal, we have simulated the algorithm proposed in the previous section. We have used Matlab [20] as a simulation tool to show the validity of our proposal. The sensor nodes are randomly placed in an area of 10x10 Km. They have radio coverage of 75 meters, so the distances between nodes, which are communicated, are always less than 75 m. The arrival times follow an exponential distribution. For each offset time, we have considered three cases: a) μ=15 and σ=3, b) μ=20 and σ=6 and c) μ=36 and σ=13. Next figures present the simulation results based on scheme shown in Fig. 5. These graphs show the offset time required between two sensor nodes. The main conclusion from these simulations is the offset time tends to be constant, but now we go to analyze them in detail in the following paragraphs.

Figure 6 shows the offset time between sensor node A and B. In this case we can see that when σ is low (between 3 and 6) the offset time is lower than with higher σ. The offset time is around 5 μs with few samples and low σ, and as the number of samples is increasing, the offset time decreases. It is around 1 μs for 40 samples. Simulation behavior is different when we have an σ=13. In this case we have a high offset time between both sensor nodes until 10 samples. Then, its behavior is more stable and it remains constant around 2 μs after 40 samples. These differences are due to variability (σ=13) in the sampling times used for the simulations.

Next, figure 7 shows us the offset time between sensor node C and n. According to our simulated topology, the most constant values are when our synchronous referenced time has an σ=6. The offset time is around 1 μs in almost all samples. When σ is 3, the offset time is 6 μs in the first samples, but then it is more stable and its fluctuations are between 0 and 2 μs. Finally, the most unstable offset time is when we have a distribution time with σ=13. In this case there are several fluctuations,

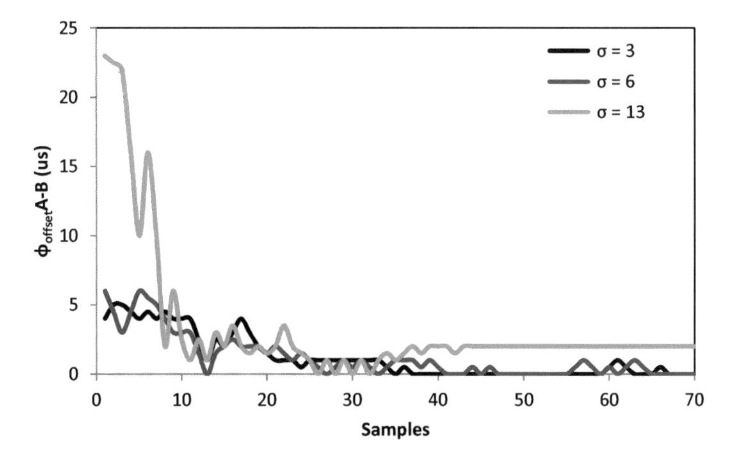

Fig. 6. Offset time between sensor nodes A and B $\left(\varphi_{offset}^{A-B}(\mu s)\right)$

which vary from 3 to 13 μs until the sample 13. However from this sample, fluctuations are decreasing and they remain between 0 and 4 μs until the 55th sample. From this point, the offset time is stable around 2 μs.

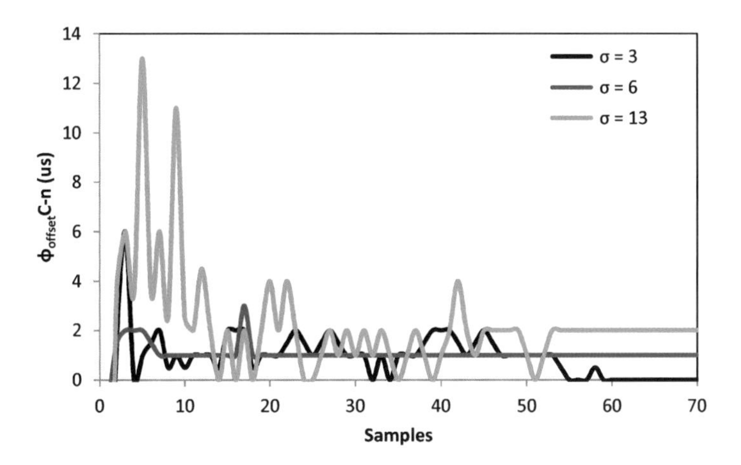

Fig. 7. Offset time between sensor nodes C and n $\left(\varphi_{offset}^{C-n}(\mu s)\right)$

Next figure (Fig. 8.) shows the offset time between sensor nodes A and 1 (φ_{offset}^{A-1}). In this case, the most stable result is when σ=3. All samples are around 10μs. For σ=6, the first samples are around 10 μs, but from the 16th sample, it changes to 20 μs. There are several fluctuations between 10 and 20 μs until the 35th sample. Then it remains stable around 20 μs. Finally, the results obtained with σ=13 are quite different on the first samples, the offset time reaches up to 60 μs. Fluctuations are so high until the 40th sample, they vary between 10 and 33 μs indeed. Next, this time is stabilized around 26 μs. The time offset variability in these simulations is the expected one, as our initial distribution has a higher σ, the results are more unstable compared to the results obtained with smaller σ's.

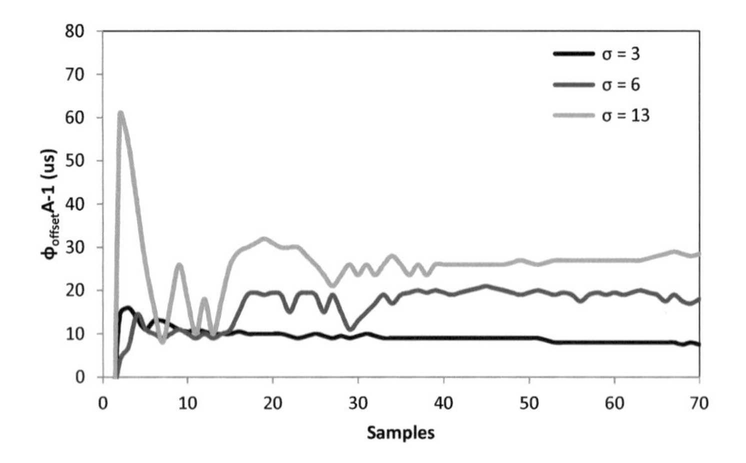

Fig. 8. Offset time between sensor nodes A and 1 $\left(\varphi_{offset}^{A-1}(\mu s)\right)$

Finally, we present the simulations of φ_{offset}^{C-B} (see figure 9). Simulations for scenarios with σ=3 and σ=6 are stable from the 15th sample. From this sample the difference between both simulations is 5 μs. When we have σ=3 the offset time is around 10 μs. It changes until 15 μs when we have an arrival time distribution with an σ=6. This offset time increases until 35 μs (it is stable from the 45th sample) when our distribution time has an σ=13. In this simulation the worst case is from 10th to 25th sample. This is due to exponential feature of our arrival time distribution. In this period of time our variance is higher than at other times.

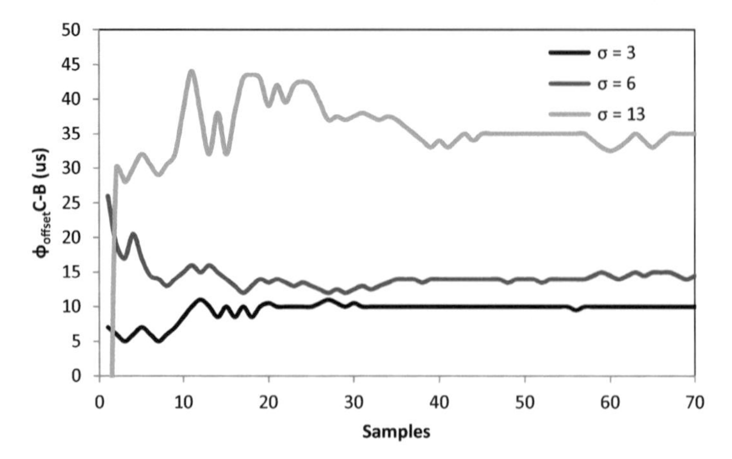

Fig. 9. Offset time between sensor nodes C and B $\left(\varphi_{offset}^{C-B}(\mu s)\right)$

But, the most important feature obtained from all simulations shown (see Fig. 6, Fig. 7, Fig. 8 and Fig. 9), is the stability achieved by our proposal after several samples. So, the proposed system is able to find a constant offset time between each pair of sensor nodes and it never exceeds 35 μs in the worst case. Moreover, the information exchanged between nodes is secure because each message is encrypted using group-based keys.

7 Conclusion and Future Work

In this paper we have done a review about clock synchronization in wireless sensor network. We have shown the main reasons to use a proper synchronization technique. Besides, the simplest method to make a protocol, which manages the offset times between sensor nodes, have been explained briefly. Lastly, we have also described the typical sources of error.

After this review we have shown our secure group-based WSN and the main issue is the key's exchange. This is an essential part, because it is essential in order to exchange encrypted messages inside a group. Next, we have explained in detail each step needed to know the offset times between sensor nodes. We have based our proposal on receiver-receiver protocol, but adapting the main idea to a group-based topology. All explanation is based on our scheme (see Fig. 5) but it could be developed for any group-based topology.

Finally, our proposal has been simulated for the same scenario changing μ and σ values of our time arrival distribution. All cases show that our proposal can resolve the offset time between nodes in any moment. But this offset time is more stable when as the number of samples is going up.

Our future work about this research line is to extend this synchronization algorithm in order to work in intergroup communications. In this way the communication between groups will be synchronized and the communication between nodes inside a group too. This extension should take into account that border nodes in neighboring groups are synchronized.

Acknowledgments. This work is supported by the "Ministerio de Ciencia e Innovación", through the "Plan Nacional de I+D+i 2008-2011", project TEC2011-27516 and by the "Universitat Politècnica de Valencia" PAID-00-11.

References

1. Garcia, M., Bri, D., Sendra, S., Lloret, J.: Practical deployments of wireless sensor networks: a survey. Journal on Advances in Networks and Services 3(1&2), 1–16 (2009)
2. Bohli, J.-M., Sorge, C., Westhoff, D.: Initial observations on economics, pricing, and penetration of the internet of things market. SIGCOMM Comput. Commun. Rev. 39(2), 50–55 (2009)
3. Lee, J.C., Leung, V.C.M., Wong, K.H., Cao, J., Chan, H.C.B.: Key management issues in wireless sensor networks: current proposals and future developments. IEEE Wireless Communications 14(5), 76–84 (2007)

4. Wu, Y.-C., Chaudhari, Q., Serpedin, E.: Clock Synchronization of Wireless Sensor Networks. IEEE Signal Processing Magazine 28(1), 124–138 (2011)
5. Mills, D.: Internet time synchronization: The network time protocol. IEEE Trans. Commun. 39(10), 1482–1493 (1991)
6. Kaplan, E.D., Hegarty, C.J. (eds.): Understanding GPS: Principles and Applications. Artech House, Norwood (2006)
7. Sadler, B.M., Swami, A.: Synchronization in Sensor Networks: an Overview. In: IEEE Military Communications Conference, MILCOM 2006, October 23-25, pp. 1–6 (2006)
8. Elson, J., Estrin, D.: Time Synchronization for Wireless Sensor Networks. In: Proc. of the 15th International Parallel & Distributed Processing Symposium, April 23-27, p. 186 (2001)
9. Elson, J., Girod, L., Estrin, D.: Fine-grained network time synchronization using reference broadcasts. In: Proc. Fifth ACM SIGOPS Operating Syst. Review, Boston, MA, pp. 147–163 (December 2002)
10. Ganeriwal, S., Kumar, R., Srivastava, M.: Timing-sync protocol for sensor networks. In: Proc. ACM SenSys 2003, Los Angeles, CA (November 2003)
11. Maroti, M., Simon, G., Kusy, B., Ledeczi, A.: The flooding time synchronization protocol. In: Proc. 2nd Intl. Conf. on Embedded Networked Sensor Systems (SenSys 2004), Baltimore, MD, USA. ACM Press (November 2004)
12. Yoon, S., Veerarittiphan, C., Sichitiu, M.L.: Tiny-sync: Tight time synchronization for wireless sensor networks. ACM Trans. Sen. Netw. 3(2), Article 8 (June 2007)
13. Kim, H., Kim, D., Yoo, S.: Cluster-based hierarchical time synchronization for multi-hop wireless sensor networks. In: 20th International Conference on Advanced Information Networking and Applications, AINA 2006, April 18-20, vol. 2, p. 5 (2006)
14. Yang, Y., Sun, Y.: Securing Time-Synchronization Protocols in Sensor Networks: Attack Detection and Self-Healing. In: IEEE Global Telecommunications Conference, GLOBECOM 2008, November 30-December 4, pp. 1–6 (2008)
15. Ganeriwal, S., Capkun, S., Han, C., Srivastava, M.B.: Secure time synchronization service for sensor networks. In: Proceedings of the 4th ACM Workshop on Wireless Security (WiSe 2005), pp. 97–106. ACM, New York (2005)
16. Li, H., Zheng, Y., Wen, M., Chen, K.: A Secure Time Synchronization Protocol for Sensor Network. In: Washio, T., Zhou, Z.-H., Huang, J.Z., Hu, X., Li, J., Xie, C., He, J., Zou, D., Li, K.-C., Freire, M.M. (eds.) PAKDD 2007. LNCS (LNAI), vol. 4819, pp. 515–526. Springer, Heidelberg (2007)
17. Ageev, A.: Time Synchronization and Energy Efficiency in Wireless Sensor Networks. PhD thesis, University of Trento (2010), `http://eprints-phd.biblio.unitn.it/260/`
18. Lloret, J., Garcia, M., Tomas, J., Boronat, F.: GBP-WAHSN: A group-based protocol for large wireless ad hoc and sensor networks. Journal of Computer Science and Technology 23(3), 461–480 (2008)
19. Garcia, M., Lloret, J., Sendra, S., Laquesta, R.: Secure Communications in Group-based Wireless Sensor Networks. International Journal of Communication Networks and Information Security (IJCNIS) 2(1), 8–14 (2010)
20. Matlab Software, `http://www.mathworks.com/products/matlab/`

Key Terms & Definitions

Clock offset: At a particular moment as the difference between the time reported by the clock and the "true" local time as defined by UTC. If the clock reports a time T_c and the true local time is T_t, then the clock's offset is T_c-T_t.

Group-based architecture: A group is defined as a small number of interdependent nodes with complementary operations that interact in order to share resources or computation time, or to acquire content or data and produce joint results. In a wireless group-based architecture, a group consists of a set of nodes that are close to each other (in terms of geographical location, coverage area or round trip time) and neighboring groups could be connected if a node of a group is close to a node of another group.

Intergroup communication: It is the communication done between nodes of different groups.

Intragroup communication: It is the communication done by nodes inside the same group.

MAC: The Media Access Control (MAC) sublayer is the part of the OSI network model data link layer that determines who is allowed to access the physical media at any one time. It acts as an interface between the Logical Link Control sublayer and the network physical layer. The MAC sublayer is primarily concerned with recognizing where frames begin and end in the bit-stream received from the physical layer (when receiving) delimiting the frames (when sending).

Proactive protocol: In networks utilizing a proactive routing protocol, every node maintains one or more tables representing the entire topology of the network. These tables are updated regularly in order to maintain a up-to-date routing information from each node to every other node.

Reactive protocol: Unlike proactive routing protocols, reactive routing protocols does not make the nodes initiate a route discovery process until a route to a destination is required. This leads to higher latency than with proactive protocols, but lower overhead.

TDMA: Time division multiple access (TDMA) is a channel access method for shared medium networks. It allows several users to share the same frequency channel by dividing the signal into different time slots. The users transmit in rapid succession, one after the other, each using its own time slot. This allows multiple stations to share the same transmission medium while using only a part of its channel capacity.

Timestamp: A timestamp is a sequence of characters or encoded information identifying when a certain event occurred, usually giving date and time of day, sometimes accurate to a small fraction of a second. The term derives from rubber stamps used in offices to stamp the current date, and sometimes time, in ink on paper documents, to record when the document was received.

Time synchronization: Also kwon as clock synchronization, it is a problem from computer science and engineering which deals with the idea that internal clocks of several computers may differ. Even when initially set accurately, real clocks will differ after some amount of time due to clock drift, caused by clocks counting time at slightly different rates. There are several problems that occur as a repercussion of clock rate differences and several solutions, some being more appropriate than others in certain contexts.

Wireless sensor network: A wireless sensor network (WSN) in its simplest form can be defined as a network of (possibly low-size and low-complex) devices denoted as nodes that can sense the environment and communicate the information gathered from the monitored field through wireless links; the data is forwarded, possibly via multiple hops relaying, to a sink that can use it locally, or is connected to other networks (e.g., the Internet) through a gateway.

Capacity-Approaching Channel Codes for Discrete Variable Quantum Key Distribution (QKD) Applications

Maria Teresa Delgado Alizo[1], Inam Bari[1], Fred Daneshgaran[2], Fabio Mesiti[1], Marina Mondin[1], and Francesca Vatta[3]

[1] Politecnico di Torino, Italy
{marina.mondin,inam.bari}@polito.it, fabio.mesiti@gmail.com, mtdelgado@hotmail.com
[2] California State University, Los Angeles, USA
fdanesh@calstatela.edu
[3] University of Trieste, Italy
vatta@units.it

Abstract. Secure communications and cryptography is as old as civilization itself. The Greek Spartans for instance would cipher their military messages and, for Chinese, just the act of writing the message constituted a secret message since almost no-one could read or write Chinese. Modern public key Cryptography until the mid 1980's was founded on computational complexity of certain trap-door one-way functions that are easy to compute in one direction, but very difficult in the opposite direction. To a large extent computational complexity is still the lynchpin of modern cryptography, but the whole paradigm was revolutionized by introduction of Quantum Key Distribution (QKD) which is founded on fundamental laws of Physics. Indeed, to date, QKD is de-facto the most successful branch of Quantum Information Science (QIS) encompassing such areas as quantum computing which is still in its infancy.

Modern QKD is fundamentally composed of a series of three steps that shall be explained later in the chapter: 1) data transmission over the error-prone quantum channel; 2) information reconciliation to allow the parties engaged in communication to have two identical copies of a message that may not be as secure as desired; and 3) privacy amplification that ensures the parties possess copies of messages about which the information that could have possibly be gleaned by the eavesdropper is below a desirable threshold. It is this sufficiently private and often much shorter message that can be used as the secret key to allow exchange of longer messages between the legitimate parties.

Step-1 must be based on the laws of quantum physics, whereas step-2 and -3 either necessitate the use of quantum error correcting codes which are often complex or as is often done in practice, based on information exchange over a classical public channel.

Objective of this chapter is to give a tutorial presentation and evaluation of QKD protocols at the systems level based on classical error-correcting codes. The QKD systems can provide perfect security (from the viewpoint of information theory) in the distribution of a cryptographic key. QKD systems and related protocols, under particular conditions, can use the classic channel

S. Khan and A.-S.K. Pathan (Eds.): *Wireless Networks and Security*, SCT, pp. 423–456.
DOI: 10.1007/978-3-642-36169-2_13

coding techniques instead of quantum error-correcting codes, both for correcting errors that occurred during the exchange of a cryptographic key between two authorized users, and to allow privacy amplification, in order to make completely vain a possible intruder attempt. The secret key is transmitted over a quantum, and thus safe channel, characterized by very low transmission rates and high error rates. This channel is safe given the properties of a quantum system, where each measurement on the system perturbs the system itself, allowing the authorized users to detect the presence of any intruder. Moreover, as shown by accurate experimental studies, the communication channel used for quantum key exchange is not able to reach high levels of reliability (the Quantum Bit Error Rate - QBER - may have a high value), both because of the inherent characteristics of the system, and of the presence of a possible attacker. In order to obtain acceptable residual error rates, it is necessary to use a parallel classical and public channel, characterized by high transmission rates and low error rates, on which to transmit only the redundancy bits of systematic channel codes with performance possibly close to the capacity limit. Furthermore, since the more redundancy is added by the channel code, the more the corresponding information can be used to decipher the private message itself, it becomes necessary to design high-rate codes obtained by puncturing a low-rate mother code, possibly achieving a redundancy such that elements of the secret message cannot be uniquely determined from the redundancy itself.

1 Introduction and Problem Setup

It has been said that the security of conventional cryptographic techniques relies on the assumption of limited advancement of mathematical algorithms and computational power in the foreseeable future, and also on limited financial resources available to a potential adversary. Computationally secure cryptosystems, no matter whether public- or secret-key, will always be at the mercy of mathematical and/or computational breakthroughs, which are difficult to predict and may even be hidden. In addition, steady progress in code-breaking allows the adversary to reach back in time and break older, earlier captured messages encrypted with weaker keys. As a consequence, periodic re-encryption or re-signing certain sensitive documents is necessary, along with the requirement to carefully sort information according to the used cryptosystem.

Another common problem of conventional cryptographic methods is the so-called side-channel cryptanalysis. Side channels are undesirable ways through which information related to the activity of the cryptographic device can leak out. The attacks based on side-channel information do not assault the mathematical structure of cryptosystems, but their particular implementations. It is possible to gain information for instance by measuring the amount of time needed to perform a certain operation, by measuring power consumption, heat or electromagnetic radiation.

Quantum mechanics offers a solution for the secure key distribution in cryptosystems. While the security of classical cryptographic methods can be undermined by advances in technology and mathematical algorithms, the quantum approach can provide *unconditional security*. In quantum mechanics the security is guaranteed by the *Heisenberg uncertainty principle*, which does not allow us to discriminate non-orthogonal states with certainty. Within the framework of classical physics, it is impossible to reveal possible eavesdropping, because information encoded into any property of a classical object can be obtained without affecting the object itself. All classical signals can be monitored passively. In classical communications, one bit of information is encoded in billions of photons, electrons, atoms or other carriers. It is always possible to passively listen in, by splitting part of the signal and performing a measurement on it. Quantum cryptosystems eliminate this side channel by encoding each bit of information into an individual quantum object, such as a single photon. Single photons cannot be split, copied or amplified without introducing detectable disturbances.

It is important to notice that quantum mechanics does not prevent eavesdropping; it only allows one to detect the presence of a possible eavesdropper. Since only the cryptographic key is transmitted, no information leakage can take place when someone attempts to listen in. Eavesdropping causes discrepancies between measurements and when discrepancies are found, the key is simply discarded and the users may repeat the procedure to generate a new key.

1.1 Quantum Key Distribution

In the early 1980s, Bennett and Brassard proposed a solution to the key distribution problem based on quantum physics [1]. They presented a protocol that allows users to establish an identical and purely random sequence of bits at two different locations, while revealing any eavesdropping with a very high probability. This idea, independently rediscovered by Ekert a few years later [2], was the beginning of quantum key distribution, which was to become the most promising element of quantum cryptography[1].

Quantum Key Distribution (QKD) is a technology to distribute, or rather generate, secure random keys between two communicating parties using optical fiber or free-space as a communication channel. It has been said that QKD has emerged in the last decades as one of the most important applications of quantum mechanics. Hence, in this paragraph the basic configuration and elements of such an important application will be introduced. Alternative introductions to this subject are available in many sources, ranging from books [3],[4],[5],[6] to other review articles [7],[8],[9].

1.2 Generalities

The general setting of QKD is shown in Figure 1. The two *authorized parties*, wishing to share a secret message are traditionally called Alice and Bob. *Alice*, the sender, is the one who starts a key transmission, while *Bob*, the receiver, is the one

[1] For some authors, quantum cryptography and quantum key distribution are synonymous. For others, however, quantum cryptography also includes other applications of quantum mechanics related to cryptography, such as quantum secret sharing or every other possible tasks related to secrecy that are implemented with the help of quantum physics.

who receives the quantum states and extracts the key sent by Alice. This is just a convention used in the field, but not a strict definition. The third important character is the eavesdropper, *Eve*, who is trying to intrude in the QKD and gain information about the key generated by Alice and Bob. Alice and Bob share a quantum *secure* channel, on which they send the *quantum* signals; and a classical public channel, on which they can send classical messages possibly back and forth. The classical channel needs to be authenticated; this means that Alice and Bob identify themselves, a third party can listen to the conversation but cannot participate in it. The quantum channel however, is open to any possible manipulation. The task of Alice and Bob is that of guaranteeing security against a possible eavesdropper that taps into the quantum channel and listens to the exchanges on the classical channel. In order to guarantee the security, either the authorized partners are able to create a secret key (a common list of secret bits known only to themselves) or they shall abort the protocol. Therefore, after the transmission of a sequence of symbols, Alice and Bob must estimate how much information about their set of bits has leaked out to Eve. In classical communications, such an estimate is obviously impossible, when Eve listens to the exchanges on the classical channel the communication goes on unmodified. This is where quantum physics comes into play: in a quantum channel, the leakage of information is directly related to the degradation of the communication quality.

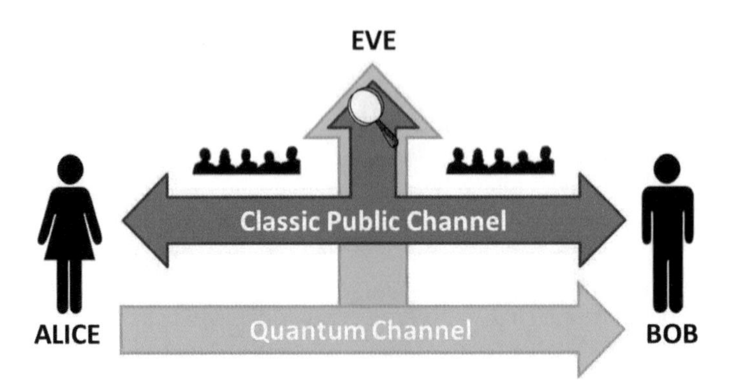

Fig. 1. Quantum key distribution comprises a quantum channel and a public classical authenticated channel. As a universal convention in quantum cryptography, *Alice* sends quantum states to *Bob* through a quantum channel. *Eve* is suspected of eavesdropping on the line.

In general, quantum information processing can be implemented with any quantum state of matter including energy state of ions, atoms, polarization states of light, electron spins, etc. Abstractly, this is also the case for QKD: one could imagine performing a QKD experiment with electrons, ions, and molecules; however, light is the only practical choice since it does not interact much with the environment leading to what is called *de-coherence*. Indeed, the task of key distribution makes sense only if Alice and Bob are separated by a macroscopic distance; if they are in the same room, there are much easier ways of generating a common secret key. Since, at any practical distance of interest, light propagates faster and with smaller de-coherence than matter, photons are the information carriers of choice. Various properties of photons can be employed to encode information for QKD, such as polarization, phase, quantum correlations of Einstein-Podolsky-Rosen pairs, wavelength or quadrature components of squeezed states of light. It is also well known that light does not interact easily with matter.

The way losses affect QKD varies with the type of protocol and its implementation. Losses impose bounds on the secret key rate and on the achievable distance and may also leak information to the eavesdropper, according to the nature of the quantum signal (for coherent pulses this is certainly the case while for single photons it is not). Another difference is determined by the detection scheme. Implementations that use photon counters rely on post-selection. If a photon does not arrive, the detector does not click and the event is simply discarded. On the contrary, implementations that use homodyne detection always give a signal, therefore losses translate into additional noise. QKD is always implemented with light and there is no reason to believe that things will change in the future. As a consequence, the quantum channel is any medium that propagates light with acceptable losses, typically either an optical fiber or just free space, provided a line of sight path exists between Alice and Bob.

1.3 The BB84 Protocol

The first and probably most famous QKD protocol is the so-called BB84 protocol, which can help one to understand the basic QKD concepts. Suppose Alice holds a source of single photons. The spectral properties of the photons are sharply defined, so that the only degree of freedom left is the polarization[2]. Alice and Bob align their polarizers[3] and agree to use either the horizontal or vertical (+) basis (*rectilinear*), or the complementary basis of linear polarizations, i.e., +45/-45 degrees (×) (*diagonal*). The transmitted bits are "prepared" at the transmitter (using the states of the selected basis) and "measured" at the receiver. The theory of quantum-mechanics states that:

- Measurements performed in the basis identical to the basis of preparation of states will produce deterministic results;
- Any measurements in the diagonal basis on photons prepared in the rectilinear basis will yield random outcomes with equal probabilities and vice-versa;

---

[2] Usually the way to encode the information being sent over the quantum channel is through the transmission of photons in some polarization states. The direction of the polarization encodes a classical bit.

[3] A polarizer is an optical filter that passes light of a specific polarization and blocks waves of other polarizations.

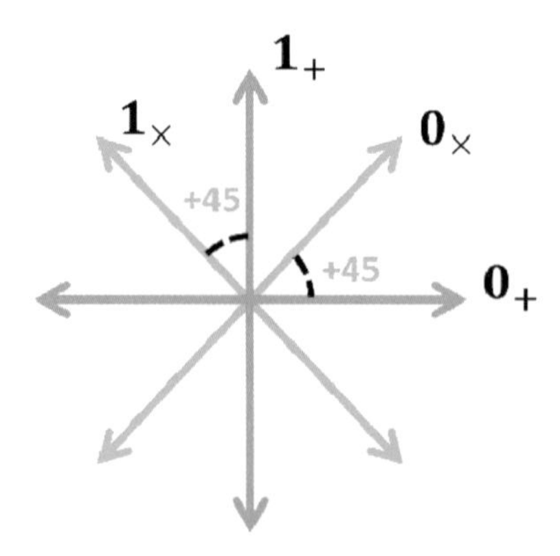

Fig. 2. The four states of the BB84 Protocol

Once Alice and Bob have agreed on the coding, the BB84 protocol (See Figure 3) can be summarized by the following steps:

1. Key Transmission: Alice, the sender, generates a sequence of N random bits for transmission and chooses the encoding basis (rectilinear or diagonal) in a random and independent way for each bit. Physically this means that she transmits photons in the four polarization states shown in Figure 2 equally frequently. Bob, the receiver, randomly and independently of Alice, chooses his measurement basis, either rectilinear or diagonal. Statistically, Alice and Bob's bases match in 50% of the cases. At the end of this stage Alice and Bob will share what is called the ***raw key.***

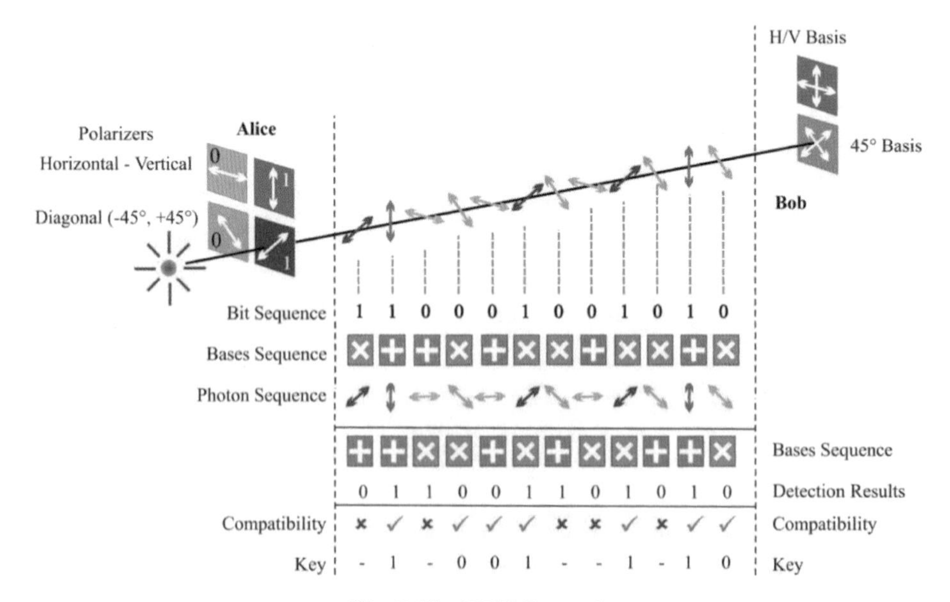

Fig. 3. The BB84 Protocol

2. Basis Announcement: Alice and Bob communicate over the classical channel and compare the basis used for each transmitted and detected photon. Whenever their bases coincide, Alice and Bob keep the bit whereupon it becomes part of the cryptographic key after reconciliation and privacy amplification. The bit is discarded when they chose different basis, when Bob's detector fails to register a photon due to the imperfect efficiency of detectors, or when the photon was lost somewhere along the way. Any potential eavesdropper, can only learn if Alice and Bob chose the same basis, but cannot determine whether Alice originally sent a "0″ or "1″. This step is called ***sifting***. At the end, Alice and Bob have a string of bits of approximately *N/2* bits, called the ***sifted key.***

3. Error Estimation: Alice and Bob disclose part of their strings, a subset of the bits of size *K*, and estimate the error rate in the quantum channel. If Eve tries to eavesdrop on the quantum channel, she cannot passively monitor the transmissions. Instead she can intercept the photons sent by Alice, perform measurements on them and resend them. However, since Alice had chosen her encoding bases randomly Eve has to guess. Half the times Eve will guess the basis right and resend correctly polarized photons, while in the other 50% of the cases, she measures in the wrong basis, producing errors. When Alice and Bob reveal a random sample of the bits of their raw keys, they discover these errors. Alice and Bob use a predetermined "failure" error threshold ($e_{\max}$) to decide whether or not an eavesdropper is present. In the literature, the most common failure error rate chosen is greater than or equal to 0.15 [10]. At 0.15 error rate, an eavesdropper could have intercepted over half of the bits transmitted. Both Alice and Bob compute the observed error-rate e and accept the quantum transmission if $e < e_{max}$. In this case they remove the K bits announced from the raw key. Otherwise if $e > e_{max}$ Eve is suspected of tampering with the channel, and the cryptographic key is thrown away. Thus, no information leak occurs even in the case of eavesdropping. It should be mentioned that no physical apparatus is perfect and noiseless. Alice and Bob will always find discrepancies, even in the absence of Eve. As they cannot tell errors stemming from eavesdropping from the noise of the apparatus, they conservatively attribute all the errors in transmissions to Eve. The actual error rate stems from both noise in the channel and possibly, interference from an eavesdropper.

4. Reconciliation and Privacy Amplification: If there are errors however, Alice and Bob have to correct them and have to eliminate the information that could have been obtained by Eve. *Information reconciliation* is a form of error correction carried out on Alice and Bob's keys, in order to ensure both keys are identical. It is conducted over the public channel and as such it is vital to minimize the information sent about each key, since any such information is totally accessible by Eve. In the earlier versions of the complete protocol, Alice and Bob perform the error correction through an interactive reconciliation protocol called *Cascade*. This is a simple protocol that leaks an amount of information close to the theoretical bound of an *almost ideal* protocol, when the error probability is below 15%. *Cascade* was presented in [11] as an improvement of the procedure suggested in [12]. *Cascade* operates in several rounds. During each round, Alice and Bob divide their raw keys into blocks, and disclose the parity of each block and compare them. If the parity bits do not match then a *binary* search is performed in order to find and correct the error. After all blocks have been compared, Alice and Bob both reorder their keys in the same random way, and a new round begins. If an error is found in a block from a previous round that had correct parity then another error must be contained in that block; this error is found and corrected as before. This process is repeated recursively, which is the origin of the name cascade. At the end of multiple rounds, Alice and Bob have identical keys with high probability, however Eve has additional information about the key from the parity information exchanged.

Once the information reconciliation has been performed, Alice and Bob share what is known as the ***reconciled key***. *Privacy amplification* is a method for reducing (and effectively eliminating) Eve's partial information about Alice and Bob's key. This partial information could have been gained both by eavesdropping on the quantum channel during key transmission (thus introducing detectable errors), and on the public channel during information reconciliation (where it is assumed Eve has access to all the parity information). Privacy amplification uses Alice and Bob's key to produce a new, condensed key, in such a way that Eve's amount of information about the new key is negligible. This can be done using universal hashing functions, chosen randomly from a publicly known set. The size r of the ***secret key*** that Alice and Bob can distill depends on the kind, as well as the amount, of information available to Eve. It is important to notice that the final distilled key has a very short length when compared to the initial key size, as shown in Figure 4.

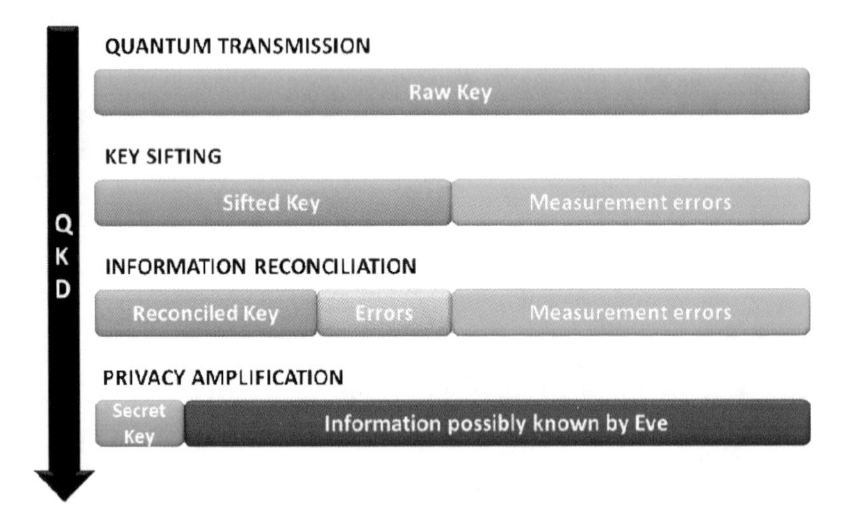

Fig. 4. Distillation process and key length in BB84 Protocol

From the description of the BB84 protocol, it can be observed that, although the security of QKD relies on the laws of quantum mechanics, a considerable part of the protocol utilizes the classical communication channel and classical techniques exclusively. Once the raw key has been transmitted over the quantum channel, a secret key is distilled using classic post-processing techniques that require interaction. In the process, some information about the key is exchanged via the public channel in order to correct the errors and eliminate the possible information that Eve may have derived. Information reconciliation is a mechanism that allows for elimination of the discrepancies between two correlated variables. It is an essential component in every key agreement protocol where the key has to be transmitted through a noisy channel. Hence, it is important to explore other classical techniques in the context of QKD systems, to minimize the information exchanged over the public channel so jeopardizing the provable security that quantum physics guarantees can be avoided.

1.4 System Model

As seen in the previous section, once the presence of eavesdropping has been excluded, the information reconciliation process can be modeled as an error correction process, which requires the transmission of additional redundancy on the public channel. If we denote as k the length of the *sifted key* (the fraction of unknown bits remaining after the presence of Eve has been excluded) and as $m=n-k$ the number of redundant bits required for error correction, the Information reconciliation process can be modeled as error correction process based on a systematic code with codeword length n and code rate $R_c = k/n$.

While the first QKD protocols were based on interactive error correction schemes [10],[11],[12] (like the Cascade algorithm), more recently Forward Error Correction (FEC) schemes have been suggested [13],[14],[15],[16],[17], which can avoid retransmission, increasing the system efficiency, and must be decoded by decoders exploiting the information available at the output of both the quantum and the public channel. In practice, the FEC block code must operate on an equivalent *composite parallel channel* formed by the quantum and the public channels, as shown in Figure 5, where the k information bits (the *sifted key*) are transmitted over the

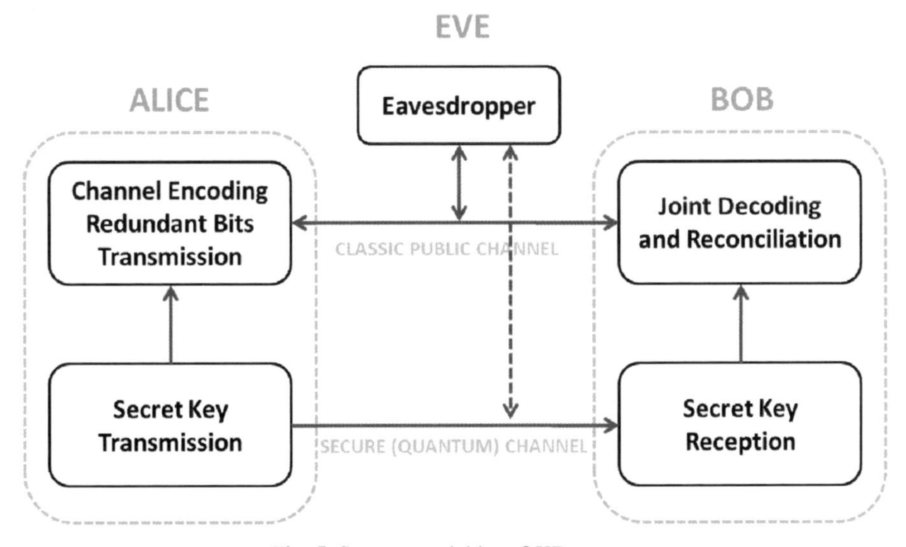

Fig. 5. System model in a QKD system

quantum *private* channel, while the $m=n-k$ redundancy bits are transmitted over the classical *public* channel. The eavesdropping on the secure channel in Figure 5 is shown as a dotted line, because if the system is properly designed the presence of Eve can be detected, and the information leaked to Eve can be made arbitrarily small, as if the eavesdropper did not exist. Given this hypothesis, we will from now on focus on the overall channel model linking Alice to Bob, neglecting the presence of Eve.

Since extremely low residual Bit Error Rate (BER) must be achieved (theoretically, error free decoding is needed), capacity achieving codes with acceptable decoding complexities and with very long code-length n have been considered in the information reconciliation literature. Furthermore, in order to minimize the quantity of information derived by Eve from the public channel, the code rate k/n must be maximized, and it must be larger than 0.5. The description of the appropriate models for all the involved channels in Figure 5 (and in particular, the private and the public channels) will be given in Section 2, while the possible capacity achieving codes for QKD applications are described in Section 3.

2 Soft-Metric QKD Schemes and Channel Models

In this section we present the system level models of the public and private channels in Figure 5. A far as the public channel is concerned, a typical Additive White Gaussian Noise (AWGN) channel model may be used, as shown in Figure 6, where n_k is a Gaussian random variable with variance σ^2 where for simplicity, a binary transmission scheme has been considered with the following association between the information bits b_k and the transmitted levels x_{r_k} :

$$x_{r_k} = \begin{cases} +\sqrt{E_b} & \text{if} \quad b_k = 1 \\ -\sqrt{E_b} & \text{if} \quad b_k = 0 \end{cases}$$

$$x_{r_k} = \pm\sqrt{E_b} \qquad \text{Classic Channel} \qquad y_{r_k} = x_{r_k} + n_k$$

Fig. 6. Classic AWGN model for the public channel

If soft decoding techniques are used in decoding the capacity achieving block codes, the channel output must be characterized with a likelihood ratio, i.e., the ratio between the likelihood (probability) of obtaining a given channel output conditioned on the possible transmitted bit. Often, the logarithm of this quantity defined as the Log Likelihood Ratio (LLR) is used. The LLR value for the public channel in Figure 6 is [18]:

$$LLR(Y_{r_k}) = log\left[\frac{P(X_{r_k} = +\sqrt{E_b}|Y_{r_k})}{P(X_{r_k} = -\sqrt{E_b}|Y_{r_k})}\right] = log\left[\frac{P(Y_{r_k}|X_{r_k} = +\sqrt{E_b})}{P(Y_{r_k}|X_{r_k} = -\sqrt{E_b})}\right] = log\left[\frac{P(Y_{r_k}|b_k = 1)}{P(Y_{r_k}|b_k = 0)}\right] = \frac{2Y_{r_k}\sqrt{E_b}}{\sigma^2}$$

As far as the private channel is concerned, when a single-photon is transmitted, the quantum channel can be modeled as a simple binary channel with error probability equal to the quantum bit error rate (QBER) Q, as shown in Figure 7. This is a very general model that can be used to model the quantum channel when polarization encoding is applied to photons in any QKD protocol.

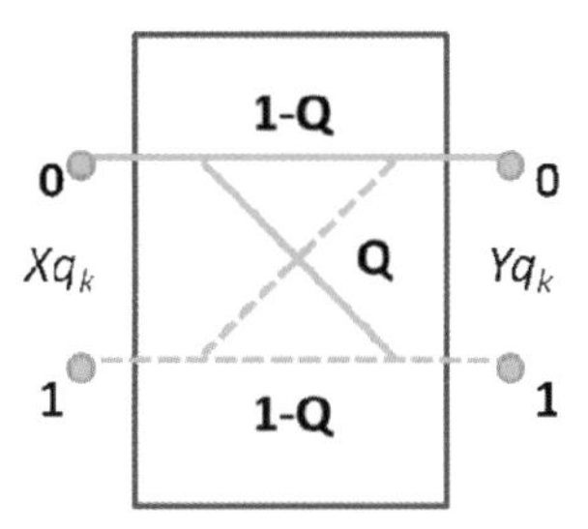

Fig. 7. Quantum channel modeled as a Binary Symmetric Channel (BSC)

The expression for the Log-Likelihood-Ratio metrics at the output of the quantum channel represented by the model of Figure 7 used as input for possible soft-metric decoding, is given by:

$$LLR(Y_{q_k}) = log\left[\frac{P(Y_{q_k} = 1|X_{q_k})}{P(Y_{q_k} = 0|X_{q_k})}\right] = \begin{cases} log\left(\frac{1-Q}{Q}\right) & if\ X_{q_k} = 1 \\ log\left(\frac{Q}{1-Q}\right) & if\ X_{q_k} = 0 \end{cases}$$

Notice that if the QBER value is not perfectly known, it should be substituted by its estimate. It is also important to notice that the log-likelihoods (metrics) $LLR(Y_{q_k})$ can only assume two values, and will therefore be referred to as *hard metrics* or *q-metrics*, while the metrics from the public channel $LLR(Y_{r_k})$ can assume any real value, and are called *soft metrics*. Since typically the raw BER value of the public channel is much lower than the QBER of the private channel, the *q-metrics* can be much less reliable than the *soft-metrics*, and it may be necessary to scale their weight in the decoding process with a proper weighting factor α_Q in order to optimize the overall performance, as we will discuss in Section 4.

The overall model for the composite channel as described above is shown in Figure 8.

Finally, since the *q-metrics* will have to be jointly used and compared in the channel decoder with the AWGN *soft metrics*, they need to be compatible and comparable. Suppose that the equivalent BSC model used for the quantum channel is obtained using an antipodal modulation scheme with transmitted levels

$\pm\sqrt{E_p} = \sqrt{E_p}(2X_{q_k} - 1)$, where X_{q_k} is the transmitted bit, Y_{q_k} is the decided raw bit, E_p is the energy per bit and σ_p^2 is the noise variance per dimension.

Denoting the equivalent received sample as Y_{p_k} , we would have $\sqrt{E_p}(2X_{q_k} - 1) + N_{p_k}$, and the theoretical quantum error probability would be:

$$P(X_{q_k} in\ error) = \frac{1}{2} erfc\left(\sqrt{\frac{E_p}{2\sigma_p^2}}\right),$$

with transmitted levels $X_{p_k} = \sqrt{E_p}(2X_{q_k} - 1) = \pm\sqrt{E_p}$ and $N_{p_k} \in \mathcal{N}(0, \sigma_p^2)$ as shown in Figure 9. Setting $P(X_{q_k} in\ error) = Q$, we would have, as a consequence:

$$\frac{E_p}{2\sigma_p^2} = \left(erfc^{-1}(2Q)\right)^2.$$

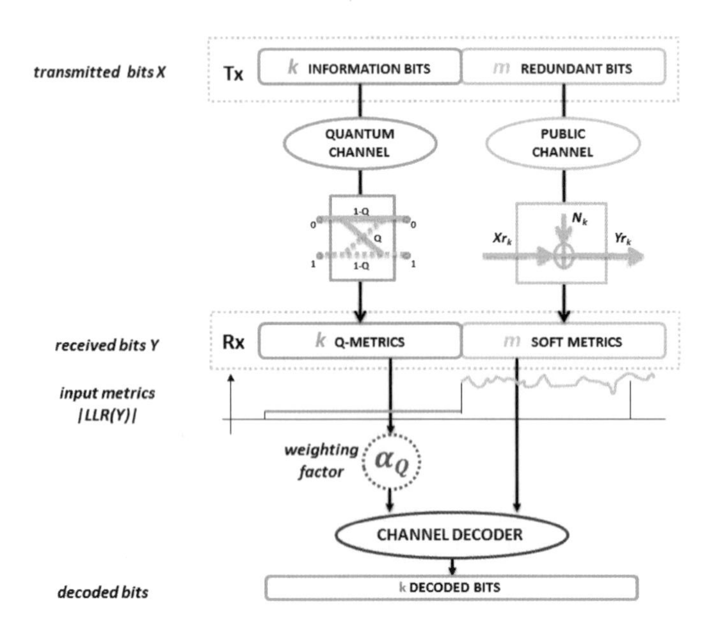

Fig. 8. The composite channel for QKD applications

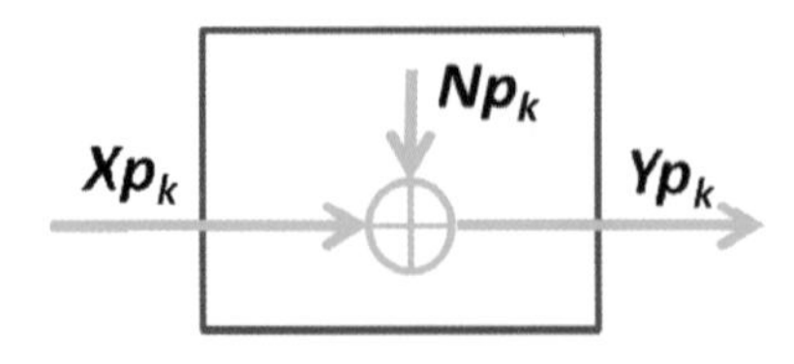

Fig. 9. Equivalent model for the single-photon quantum channel when using an equivalent antipodal modulation scheme over an equivalent (and fictitious) AWGN channel

2.1 Theoretical Limits of Information Reconciliation

We discuss now the theoretical limits of Information reconciliation, using the reference system model shown in Figure 10, which is an alternative version of the model of Figure 5, where "channel 1" represents the private secure (quantum) channel, while "channel 2" represents the public channel. Furthermore, X denotes the random variable transmitted by Alice and Y the random variable received by Bob during the transmission phase, i.e. during the transmission of the random raw key on the quantum channel (or, more precisely, of the sifted key, i.e. the fraction of key that remains after the presence of Eve has been excluded). As noted above, during this process (performed in the upper part of Figure 10), "channel 1" is typically modeled as a BSC channel with transition probability Q (Q is the quantum bit error rate - QBER - of the quantum channel). We denote as k the length of the information block, as n the encoded codeword length, as $m=n-k$ the number of redundant bits and as $R_c=n/k$ the code rate. The amount of information that needs to be transmitted is $H(X)=k$ bits, and the amount of information actually transmitted during the "transmission phase" is $H(X)-H(X|Y)=k[1-H_2(Q)]$ bits. In fact, Bob receives Y, which is an imperfect version of X, and is therefore left with some uncertainty about X, and in particular he is left with an uncertainty equal to $H(X|Y)=kH_2(Q)$, where $H_2(Q) = -Qlog(Q) - (1-Q)log\,(1-Q)$ is the binary entropy function.

The two main requirements for the information reconciliation phase are correction performance (all the errors must be eliminated with very high probability), and efficiency (the smallest possible redundancy needed for error correction must be transmitted, in order to maximize the effective rate).

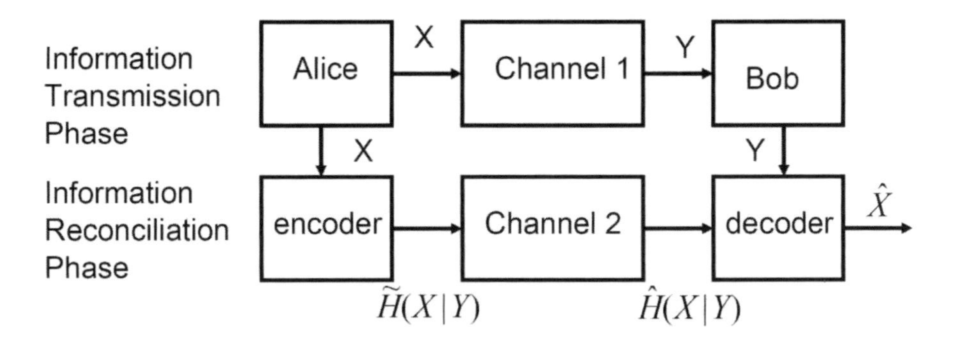

Fig. 10. Model of the information exchange occurring in QKD schemes

The system efficiency is defined modeling the information reconciliation problem as a compression with side information problem, and using as reference the system depicted in Figure 10. In this model, once the physical layer has taken care of the information transmission phase, in order to remove the residual uncertainty during the information reconciliation process, Alice performs channel encoding of X and transmits the redundancy $\tilde{H}(X|Y)$ through the information reconciliation channel

("channel 2" in Figure 10), generating the received uncertainty $\hat{H}(X \mid Y)$. Notice that we must have,

$$\hat{H}(X \mid Y) \geq \tilde{H}(X \mid Y) \geq H(X \mid Y).$$

The scheme in Figure 10 can be interpreted as a source with side information coding problem, schematically depicted in Figure 11, where the decoder at Bob side recovers an estimate $\hat{X}$ of X starting from the knowledge of Y and of the estimate $\hat{H}(X \mid Y)$ of the uncertainty $H(X \mid Y)$.

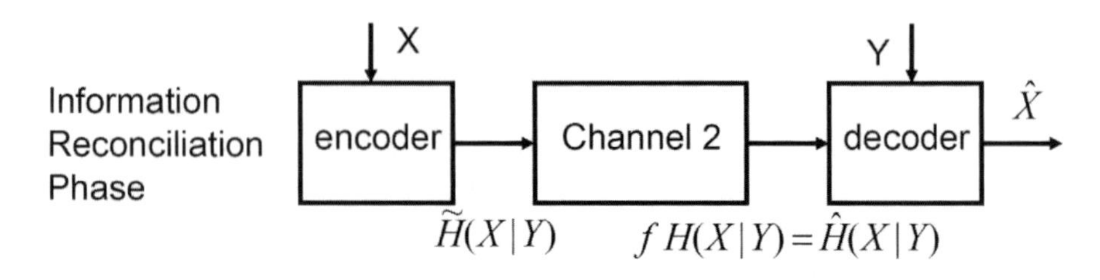

Fig. 11. Information Reconciliation process modeled as a source coding with side information problem

Since $\hat{H}(X \mid Y)$ is always larger than $H(X \mid Y)$ we can write it as $f\ H(X|Y)$, where the parameter f is what we define as *efficiency*, and is always larger than 1 (and equal to 1 only in the case of "perfect encoding"), i.e.

$$f = \frac{\hat{H}(X \mid Y)}{H(X \mid Y)} \geq 1.$$

If prior to the reconciliation Alice and Bob apply a random permutation, the errors introduced on the quantum channel can be considered uncorrelated and symmetric, which allows us to express the uncertainty on X given Y as,

$$H(X \mid Y) = kH_2(q).$$

The received uncertainty $\hat{H}(X \mid Y)$ is defined as "dimension M of the received message", the quantity of information transferred during the information reconciliation phase, or, in practice, the average number of correct bits transferred during the information reconciliation. Efficiency can be experimentally measured by simulating the considered encoding schemes, leading to the definition of experimental efficiency as,

$$f = \frac{M}{kH_2(Q)}$$

The system efficiency can also be defined by simulating the BER performance of the considered QKD scheme. The received uncertainty $\hat{H}(X|Y)$ would in fact be equal to the number of redundant bits, i.e. $m=n-k$ bits, in case of perfect transmission through "channel 2". Notice that the number of redundant bits can also be expressed as $m=n-k=\frac{k}{R_c}-k=\frac{k(1-R_c)}{R_c}$, where n is the block length of the information reconciliation code, k the number of information bits, and $R_c=\frac{k}{n}$ the code rate.

However, the final decoding may contain errors, and the effective rate of information transfer is not R_c, but lower. At this point we must recall the considered composite channel model (already described in the previous section, and recalled in the upper part of Figure 12), and the fact that, after proper decoding has taken place, the overall effect of the QKD transmission plus Information reconciliation can be modeled with an equivalent BSC channel operating on the k information bits with a crossover probability equal to the residual error probability after decoding, that we will denote as p_{DEC}. The considered equivalent scheme is shown in Figure 12 (lower part). Since the final error probability after information reconciliation must be 0 with high probability, we assume in our simulation that the bits in error can be discarded, leading to a rate loss (note that in a practical system, a further CRC based control scheme is often present, capable of discarding wrong frames).

Given this setup, where we only consider the correctly received bits, and we denote as p_{DEC} the equivalent bit error rate of the encoded system, we can define as k_{eff} and r_{eff} the effective number of information bits and redundancy bits, and as R_{eff} the effective rate, which can be expressed as:

$$k_{eff}=k(1-p_{DEC})$$

$$R_{eff}=\frac{k_{eff}}{n}=R_c(1-p_{DEC})$$

$$r_{eff}=n-k_{eff}=\frac{k}{R_c}-k(1-p_{DEC})=k\frac{[1-R_c(1-p_{DEC})]}{R_c}=k\frac{(1-R_{eff})}{R_c}$$

We can now estimate the efficiency of an error correction scheme operating on a composite BSC quantum channel with QBER Q (channel 1) and a possibly different channel on which the redundancy is transmitted (channel 2), as depicted in Figure 12. In particular, we need to experimentally estimate the overall (after decoding) BER $p_{DEC}(Q)$ as a function of the QBER parameter Q, so that the experimental efficiency as a function of Q will be,

$$f(q) = \frac{\hat{H}(X \mid Y)}{H(X \mid Y)} = \frac{1 - R_c[1 - p_{DEC}(Q)]}{R_c H_2(Q)}.$$

The $f(Q)$ function generally shows a minimum, in correspondence with the optimal QBER value for the considered rate. The selected code must in fact be matched to the channel, and, for a given QBER Q, the code with optimal rate should be selected (in order to find an optimal compromise between the loss due to the code redundancy and that due to its residual BER).

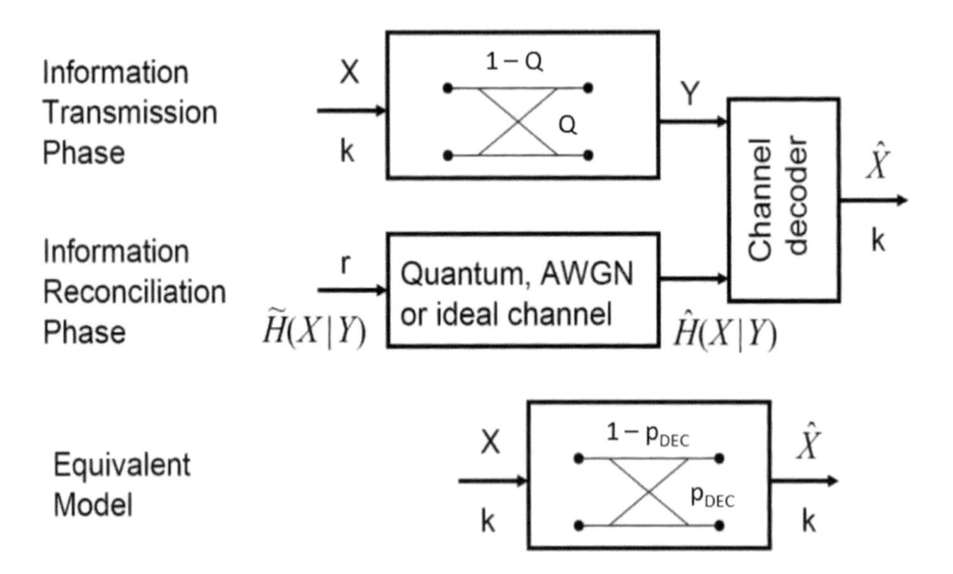

Fig. 12. The composite channel (composed of the parallel secure and public channels) linking Alice and Bob in QKD applications. Notice that different channel model can be used for the public channel.

2.2 Secret Key Rates

Classical information theory tells us that if $P(A,B,E)$ is the joint probability density function (pdf) of Alice, Bob and Eve's data, then the lower-bound on maximum secure key rate R is given by:

$$R \geq max\{I(A;B) - I(A;E), I(A;B) - I(B;E)\}$$

where, $I(X;Y)$ is the mutual information between X and Y (i.e., the amount of information provided about X by Y , which by symmetry is the same as the amount of information provided about Y by X). This result by Csiszár and Körner is only valid in the case of independent identically distributed random variables that can only hold true in the case of individual attacks on the quantum system. A classical upper-bound on secret key rate is due to Maurer and is given by:

$$R \leq \min_{E \rightarrow \hat{E}} \{H(A|\hat{E}) + H(B|\hat{E}) - H(a, B|\hat{E})\}$$

where, $E \rightarrow \hat{E}$ represents a mapping from random variable E to random variable $\hat{E}$ (i.e., we require the joint pdf $P(A,B,\hat{E})$ to be consistent in the sense that it should lead to the proper marginal $P(A,B)$). Above is simply the expression for the minimum of the conditional mutual information between Alice and Bob, given Eve's data (the expression for the minimum is called the intrinsic information and is interpreted as a measure of how much Bob learns about Alice's data by looking at his own data after Eve announced her data or a function of it).

It turns out that the gap between the lower and upper bounds is quite wide. There are no known protocols that achieve the upper-bound. The method of advantage distillation taps into the gap region. We note that most practical QKD schemes operate in two phases. In the first phase, a physical setup generates quantum states which are distributed and subsequently measured. In phase two, Alice and Bob use a classical channel and a one or two way communication to sift their data, perform error correction and privacy amplification and end up with a secret key. The computation of bounds on secret key rate should really look at all the phases.

We will now develop an example of secret key rate evaluation as a function of the receiver parameters, focusing on the BB84 protocol, previously described.

BB84 with Single Photon Source: Assuming that an on-demand Single Photon Source (SPS) did exist, what would be the secure key rate? Shor and Preskill have provided the following expression for secret key rate (all results are stated without proof. The interested reader can look at the excellent references on the subject for derivations of the stated results):

$$R = \frac{P_{det}}{2}[1 - 2H_2(Q)],$$

where, P_{det} is the probability that a signal is detected (including the possibility of dark counts whereby the detector clicks even when there is no signal), Q is the quantum bit error rate and $H_2(.)$ is the binary entropy function.

Suppose that the single photon source produces n photons with probability P_n. Now, any real source has non-zero probability values at least for P_0, P_1, and P_2. One would hope that for $n \geq 2$, P_n is so small that the effect can be neglected. An ideal source would have $P_1 = 1$ and $P_0 = 0$; $P_n = 0$ for $n \geq 2$.

Consider finding the probability of detection error at Bob. To find this probability, we need to condition on several other events, and then remove the conditioning. In particular, we need to condition on the number of photons send by Alice, and note that for Bob to have detection errors, he must first detect a signal:

*P(*Bob detects a signal and Bob has detection error for n-photons|Alice sends an n-photon*)* =
*P(*Bob has detection error for n-photons|Bob detects a signal and Alice sends an n-photon*)* ×
*P(*Bob detects a signal|Alice sends an n-photon*)*

To simplify the expressions, let us define the following events:

$Ev1$ = {Bob has detection error for n-photons}
$Ev2$ = {Bob detects a signal}=*Detection probability* Q_μ
$Ev3$ = {Alice sends an n-photon}

Then, we can write (below we use a comma to denote the AND operation):

$$P(\text{Bob has detection error}, Ev2) = \sum_n P(Ev1, Ev2|Ev3)P(Ev3).$$

Now, Bob may detect a signal even when there is none, due to the detector dark counts. Hence, we can write (we note that $P($Bob detects an n-photon or there is dark count$|Ev3)$ is called the Yield and denoted Y_n, and $P(Ev3)$ is simply P_n):

$$P(Ev2) = \sum_n P(\text{Bob detects an } n\text{-photon or there is dark count}|Ev3)P(Ev3).$$

$$P(Ev2) = Q_\mu = \sum_n Y_n P_n$$

Finally:

$$P(Bob\ has\ detection\ error|Ev2) = \frac{P(Bob\ has\ detection\ error|Ev2)}{P(Ev2)}$$

For an ideal SPS, in order for Bob to detect a signal, either Alice's photon gets to Bob, which happens with probability η (i.e., the communication link transmissivity), or the detector has a dark count, which happens with probability P_{dark}. It is straightforward to show that:

$$P(Bob\ has\ detection\ error|Ev2) = \frac{e_{det} \cdot \eta + 0.5 \cdot P_{dark}}{\eta + P_{dark}}$$

where, e_{det} is the detector error rate which is independent of the number of photons. Putting the pieces together, we find the final expression:

$$R = \frac{\eta + P_{dark}}{2} \cdot \left[1 - 2H_2\left(\frac{e_{det} \cdot \eta + 0.5 \cdot P_{dark}}{\eta + P_{dark}}\right)\right].$$

Note that if the detector dark count probability was to approach zero and the detector error rate was small (on the order of 10^{-3} or less), then the secret key rate would be $R \sim \eta$. This is the best that can be achieved with BB84, ideal SPS and almost ideal detector. The point where the rate tends to zero as a function of transmissivity (hence establishing the communication range) can be easily obtained by setting

$$H_2\left(\frac{e_{det} \cdot \eta + 0.5 \cdot P_{dark}}{\eta + P_{dark}}\right) = 0.5$$

which occurs when

$$\frac{e_{det} \cdot \eta + 0.5 \cdot P_{dark}}{\eta + P_{dark}} = 0.11 \ .$$

This leads to the following relation:

$$\eta^* = \frac{0.39 \cdot P_{dark}}{0.11 - e_{det}} \ .$$

The interpretation of this expression is that for a given detector performance as measured by P_{dark}, e_{det}, there is a lower-bound on transmissivity η^* below which it is impossible to have secure key exchange.

3 Capacity Achieving Codes (Turbo and LDPC) and Asymptotic Performance Evaluation

The fact that the use of classical error correction techniques entails no loss in security of the underlying scheme has been well demonstrated in the literature. In [19] the authors present a proof of security of QKD showing the connection between the Calderbank-Shor-Steane (CSS) Quantum Error Correction Codes (QECCs) and QKD and demonstrate the security of the BB84 protocol [20] and other QKD schemes. The proof presented in [21] is much less complicated than other approaches (see for example, [22]). Furthermore, the QKD protocols presented in [19] do not require a quantum computer for their implementation, as is required for instance by the protocols presented in [23]. In [22] it is demonstrated that despite the introduction of quantum error correction, the proposed schemes do not require the entanglement of photon states, but rather, it is sufficient to have separable photon states. Starting from this key result, in [24] it has been shown that simpler protocols in which classical error correction replaces quantum error correction suffice to prove the security of QKD.

The main ingredients that allow the simplification of using classical error-correcting codes in place of quantum error-correcting codes are the separability of Alice's state and Bob's knowledge of the basis used by Alice. On the other hand, this simplification cannot be applied if Alice sends entangled states because, in this case the state sent by Alice consists of a superposition of states that are non-separable regardless of the basis used. Note that even if the protocol used was not BB84 and Bob did not know the basis used by Alice, the problem would still be treatable with classical error-correcting codes, but in this case out of a block of N transmitted qubits, approximately $N/2$ additional errors that are due to a measurement made in the wrong basis should be taken into account.

Classical channel coding got a significant boost in 1993 with the introduction of Turbo codes and the subsequent developments of iterative decoding techniques [25]. With Turbo codes, Bit Error Rate (BER) of 10^{-5} at SNRs within 1dB of the Shannon's limit was readily achieved. For iterative decoding of Turbo codes, a practical implementation of the Maximum A-Posteriori Probability (MAP) decoding of convolutional codes known as the BCJR algorithm [26] is typically used.

As research progressed, it was realized that other capacity approaching codes once considered to be too complex to decode in the past, could be iteratively decoded achieving significant improvements in performance. In particular, Low-Density Parity-Check (LDPC) codes originally invented in 1960 by Robert Gallager in his PhD dissertation [27], which in spite of the excellent performance, were ignored for a long period because of the huge amount of hardware complexity involved in their decoding and the limitation of technology of that time, were resurrected. Tanner [28] provided a graphical representation denoted by Tanner graph, that is well suited to decoding of LDPC codes. Subsequently, LDPC codes began to be recognized as strong competitors to turbo codes in late 1990's, when Shokrollahi, MacKay, and others rediscovered LDPC codes and showed that they have excellent theoretical performance [29],[30]. Like turbo codes, iteratively decoded LDPC codes can get very close to the Shannon capacity limit. The asymptotic performance of LDPC codes, when the codeword length tends to infinity, has been studied using an analytical technique called density evolution or Gaussian approximation [29],[31],[32] and show close proximity to theoretical limits.

In the following two subsections we present material on analysis and design of Turbo and LDPC codes for correction of errors that occur during the exchange of a cryptographic key.

3.1 Analysis and Design of Classical Turbo Coding Techniques to Correct Errors That Occurred during the Exchange of a Cryptographic Key

To introduce Turbo codes, consider a rate R_c=1/3 eight state Parallel Concatenated Convolutional Code (PCCC). The encoder consists of a parallel interleaved (with an interleaver of length k) concatenation of a rate-1/2 systematic recursive convolutional upper code and a rate-1 recursive convolutional lower code both with generators $(g_1,g_2)=(13,17)_8$, where g_1 and g_2 are the feedback and the output polynomials in octal notation, respectively. The feedback polynomials specify the connections in the XOR sum of the state variables at the output of each constituent encoder memory cells. The design of concatenated codes with interleavers involves the choice of the interleaver and the constituent encoders. Unfortunately, a joint optimization is prohibitively complex.

In [33] Benedetto and Montorsi proposed a method based on random interleaving arguments to evaluate the error probability of PCCCs independently from the interleaver used. The method consists in a decoupled design, in which one first designs the constituent encoders, and then tailors the interleaver to their characteristics. To achieve this goal, the notion of *uniform interleaver* was introduced in [33]. The use of the uniform interleaver drastically simplifies the performance evaluation of turbo codes.

Denote by w_m the minimum weight of an input sequence generating an error event of the parallel concatenated code C, and by h_m and h_M the minimum and maximum weight, respectively, of the codewords of C. Also, let $A^{C}_{w,h}$ denote the Input-Output Weight Enumerating Function (IOWEF), that is, the number of codewords in code C with input weight w and output weight h. Similarly, we define $A^{C_U}_{w,h_U}$ and $A^{C_L}_{w,h_L}$ for the upper constituent code C_U and the lower constituent code C_L, respectively. The bit error probability $P_b(e)$ of a PCCC over an additive white Gaussian noise channel can be upper bounded by [33]:

$$P_b(e) \leq \frac{1}{2} \sum_{h=h_m}^{h_M} \sum_{w=w_m}^{k} \frac{w}{k} A^{C}_{w,h} \operatorname{erfc}\left(\sqrt{\frac{hR_c E_b}{N_0}} \right) \tag{1}$$

where $N_0/2$ is the two-sided noise power spectral density and E_b is the energy per information bit.

Likewise, the frame error probability $P_w(e)$ is upper bounded by:

$$P_w(e) \leq \frac{1}{2} \sum_{h=h_m}^{h_M} A_h^C \operatorname{erfc}\left(\sqrt{\frac{hR_c E_b}{N_0}}\right) \tag{2}$$

where $A_h^C = \sum_{w=w_m}^{k} A_{w,h}^C$

$A^C{}_{w,h}$ can be calculated by replacing the actual interleaver with the uniform interleaver [33] and exploiting its properties. The uniform interleaver of length k transforms an input sequence of weight w at the input of the upper constituent encoder into all its distinct $\binom{k}{w}$ permutations. As a consequence, each input sequence of the upper code of weight w, through the action of the uniform interleaver, enters the lower constituent encoder generating $\binom{k}{w}$ codewords of the lower code. The IOWEF of the overall PCCC can then be evaluated from the knowledge of the IOWEFs of C_U and C_L [33]:

$$A_{w,h}^C = \frac{A_{w,h_U}^{C_U} A_{w,h_L}^{C_L}}{\binom{k}{w}} \tag{3}$$

where h, h_U, and h_L are related by the equation $h=h_U+h_L$.

This general result needs to be modified to take into account the fact that in the context of the QKD application, we have two parallel channels to consider, the quantum channel modelled as a BSC and the classical public channel modelled as an AWGN.

Assume that the systematic part of the PCCC described above is transmitted over a quantum channel and the redundancy at the output of the two constituent codes on an AWGN channel. As stated earlier, the quantum channel can be modelled as an equivalent binary symmetric channel (BSC) with error probability Q [18]. To generate soft metrics that can be combined with the metrics generated at the output of the classic AWGN channel, we need to convert the error probability Q into an effective signal to noise ratio associated with a fictitious AWGN channel, as shown in Section 2:

$$\mathrm{SNR_Q} = \left.\frac{E_b}{N_0}\right|_Q = \left(\operatorname{erfc}^{-1}(2Q)\right)^2 \tag{4}$$

Since the communication system consists in two parallel channels, namely a quantum channel with signal-to-noise ratio $\mathrm{SNR_Q}$, on which to transmit the information of weight w, and a parallel classical public channel, with signal-to-noise-ratio $\mathrm{SNR_P}$ on which to transmit the redundancy of weight $h-w$, the bit error probability of a PCCC over this system can be upper bounded by [33]:

$$P_b(e) \leq \frac{1}{2} \sum_{h=h_m}^{h_M} \sum_{w=w_m}^{k} \frac{w}{k} A_{w,h}^C \operatorname{erfc}\left(F\left(R_c, w, h, \mathrm{SNR}_\mathrm{Q}, \mathrm{SNR}_\mathrm{P}\right)\right) \tag{5}$$

Where,

$$F\left(R_c, w, h, \mathrm{SNR}_\mathrm{Q}, \mathrm{SNR}_\mathrm{P}\right) = \sqrt{R_c\left[w\frac{E_b}{N_0}\bigg|_Q + (h-w)\frac{E_b}{N_0}\bigg|_P\right]} \tag{6}$$

Likewise, the frame error probability is upper bounded by,

$$P_w(e) \leq \frac{1}{2} \sum_{h=h_m}^{h_M} \sum_{w=w_m}^{k} A_{w,h}^C \operatorname{erfc}\left(F\left(R_c, w, h, \mathrm{SNR}_\mathrm{Q}, \mathrm{SNR}_\mathrm{P}\right)\right) \tag{7}$$

Figure 13 depicts the union bounds on the bit error probability (BER) for the rate-1/3 PCCC with generators $(g_1, g_2) = (13, 17)_8$. Here, the block length is $k = 250$ bits; this choice was made following the assumptions in [18]. The black curve with no markers

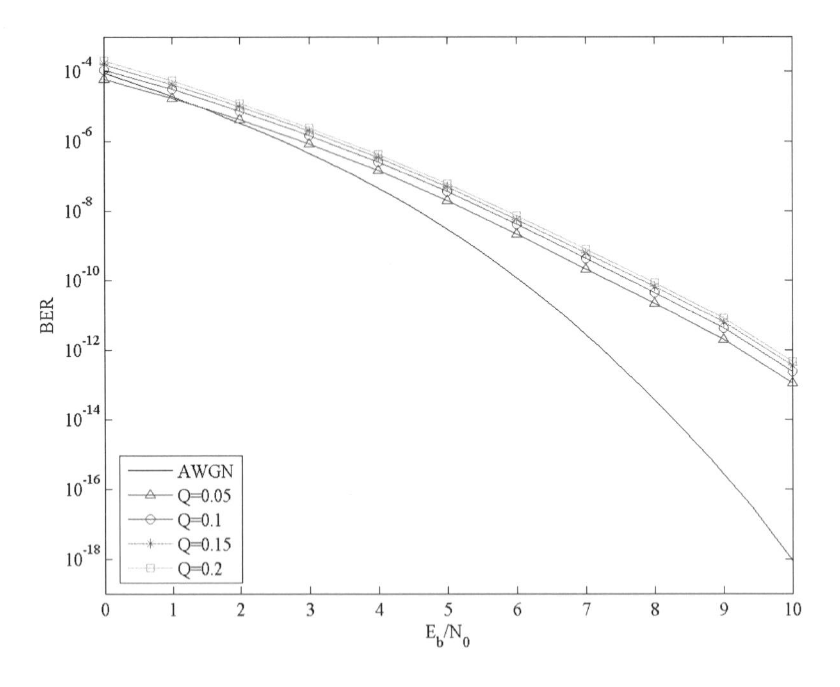

Fig. 13. Bit Error Rate vs. Signal-to-Noise Ratio E_b/N_0 for the rate-1/3 PCCC with generators $(g_1,g_2)=(13,17)_8$. Union bound results are reported for different Q values: $Q=\{0.05, 0.1, 0.15, 0.2\}$ and also for the classical AWGN case (curve with no markers). The block length is $k=250$ bits.

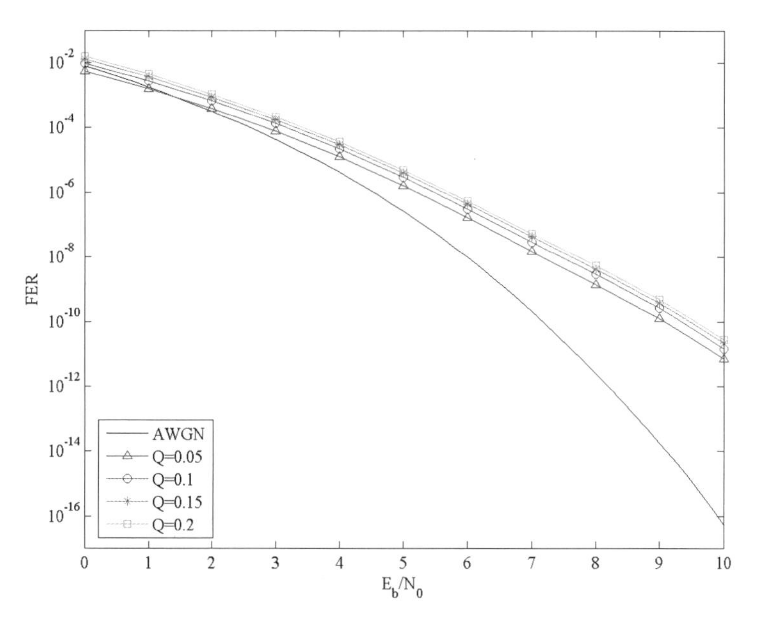

Fig. 14. Frame Error Rate vs. Signal-to-Noise Ratio E_b/N_0 for the rate-1/3 PCCC with generators $(g_1,g_2)=(13,17)_8$. Union bound results are reported for different Q values: $Q=\{0.05, 0.1, 0.15, 0.2\}$ and also for the classical AWGN case (curve with no markers). The block length is k=250 bits.

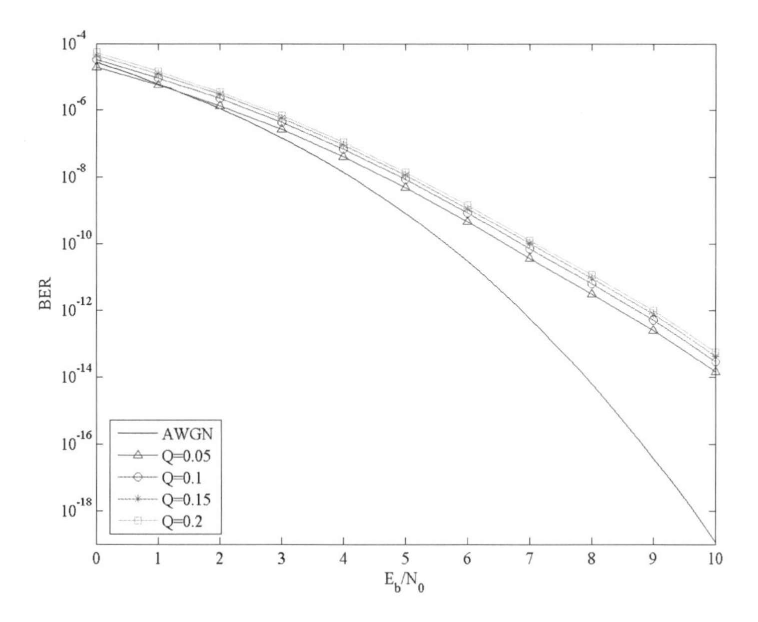

Fig. 15. Bit Error Rate vs. Signal-to-Noise Ratio E_b/N_0 for the rate-1/3 PCCC with generators $(g_1,g_2)=(13,17)_8$. Union bound results are reported for different Q values: $Q=\{0.05, 0.1, 0.15, 0.2\}$ and also for the classical AWGN case (curve with no markers). The block length is $k = 500$ bits.

is the union bound results assuming the transmission of both the secret key and the redundancy on the classical AWGN channel. The coloured curves with markers report the union bound results assuming the transmission of the secret key on a quantum channel with fixed error probability Q ranging from 0.05 to 0.2, and of the redundancy on the classical AWGN channel. Figure 14 provides the union bounds on the frame error probability (FER) for the same code.

In Figure 15 we provide the union bounds on the bit error probability (BER) with a block length of k=500 bits.

3.2 Analysis and Design of Classical LDPC Coding Techniques to Correct Errors That Occurred during the Exchange of a Cryptographic Key

This section describes the characteristics of another class of capacity achieving block codes, the Low-Density Parity-Check (LDPC) codes. LDPC codes are a class of linear block codes whose name comes from the characteristic of their parity-check matrix which contains few ones in comparison to the number of zeros. Their main advantage is that they provide a performance which is very close to the capacity for a lot of different channels and there are linear time complexity decoding algorithms available for them. Furthermore, they are suited for implementations that make heavy use of parallelism. LDPC codes can be represented through a matrix as well as a graph.

An (n, k) LDPC code is represented by a parity check matrix which consists of m=n-k rows and n columns, where n is the codeword length, k the number of information bits and m the number of redundant bits. For example, the matrix defined in Equation (8) is a parity check matrix H with dimension n×k for a (8,4) code.

$$H = \begin{bmatrix} 0\,1\,0\,1 & 1\,0\,0\,1 \\ 1\,1\,1\,0 & 0\,1\,0\,0 \\ 0\,0\,1\,0 & 0\,1\,1\,1 \\ 1\,0\,0\,1 & 1\,0\,1\,0 \end{bmatrix} \tag{8}$$

In a Tanner Graph representation of a (n, k) LDPC code, the n nodes related to the rows of the parity check matrix are denoted as Variable Nodes or Bit Nodes (*V-nodes*). On the other hand there are m nodes, called Check Nodes (*C-nodes*), that are related to the rows of the H matrix, i.e., the m parity check equations of the code. An edge on the Tanner Graph connects a *V-node* to a *C-node* only if the corresponding element is a “1” in the parity check matrix H. From the parity check matrix H of Equation (8), we have n = 8 *V-nodes* connected to m = 4 *C-nodes*. Figure 16 shows the Tanner graph representation of the parity check matrix of Equation (8). Notice that the bit nodes values connected to same check node must sum to zero. Similarly, a Tanner graph can also be constructed from the columns of H.

We can define two numbers describing the matrix H, w_r for the number of 1’s in each row and w_c for the number of 1’s in each column. For a matrix to be called low-density the two conditions $w_c \ll n$ and $w_r \ll m$ must be satisfied. In order to achieve this, the parity check matrix should usually be very large, so the example matrix presented above is not really low-density.

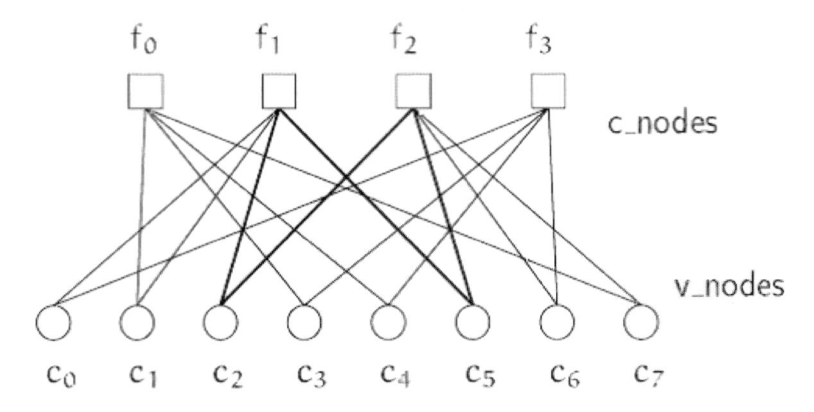

Fig. 16. Tanner Graph representation of the LDPC code corresponding to the parity check matrix ***H***. The marked path c2 →f1 → c5 → f2 → c2 is an example of a short cycle. Those should usually be avoided since they are bad for decoding performance.

In an (*n*, *k*) LDPC code, if the rank of the *H* matrix is *r*, then *n-r* information bits can be transmitted per codeword. Accordingly the code rate is given by,

$$R_c = k/n = (n-m)/n \leq (n-r)/n \tag{9}$$

where the inequality holds when all *m* rows are linearly independent.

LDPC codes are encoded using the generator matrix *G* spanning a space which is the orthogonal complement of the space spanned by the parity-check matrix *H*,

$$GH^{T} = 0 \tag{10}$$

G and *H* can be derived from each other using Gaussian elimination. If the code is systematic, the matrix *H* can be expressed as,

$$H = \left[I_{n-k} | P \right] \tag{11}$$

where *I* is a $(n-k) \times (n-k)$ identity matrix and *P* is the $(n-k) \times k$ parity matrix. The generator-matrix *G* can be written in the systematic form as

$$G = [\, P^T \mid I_k \,] \tag{12}$$

where P^T represents the $k \times (n-k)$ transposed parity matrix. If we consider a sequence of information bits x that contains *k* bits, the encoding process is achieved by simply multiplying this sequence by the generator matrix to get the codeword,

$$c = x\,G \tag{13}$$

The iterative decoding algorithm for LDPC codes iteratively computes the distribution of variables and is well known as the Sum Product Algorithm (SPA), Belief Propagation Algorithm (BPA) or Message Passing Algorithm (MPA). The term *message passing* refers to the fact that during each round messages are passed from *V-nodes* to *C-nodes* and vice versa. An important aspect of iterative decoding is that message to be sent from the ith *V-node* V_i to the jth *C-node* C_j must not take into account the message sent in the previous iteration from C_j to V_i. The same rule holds for messages to be sent from C_j to V_i.

The Belief Propagation Algorithms (BPA) is an important class of message passing algorithm where the messages passed along the edges of a Tanner Graph are probabilities (or beliefs) [33]. More precisely, the message passed from the *V-node* V_i to the *C-node* C_j is the probability that V_i has a certain value, given its own noisy observed value, and all the values received in the previous iteration from its neighboring *C-nodes* (two nodes are said to be neighbors if they are connected to the same edge of Tanner graph) excluding C_j. Similarly, the message passed from C_j to V_i is the probability that V_i has a certain value given all the messages passed to C_j in the previous iteration from neighboring *V-nodes* other than V_i.

The aim of the belief propagation algorithm is to compute the A-Posteriori Probability (APP) that a given bit in the transmitted codeword $C = [\, c_0\, c_1 \dots c_{n-1}\,]$ equals 1, given the received sufficient statistic samples $Y = [\, y_0\, y_1 \dots y_{n-1}\,]$, i.e., the APP probability

$$p_i = P_r\,(c_i = 1|Y) \tag{14}$$

or the APP ratio (also called Likelihood Ratio (LR)),

$$l(c_i) = \frac{P_r\,(c_i = 0|Y)}{P_r\,(c_i = 1|Y)} \tag{15}$$

The LR can be iteratively computed exploiting the code's Tanner graph. In one half iteration, each *V-node* processes its input messages (probabilities or LLRs) and passes its resulting output messages to the neighboring *C-nodes*. In the other half iteration the *C-node* passes its messages to the *V-nodes*. After a pre-defined number of iterations, or after some stopping criteria have been met, the decoder computes the APP (A-Posteriori Probabilities), or LLR (Log Likelihood Ratios) from which decisions on the bits can be taken.

Let f_{ij}^k and g_{ji}^k be the messages from V_i to C_j and C_j to V_i in k^{th} iteration, respectively. The belief propagation algorithm in probability domain can be described as,

1) Initialization:

$$f_{ij}^0(0) = 1 - p_i \tag{16}$$

$$f_{ij}^0(1) = p_i \tag{17}$$

2) *C-node* update:

$$g_{ji}^k(0) = \frac{1}{2} + \frac{1}{2}\prod_{i\in f_{j/i}} (1 - 2f_{ij}^{(k-1)}(1)) \tag{18}$$

where $f_{j/i}$ is the set of all *V-nodes* connected to C_j excluding V_i, and,

$$g_{ji}^k(1) = 1 - g_{ji}^k(0) \tag{19}$$

3) *V-node* update:

$$f_{ij}^{k}(0) = 1 - A_{ij} f_{ij}^{0}(0) \prod_{j \in g_{i/j}} g_{j\,i}^{k}(0) \tag{20}$$

where $g_{i/j}$ is the set of all *C-nodes* connected to V_i excluding C_j, and,

$$f_{ij}^{k}(1) = A_{ij} f_{ij}^{0}(1) \prod_{j \in g_{i/j}} g_{j\,i}^{k}(1) \tag{21}$$

where A_{ij} are constants, which satisfy

$$f_{ij}^{k}(0) + f_{ij}^{k}(1) = 1 \tag{22}$$

4) Soft Decision:

$$F_{i}^{k}(0) = A_i f_{ij}^{0}(0) \prod_{j \in g_i} g_{ji}^{k}(0) \tag{23}$$

$$F_{i}^{k}(1) = A_i f_{ij}^{0}(1) \prod_{j \in g_i} g_{ji}^{k}(1) \tag{24}$$

where g_i is the set of all *C-nodes* connected to V_i, and A_i is chosen to satisfy,

$$F_{i}^{k}(0) + F_{i}^{k}(1) = 1 \tag{25}$$

5) Hard Decision:

$$\tilde{c}_i = \begin{cases} 1 & \text{if } F_i^k(1) \gg 0 \\ 0 & \text{otherwise} \end{cases}$$

If $\tilde{c}H^T = 0$, or maximum number of iterations is reached, stop, else go back to step 2), where $\tilde{c}$ is the decoded codeword.

As it can be seen above, the decoding process involves the multiplication of probabilities, which have high computational complexity. With the increase in number of iterations a log domain manipulation is required to decrease the complexity, by converting multiplications to additions.

In log domain the algorithm can be described as follows, first we define:

$$L(c_i) = log\left(\frac{P_r\ (c_i = 0|Y)}{P_r\ (c_i = 1|Y)}\right) \tag{26}$$

$$L(f_{ij}) = log\left(\frac{f_{ij}(0)}{f_{ij}(1)}\right) \tag{27}$$

$$L(g_{ij}) = log\left(\frac{g_{ij}(0)}{g_{ij}(1)}\right) \tag{28}$$

$$L(F_i) = log\left(\frac{F_i(0)}{F_i(1)}\right) \tag{29}$$

1) Initialization:

$$L^0(f_{ij}) = L^0(c_i) \tag{30}$$

2) C-node update:
From (18) and (19) we get

$$1 - 2g_{ji}(1) = \prod_{i \in f_{j/i}} (1 - 2f_{ij}(1)) \quad (31)$$

Now since $\tanh\left[\frac{1}{2} log\left(\frac{a}{b}\right)\right] = 1 - 2b$ and using (27) and (28), (31) can be written as,

$$\tanh\left[\frac{1}{2} L^k(g_{ij})\right] = \prod_{i \in f_{j/i}} \tanh\left[\frac{1}{2} L^{k-1}(f_{ij})\right] \quad (32)$$

Equation (32) still involves multiplication and a complex tanh(.) function that needs to be simplified. Let us represent $L(f_i)$ in its sign and magnitude form; in particular, let Θ_{ij} represent the sign of $L(f_{ij})$, and δ_{ij} represent the magnitude of $L(f_{ij})$. Using these, (32) becomes,

$$\tanh\left[\frac{1}{2} L^k(g_{ij})\right] = \prod_{i \in f_{j/i}} \Theta^{k-1}{}_{ij} . \prod_{i \in f_{j/i}} \tanh\left[\frac{1}{2} \delta^{k-1}{}_{ij}\right] \quad (33)$$

Then,

$$L^k(g_{ij}) = \prod_{i \in f_{j/i}} \Theta^{k-1}{}_{ij} . 2\tanh^{-1} . \log^{-1} . \log\left[\prod_{i \in f_{j/i}} \tanh\left[\frac{1}{2} \delta^{k-1}{}_{ij}\right]\right] \quad (34)$$

$$= \prod_{i \in f_{j/i}} \Theta^{k-1}{}_{ij} . 2\tanh^{-1} . \log^{-1} . \sum_{i \in f_{j/i}} \log\left[\tanh\left[\frac{1}{2} \delta^{k-1}{}_{ij}\right]\right] \quad (35)$$

Let γ be a map from the real numbers $[-\infty, \infty]$ to $F2 \times [0, \infty]$ defined by $\gamma(\mathrm{x}) := (\mathrm{sgn}(x)\text{-}\log(\tanh(\frac{|x|}{2})))$, whereby, $\mathrm{sgn}(x) = 1$ if $x \geq 1$ and 0 otherwise. Equation (35) can be written as,

$$L^k(g_{ij}) = \gamma^{-1}\left(\sum_{i \in f_{j/i}} \gamma(\delta^{k-1}{}_{ij})\right) \quad (36)$$

3) V-node update:
Dividing (21) by (20) and taking log, we have,

$$L^k(f_{ij}) = L^0(c_i) + \sum_{j \in g_{i/j}} L^k(g_{ji}) \quad (37)$$

4) Soft Decision:

$$L(F_{ij}) = L^0(c_i) + \sum_{j \in g_i} L(g_{ji}) \quad (38)$$

5) Hard decision:

$$\tilde{c}_i = \begin{cases} 1 & \text{if } L_i^k(1) < 0 \\ 0 & \text{otherwise} \end{cases}$$

If $\tilde{c}H^T = 0$, or maximum number of iterations is reached, stop, else go back to step 2), where $\tilde{c}$ is the decoded codeword.

As stated previously, the asymptotic performance of LDPC codes, when the codeword length tends to infinity, has been studied using an analytical technique called density evolution (DE) or Gaussian approximation [29],[31],[32].

The density evolution computes the probability density function (PDF) of the messages defined by the message-passing algorithm on Tanner graphs at any iteration. From [29], "Asymptotically, the actual density of the messages passed is very close to the expected density. Tracking the expected density during the iterations thus gives a very good picture of the actual behavior of the algorithm". Two assumptions are made for the calculation of density evolution [32],

- The independence condition assures that the messages passed on the Tanner graph are statistically independent;
- For infinite code length, the factor graph can be viewed as a cycle free graph.

In general, an LDPC code ensemble is specified by a degree profile (λ, ρ). Its corresponding generating functions are $\lambda(x) = \sum_{i=2}^{d_{v\max}} \lambda_i x^{i-1}$ and $\rho(x) = \sum_{i=2}^{d_{c\max}} \rho_i x^{i-1}$ where $\lambda_i(\rho_i)$ is the fraction of edges with variable (check) node of degree i and $d_{v\max}$ ($d_{c\max}$) is the maximal variable (check) node degree (number of edges connected to it), respectively.

Let ∂_{c_k} denote the common density function of the messages g_{ji}^k sent from C-nodes to V-nodes at round k and let ∂ denote the density of the messages f_{ij}^0, i.e., the likelihood of the messages sent at iteration 0 of the algorithm. Then the update rule for the densities in (36) implies that the common density $\partial_{v_{k+1}}$ of the messages sent from V-nodes to C-nodes at round $k+1$ conditioned on the event that the degree of the node is d, equals $\partial * \partial_{c_k}{}^{d-1}$, where $(\partial * \partial_{c_k})(\tau) = \int (\partial(\sigma)\partial_{c_k}(\tau - \sigma))\, d\tau$ is the convolution over some group G of ∂ and ∂_{c_k} [32].

Using $\gamma(x)$ defined above, let $\Gamma(\gamma(x))$ be the density of $\gamma(x)$. Using Eq. (36) and the independence assumption, it can be shown that

$$\partial_{c_k} = \Gamma^{-1}(\rho(\Gamma(\partial_{v_k}))) \tag{39}$$

where Γ is the Laplace transform of the expected densities derived in [31],[32]. From this, the following recursion formula can be obtained for density evolution (DE):

$$\partial_{v_{k+1}} = \partial * \lambda(\Gamma^{-1}(\rho(\Gamma(\partial_{v_k})))) \tag{40}$$

The convolution can be efficiently computed using Fourier transform $\mathcal{F}$, so the DE can be expressed as

$$\partial_{v_{k+1}} = \Gamma \mathcal{F}^{-1}(\mathcal{F}(\partial)\,\lambda(\mathcal{F}(\partial_{c_k}))) \tag{41}$$

From density evolution together with Fourier transform techniques, asymptotic thresholds below which belief propagation decodes the code successfully, and above which belief propagation does not decode successfully, can be derived [29],[31],[32].

The asymptotic performance of LDPC is characterized by finding the maximum channel parameter (threshold σ^*) such that if $\sigma < \sigma^*$ then $\lim_{k\to\infty} P_e{}^k = 0$, and $P_e{}^k$ is the expected fraction of incorrect messages at the kth iteration.

4 Simulation Results

In order to exemplify the performances of capacity achieving iteratively decoded codes for QKD applications, this section presents simulation results for the error rates achievable with LDPC code of various rates on the composite channel model of Figure 8 as a function of various system parameters, such as the quantum bit error rate Q and the weighting factor α_Q used to properly scale the somewhat unreliable q-metrics. The public channel will be modeled as an AWGN channel with a high signal-to-noise ratio (E_b/N_o) (since powerful coding is allowed on the public link and therefore possible errors on this channel may be considered extremely rare).

For the presentation of the simulation results, two figures will be used: the Bit Error Rate (BER) and the Frame Error Rate (FER) of the sifted key (where one frame is equivalent to one decoded code block).

In Figure 17 and 18 the simulated BER and FER is reported for an LDPC code with n =504, and R_c=0.5 for different values of Q as a function of the weigh parameter α_Q. It can be observed how an optimal value of α_Q in the order of 0.5-0.6 can be identified (depending on the actual QBER of the private channel), which shows how proper weighting of the q-metric may actually improve the system performance at no extra cost.

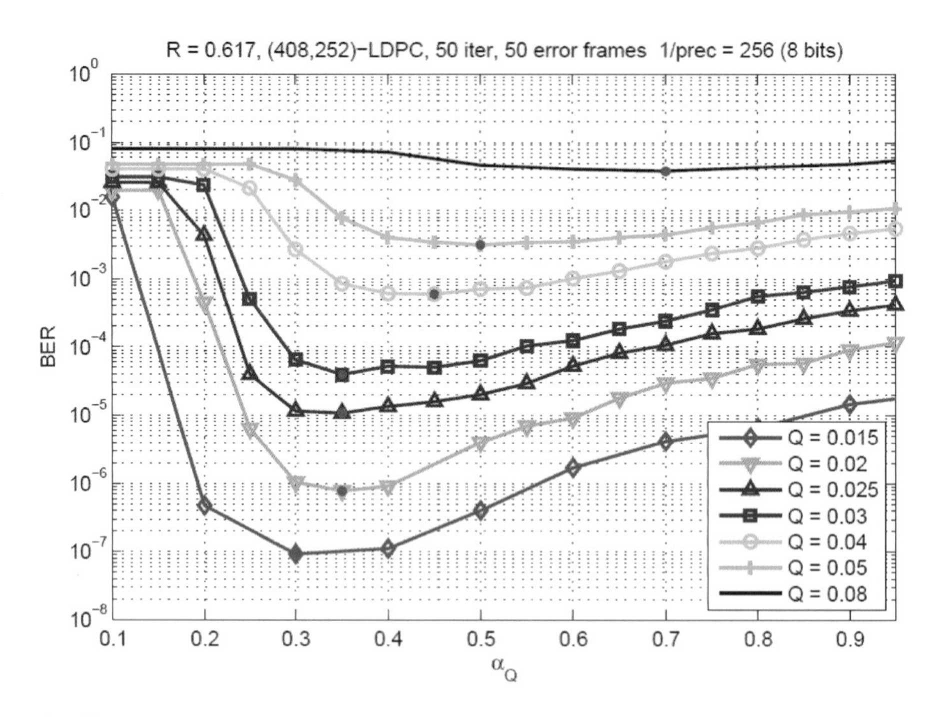

Fig. 17. BER performance of a LDPC code with block length n=408 and rate R_c=0.61, as a function of Q and α_Q (50 decoding iterations have been used)

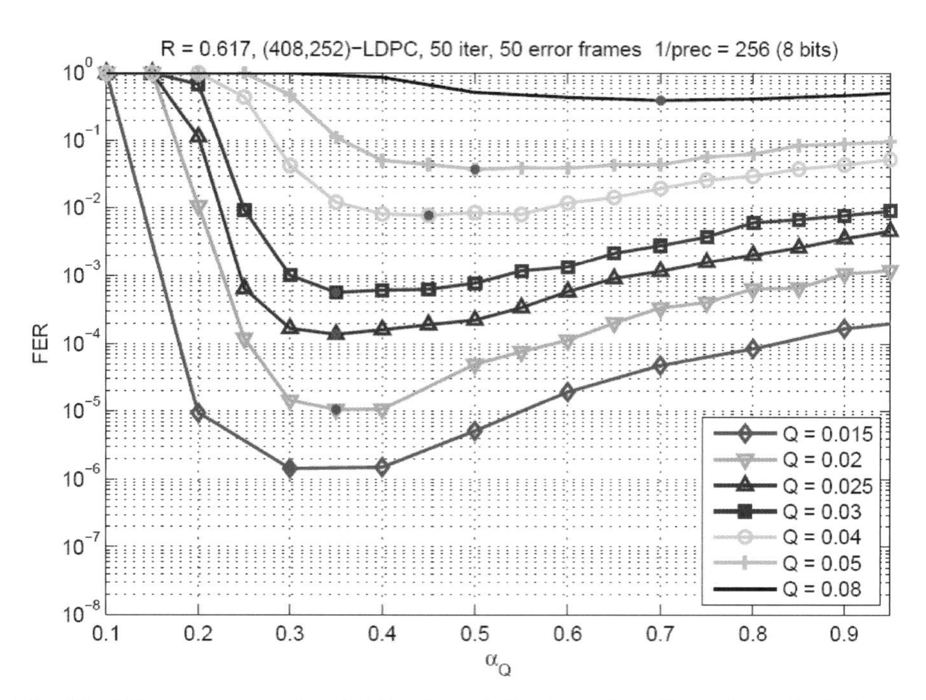

Fig. 18. FER performance of a LDPC code with block length n=408 and rate R_c=0.61, as a function of Q and α_Q (50 decoding iterations have been used)

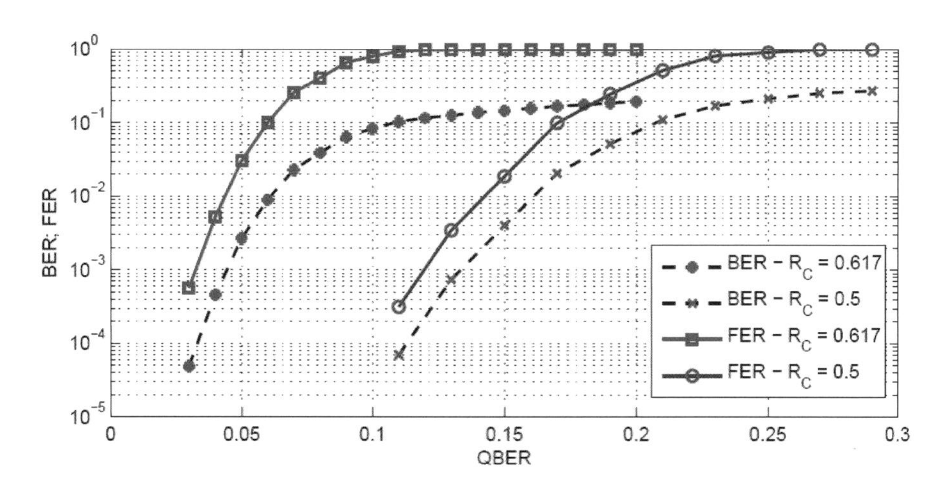

Fig. 19. BER and FER performance of two LDPC codes with block length n=*504* and R_c=*0.5* and n=408 and R_c=0.61, respectively, as a function of the private channel QBER

In Figure 19, the performance of two codes with rates 0.5 and 0.61 are compared, showing that a higher coding rate leads to a worst BER performances as expected, but higher security, since a lower fraction of bits are revealed on the public channel.

Finally, Figure 20 depicts the number of iterations needed to achieve the BER and FER values of Figures 17 and 18, for an LDPC code with rate *Rc*=0.61, showing the strong correlation between BER (and FER) of the decoded sequence and number of decoding iterations.

Since the FER/BER of the decoded sequence is typically correlated to the channel Q parameter, the behavior of Figure 20 shows that the number of decoding iterations can be used as a valid indicator of the actual channel QBER. Being able to estimate the parameter Q from the decoded bits may allow the detection of a possible eavesdropper without wasting additional bits.

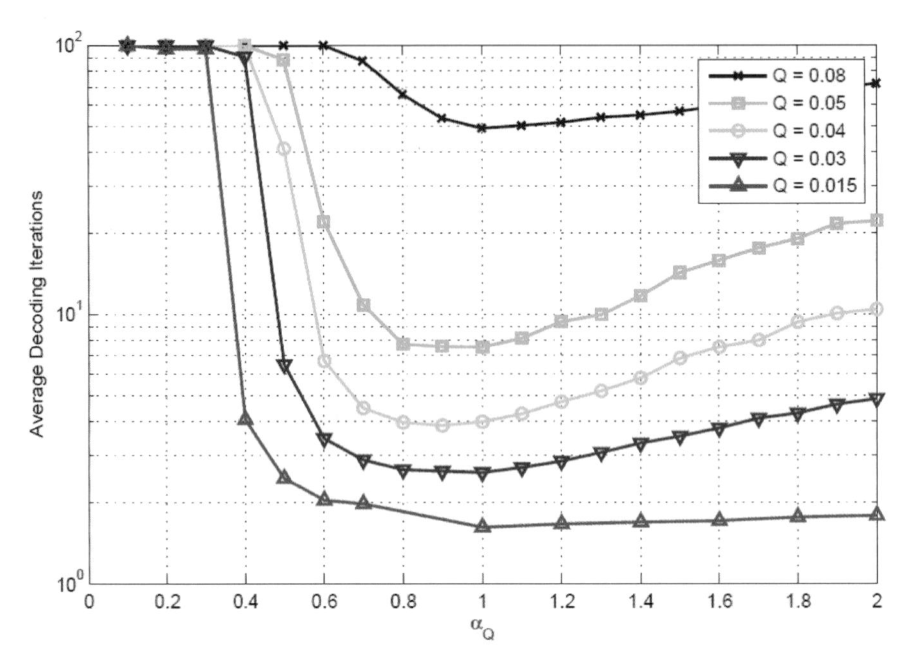

Fig. 20. Average number of iterations for the LDPC code with R_c= 0.61 as a function of α_Q and for several values of Q

References

[1] Bennett, C.H., Brassard, G.: Quantum cryptography: Public key distribution and coin tossing, Bangalore, India, vol. 175, pp. 175–179 (1984)

[2] Bouwmeester, D., Ekert, A., Zeilinger, A.: The Physics of Quantum Information. Springer (2000) ISBN: 978-3-540-66778-0

[3] Lo, H.-K., Spiller, T., Popescu, S.: Introduction to Quantum Computation and Information, vol. 399. World Scientific (1998)

[4] Ilic, N.: The Ekert Protocol. Quantum (1991)

[5] Bellac, M.L.: A Short Introduction to Quantum Information and Quantum Computation, vol. 60. Cambridge University Press (2006)

[6] Scarani, V.: Quantum Physics A First Encounter: Interference, Entanglement, and Reality. Oxford Univ. Press, Oxford (2006)
[7] Dusek, M., Lutkenhaus, N., Hendrych, M.: Quantum cryptography. Progress in Optics 18(8), 51 (2006)
[8] Lo, H.-K., Zhao, Y.: Quantum cryptography. Encyclopedia of Complexity and Systems Science 8, 7265 (2009)
[9] Gisin, N., Ribordy, G., Tittel, W., Zbinden, H.: Quantum cryptography. Reviews of Modern Physics 74(1), 145–195 (2002)
[10] Nakassis, A.: Expeditious reconciliation for practical quantum key distribution. In: Proceedings of SPIE, vol. 5436, pp. 28–35 (2004)
[11] Brassard, G., Salvail, L.: Secret-key reconciliation by public discussion. Advances 765, 410–423 (1994)
[12] Bennett, C.H., Bessette, F., Brassard, G., Salvail, L., Smolin, J.: Experimental quantum cryptography. Journal of Cryptology 5(1), 3–28 (1992)
[13] Martinez-Mateo, J., Elkouss, D., Martin, V.: Blind Reconciliation. Quantum Information and Computation (2003)
[14] Pearson, D.: High-speed QKD Reconciliation using Forward Error Correction. In: 7th International Conference on Quantum Communication, Measurement and Computing, vol. 734, pp. 299–302 (2004)
[15] Elkouss, D., Leverrier, A., Allaume, R., Boutros, J.: Efficient reconciliation protocol for discrete-variable quantum key distribution. In: 2009 IEEE International Symposium on Information Theory vol. (1), pp. 1879–1883 (2009)
[16] Mondin, M., Daneshgaran, F., Delgado, M.T., Mesiti, F.: Soft-metric-based information reconciliation techniques for QKD. In: SPIE Optics + Photonics 2010, San Diego, USA, August 1-5 (2010)
[17] Mesiti, F., Daneshgaran, F., Delgado, M.T., Mondin, M.: Sparse-graph codes for information reconciliation in QKD applications. In: ISABEL 2010, Roma, Italy, Novembre 7-10, pp. 1–5 (2010)
[18] Mondin, M., Daneshgaran, F., Delgado, M., Mesiti, F.: Novel Techniques for Information Reconciliation, Quantum Channel Probing and Link Design for Quantum Key Distribution. In: Sithamparanathan, K., Marchese, M., Ruggieri, M., Bisio, I. (eds.) PSATS 2010. LNICST, vol. 43, pp. 305–316. Springer, Heidelberg (2010)
[19] Shor, P., Preskill, J.: Simple proof of security of the BB84 quantum key distribution protocol. Physical Review Letters 85, 441–444 (2000)
[20] Bennett, C.H., Brassard, G.: Quantum Cryptography: Public Key Distribution and Coin Tossing. In: Proceedings of IEEE International Conference on Computers Systems and Signal Processing, Bangalore, India, pp. 175–179 (1984)
[21] Bennett, H., et al.: Experimental quantum cryptography. Journal of Cryptology 5(1), 3–28 (1992)
[22] Biham, E., Boyer, M., Boykin, P.O., Mor, T., Roychowdhury, V.: A proof of the security of quantum key distribution. In: Proceedings of the 32nd Annual ACM Symposium on Theory of Computing, p. 175. ACM Press, New York (2000)
[23] Lo, H.-K., Chau, H.F.: Unconditional security of Quantum Key Distribution over arbitrarily long distances. Science 283, 2050–2056 (1999)
[24] Srikanth, R., Pasupathy, J.: BB84 Quantum Key Distribution protocol based on classical error correction, TR-PME-2002-11, Center for Theoretical Studies, Indian Institute of Science, Bangalore, August 22 (2002)

[25] Berrou, C., Glavieux, A., Thitimajshima, P.: Near Shannon limit error-correcting coding. In: Proceedings of the 1993 IEEE International Conference in Communications, pp. 1064–1070 (1993)
[26] Jelinek, F., Bahl, L., Cocke, J., Raviv, J.: Optimal decoding of linear codes for minimizing symbol error rate. IEEE Trans. Inform. Theory 20(3), 284–287
[27] Gallager, R.G.: Low Density Parity Check Codes. IEEE Trans. Inform. Theory 8(1), 21–28 (1962)
[28] Tanner, R.M.: A recursive approach to low complexity codes. IEEE Trans. Inform. Theory 27, 533–547 (1981)
[29] Richardson, T.J., Shokrollahi, M.A., Urbanke, R.L.: Design of capacity-approaching irregular low density parity-check codes. IEEE Trans. Inform. Theory 47(2), 619–637 (2001)
[30] MacKay, D.J.: Good error-correcting codes based on very sparse matrices. IEEE Trans. Inform. Theory 45(2), 399–431 (1999)
[31] Richardson, T.J., Urbanke, R.L.: The capacity of low-density parity-check codes under message-passing decoding. IEEE Trans. Inform. Theory 47(2), 599–618 (2001)
[32] Urbanke, S.-Y.C.: Analysis of sum-product decoding of low-density parity-check codes using a Gaussian approximation. IEEE Trans. Inform. Theory 47(2), 657–670 (2001)
[33] Benedetto, S., Montorsi, G.: Design of parallel concatenated convolutional codes. IEEE Transactions on Communications 44(5), 591–600 (1996)
[34] Bruss, D.: Optimal eavesdropping in quantum cryptography with six states. Physical Review Letters 81, 3018–3021 (1998)
[35] Ekert, K.: Quantum Cryptography based on Bell's theorem. Physical Review Letters 67, 661–663 (1991)
[36] Shannon, C.: A Mathematical theory of communication: Part 1. Bell System Technical Journal 27, 379–423 (1948)

A Comparative Study on Security Implementation in EPS/LTE and WLAN/802.11

Siwar Ben Hadj Said[1], Karine Guillouard[2], and Jean-Marie Bonnin[1]

[1] TELECOM Bretagne, France
[2] France Telecom R&D
{siwar.benhadjsaid,karine.guillouard}@orange.com,
{siwar.benhadjsaid,jm.bonnin}@telecom-bretagne.eu

Abstract. Security in the wireless access network gained increasing interest over the last years. Its implementation varies from one access network to another. The current trend in wireless access network is towards implementing mechanisms for mobility management namely handoff process and quality of service control. Consequently, security should be taken into consideration at each handoff process which may occur between different technologies (inter-technology) or within the same (intra-technology). At the same time, security provisioning impact on network performances (e.g. end-to-end delay, throughput) should be controlled. This chapter aims to give a better understanding of security measures and protocols available in two distinct wireless network families, namely the Wireless Wide Area Network (WWAN) and the Wireless Local Area Network (WLAN). WWAN family includes the wide coverage area technologies such as the Long-term Evolution (LTE), also named as Evolved Packet System (EPS). On the other hand, WLAN are characterized by having a small coverage area. It includes the WiFi (802.11) technology. Each time, the chapter highlights the mechanisms employed by access network to ensure the trade-off between secured mobility and application requirements in terms of delay and throughput.

Introduction

Over the last few decades, we have witnessed the emergence of several wireless accesses which cover various requirements. The cellular access represents the first category of wireless access. The Global System for Mobile communication (GSM) was the first standard developed for the cellular access. Initially, The GSM was dedicated for voice traffic. The General Packet Radio Service (GPRS) extends the GSM capabilities to support the IP packet transfer. Several security problems were detected in this original standard. One decade later, the Universal Mobile Telecommunication System (UMTS) was introduced as an evolution of the GSM access. It provides a better data rate and improves user's security. One more decade later, the new radio technology Long Term Evolution (LTE) appeared as the solution that will increase the capacity and the speed of wireless access. This paves the way to the Evolved Packet System (EPS) emergence [1]. The mobility management mechanisms were natively implemented in cellular access.

S. Khan and A.-S.K. Pathan (Eds.): *Wireless Networks and Security*, SCT, pp. 457–489.
DOI: 10.1007/978-3-642-36169-2_14

The fixed wireless access represents the second category of wireless access. It is called fixed access because the mobility was not natively considered. The Wireless Local Access Network (WLAN) was the early fixed wireless access. It is seen as the extension of the wired LAN. The IEEE 802.11 [2] is the most known WLAN and based on the WiFi technology. It is obvious that the low cost of the IEEE 802.11 attracts a large number of users. Therefore, the WiFi access points become pervasive. The current trend in WLAN access is towards implementing mechanisms for mobility management namely handoff process and location update [3]. This enables the users to move from an access to another while keeping the session continuity. Moreover, the initial IEEE 802.11 standard defined a poor security implementation. The IEEE 802.11i [4] standard was introduced to improve the security features.

Security in wireless access network is needed to prevent the misuse. Actually, unlike wired networks, where data transit in cables connecting communicating entities, transmission in wireless networks uses the open air as a medium. The wireless medium has several characteristics that affect the security implementations: it is broadcast medium; it has a limited bandwidth; and the computational power of subscriber's device is limited. For instance, the broadcast nature of wireless network exposes subscribers to a greater risk from intruders who may eavesdrop and potentially alter transmitted messages, impersonate a legitimate subscriber, and therefore gain unauthorized access to network equipments. Security mechanisms are of key importance for establishing trust relationship between communicating entities. Users should be able to insure that network equipments are not compromised or impersonated. On the other hand, network equipments should be able to check whether the requestor of network resources is indeed a reliable subscriber.

On the other hand, the number of mobile users keeps growing and the amount of traffic related to real-time application keeps increasing. Therefore, making the handoff seamless and the user reachable is the key requirement for the mobility management in future wireless access networks. The seamless handover ensures the session continuity for real-time application during the subscriber movement. In addition, the device location should be known by the access provider so that it stays reachable at any time.

The wireless access network faces dual challenge of mobility management and security provisioning. In fact, during each handoff process, the security procedures should take place so that an intruder cannot produce undesirable effects within the access network. At the same time, the handoff latency including the security establishment delay should not impact the seamless session continuity. However, it was shown that the authentication delay represents approximately 64% of the total handoff latency [5]. As a result, the security fast reestablishment during the handoff process is required.

The emerging of heterogeneous wireless access technologies, the disparity of the application requirements, and the urge to offer the mobility service has driven the wireless industry to evolve towards a convergent access network. For instance, the Third Generation Partnership Project (3GPP) specifications [6] defined a common IP core network that is able to interconnect heterogeneous access technologies including the cellular access (EPS/LTE) and the fixed access (WLAN/802.11). The mobility management and the security provisioning are the major differences between the cellular access and the WLAN access.

Within this context, we conduct a comparative study between the cellular access EPS/LTE and the WLAN/802.11 access. The goal of this chapter is to cover the EPS and WLAN security aspects.

The chapter consists of four main parts. In the first part (Part I), we give an overview of the relevant security concepts such as the security services and the potential security attacks related to the wireless access network. We present also the security issues associated with the mobility management. In the second part (Part II), we discuss the security implementation in the EPS system. Specifically, we start with an overview of the network components and architectural model of the EPS system. Then, we analyze the way security mechanisms are implemented and the kind of security services the EPS system is intended to ensure. This part ends with a survey of the mechanisms implemented by EPS to ensure secured and seamless mobility. In the third part (Part III), we present the security features of WiFi along with its architectural model description. After that, we describe the security implementation evolution from unprotected to highly protected access points. This part concludes with an outline of mechanisms for security setup optimization during handoff process in WLAN network. The purpose of the last part (Part IV) is to compare the security scheme weakness and strength points between the two studied access networks, with due consideration for mobility. The conclusion includes security recommendation for LTE and WiFi interworking.

1 Part I: Security Concepts

In this section, we provide the basic concepts in security implementation. After giving the essential terminologies that we will use along this chapter, we present an overview of the main threats related to wireless access network. Then, the security services or properties that should be present in access networks are described. A mapping between each security service and some security mechanisms is provided at the end of the section.

1.1 Terminologies

Access Provider: represents an organization that arranges for a subscriber to have access to the Internet and potentially to additional services platforms. The cellular network operator is an example of the access provider.

Subscriber: represents the user that requests resources from an access network such as IP address. It has already a subscription with the access provider in use.

Intruder: refers to an attacker that would like to listen or alter messages of a legitimate subscriber. He may intend to alter network equipments.

1.2 Threats in Wireless Access Networks

In general, security threats are classified into two categories: passive and active attacks [7]. A passive intruder seeks to learn information from the communication. He may even make use of the obtained information. However, he does not try to affect communicating entities or their communication. In the other side, an active intruder attempts to alter exchanged messages, abuse of network resources or affect the functioning of communicating entities.

We propose to classify security threats into 4 main categories:

1.2.1 Threats Related to Subscriber Privacy

It includes the disclosure of information related to subscriber profile such as his identity, location, etc. Subscriber identity theft or subscriber tracking are examples of threats targeting subscriber privacy. Even in case of message encryption, the intruder might observe the pattern of these messages and determine the location and identity of the communicating hosts. Also, the intruder might guess the nature of the communication that was taking place by observing the frequency and length of exchanged messages.

1.2.2 Threats Related to Communication

Those threats target subscriber communication. The following attacks are example of those threats:

- *Message contents disclosure:* An intruder may listen to a legitimate subscriber communication without his consent. A communication may have sensitive or confidential information. If the communication is not protected, the intruder may learn the content of these transmissions.
- *Message modification:* An intruder may capture messages of a legitimate subscriber, alter, delay or reorder them in order to produce an unauthorized effect.

1.2.3 Threats Related to Access Network Resources

Those threats aim to open access network for an unauthorized subscriber, to consume network resources in inappropriate way or to make it possible for access network to serve legitimate users. It includes the following attacks:

- *Masquerade:* An intruder may pretend to be a legitimate subscriber. It may steal the real subscriber identity and use it to get an unauthorized access. As a result of this attack, the intruder may get extra privileges by impersonating the legitimate subscriber that possesses those privileges.
- *Replay:* An intruder may capture an authentication sequence and retransmit it in order to produce an unauthorized effect such as unauthorized resource usage.
- *Denial of service:* An intruder may engage network equipment with excessive and unnecessary processing, thus denying access to authorized subscribers. The intruder may overload a network equipment leading to reduced availability. He may also prevent signaling from being transmitted inducing protocol failure.

1.2.4 Threats Related to Mobility Management

Thanks to mobility management, the subscriber may move from an access to another while maintaining session continuity. Moreover, the subscriber may still reachable as long as the mobility management service is active. There are two categories of mobility: Network-based mobility and Host-based mobility. We are interested in the first category where the network is responsible for ensuring seamless mobility while it moves from an access network to another.

There are two categories of traffic that should be distinguished in mobility management: signaling traffic (that is messages for location update or handover command) and user traffic.

An intruder may be merely a passive attacker by observing the signaling traffic related to the mobility management. With this kind of attack, the intruder may derive sensitive information such as subscriber location leading to a privacy violation.

An intruder may choose to be an active attacker by manipulating the mobility management messages when they are carried unprotected. It may replay signaling messages or even flood network equipments. This may lead to unauthorized access to network services, denial of service, or even redirection of other user and control traffic.

1.3 Security Services

RFC 4949 [8] defines a security service as "a processing or communication service that is provided by a system to give a specific kind of protection to system resources". Therefore, network security services may be defined as the security services ensured by the access provider in order to avoid potential threats such as user's identity theft, replay attack, and denial of service. Network security services may be implemented by several security mechanisms.

In the following, we address an outline of the main security services.

1.3.1 Identification

Identification service is a crucial as it helps access providers to identify the subscribers that use their network. In general, there are two kinds of identity:

- *Subscription identity:* is allocated to the subscriber by the access provider upon signing a subscription contract. It serves to identify subscribers when they use the access network. This parameter enables the access provider to determine the subscriber profile and the services to which he subscribed.
- *Device identity:* is more related to the device and is specific to vendors.

The type of the subscriber identity that shall be presented to an access network varies according to the method of authentication implemented in this access. For instance, the login represents the subscriber identity in the login/password authentication mechanism.

1.3.2 Subscriber Authentication

The authentication service consists in validating the subscriber identity. The fact that the subscriber claims an identity does not necessarily mean that he is the holder of that identity. The subscriber must provide evidence to prove its identity to access the network. This proof can be a certificate or a response to a challenge. For security reasons, the authentication should be mutual in the access network, i.e., not only the user's identity is verified by the network but also the network credibility is checked by the user. Several authentication methods exist.

There are three components involved in the subscriber authentication process:

- *Subscriber:* represents the party in authentication process that will provide its identity and evidences that proves its veracity.
- *Authenticator:* represents the party in authentication process that is providing resources only for allowed subscribers. It needs to ascertain subscriber identity and profile before allocating resources.
- *Security authority:* represents the party in authentication process that is responsible for subscriber identity check.

There are mainly three kinds of identity check mechanism: challenge/response authentication, mutual public key authentication, and tunneled authentication (i.e. first, the security authority is authenticated using its certificate, establishes encrypted channel with the subscriber and then authenticates the subscriber using the login/password or the challenge/response).

1.3.3 Access Control

The access control service prevents either the unauthorized use of the network resources or the use of network resources in unauthorized manner. Therefore, it ensures two main properties:

- *Authorized network resource usage:* In general, the access network is intended to be used only by authorized subscribers.
- *Appropriate network resource usage:* Once authorized, the subscriber is able to use network resources according to the restrictions that are defined in his profile.

Therefore, the access control service should include two mechanisms:

- *Authorization mechanism:* is intended to verify whether the subscriber has permission to use network resources.
- *Resource usage control mechanism:* is intended to control the manner with which the network is used.

1.3.4 Confidentiality and Integrity Protection

The confidentiality service has the role of preventing the disclosure of the information contained in user's communication. The communication between the user and the network should be unintelligible to a third-party. It may be ensured by the ciphering operations.

The integrity protection service has the role of preventing the modification of messages during their trip from the sender to the receiver. It prevents also from message replay attacks. In fact, with this service, any alteration of the original message by an unauthorized party will be notable at the receiver side.

1.3.5 Privacy

The access provider tends to maintain some basic information about subscribers such as subscription identities, devices identities, subscriber locations, etc. From privacy viewpoint, those information should be protected from other parties. In general, access providers should mainly keep confidential the following information:

- *Identity confidentiality:* is defined as the act of keeping confidential the subscriber's identity. The identity privacy becomes crucial as it prevents identity theft, and therefore user impersonation attacks.
- *Location confidentiality:* is defined as the act of keeping confidential subscriber's locations. In fact, an intruder may observe and analyze location update messages sent by the subscriber each time he moves to a new access. This could lead to the disclosure of the subscriber location and therefore to privacy violations.

1.4 Security Mechanisms

Each network security service may be ensured with various security mechanisms. For instance, the authentication service may be ensured with either challenge/response or certificate-based mechanisms. In the following table, we present the mapping between the network security services and examples of the corresponding mechanisms.

Network security service	**Network security mechanism**
Identification	Permanent identity provisioning (device identity, subscription identity)
Authentication	Login/password Challenge/response Certificate-based authentication
Access control	Access control list, port-based control
Confidentiality	Ciphering
Integrity protection	Message digest (hashing) Use of the sequence number
Privacy	Binding with a temporary identity

2 Part II: Security Implementation in EPS

In this section, we will present the overall picture of the design of security implementation in the 3GPP Evolved Packet System (EPS) architecture. We start by describing the EPS architectural model. An in-depth description of security implementation appears in the subsequent sections of this part.

2.1 Architectural Model

The EPS architectural model is depicted in Figure 1. It is composed of two parts: a radio access network (RAN) and core network (CN) [1]. Further, in EPS system, there is a control plane for signaling traffic exchange and a data plane for data traffic transmission. Dotted lines in Figure 1 are for signaling plane messages and solid lines are for user data traffic.

User Equipment (UE): represents the subscriber in EPS world. It consists of a smart card called Universal Integrated Circuit Card (UICC) and a terminal called Mobile Equipment (ME). UICC houses the UMTS Subscriber Identity Module (USIM application). In general, UICC is known as the USIM card. The subscriber identity and the corresponding cryptographic key are contained in the USIM card.

Radio access network (RAN): consists of several eNodeBs (eNB). The eNB represents the first contact point of the UE with the EPS system. It acts as a gatekeeper opening the door only for data traffic related to authenticated and authorized subscribers and forwarding signaling traffic related to unauthenticated subscribers to the equipment responsible for user's authentication.

Core Network (CN): consists of three main entities namely Mobility Management Entity (MME), Serving Gateway (S-GW) and PDN Gateway (P-GW).

- **Mobility Management Entity (MME):** acts as the manager of network connectivity. It is responsible for the subscriber authentication and authorization. It generates the required keying material to secure traffic over the wireless link (i.e. between UE and eNB). It selects the adequate S-GW and P-GW and then initiates resources allocation procedure. It manages subscriber mobility. For instance, it ensures session continuity while the subscriber moves from an eNB to another. It ensures subscriber reachability by tracking his movement even when no session is going on.
- **Serving Gateway (S-GW):** acts as a demarcation point between the eNodeB and the core network. It anchors UE data traffic for local mobility (i.e. packets are still routed through this point while UE moves between eNBs connected to this S-GW). For each UE, it maintains an always-on connection with the P-GW to ensure its reachability. It triggers the UE paging operation in the MME and buffers packets when downlink data arrive for an idle UE.
- **Packet Data Network (PDN) Gateway (P-GW):** is the termination point of the core network towards one or more IP networks. It anchors UE data traffic for intra-LTE mobility (i.e. packets are still routed through this point while UE moves between several S-GWs) or inter-technology mobility (i.e. packets are still routed through this point while UE changes the access technology). It is responsible for IP address allocation from either its own pool or the target IP networks.

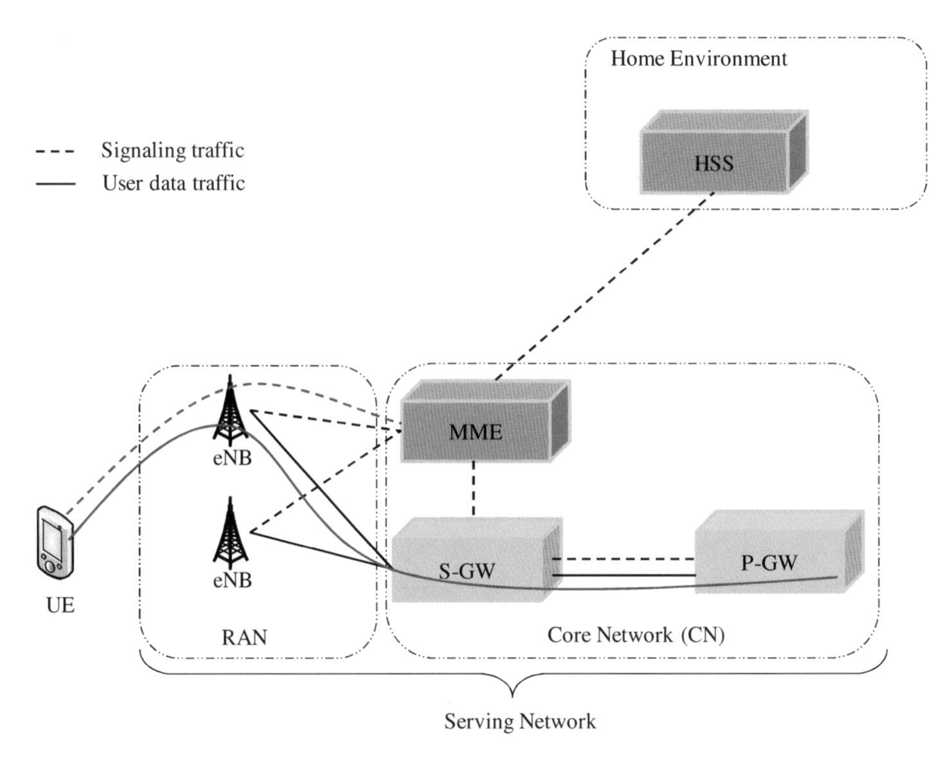

Fig. 1. EPS/LTE architectural model

In Figure 1, a distinction between serving network and home environment is made. This distinction is relevant for security description in the following.

- *Serving network*: represents the network equipments (in CN and RAN) that presently serve the subscriber.
- *Home Environment:* includes the Home Subscriber Server (HSS) which acts as a central repository of all subscriber-specific authorizations and service profiles and preferences. In particular, the HSS contains the subscriber identity and the long-term subscriber key for each subscriber. The HSS is an evolution of the HLR of the original GSM and 3G architectures.

The 3GPP specification [1] describes several procedures that are required to establish and maintain UE connection to the network. The basic procedures are:

- *Network attachment:* is defined as the registration procedure. The UE needs this procedure to gain services from the network. For instance, due to the registration procedure, the MME is able to locate the subscribers and to deliver to them any received calls or messages. Upon receiving an initial attach request, the MME connects the UE to a default P-GW. A tunnel between the S-GW and P-GW is established and maintained as long as the UE is registered to the network.
- *Service Request:* is performed when the UE is already attached and needs to send data traffic. This procedure may be initiated by the UE to respond to a paging request.

- *Tracking Area Update (TAU):* is defined as the procedure that helps the network to be up-to-date of the latter UE location. It is called also UE idle mobility. In fact, the cellular network is divided in different areas called tracking areas (TA). An area consists of several cells. The TAU procedure is initiated by the UE when a change of TA is detected.
- *Handover:* is defined as the procedure that enables session continuity when the UE moves from an eNB to another. The UE mobility may include S-GW and MME changes. The handover principle is to hide the UE movement from the serving P-GW in order to avoid breaking the session.

Each time the UE initiates one of this procedures with the network, security mechanisms are invoked. For instance, the MME should verify the identity included in the attach request and download the UE profile. Therefore, MME may decide whether the UE is authorized to use network resources. In addition, an agreement on security parameters such as cryptographic keys and algorithms is performed between the UE and the MME to protect their signaling exchanges. A second agreement is performed between the UE and the eNB to protect the traffic at the radio level.

2.2 Security Implementation

The EPS requirements in terms of security and the various alternatives for security implementation were discussed in the Technical Report [9]. The final version of security implementation in EPS is described in the Technical Specification [10]. It is worth mentioning that a trust relationship is assumed between the network equipments namely the eNB, the MME, the S-GW and the P-GW. [10] specifies only the way to authenticate subscribers, secure the signaling traffic exchange between UE and MME, and secure signaling and traffic exchange between UE and eNB.

In the following, we map each security service presented in Part I with the corresponding security mechanism implemented in the EPS architecture. The network attachment Figure 2 in EPS seems to be the best procedure that may illustrate our mapping. In the following, we give the attachment phases in the EPS architecture.

- *Phase A:* The UE gets access to the radio level by requesting a radio control channel from the eNB. After that, the UE sends an Attach request message including its identity. As a result, the MME starts a number of security procedures namely UE identification and authentication. We assume that the UE indicates in the attach request message that it has data to be sent.
- *Phase B:* As the signaling exchange between UE and MME should be protected, an agreement on security parameters should happen. This includes cryptographic algorithm selection and cryptographic keys generation. The MME may initiate optionally the device identity check procedure.
- *Phase C:* The MME downloads the UE profile from HSS so it may verify whether the UE is authorized to use the current eNB.
- *Phase D:* Due to the always-on IP connectivity property of EPS access, MME requests from the S-GW and P-GW to establish a permanent tunnel dedicated for this UE.

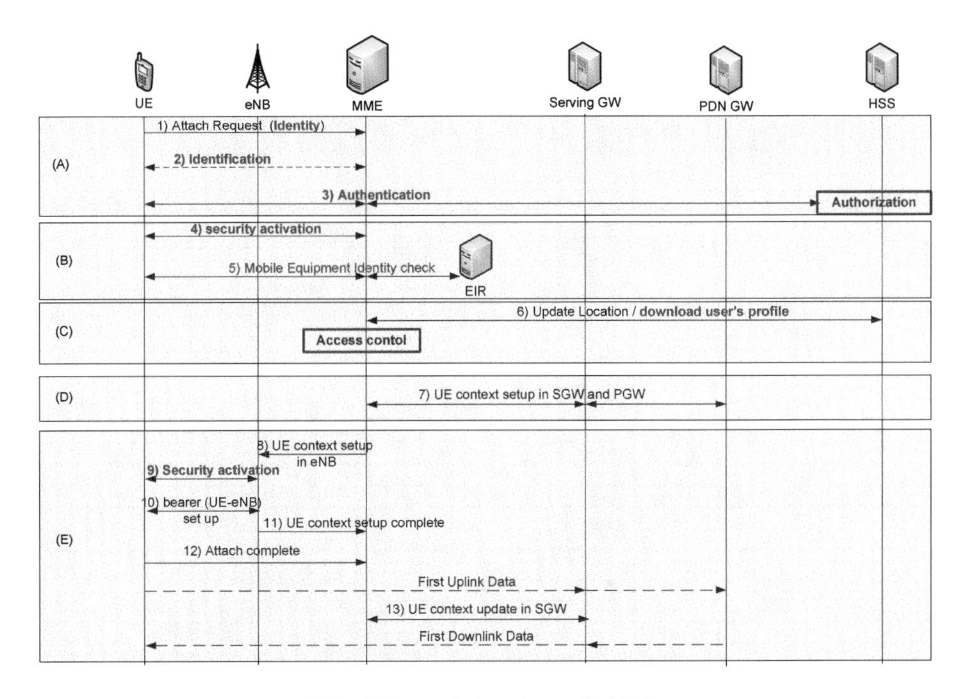

Fig. 2. Network Attachment in EPS

- *Phase E:* The MME generates a temporary identity and delivers it to the UE within the attach answer message. This message is not shown in the flow chart. Actually, the MME sends it to the eNB during the UE context setup message (procedure number 8 in the flow chart). This initiates the cryptographic keys generation and security algorithms selection in eNB and UE to secure their exchange at radio level (procedure number 9 in the flow chart). The eNB forwards the attach answer message to the UE during the bearer UE-eNB set up (procedure number 10 in the flow chart). The UE will use this temporary identity each time it sends a signaling message to MME. Upon receiving the attach complete from the UE, the MME updates the context maintained within the S-GW by indicating the eNB to which the UE is connected. This will enable the activation the data tunnel between the serving eNB and the S-GW.

2.2.1 Identification

In EPS, each UE possess two identities:

- *Subscription identity:* IMSI (International Mobile Subscriber Identity). It has three main components:
    - o Mobile Country Code (MCC): identifies the country of the subscriber
    - o Mobile Network Code (MNC): identifies the access provider of the subscriber within that country.
    - o Mobile Subscriber Identification Number (MSIN): identifies the subscriber within the home network.
- *Device identity:* IMEI (International Mobile Equipment Identity).

Whenever an UE sends a signaling message to the MME, it must include his subscription identity (IMSI) within this message. Therefore, the MME may fetch from the HSS the profile corresponding to this identity and check whether this subscriber is authorized to use the network resources. As the HSS contains all subscribers' profiles, it is able to verify whether the identity presented to the MME corresponds to an existing subscription. When the MME cannot determine the subscriber identity from the signaling message, it sends back to the subscriber an Identity Request message Figure 3.

In emergency cases, the UE may present its device identity instead of its subscription identity.

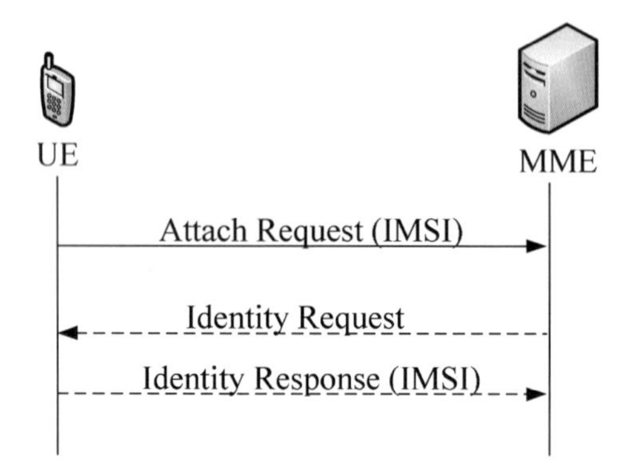

Fig. 3. Subscriber Identity Request procedure [10]

EPS may propose the device identity check procedure as an optional service Figure 4. In this procedure, the MME requests device identity (IMEI) from the UE. Then, it checks whether the received identity exists in a central database, called Equipment Identity Register (EIR). This database contains the IMEI of stolen or blacklisted devices. The EIR database belongs to the Home environment part (see Figure 1). Two EIRs of different access providers may communicate with each other in order to check the device validity.

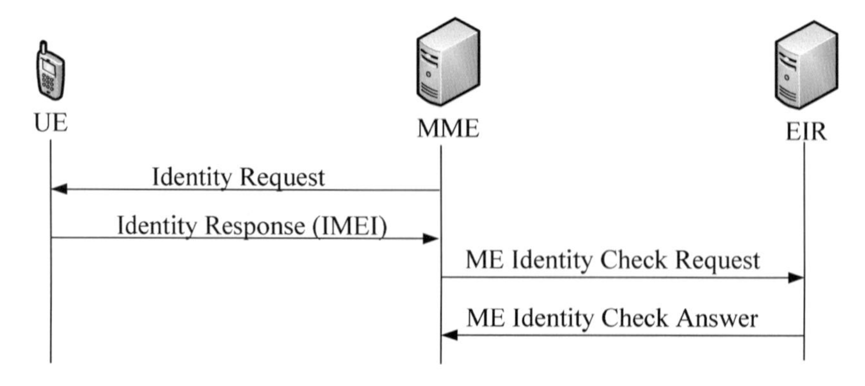

Fig. 4. Device Identity Check procedure [10]

At the access network level, network equipments generate several identifiers for addressing purpose. For instance, the UE is uniquely identified by the eNB within a cell using the C-RNTI identity. The C-RNTI denotes the cell radio network temporary identifier and is used to identify a user's radio connection within a cell radio coverage area. The C-RNTI is allocated by the eNB and transmitted to the UE.

2.2.2 Authentication

In EPS access, the challenge/response authentication mechanism, called EPS-AKA (EPS-Authentication and Key Agreement), has been adopted. It is based on a permanent pre-shared secret between UE and HSS database. In fact, at the time of signing the subscription contract, the access provider gives a USIM card to the subscriber where the subscriber identity and the permanent pre-shared key are stored.

With AKA, the network is able to firstly authenticate the subscriber and secondly agree with him on a common security key set.

The network equipments involved in the authentication procedure are:

- UE: acts as the subscriber in the authentication mechanism.
- MME: acts as the authenticator and the security authority in the authentication mechanism.
- HSS: acts as a database that provides for the security authority the information needed for the smooth running of the authentication mechanism.

The flow chart of the authentication mechanism in EPS is shown in Figure 5.

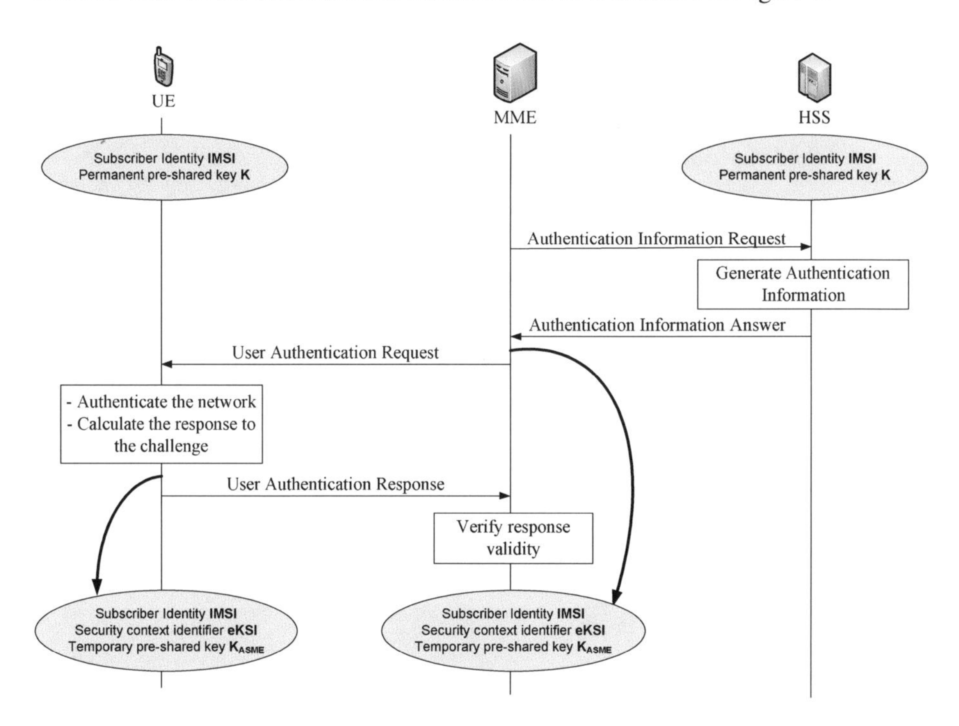

Fig. 5. Authentication mechanism in EPS

Upon receiving the subscriber identity, the MME requests authentication information, called also authentication vectors, from the HSS database [11]. This triggers the authentication vector generation procedure in the HSS using the permanent pre-shared key K. An authentication vector contains four components namely a temporary pre-shared key (K_{ASME}) called also master key, a challenge (RAND), an expected response to the challenge (XRES), and an authentication token (AUTN) that serves to authenticate the home network. The number of authentication vectors generated for an MME depends on the access provider policies. Then, the authentication vectors are sent back to the MME within the Authentication Information Answer message. After that, the MME chooses one of the received authentication vectors and sends the challenge and the authentication token to the subscriber within the User Authentication Request message. As a result, the subscriber uses its permanent key K, the parameters AUTN and RAND to authenticate the home network and calculate the response to the challenge. It sends back the response to the MME within the User Authentication Response message. The last phase of this mechanism consists in verifying the received response with the expected response. At the end of a successful authentication, the MME and the UE maintain each one a security context composed of the following parameters: the subscriber identity IMSI, temporary pre-shared secret K_{ASME}, and an identifier of this context eKSI (Key Set Identifier in EPS). This identifier serves as a pointer to the security context that should be used. The MME generates this parameter, checks its uniqueness and communicates it to the UE in the same message of challenge message.

2.2.3 Access Control

The authorized use of network resources represents the first property of the access control service. The authorization to use the current MME is checked by the HSS upon receiving the Authentication Information Request message. In fact, one subscriber may be allowed to use only the UMTS access. Therefore, when the HSS receives the Authentication Information Request message for the subscriber, it should inform the MME that the corresponding subscriber is not authorized to use the EPS access.

After a successful authentication, the MME fetches the subscriber profile from the database HSS (see Figure 6). In this profile, the network indicates the network services to which the user is subscribed to such as multimedia broadcast (MBMS) and local IP access (LIPA) services. It indicates also the quota of resources affected to this subscriber such as the maximum bandwidth.

Each time the subscriber requests a particular network service, the MME checks whether this subscriber is authorized to gain access to this service.

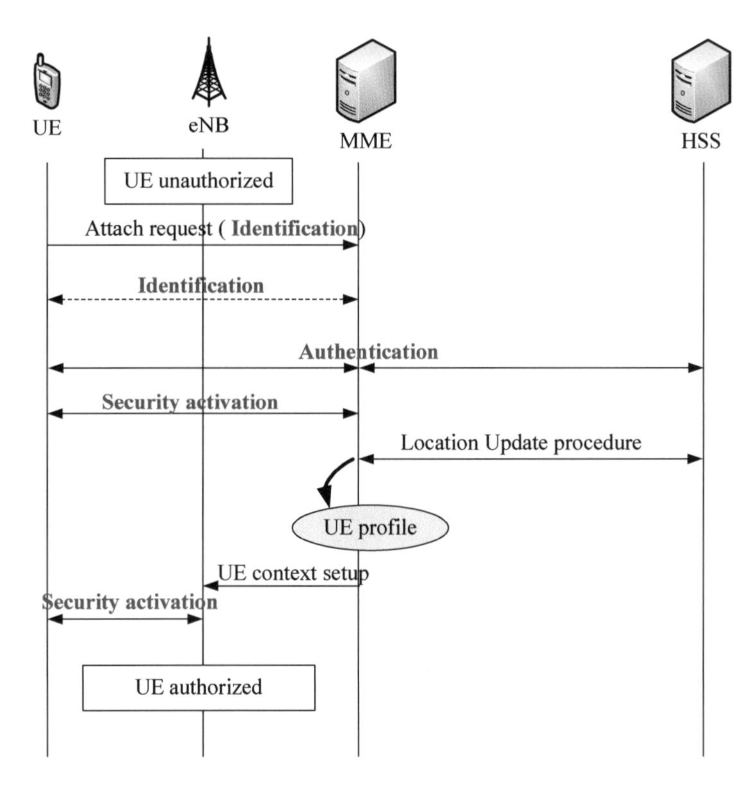

Fig. 6. Subscriber Profile download

2.2.4 Confidentiality and Integrity Protection

The 3GPP specification distinguishes between two categories of traffic: the signaling and user exchanges. The first category requires total protection as it contains sensitive data. For instance, the confidentiality and integrity protection for the signaling exchange between the UE and the MME is mandatory. Any signaling messages received in clear are rejected by the MME except the Attach Request and TAU requests.

The protection for the second category of traffic is optional [10]. In this case, the integrity protection is not provided for the data traffic at the radio level. It was established that the integrity protection is not required for user traffic [9]. Essentially, the integrity protection mechanism generates additional overhead if it is applied to each user packet. Therefore, integrity protection mechanism may reduce the available bandwidth. In addition, this mechanism may add an intolerable delay in the end-to-end communication latency.

The Figure 7 shows how security contexts are setup in both of MME and eNB.

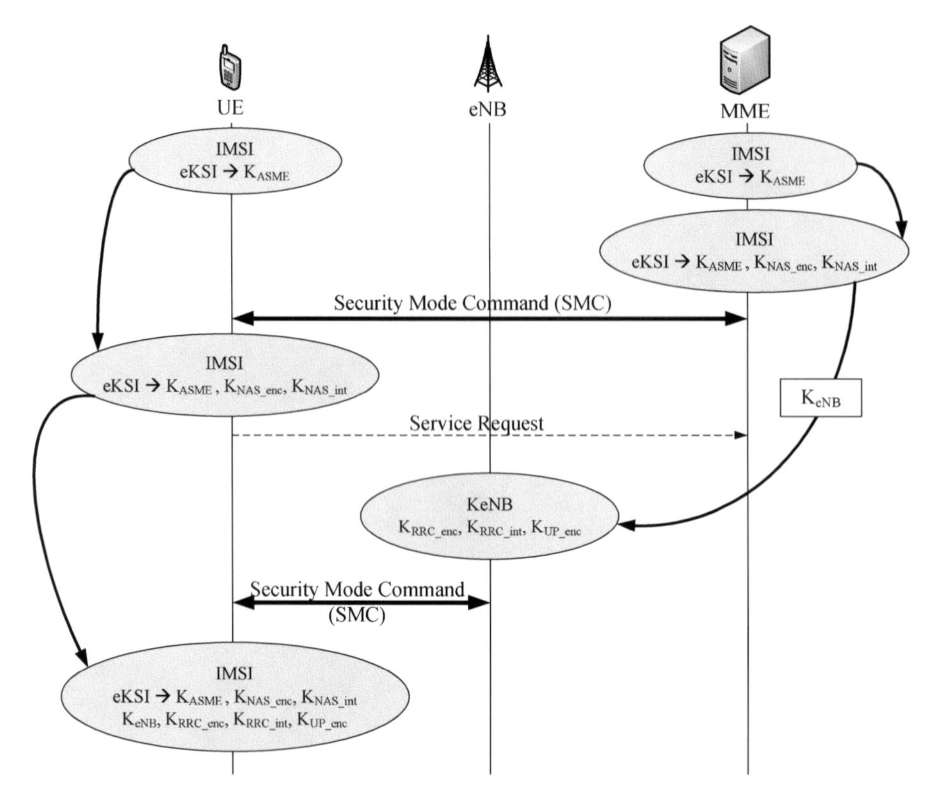

Fig. 7. Confidentiality and integrity protection setup procedure

The MME uses the security context resulting from the authentication mechanism (IMSI, UE security capabilities, eKSI, K_{ASME}) to derive two new cryptographic keys (K_{NAS_enc} and K_{NAS_int}). These keys are required to ensure the confidentiality (K_{NAS_enc}) and integrity protection (K_{NAS_int}) of the signaling exchange between the UE and the MME.

The keys derivation is based on the temporary pre-shared key (K_{ASME}) and the identifier of the cryptographic algorithms that will be used in ciphering and integrity protection. It is worth mention here that the UE communicates its cryptographic capabilities (i.e. the identifiers of the cryptographic algorithms supported by the UE) to the MME within the attach request. Therefore, the MME choose the strongest security algorithms that are supported by both of them. After that, the MME initiates a Security Mode Command (SMC) to announce to the UE the selected algorithms.

Upon receiving a service request or if the subscriber indicates in the attach request that he has data to be sent, the MME generates an additional key (K_{eNB}). This key will be forwarded to the eNB and serves as the root for other keys derivation at eNB level. In fact, eNB derives from K_{eNB} three new keys (K_{RRC_enc}, K_{RRC_int} and K_{UP_enc}). The MME forwards also UE capabilities so the eNB may choose an algorithm supported by the two parties (eNB and UE). Likely to the MME, the eNB initiates the Security Mode Command (SMC) procedure to announce to the UE the selected algorithms and to trigger cryptographic keys derivation.

2.2.5 Privacy

From subscriber privacy points of view, subscriber and device identities should be confidentially protected. Therefore, the EPS access takes any precaution to prevent the subscriber from sending identities in clear. For instance, it is established that the MME requests the device identity only after setting up security parameters for UE-MME signaling protection.

Concerning the subscriber identity, the UE should include it in each signaling message so the MME may identify the corresponding parameters. In mobility cases, the new MME requires this parameter to fetch the corresponding authentication data from the HSS or from the old MME. Therefore, sending this identity in clear is unavoidable. As a countermeasure, MME allocates a temporary identity called GUTI to the UE. The purpose of this identity is to provide to the UE an alternative identity that does not reveal the subscriber identity. Consequently, the UE may communicate with any MME in EPS access with this new identity.

The MME uses the Attach Accept message to announce the temporary identity to the UE.

GUTI is the abbreviation of Globally Unique Temporary UE Identity. It has two main components:

- Globally Unique MME Identifier (GUMMEI): identifies the MME that allocated the GUTI in a global and unique way. It is constructed from the MCC, the MNC and the MME Identifier (MMEI).
- M-TMSI: uniquely identifies the UE within the MME that allocated the GUTI.

The use of the temporary identity ensures also the location confidentiality as it prevents the intruder from matching the TAU messages with the subscriber identity.

2.3 Security and Mobility

Initially, EPS was conceived with the mobility consideration in mind. Whenever the UE moves from an access point to another, EPS ensures the seamless continuity of active sessions by hiding UE location change to the other end-party of the communication. Even in case of mobility with no active sessions, EPS provides subscriber's reachability by enabling the TAU mechanism.

For each mobility management procedures, the security should be provided to avoid unauthorized use of resources or denial of service attacks. For instance, EPS should run authentication mechanism upon receiving a TAU request in order to make sure of the UE's reliability. Thus, the unauthorized use of network equipment is warded off. Moreover, the TAU request sent by the UE should be integrity protected so an intruder cannot modify the message content.

The security implementation may impact network performances. Firstly, the authentication method AKA and the SMC procedures are time-consuming procedure [12]. Therefore, running these procedures during each handover may impact session latency. Secondly, running AKA for each location update generates additional overhead on network and may overload network equipment [13][14]. A mechanism for fast security re-establishment is required to support the mobility management mechanisms.

In order to carry out authentication, the MME obtains authentication data from the HSS. It was established that the MME may obtain an array of AV so that the authentication delay is reduced [10]. The 3GPP specification left the choice of the AV array length to the access provider policies. Both of [13] and [14] propose a mechanism that dynamically selects the length of the AV array. This reduces the authentication signaling traffic only between the serving network and the home environment. However, no mechanism was proposed in these papers to reduce the signaling traffic between the UE and the serving network.

In the 3GPP specification, the context transfer protocol is used to rapidly re-establish the security parameters in the new access point and to reduce the authentication signaling exchange during the UE mobility. In general, the context protocol consists in transferring session state from one network equipment to another contributing therefore to the enhancement of handoff performance.

In the following we describe the way the context transfer is implemented in the EPS network. We note that a trust relationship is assumed between the network equipments that belongs to the same serving network.

The running of security mechanisms described earlier results in the creation of two security contexts (Figure 8). The first context is maintained in MME as long as the UE is registered to the network and consists in the permanent identity, the temporary identity, the cryptographic keys and algorithms in use, and the unused authentication vectors (AV). The second one is maintained in eNB as long as the UE has an active session and consists in the cryptographic keys and algorithms in use.

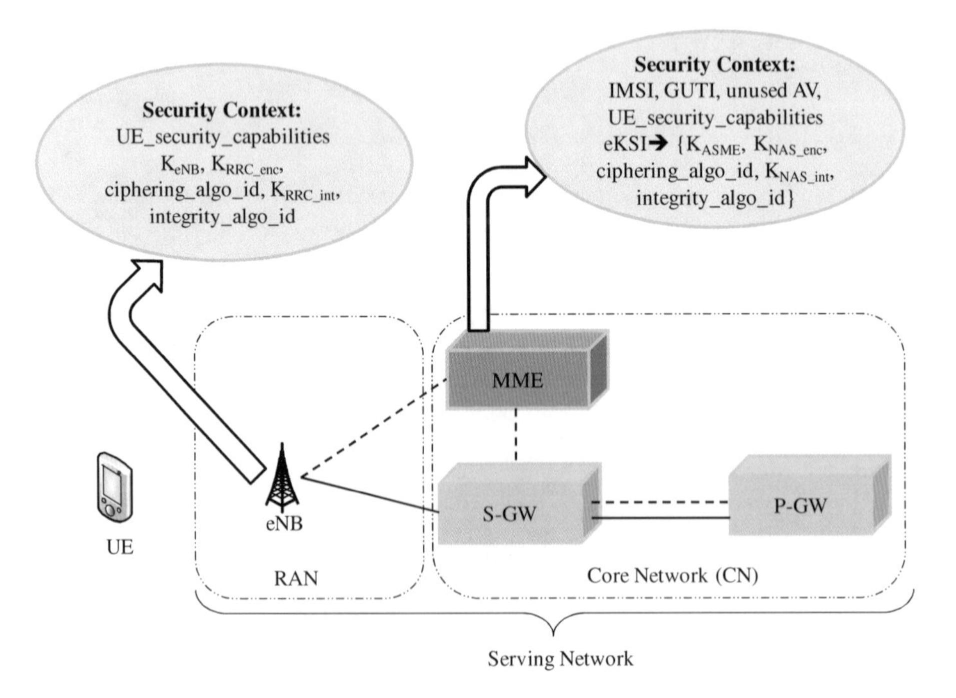

Fig. 8. Security contexts in EPS

The security parameters that should be transferred as well as the network equipment concerned with this transfer depend on the mobility level. In fact, there are two levels of mobility. In the first level, the UE changes the eNB but keeps the same MME. It is called *intra-MME mobility*. In the second one, the UE moves from the old MME to a new one. It is called *inter-MME mobility*.

- *Intra-MME mobility*: There are two kinds of handover: the X2-based handover and the S1-based handover. In X2-based handover, the context transfer takes place directly between eNBs. The old eNB derives a new key, called K^{*}_{eNB}. Then, the K^{*}_{eNB} and the UE security capabilities are sent to the new eNB within the Handover Command message. Therefore, the new eNB can select the appropriate cryptographic algorithms and generates new cryptographic keys for radio link level security. Then the new eNB communicates its choices to the UE through the old eNB. In S1-based handover, the context transfer takes place between the MME and the new eNB. The MME sends the UE security capabilities and a couple of parameters for cryptographic key generation to the new eNB using the Handover Command message. Once the cryptographic algorithms are selected and the cryptographic key generated, the new eNB informs the MME about the selected algorithm. The MME communicates this information to the UE through the old eNB.
- *Inter-MME mobility*: the context transfer takes place between MMEs. Upon a mobility signaling message (TAU Request or Attach Request with Handover indictor) from an UE, the new MME uses the GUTI parameter to identify the old one. Then, it forwards the received message to the old MME. After checking message integrity, the old MME sends back the UE security context. For security reasons, the 3GPP specification defined several rules about the content of the UE security context [10]. If the old and new MMEs belong to the same serving network, the UE security context includes the permanent identity, the unused authentication vectors and potentially the cryptographic keys in use. This avoids the signaling exchange with the HSS for authentication information retrieval. If the MMEs belong to different serving network, the UE security context includes only the permanent identity. In this case, the new MME should rerun the complete authentication mechanism..

3 Part III: Security Implementation in WLAN

Before describing security implementation in WLAN access, we need to briefly preview WLAN architectural model [2]. An overview of the techniques for fast reestablishment security during handoff process is given at the end of this part.

3.1 Architectural Model

The wireless LAN standard IEEE 802.11 defines two basic modes of operation: the ad-hoc network and the infrastructure network (see Figure 9). In ad-hoc mode, wireless devices in the same area may directly interconnect, for example laptops in a meeting room. In the infrastructure mode, wireless devices may connect to a backbone, called Distribution System, through an access point (AP). The laptop can move from cell to cell while keeping access to resources on the wireless LAN.

The term *Station (STA)* is specific to the terminology used in IEEE 802.11 Standard. It refers to the subscriber that supports the IEEE 802.11 protocol. A laptop, a PDA, or a smartphone are examples of STA.

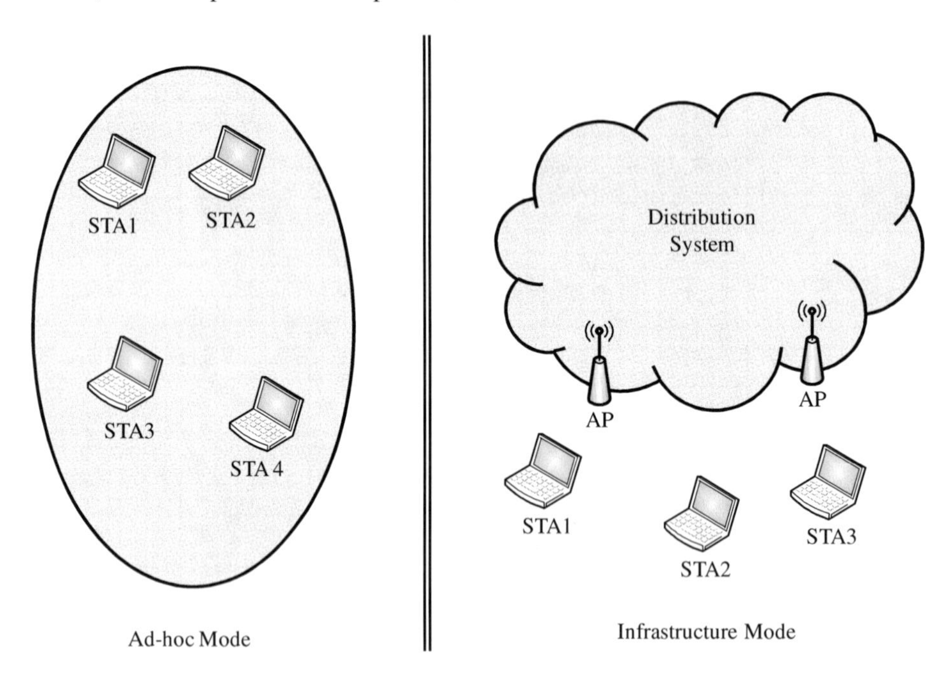

Fig. 9. WLAN modes of operation

As the document aims at comparing the EPS with the WLAN architecture, we are mainly interested in the WLAN infrastructure mode.

The basic architectural model of the WLAN infrastructure mode, depicted in Figure 10, is made of three main equipments:

- *Access Point (AP):* represents the first point of contact of STA with the WLAN network. It represents the boundary between the wireless coverage and the wired backbone. Depending on security configuration, AP may behave differently. For instance, an open AP with no security configuration serves any STA within its coverage area as long as the maximum number of STAs that it can serve immediately is not reached. Whenever the access provider decides to protect his access network, the AP acts as a gatekeeper. In this case, it opens the door only for data traffic related to authenticated and authorized STAs. Otherwise, the AP rejects association request as long as STA authentication procedure fails. Several security configurations will be described in the following.

- *Access Router (AR):* represents the first hop router in WLAN networks. It provides basic connectivity functions such as packet routing, IP address allocation. It potentially participates in Authentication, Authorization and Accounting (AAA) procedure.
- *Wireless Access Gateway (WAG):* is the termination point of the WLAN network. It acts as the point of exit of subscriber's packets towards the Internet, the services platforms, and other networks (e.g. domains managed by other operators)[28].

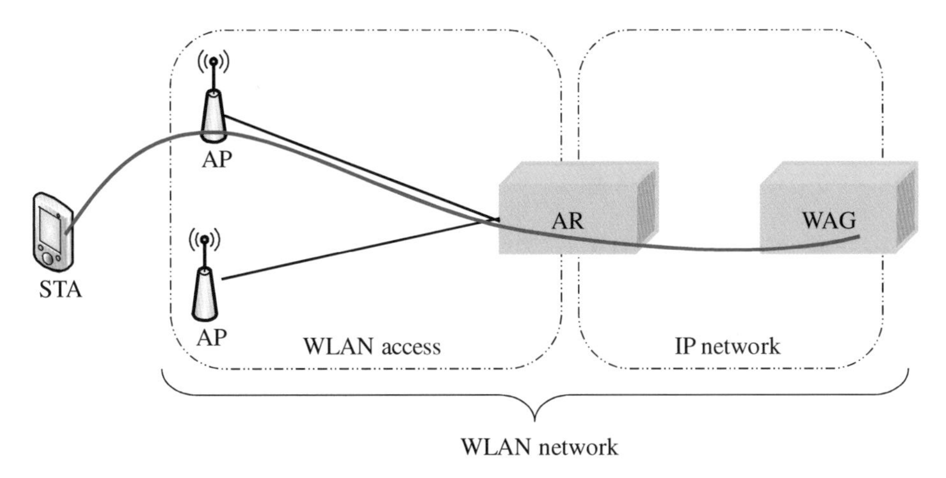

Fig. 10. WLAN architectural model

As the security configuration depends on the authentication mechanism in use, security components are not shown in Figure 10.

Unlike the EPS architecture, no separation between the control and user plane is made in the WLAN architecture.

3.2 Security Implementation

Security implementation in WLAN access is undoubtedly continuously improving. The Wired Equivalent Privacy (WEP) was the first security algorithm specified in IEEE 802.11 standards. The aim of this algorithm was to provide a level of security comparable to that of wired network [2]. WEP ensures the confidentiality protection for subscriber's messages. However, the secret key (WEP key) of the AP should be configured in each subscriber that connects to this AP. Security analysis showed that security implementation in the original IEEE 802.11 is vulnerable to several attacks capable of WEP key cracking [15][16]. The basic goal of the IEEE 802.11i was to overcome the WLAN security weakness. The WiFi alliance introduced the WiFi Protected Access (WPA) as a replacement of WEP in order to improve security in IEEE 802.11 networks. WPA was just an intermediary version of the security implementation in WLAN. Finally, IEEE 802.11i standard [4] was proposed as an amendment of the original IEEE 802.11 and is referred to as Robust Security Network (RSN). It relies on the IEEE 802.1X protocol to ensure authentication and access control services. It specifies the cryptographic algorithm to be used for confidentiality and integrity protection.

The Figure 11 shows the main phases of the association procedure in IEEE 802.11i [7].

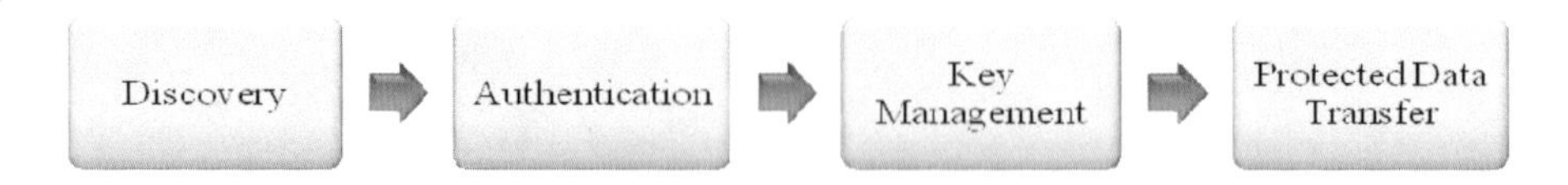

Fig. 11. Association procedure in WLAN

Discovery: represents the first phase in the association procedure in WLAN. Upon exchanging the security capabilities, the STA and the AP decides on several security procedures such as the authentication method to be used, confidentiality and integrity protection protocols, and cryptographic key management approach. At the end of this phase, the STA is just associated to the AP and needs to be authenticated and authorized before getting the complete access to WLAN network.

Authentication: relies on IEEE 802.1x standard which was designed to provide access control function in LANs. This standard introduces a new entity called Authentication Server (AS) in WLAN architectural model. Therefore, the AP delegates the authentication task to the AS. Upon a successful authentication, the AS may generate and forward the cryptographic key associated with the authenticated STA to the AP.

Key Management: represents the security activation phase. During this phase, a 4-way handshake between the STA and the AP takes place. The aim of this exchange is to establish fresh keys to protect link-layer frames. Therefore, the confidentiality and integrity protection can be ensured.

Protected Data Transfer: represents the last phase of the association procedure. During this phase, the STA and the AP apply the cryptographic algorithm for confidentiality and integrity protection. The STA traffic is now protected at the radio level.

In the following, we present the security mechanism implemented by the IEEE 802.11i for security services.

3.2.1 Identification

Identification in WLAN is still a matter of concern. Several types of subscriber identities exist. For instance, the subscriber uses the login as a subscriber identity in WiFi hotspots. In I-WLAN access (i.e. the Interworking between the WLAN and 3GPP access [28]), the subscriber presents its IMSI as a subscriber identity. In other examples, the subscriber may use the MAC address of its device as a subscriber identity. However, this is in general not recommended if the authentication mechanism is not strong enough; The MAC address represents the device identity.

For data transfer purpose, the subscriber is identified with the MAC address within the AP coverage area. Further, a binding between the MAC and IP addresses is required at the AR level for identification purpose.

3.2.2 Authentication

In WEP, two authentication methods are defined:

- Default authentication: In this mode, no authentication is carried out. Upon receiving an association request, the AP approves it without requesting a proof of the identity presented by the subscriber. However, the subscriber should be configured with the security key (WEP key) of the AP in use. This key is required for data encryption.
- Shared Key authentication: The subscriber and the AP should share the security key (WEP key). In this mode, the AP authenticates the subscriber in a four step challenge-response handshake.

The IEEE 802.11i relies on the 802.1x standard to perform authentication mechanism. The IEEE 802.1x is known as port-based network access control. It introduces three main functional entities in WLAN architecture (see **Figure 12**):

- *Supplicant:* represents the party that needs to be authenticated. It is located at the subscriber side.
- *Authenticator:* is defined as the gatekeeper of the access network. It delegates the authentication service to the authentication server and waits for its approval to open the access for the subscriber. It is located in the AP.
- *Authentication Server (AS):* is defined as the entity that provides authentication service to the authenticator. Therefore, the AS task is to check the supplicant credentials and to give the supplicant authorizations to the authenticator. There are two possibilities for the AS location in WLAN network. The first possibility consists of a standalone mode where the AS serves a number of authenticators. In this case, it needs an AAA protocol, namely RADIUS and Diameter to communicate with the authenticator. Co-locating the AS with the authenticator in the same AP represents the second possibility.

The authentication exchange between the AS and the supplicant relies on the Extensible Authentication Protocol (EAP).

EAP is a generic authentication framework that implements diverse authentication methods, called EAP methods, such as EAP-TLS, EAP-AKA, etc. In addition, EAP messages are easily transported over several protocols. For instance, any EAP method may run over wired links as well as wireless links. Moreover, any EAP method may run at the link-layer as well as at the IP-layer. The number of messages exchanged during an authentication procedure depends on the selected authentication method.

EAP-TLS is a mutual authentication method where the supplicant and the server use certificates as credentials. A PKI infrastructure is required for the smooth running of the method. However, the disadvantages of PKI infrastructures reside in their expensive cost and implementation complexity. Therefore, access providers will use the vendor certificate associated with the MAC address at the subscriber side and provide the AS with a certificate. Upon a successful authentication, session key is generated at both sides (supplicant and AS).

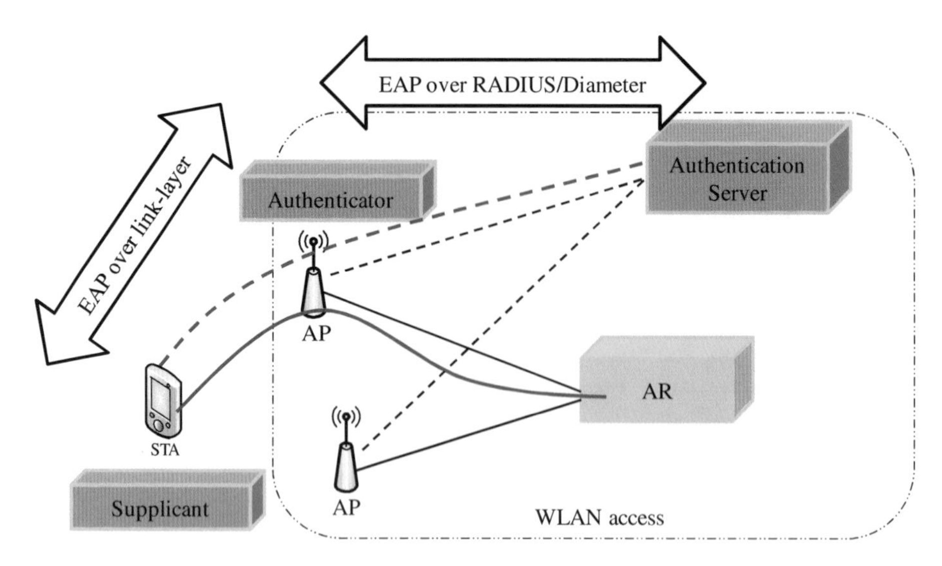

Fig. 12. Authentication components in WLAN

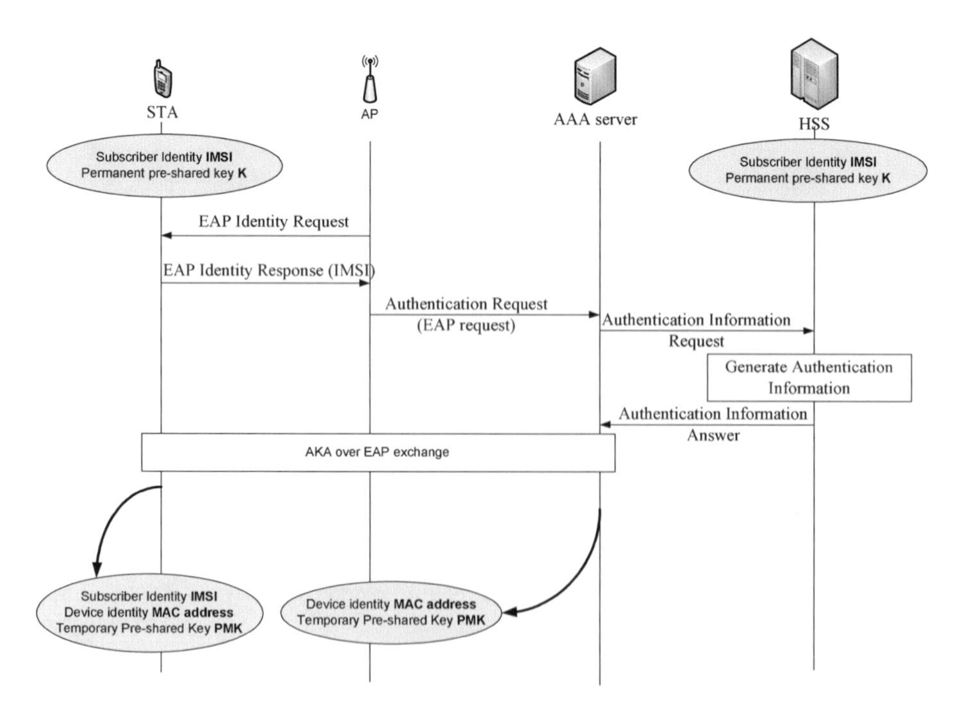

Fig. 13. EAP-AKA in 802.1x context

EAP-AKA is a well-known mutual authentication method. Likewise the EPS – AKA authentication mechanism used in the 3GPP cellular access, it is based on a pre-shared key authentication method and requires a USIM card at the subscriber side. The 3GPP specification [6] considers any WLAN implementing EAP-AKA as a trusted access. Figure 13 depicts the authentication in trusted access as it was defined in 3GPP specifications.

3.2.3 Access Control

Several mechanisms for the access control service exist [18]. For instance, the AP may be pre-configured with a list of authorized MAC addresses. Moreover, the access control may be implicit. That is the case of WEP. In fact, receiving data encrypted with WEP key, assures the AP ensures that the sender possesses the pre-shared key. Therefore, it is authorized to gain access to network resources. These mechanisms are vulnerable and static. A dynamic and stringent access control is needed.

The 802.1x protocol ensures the network access control service by maintaining two kinds of ports at the authenticator side. The first port is always open. However, it only transfers traffic related to the authentication exchange towards the AS. This enables the subscriber to present its identity and its credentials to the AS. The second port is dedicated to the data transfer. It is maintained blocked until the subscriber is successfully authenticated (see Figure 14).

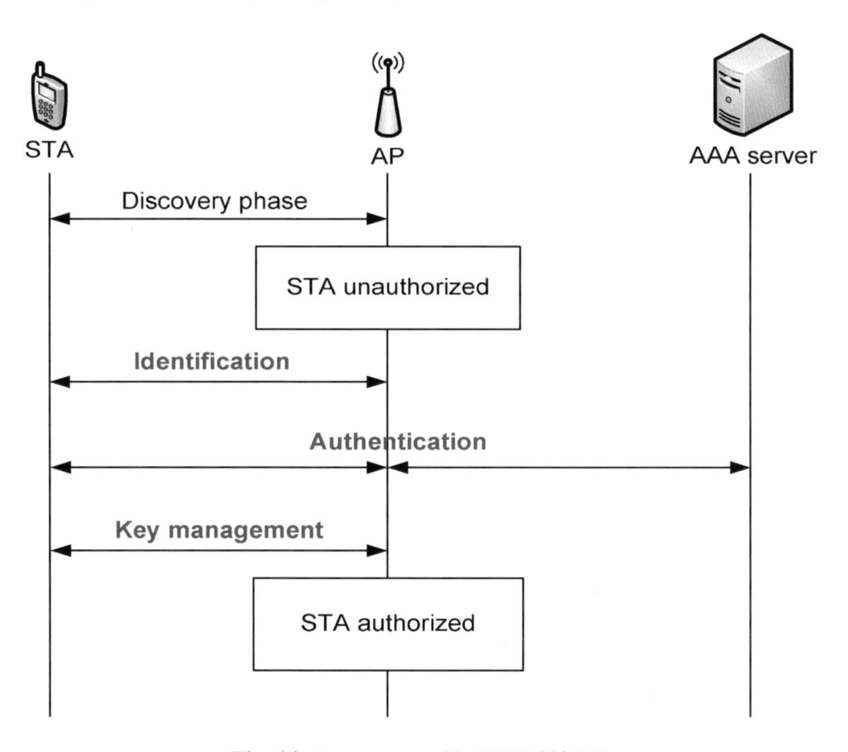

Fig. 14. Access control in IEEE 802.11i

3.2.4 Confidentiality and Integrity Protection

At the end of a successful authentication, common temporary key called Pairwise Master Key (PMK) is setup in the subscriber and the AP. From this key, a number of dynamic and short-lived keys are derived. A 4-way handshake takes place between the subscriber and the AP to confirm the presence of the PMK, to select the cipher suite and to derive the other keys.

Actually, AP triggers the key derivation by sending its MAC address to STA. Therefore, the Pairwise Transient Key (PTK) is derived from the PMK using the AP and STA MAC addresses at STA side. The use of MAC address in key derivation provides a protection from impersonation attacks. After that, STA sends back its MAC address to the AP. The sent message is integrity protected. At this time, the AP can derive the PTK key using both of the MAC address and the PMK key. An acknowledge message concludes the 4-way handshake.

The PTK key has 3 main components:

- EAPOL-KCK (EAP over LAN - Key Confirmation Key): is used to ensure the integrity protection of the control frame transporting key material during the 4-way handshake.
- EAPOL-KEK (EAP over LAN - Key Encryption Key): is intended to ensure the confidentiality protection for the control frame transporting key material during the 4-way handshake.
- TK (Temporal Key): It enables user traffic protection at the Radio link level.

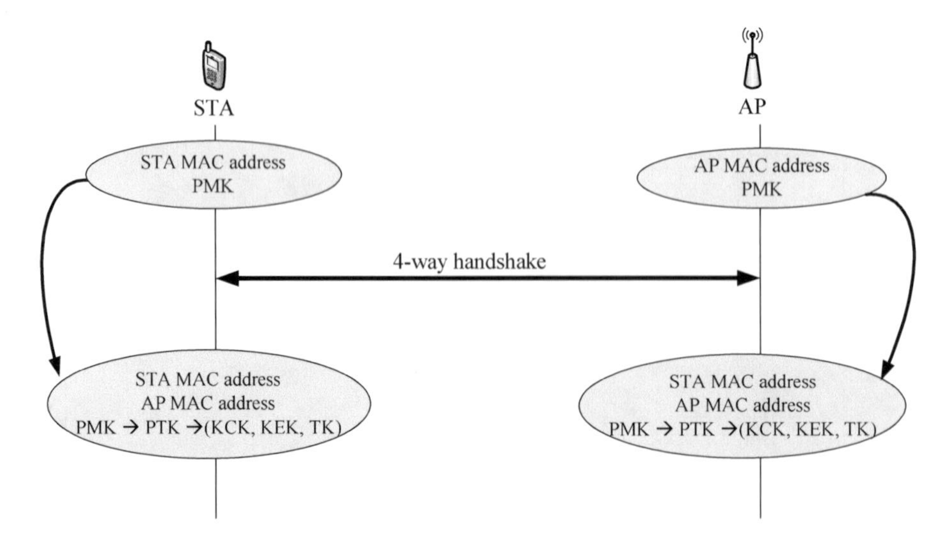

Fig. 15. Cryptographic key management in 802.11i

3.2.5 Privacy

Regarding identity privacy, the EAP framework generates temporary identity that can be used in the next authentication request. However, the subscriber should send its identity in clear whenever it changes the authenticator. Moreover, the MAC addresses of both of the subscriber and the AP respectively are exchanged in clear during the 4-way handshake for the security keys setup. Thus, we may conclude that identity and location privacy is not ensured.

3.3 Security and Mobility

The security implementation in WLAN access and its impact on network performances were extensively studied. For instance, [19] carried out an analytical and experimental study about the impact of several security levels on WLAN performance. It showed that the major contributor in the authentication delay is the time needed to detect the surrounding AP. [20] showed that the security may add a slight degradation on the throughput. However, no mobility considerations were made in these papers.

The original IEEE 802.11 standard specified no mobility management mechanism leading to session broken whenever the AP change. Several approaches have been proposed to encounter this problem. At link-layer, the IEEE 802.11r supports the fast-handover by enabling the STA to change the access point seamlessly while moving. At IP-layer, the Mobile IP (MIP) and Proxy Mobile IP (PMIP) are examples of IETF protocols for the mobility management. Both of these approaches define no mechanism to rapidly reestablish the security parameters. Even when the EAP framework was introduced in IEEE 802.11i, the fast reestablishment of the security parameters during the mobility event without generating a lot of signaling was not addressed. In fact, the mobility has not been considered at the EAP framework design time. In fact, [21] states that the EAP authentication should be performed each time the user changes the EAP authenticator regardless of whether the user is already authenticated by an old EAP authenticator or has valid keying materials.

Maintaining seamless mobility and ensuring security at the same time are ones of major challenges in WLAN. Actually, the aim of the mobility management is to provide the subscriber with the same Quality of Service (QoS) while it moves from an access point to another. The security procedures may generate additional overhead. Moreover, it may make the network association longer. For instance, **[5]** mentioned that the 802.11i authentication latency represents 65% of the total handover latency. It was shown in [22] that the IEEE 802.1X authentication is a time-consuming mechanism within the handover process. Hence, each time the STA changes the AP, the authentication delay affects the real-time application. A study was carried out about the effect of the mobility on four different security levels [23]. This study shows that the higher the STA mobility rate, the greater the authentication signaling and processing cost. This highlights the need for mechanisms to reduce authentication delay during handover.

In literature, several mechanisms were proposed to enable the fast security reestablishment:

- *Context transfer*: consists in transferring the security parameters created by the STA at the current AP to a new one. The IEEE 802.11f standard specified the way to exchange context between APs during the handoff process. When the mobility is managed at IP layer, the IEEE 802.11f proposal works separately from the mobility protocol. This may introduce an additional delay to the handoff process. Consequently, the [24] proposes extensions to the mobility messages to carry the security context. From the network performances viewpoint, the context transfer technique reduces the handoff latency and the authentication signaling traffic. However, from the security viewpoint, this mechanism raises some security concerns such as the domino effect [25]. Actually, a strong trust relationship between the involved APs is required.

- *Pre*-authentication: consists in pre-installing the security context into the next AP before the STA moves. The IEEE 802.11i introduced the pre-authentication mechanism to reduce the handoff delay. Actually, the new authentication mechanism takes place between the STA and the next AP through the current AP. Once a successful authentication, a new PMK is generated and maintained in both of new AP and STA. When the STA moves to the new AP, the 6-way handshake takes place using the new PMK. As the 802.11i is based on link-layer frames, the proposed mechanism for fast security reestablishment cannot work when the involved APs belong to different Distribution Systems (DSs). To overcome this issue, [26] proposed a mechanism based on IP-layer that assists the handover process between different DSs. Thanks to the pre-authentication technique, the handoff delay is reduced. However, the AS is significantly loaded.
- *Proactive Key distribution*: consists in distributing cryptographic keys into several APs that will probably host the STA during its mobility [27]. A trusted third-party such as the AS will generate keys and pre-distribute them to the corresponding APs. The proactive key distribution requires the knowledge about the network topology such as AP location. Therefore, it cannot work when the STA moves between different DSs. This technique enables a reduction in handoff delay. However, it generates a significant network load upon receiving the initial authentication procedure. Moreover, it keeps an AP busy as long as the STA has not reached this AP.

4 Part IV: Comparison between EPS and WLAN

In this section, we address an overall comparison between the EPS and the WLAN accesses.

1. From the Architecture Viewpoint

In EPS architecture, a distinction is made between the signaling plane and the data plane. In fact, a number of protocols were specified to only carry the signaling traffic. For instance, the Non-Access Stratum (NAS) protocol carries the security and mobility signaling traffic between the UE and the MME.

In the original IEEE 802.11standard, the signaling plane is not separated from the data plane. Then, the IEEE 802.11i amendment introduced the IEEE 802.1x protocol which separates the security messages from the data message (i.e. through the use of the EAPoL messages). This may be seen as an initiative to separate the signaling exchange from the data exchange.

2. From the Security Implementation Viewpoint

- *Identification:* In EPS/LTE architecture, only the IMSI should be used as the subscriber identity. Whereas in WLAN/802.11, the subscriber identity depends on the authentication method in use. Where the addressing plane in EPS is based on multiple temporary generated identities, it relies on the use of the device identity namely the MAC address in the IEEE 802.11.
- *Authentication:* The 3GPP standard specified the EPS-AKA as the unique authentication method that should be used. It is based on a pre-shared key which is maintained in both of the HSS database and the subscriber terminal.

  In the IEEE 802.11i, the subscriber negotiates with the AP the authentication suite to be used. For instance, it may choose to use the pre-shared key method. It may also choose the IEEE 802.1X which relies on the EAP framework. Even within the EAP framework, the subscriber can negotiate the authentication method to be used.
- *Access control:* In EPS, the access control is distributed between the HSS database, the MME and the eNB. Actually, the HSS decides whether the subscriber is authorized to use the current MME. According to the subscriber profile, the MME decides whether the subscriber is authorized to use the current eNB. Finally, the eNB enforces the MME authorization decision by opening the door for the subscriber data. When the IEEE 802.11 uses the pre-shared secret, the access control is performed implicitly. When the IEEE 802.11 uses the EAP framework, the access control is distributed between the AS and the AP. In fact, the AS decides whether the subscriber is authorized to use the current AP. The latter enforces the AS authorization decision.
- *Confidentiality and integrity protection:* The security in EPS is layered. Actually, the first security layer aims at protecting the exchange between the UE and the MME. The second one aims at protecting the exchange between the UE and the eNB.

  A unique security layer was specified in IEEE 802.11i. This aims at protecting the exchange between the STA and the AP.
- *Privacy:* In EPS, the identity confidentiality is ensured thanks to the temporary identity allocation. The location confidentiality is ensured thanks to the use of the temporary identity and the ciphering of the signaling exchange. In fact, whenever the MME changes it is possible guess the old MME through the temporary identity. Therefore, an intruder cannot link the subscriber to its location.

  The EAP framework defined a temporary identity for the re-authentication. However, this identity becomes useless when the authenticator change. Therefore, the subscriber should send its permanent identity in clear at each authenticator change. The tunneled authentication method can preserve the identity confidentiality.

The following table summarizes the comparison between the EPS/LTE and the WLAN/802.11 access from the security implementation viewpoint.

Security service	EPS access	WLAN access (802.1x)
Identification	*Subscriber identity :* IMSI *Device identity :* IMEI The addressing plane relies on multiple temporary identities	Subscriber identity: Multiple subscription identities *Device identity:* MAC address The addressing plane uses the MAC address
Authentication	EPS-AKA	Several authentication method
Access control	Distributed between the eNB, the MME and the HSS.	Distributed between the AP and the AS.
Confidentiality and integrity protection	Layered security: - confidentiality and integrity protection for the UE-MME exchange - confidentiality and integrity protection for the UE-eNB exchange	One security layer: - Security and integrity protection for the STA-AP exchange.
Privacy	*Identity confidentiality:* Temporary identity allocation *Location confidentiality:* the use of temporary identity, encryption of the location update messages	*Identity confidentiality:* the temporary identity allocation for re-authentication

3. From the Mobility Viewpoint

The 3GPP specification selected the context transfer technique for the fast security reestablishment. Since the mobility management and the security mechanisms are natively implemented in the EPS architecture, the context transfer technique relies on the mobility management messages for security parameters transfer between either the eNBs or the MMEs. This enables the handoff latency and signaling reduction. However, it is established in the 3GPP specification that the context transfer should be used carefully. For instance, two MMEs belonging to different serving network cannot exchange authentication information.

The IEEE 802.11i specified the pre-authentication technique as a support for the fast handoff. This technique has reduced the handoff delay. However, it has generated a significant load at the AS. In the literature, multiple alternative techniques were proposed for the fast security reestablishment.

Conclusion and Future Work

Throughout this chapter, we reviewed the security aspects of two distinct wireless access network namely the EPS/LTE and the WLAN/802.11. We described the architectural model of each wireless access. In more details, we focused on the security implementation in both of them. We presented the impact of the security on the mobility management performances. This was followed by an overview of the techniques that was proposed in the literature for the fast security reestablishment during the handoff process. At the end of the chapter, we compared the EPS/LTE access and the WLAN/802.11 access.

Under the interworking between the LTE and 802.11 access technology, optimizing the authentication schemes becomes a key challenge. In fact, the comparison between the EPS/LTE and the WLAN/802.11 conducted within this chapter reveals some point of differences. Where the authentication method is systematic in the first access, the authentication method can be negotiated in the second one. Therefore, when the subscriber moves from the EPS/LTE access to the WLAN/802.11 access, the authentication negotiation phase may increase the handoff delay leading to performance degradation. Moreover, the security level at the target access should be the same as the level at the old access. For instance, the subscriber has a high security level when he is connected to the EPS/LTE access. He should obtain the same security level when he moves to the WLAN/802.11 access.

Currently, the research community is working on improving the handover delay between heterogeneous accesses by making the authentication automatic and optimizing therefore the association phase. In fact, the network may prepare the target access point by choosing the adequate authentication mechanism that should be run. For instance, the Access Network Discovery and Selection Function (ANDSF) entity, which is specified by the 3GPP standards group, may inform the subscriber about the authentication mechanism and the credentials that should be used in the next hotspot. Hence, the association phase in the next hotspot may be optimized.

The dynamic nature of the current environment requires a very active flexible and adaptable security implementation. Actually, the future security implementation should be able to adapt to situations with a scarce resources. At the same time, it should be able to evolve and provide a strong security guarantees when more resources are available. Moreover, the future security implementation should be able to cope with several categories of devices. Consequently, a research challenging task is to make the future access network more flexible without adding complexity. In fact, everytime we add extra network mechanisms into our architecture to make it more flexible, we are usually adding more complexity.

References

[1] 3GPP TS 23.401 (v10.3.0), General Packet Radio Service (GPRS) enhancements for Evolved Universal Terrestrial Radio Access Network (E-UTRAN) access (release March 10, 2011)

[2] IEEE 802.11, Number Part 11: Wireless LAN Medium Access Control (MAC) and Physical Layer (PHY) Specifications: Specific Requirements, IEEE Std. 802.11-1997

[3] Wireless LAN Medium Access Control (MAC) and Physical Layer (PHY) Specifications: Amendment 8: Fast BSS Transition, IEEE std., 802.11r

[4] Wireless Medium Access Control (MAC) and Physical Layer (PHY) Specifications: Specification for enhanced security, IEEE Std. 802.11i-2004

[5] Komarova, M., Riguidel, M.: Secure User's Mobility: the current situation. China Communication Journal. Special Issue on Wireless Communications 4(1), 95–104 (2007)

[6] 3GPP TS 24.302 (v10.2.0), Access to the 3GPP Evolved Packet Core (EPC) via non-3GPP access networks (release December 10, 2010)

[7] Stallings, W.: Network security essentials, 4th edn. Pearson Education (2011) ISBN: 978-0-13-706792-3

[8] Shirey, R.: Internet Security Glossary, Version 2. IETF RFC 4949 (Informational) (August 2007)

[9] 3GPP TR 33.821 (v9.0.0), Rationale and track of security decisions in Long Term Evolved (LTE) RAN / 3GPP System Architecture Evolution (SAE) (release June 9, 2009)

[10] 3GPP TS 33.401 (v11.0.0), "3GPP System Architecture Evolution (SAE); Security architecture", (release June 11, 2011)

[11] 3GPP TS 29.272 (v11.0.0), "Evolved Packet System (EPS); Mobility Management Entity (MME) and Serving GPRS Support Node (SGSN) related interfaces based on Diameter protocol," (release September 11, 2011)

[12] Hu, Y.-Z., Ma, D.-W., Li, X.-F.: An Improved Authentication Protocol with Less Delay for UMTS Mobile Networks. In: International Conference on Networking and Digital Society, ICNDS 2009 (2009)

[13] Lin, Y.-B., Chen, Y.-K.: Reducing Authentication Signaling Traffic in Third-Generation mobile Network. IEEE Transactions on Wireless Communications 2(3) (May 2003)

[14] Al-Saraireh, J., Yousef, S.: Analytical model for authentication transmission overhead between entities in mobile networks. Computer Communications 30, 1713–1720 (2007)

[15] `http://www.trainingcamp.com/usa/preclass/ceh/Reading/WEP.pdf`

[16] Bulbul, H.I., Batmaz, I., Ozel, M.: Wireless Network Security: Comparison of WEP (Wired Equivalent Privacy) Mechanism, WPA (Wi-Fi Protected Access) and RSN (Robust Security Network) Security Protocols. In: Proceedings of the 1st International Conference on Forensic Applications and Techniques in Telecommunications, Information, and Multimedia and Workshop, e-Forensics 2008 (2008)

[17] Institute of Electrical and Electronics Engineer, "IEEE standards for Local and Metropolitan Area Networks: Port-based Network Access Control", IEEE Std. 802.1X-2004

[18] `http://www.interlinknetworks.com/whitepapers/WLAN_Access_Control.pdf`

[19] Fathi, H., et al.: On the impact of security on the latency in WLAN 802.11b. IEEE Globecom (2005)

[20] Boulmalf, M., Barka, E., Lakas, A.: Analysis of the effect of security on data and voice traffic in WLAN. Computer Communications 30, 2468–2477 (2007)

[21] Marin-Lopez, R., Ohba, Y., Pereniguez, F., Gomez, A.F.: Analysis of Handover Key Management schemes under IETF perspective. Computer Standards & Interfaces 32(5-6), 266–273 (2010) ISSN 0920-5489
[22] Bargh, M.S., et al.: Fast Authentication Methods for handovers between IEEE 802.11 Wireless LANs. In: ACM WMASH 2004 (October 2004)
[23] Liang, W., Wang, W.: On performance analysis of challenge/response based authentication in wireless networks. Computer Network 48(2), 267–288 (2005)
[24] Georgiades, M., Akther, N., Politis, C., Tafazolli, R.: Enhancing mobility management protocols to minimize AAA impact on handoff performance. Science Direct Computer Communication 30, 608–618 (2007)
[25] Housley, R., Aboda, B.: Guidance for authentication, Authorization, and Accounting (AAA) key management. IETF RFC 4962 (July 2007)
[26] Lopez, R.M., Dutta, A., Ohba, Y., Schulzrinne, H., Skarmeta, A.F.G.: Network-Layer Assisted Mechanism to Optimize Authentication Delay during Handoff in 802.11 Networks. In: Fourth Annual International Conference on Mobile and Ubiquitous Systems: Networking & Services, MobiQuitous 2007, pp. 1–8 (August 2007)
[27] Mishra, A., Shin, M., Petroni, N., Clancy, C., Arbaugh, W.: Proactive key distribution using neighbor graphs. IEEE Wireless Communications 11, 29–36 (2004)
[28] 3GPP TS 33.234 (11.3.0), 3G security; Wireless Local Area Network (WLAN) interworking security (release March 11, 2012)

Biography of Authors

Miroslav Škorić

YT7MPB, has been a licensed radio amateur since 1989. During the two decades he administered various types of amateur radio bulletin board systems (based on MS DOS^{TM}, Windows^{TM} and Linux platforms) with VHF/HF radio frequency and Internet inputs/outputs. He voluntarily served as the information manager and secretary in the Amateur radio union of Vojvodina province in Serbia, where he compiled technical and scientific information for broadcasting via local amateur radio frequencies. His teaching experience includes classes in a local high-school amateur radio club, tutorials on the amateur radio in engineering education, visiting lectures and paper presentations in domestic and international conferences. He authored two book chapters, several magazine and journal articles related to amateur radio, as well as a dedicated web page http://tldp.org/HOWTO/FBB.html, which includes user manuals for amateur radio e-mail servers. He is a member of IEEE (Computer Society, Communications Society, and Education Society), ACM and IAENG.

Noman Islam

Noman Islam received his BS degree in Computer Science from University of Karachi, Pakistan in 2002 and MS in Computer Science from National University of Computer and Emerging Sciences, Pakistan in 2006. He is pursuing his Ph.D. in Computer Science from National University of Computer and Emerging Sciences, Pakistan. He has around one and half dozen publications in various internal journals and conferences. His research interests include wireless ad hoc networks, ubiquitous computing and semantic web.

Zubair Ahmed Shaikh

Zubair Ahmed Shaikh received M.E. in Artificial Intelligence and the Ph.D. degree from Polytechnic Institute, Florida, USA, in 1992 and 1996 respectively. He is Professor and Dean of Engineering and Management Sciences at DHA Suffa University, Karachi,

Pakistan. He has published more than 70 papers in international conferences and peer reviewed journals. His research interest includes Artificial Intelligence, Social Network, Human Computer Interaction, Wireless Sensor Networks and Data Provenance.

Harris Simaremare

Harris Simaremare received the B.Sc. and Master degrees in Electrical Engineering from Universitas Gadjahmada. He is currently pursuing his PhD research on security in wireless ad-hoc networks. His research interest includes communication and mobile ad hoc networks, security and pervasive computing.

Abdelhafid Abouaissa

Abdelhafid Abouaissa is an associate professor at the University of Haute-Alsace, in Colmar, France. He received a bachelor's degree from the Technical University of Wroclaw, Poland, in 1995, and a master's degree from Franche-Comté University of Besançon, France, in 1996. He obtained his PhD at the Technical University of Belfort, France, in January 2000. His interests include multimedia synchronization, group communication systems, QoS routing in ad-hoc, Mesh networks, sensor networks, MPLS, DiffServ, and QoS management.

Riri Fitri Sari

Riri Fitri Sari, PhD. is a Professor at Electrical Engineering Department of Universitas Indonesia. She received her Bsc degree in Electrical Engineering from Universitas Indonesia. She receives her MSc in Computer Science and Parallel Processing from University of Sheffeld, UK. And she received her PhD in Computer Science from University of Leeds, Leeds. Riri Fitri Sari is a senior member of the Institute of Electrical and Electronic Engineers (IEEE).

Pascal Lorenz

Pascal Lorenz received a PhD degree from the University of Nancy, France. Between 1990 and 1995 he was a research engineer at WorldFIP Europe and at Alcatel-Alsthom. He is a professor at the University of Haute-Alsace and responsible of the Network and Telecommunication Research Group. His research interests include QoS, wireless networks and high-speed networks. He was the Program and Organizing Chair of the IEEE ICATM'98, ICATM'99, ECUMN'00, ICN'01, ECUMN'02 and ICT'03, ICN'04, PWC'05 conferences, symposium co-chair of ICC'06, Globecom'07, ICC'08, Globecom'08, ICC'09 and co-program chair of ICC'04. Between 2000 and 2006, he was Technical Editor of the IEEE Communications Magazine Editorial Board. He is the vice-chair of the IEEE ComSoc Communications Software Technical Committee and chair of the IEEE ComSoc Communications Systems Integration and Modelling Technical Committee. He is senior member of the IEEE, member of many international program committees and he has served as a guest editor for a number of journals including Telecommunications Systems, IEEE Communications Magazine and LNCS. He has organized and chaired several technical sessions and gave tutorials at major international conferences. He is the author of 3 books, 2 patents and 190 international publications in journals and conferences.

M.A. Razzaque

M.A. Razzaque is Senior Lecturer in the Faculty of Computer Science and Information Systems, UTM, Malaysia. His research interests centered on the area of wireless and mobile computing and communications. Subtopics of focus include cross-layer design, autonomic communications, wireless sensor networks, wireless security, Vehicular Ad-hoc Networks, computational intelligence, bioinformatics, etc. His current research works centered on data compression, data reliability and data management in wireless sensor network, Intelligent Transport System, Computational Intelligence in WSNs, etc.. He is an expert in the fields of cross-layer design for Wireless Networks (MANETs, Wireless Sensor Networks), a reputation supported by number of internationally peer-reviewed publications. He holds a BSc and M.Sc. in Applied Physics, Electronics and Communication Engineering, PhD in computer science, is a Member of the IEEE and ACM.

Ahmad Salehi S.

Ahmad Salehi S. received B.Sc. degree in Software Engineering in 2008 from University of Azad Islamic, Iran. Currently, he is doing his M.Sc. degree on Information Security in the Faculty of Computer Science and Information Systems, UTM, Malaysia. His research interests are Wireless and Mobile Computing, Wireless Sensors Networks (WSN) and Vehicular Adhoc Networks with emphasis on quality of service, security and privacy issues.

Seyed M. Cheraghi

Seyed M. Cheraghi received B.Sc. degree in Software Engineering in 2008 from University of Azad Islamic, Iran. Currently, he is doing his M.Sc. degree on Information Security in the Faculty of Computer Science and Information Systems, UTM, Malaysia. His research interests are Wireless and Mobile Computing, Wireless Sensors Networks (WSN) and Vehicular Adhoc Networks with emphasis on routing, security and privacy issues.

Jorge Granjal

Jorge Granjal is an Invited Assistant Professor at the Department of Informatics Engineering of the University of Coimbra, Portugal, and a Researcher of the Laboratory of Communications and Telematics of the Centre for Informatics and Systems of the University of Coimbra, Portugal. His main research interests are Computer Networks, Network Security and Wireless Sensor Networks. He is a member of IEEE and ACM Communications groups, and is currently pursuing a PhD in the area of security in Wireless Sensor Networks and the Internet of Things.

Edmundo Monteiro

Edmundo Monteiro is Full Professor at the University of Coimbra, Portugal, from where he got a PhD in Electrical Engineering, Informatics Specialty, in 1995. His research interests are Computer Communications, Wireless Communications, Service Oriented Infrastructures and Security. He is author of several publications including books, patents, and over 200 papers in national and in international refereed books, journals and conferences. He is member of the Editorial Board of Elsevier Computer Communication and Springer Wireless Networks journals and he has been involved in the organization of many international conferences and workshops (e.g. QoFIs2001, IDMS2002, Networking2006, INFOCOM2006, WWIC2007, BWA2008/2011, IWQoS2012).

He participated in several research projects and research programs of the European Union, various European countries, US, Latin America, and national organizations and companies, in the areas of Information and Communication Technologies (ICT), Critical Infrastructures Protection, ICT systems for Energy Efficiency, Research and Innovation, and Human Resources and Mobility. Edmundo Monteiro is member of IEEE Communications, IEEE Computer, and ACM Communications groups.

Jorge Sá Silva

Jorge Sá Silva received his PhD in Informatics Engineering in 2001 from the University of Coimbra, where is an Assistant Professor at the Department of Informatics Engineering of the University of Coimbra and a Senior Researcher of Laboratory of Communication and Telematics, Portugal. His main research interests are Mobility, Network Protocols and Wireless Sensor Networks. He has been serving as a reviewer and publishing in top conferences and journals in his expertise areas. His publications include 2 book chapters and over 70 papers in refereed national and international conferences and magazines. He is a member of IEEE, and he is a licensed Professional Engineer.

Saeideh Sadat Javadi

Saeideh Sadat Javadi received B.S. degree in computer engineering in 2008 from Shiraz University, Shiraz, Iran. Currently, she is doing her M.S. degree on Information Technology (IT) in the Faculty of Computer Science and Information Systems, UTM, Malaysia. Her master project is centered on development of Quality of Service (QoS) aware Medium Access Control (MAC) protocol for Wireless Body Area Networks (WBANs). Her research interests are Wireless and Mobile Computing, Wireless Sensors Networks (WSN) and Wireless Body Area Networks with emphasis on quality of service, security and privacy issues.

Jaydip Sen

Jaydip Sen has 18 years of experience in the field of networking, communication and security. He has worked in reputed organizations like Tata Consultancy Services, India, Oil and Natural Gas Corporation Ltd., India, Oracle India Pvt. Ltd., and Akamai Technology Pvt. Ltd. His research areas include security in wired and wireless networks, intrusion detection systems, secure routing protocols in wireless ad hoc and sensor networks, secure multicast and broadcast communication in next generation broadband wireless networks, trust and reputation based systems, quality of service in multimedia communication in wireless networks and cross layer optimization based resource allocation algorithms in next generation wireless networks, sensor networks, and privacy issues in ubiquitous and pervasive communication. He has more than 90 publications in reputed international books, journals and referred conference proceedings. He is a member of ACM and IEEE.

Jyoti Grover

Jyoti Grover received her B.E in Computer Science and Engineering from M.D. University, Rohtak, India in 2002, and M.Tech degree in Computer Science and Engineering from G.J. University, Hisar, India in 2004 with specialization in "Mobile Ad hoc Network Routing Protocols". She is currently pursuing her PhD in Computer Engineering from Malaviya National Institute of Technology, Jaipur, India. The main focus of her research has been on modeling various attacks in VANET scenario and formulates detection approaches for these attacks.

Manoj Singh Gaur

Dr. Manoj Singh Gaur is Professor in Computer Engineering Department at Malaviya National Institute of Technology, Jaipur (India). He received his bachelor degree in Electronics and Communication Engineering from J.N.V University, Jodhpur and master's degree in Computer Science and Engineering from Indian Institute of Science, Bangalore (India). He received his Ph.D. in Department of Electronics and Computers from University of Southampton, UK. His research interests are Network Security, Embedded Systems (Network-on-Chip), Information Security, Cloud Computing, QoS in wireless communications.

Vijay Laxmi

Dr. Vijay Laxmi is currently an Associate Professor in the Computer Engineering Department at Malaviya National Institute of Technology, Jaipur (India). She received her bachelor's degree in Electronics and Communication Engineering from J.N.V University, Jodhpur and master's degree in Computer Science and Engineering from Indian Institute of Technology, Delhi (India). She completed her Ph.D from Department of Electronics and Computers from University of Southampton, UK. The main focus of her research has been on the Information Security, Machine Vision, Embedded Systems, Multi-core QoS issues during transmission of multimedia traffic over Wireless Networks and Cloud Computing.

Peter Langendörfer

Prof. Dr. Peter Langendörfer holds a diploma and a doctorate degree in computer science. Since 2000 he is with the IHP in Frankfurt (Oder). There, he is leading the sensor networks and mobile middleware group. Since 2012 he has his own chair for security in pervasive systems at the Technical University of Cottbus. He has published more than 90 refereed technical articles, filed ten patents in the security/privacy area and worked as guest editor for many renowned journals e.g. Wireless Communications and Mobile Computing (Wiley). He was chairing International conferences such as WWIC and has served in many TPC for example at Globecom, VTC, ICC and SECON. His research interests include wireless sensor networks and cyber physical systems, especially privacy and security issues.

Dr. Zoya Dyka

Dr. Zoya Dyka holds a diploma in radio-physics from the Taras Shevchenko University Kiew and received her doctorate degree from the BTU Cottbus in 2012. Zoya has published 7 refereed technical articles and filed three patents in the security area. She has served as reviewer for highly selective journals such as IEEE Transactions on Computers, and is currently organization chair of the IEEE ISSSE conference. Her research interests include design of efficient hardware accelerators for crypto-operations and anti-tampering means.

Amrita Ghosal

Amrita Ghosal is presently Assistant Professor in the Department of Computer Science and Engineering, Dr. B. C. Roy Engineering College, Durgapur, India. She received B.Tech in Electronics and Communication Engineering from Asansol Engineering College, and M.Tech in Computer Science and Engineering from Kalyani Government Engineering College, Kalyani, India in 2003 and 2006 respectively. She is currently pursuing PhD at the Department of Computer Science and Technology, Bengal Engineering and Science University. Her current area of research is wireless sensor networks. Her main areas of interest include different attacks and security in wireless sensor networks. She has published research works in reputed conference proceedings and journals in her field. She has been a technical program committee member of many international conferences and reviewer of journals such as Wireless Personal Communication, Wireless Networks in her field. She is a member of CRSI.

Subir Halder

Subir Halder is currently Assistant Professor in the department of Computer Science and Engineering at Dr. B. C. Roy Engineering College, Durgapur under West Bengal University of Technology, India. He received B.Tech degree in Electronics and Communication Engineering and M.Tech degree in Computer Science and Engineering from Kalyani Government Engineering College, Kalyani, India in 2003 and 2006 respectively. He is currently pursuing PhD at the Department of Computer Science and Technology, Bengal Engineering and Science University. His current research interests include Network Modeling and Analysis, Performance Evaluation and Optimization, Wireless ad hoc and Sensor Networks. He has published research works in reputed conference proceedings and journals in his field. He has been a technical program committee member of many international conferences in his field. He is a member of ACM.

Rafaela Villalpando-Hernández

Rafaela Villalpando Hernández received her Ph.D. in Engineering of Electronics and Telecommunications from Instituto Tecnológico y de Estudios Superiores de Monterrey (ITESM), Campus Monterrey, in December 2008. Thereafter, she joined the Engineer Center at the ITESM, Campus Laguna, Torreón México. She has participated in several research projects involving network coding, position location in wireless networks and implementation of sensor networks. Her research interests include wireless and sensor networks.

Cesar Vargas-Rosales

Cesar Vargas-Rosales received a Ph.D. in electrical engineering from Louisiana State University in 1996. Thereafter, he joined the Center for Electronics and Telecommunications at Instituto Tecnológico y de Estudios Superiores de Monterrey (ITESM), Campus Monterrey, Mexico. Dr. Vargas is the coauthor of the book Position Location Techniques and Applications. He has carried out research in the area of personal communication systems on CDMA, smart antennas, adaptive resource sharing, location information processing, and multimedia services. His research interests are personal communications networks, position location, mobility and traffic stochastic modeling, intrusion detection, and routing in reconfigurable networks. Dr. Vargas was the co-Technical Program Chair of the IEEE Wireless Communications and Networking Conference (WCNC) 2011. He is the IEEE Communications Society Monterrey Chapter Head and has been a Senior Member of the IEEE since 2001.

David Muñoz-Rodríguez

David Muñoz Rodríguez received a B.S. in 1972, an M.S.in 1976, and a Ph.D. in 1979 in electrical engineering from the Universidad de Guadalajara, México, Cinvestav, México, and University of Essex,Colchester, England, respectively. He is Senior Member of the IEEE and was formerly Chairman of the Communication Department and Electrical Engineering Department at Cinvestav, IPN. In 1992, he joined the Instituto Tecnológico y de Estudios Superiores de Monterrey (ITESM), Campus Monterrey, México,where he became the Director of Center for Electronics y Telecommunications,. And currently he holds the Telecommunications and Mobility Chair. His research interests include wireless systems and performance analysis.

Fernando Ruiz-Trejo

Fernando Ruiz Trejo was born in Cd. Victoria, Tamaulipas, México, on August 5th, 1985. He received the B.S. degree in Electronic Engineering from the Instituto

Tecnológico de Nuevo Léon (ITNL), Nuevo Léon, México in 2008. Currently he is pursuing a Master of Science in Electronic Engineering (Telecommunications) in the Instituto Tecnológico y de Estudios Superiores de Monterrey (ITESM), Campus Monterrey, México. He is a member of the research chair of Mobility and Wireless Networks in the Department of Electrical and Computer Engineering at ITESM. His research interests include Wireless Networks and Network Coding.

Miguel Garcia

Miguel Garcia (migarpi@upvnet.upv.es) received his M.Sc. in Telecommunications Engineering in 2007 at "Universitat Politecnica de Valencia" and a postgraduate degree called "Master en Tecnologías, Sistemas y Redes de Comunicaciones" in 2008. He is currently a Ph.D. student in the Department of Communications of the same university. He is a CCNA Instructor. Miguel is working as a researcher in the research group "Communications and Remote Sensing" of the IGIC. He is involved in several R&D projects. Up to now, he has more than 45 scientific papers published in international conferences; moreover he has several educational papers. He has more than 30 papers published in international journals (most of them with I.F.). Mr Garcia has been TPC member in several conferences and journals. He has been in the organization committee of several conferences. Miguel is associate editor of International Journal Networks, Protocols & Algorithms. He is IEEE graduate student member.

Diana Bri

Diana Bri (diabrmo@upvnet.upv.es) received the M.Sc. Telecommunications Engineering degree in 2007 from the "Universitat Politecnica de Valencia" and the master's degree "Tecnologías, Sistemas y Redes de Comunicaciones" in 2010 from the same university. Since then, she is doing her Ph.D. in the Department of Communications at this university. Moreover, she has become a CCNA Instructor and she works as a researcher in the Research Institute for Integrated Management of Coastal Areas (IGIC). From this researching activity, she is involved in some R&D projects and she has several scientific papers published in international conferences, journals and books. Finally, Diana has been TPC member in several conferences and journals and she has been in the organization committee of several conferences. Today, she is associate editor of International Journal Networks, Protocols & Algorithms.

Jaime Lloret

Dr. Jaime Lloret (jlloret@dcom.upv.es) is currently Associate Professor in the Polytechnic University of Valencia. He is the head of the research group "communications and remote sensing" of the IGIC Research Institute. He is the director of the University Expert Certificate "Redes y Comunicaciones de Ordenadores" and the University Master "Digital Post Production". He is currently Vice-chair of the Internet Technical Committee and the Vice-chair for the Europe/Africa Region of Cognitive Networks Technical Committee. He has authored 12 books and has more than 220 research papers published in national and international conferences, international journals (most of them with I.F.). He is editor-in-chief of the international journal "Networks Protocols and Algorithms ", IARIA Journals Board Chair (8 Journals). He led many national and international projects. He has been the general chair (or co-chair) of 16 international conferences and workshops. He is IEEE Senior and IARIA Fellow.

Pascal Lorenz

Pascal Lorenz (lorenz@ieee.org) received his M.Sc. (1990) and Ph.D. (1994) from the University of Nancy, France. Between 1990 and 1995 he was a research engineer at WorldFIP Europe and at Alcatel-Alsthom. He is a professor at the University of Haute-Alsace, France, since 1995. His research interests include QoS, wireless networks and high-speed networks .He is the author/co-author of 3 books, 3 patents and 200 international publications in refereed journals and conferences.

Marina Mondin

Marina Mondin is Associate professor at the Department of Electronics and Telecommunications, Politecnico di Torino. Her current interests are in the area of signal processing for telecommunications, coding, simulation of communication systems and quantum communications. She has been project coordinator or principal investigator for the Politecnico di Torino unit for several international and national research projects financed by national and international funding agencies and by private companies. Professor Mondin has authored and co-authored more than 50 articles on international journals, more than 100 contributions to international conferences, and two books (in Italian). She has been acting as a reviewer for several international scientific IEEE and IEE journals, has been Guest Editor for two Special issues (of the EURASIP Journal on Wireless Communications and Networking and of the International Journal of Digital Multimedia Broadcasting). She is currently Associate Editor for the IEEE Transactions on Circuits and Systems I.

Fred Daneshgaran

Fred Daneshgaran received the B.S. degree in electrical and mechanical engineering from California State University, Los Angeles (CSLA) in 1984, the M.S. degree in electrical engineering from CSLA in 1985, and the Ph.D. degree in electrical engineering from University of California, Los Angeles (UCLA), in 1992. Since 1997, he has been a full Professor with the ECE Department at CSLA and since 2006 he serves as the chairman of the department.

From 1999 to 2001, he was Chief Scientist for TechnoConcepts, Inc., where he directed the development of a prototype software-defined radio system and managed the hardware and software teams. In 2000, he co-founded EuroConcepts s.r.l., an R&D company specializing in the design of advanced communication links and software radio.

He served as the editor of IEEE Trans. On Wireless Comm. from 2003 to 2009. He is active in R&D and has numerous grants and published articles to his credit.

Francesca Vatta

Francesca Vatta received a Laurea in Ingegneria Elettronica in 1992 from University of Trieste, Italy. From 1993 to 1994 she has been with the Olivetti group, Milano, Italy, as system engineer working on design and implementation of Computer Integrated Building architectures. Since 1995 she has been with the University of Trieste where she received her Ph.D. degree in Telecommunications in 1998 and became assistant professor in 1999. Starting in 2002, she spent several months as visiting scholar at the University of Notre Dame, IN, U.S.A., cooperating with the Coding Theory Research Group under the guidance of Prof. D. J. Costello, Jr. She is an author of more than 70 papers published on international journals and conference proceedings. Her research interests are in the area of channel coding techniques concerning the analysis and design of capacity achieving coding schemes for different applications, including, more recently, also Quantum Key Distribution (QKD) applications.

Fabio Mesiti

Fabio Mesiti received his MSc degree in Telecommunication Engineering in 2005, and the PhD in Electronics and Communications Engineering in 2009 from Politecnico di Torino, Italy. He is currently working as Post-doc fellow at Norwegian University of Science and Technology (NTNU) in Trondheim (Norway). His research interests mainly include cross-layer modeling, optimization and performance analysis of wireless communications systems and channel coding for quantum communication systems. He is also involved in the investigation of biological effects induced by electromagnetic exposure, with possible applications at nanoscale for sensing and stimulation of the neuronal activity. Fabio Mesiti has been acting as a reviewer for several international scientific IEEE and Hindawi journals and as session chair of SPACOMM 2009 - (Signal Processing in Telecommunications). He also served the scientific community as a member of the technical program committee of international conferences, ISCC 2011/2012 and CogArt 2011.

Maria Teresa Delgado Alizo

Maria Teresa Delgado received her M.S. Degree in Electrical Engineering from the Universidad Central de Venezuela in Caracas, Venezuela in 2006. Later, she received her M.S. degree in Telecommunication Engineering in 2008, and the Ph.D. in Electronics and Communications Engineering in 2012 from Politecnico di Torino, in Turin, Italy.

Her interests are in the area of signal processing for telecommunications, coding, simulation of communication systems and quantum communications. During her Ph.D., her research area focused on the study and modeling of quantum key distribution systems, using innovative techniques for error correction and privacy amplification employing capacity achieving codes, minimizing the interaction between the parties involved in order to guarantee the security of the communication.

Inam Bari

Inam Bari did his Bachelor's degree (BS in Telecommunication Engineering) from National University of Computer and Emerging Science (NUCES-FAST), Pakistan in 2007. He served in NUCES-FAST Peshawar campus as a Lab Engineer for 8 months. In 2008, He was awarded a full 5 years MS leading to PhD scholarship By Higher Education Commission Pakistan. He did his Master's degree (MS in Telecommunication Engineering) from Politecnico di Torino, Italy in 2010. His MS thesis consist of "Assessing techniques for Scalable Optical Switching Fabrics based on Optical Code division Multiplexing (OCDMA)". Currently, he is doing his PhD in Electronics and Telecommunication Engineering from Politecnico di Torino, Italy. His research area is quantum communication. He is currently working on Capacity achieving codes for Quantum Key Distribution (QKD) and BIMO communication.

Siwar Ben Hadj Said

Siwar Ben Hadj Said received her Engineering degree in Telecommunication in January 2010, from both Telecom Paris'Tech inFrance and Sup'Com communication engineering school in Tunisia. From March 2010 to September 2010, she was a trainee in the research laboratory IRISA, France. She is now a Ph.D. student in Orange Labs in conjunction with the "Network, Security and Multimedia" (RSM) department in TELECOM Bretagne, France. Siwar's current research interests include: wireless networks, signaling plane performance, mobility management, and network security.

Karine Guillouard

Karine Guillouard graduated from INSA (National Applied Science Institute), France, in 1994, and received the Radiocommunications PhD degree, in 1997. She joined France Telecom R&D in 1997. She was first involved in techno-economic studies of radio access networks and feasibility studies of wireless home networks. Her focus shifted to IP based mobility management within wireless broadband IP accesses. She particularly contributed to European collaborative projects (Eurescom, IST). She was also in charge of the innovative RNRT Cyberté project (2001-2004), in the field of the multi-access IPv6 terminal architecture. Her research activities currently evolve towards dynamic and adaptive network architectures applied in the context of fixed and mobile convergence.

Jean-Marie Bonnin

Professor Jean-Marie Bonnin got his PhD degree in Computer Science at the University of Strasbourg, France in 1998. He has been with TELECOM Bretagne since 2001, where he is currently the head of the "Networks, Security and Multimedia" (RSM) department. He also leads the Mobility research team in the RSM dept. His main research interests lie in the convergence between IP networks and mobile telephony networks, and especially in heterogeneous handover issues. Recently, he has been involved in projects dealing with network mobility and its application to ITS (Intelligent Transportation Systems). He is involved in several collaborative research projects at the French and European levels and through international academic collaborations (mainly with Asia and North Africa).

Author Index

Subject Index

I

J

K

L

M

S

T

Printed by Printforce, the Netherlands